BIOMEDICAL APPLICATIONS OF THE HORSESHOE CRAB (LIMULIDAE)

PROGRESS IN CLINICAL AND BIOLOGICAL RESEARCH

Vol 1: **Erythrocyte Structure and Function,** George J. Brewer, *Editor*

Vol 2: **Preventability of Perinatal Injury,** Karlis Adamsons and Howard A. Fox, *Editors*

Vol 3: **Infections of the Fetus and the Newborn Infant,** Saul Krugman and Anne A. Gershon, *Editors*

Vol 4: **Conflicts in Childhood Cancer,** Lucius F. Sinks and John O. Godden, *Editors*

Vol 5: **Trace Components of Plasma: Isolation and Clinical Significance,** G.A. Jamieson and T.J. Greenwalt, *Editors*

Vol 6: **Prostatic Disease,** H. Marberger, H. Haschek, H.K.A. Schirmer, J.A.C. Colston, and E. Witkin, *Editors*

Vol 7: **Blood Pressure, Edema and Proteinuria in Pregnancy,** Emanuel A. Friedman, *Editor*

Vol 8: **Cell Surface Receptors,** Garth L. Nicolson, Michael A. Raftery, Martin Rodbell, and C. Fred Fox, *Editors*

Vol 9: **Membranes and Neoplasia: New Approaches and Strategies,** Vincent T. Marchesi, *Editor*

Vol 10: **Diabetes and Other Endocrine Disorders During Pregnancy and in the Newborn,** Maria I. New and Robert H. Fiser, *Editors*

Vol 11: **Clinical Uses of Frozen-Thawed Red Blood Cells,** John A. Griep, *Editor*

Vol 12: **Breast Cancer,** Albert C. W. Montague, Geary L. Stonesifer, Jr., and Edward F. Lewison, *Editors*

Vol 13: **The Granulocyte: Function and Clinical Utilization,** Tibor J. Greenwalt and G.A. Jamieson, *Editors*

Vol 14: **Zinc Metabolism: Current Aspects in Health and Disease,** George J. Brewer and Ananda S. Prasad, *Editors*

Vol 15: **Cellular Neurobiology,** Zach Hall, Regis Kelly, and C. Fred Fox, *Editors*

Vol 16: **HLA and Malignancy,** Gerald P. Murphy, *Editor*

Vol 17: **Cell Shape and Surface Architecture,** Jean Paul Revel, Ulf Henning, and C. Fred Fox, *Editors*

Vol 18: **Tay-Sachs Disease: Screening and Prevention,** Michael M. Kaback, *Editor*

Vol 19: **Blood Substitutes and Plasma Expanders,** G.A. Jamieson and T.J. Greenwalt, *Editors*

Vol 20: **Erythrocyte Membranes: Recent Clinical and Experimental Advances,** Walter C. Kruckeberg, John W. Eaton, and George J. Brewer, *Editors*

Vol 21: **The Red Cell,** George J. Brewer, *Editor*

Vol 22: **Molecular Aspects of Membrane Transport,** Dale Oxender and C. Fred Fox, *Editors*

Vol 23: **Cell Surface Carbohydrates and Biological Recognition,** Vincent T. Marchesi, Victor Ginsburg, Phillips W. Robbins, and C. Fred Fox, *Editors*

Vol 24: **Twin Research,** Proceedings of 2nd International Congress on Twin Studies
Walter E. Nance, *Editor*
Published in 3 Volumes:
Part A: Psychology and Methodology
Part B: Biology and Epidemiology
Part C: Clinical Studies

Vol 25: **Recent Advances in Clinical Oncology,** Tapan A. Hazra and Michael C. Beachley, *Editors*

Vol 26: **Origin and Natural History of Cell Lines,** Claudio Barigozzi, *Editor*

Vol 27: **Membrane Mechanisms of Drugs of Abuse,** Charles W. Sharp and Leo G. Abood, *Editors*

Vol 28: **The Blood Platelet in Transfusion Therapy,** G.A. Jamieson and Tibor J. Greenwalt, *Editors*

Vol 29: **Biomedical Applications of the Horsehoe Crab (Limulidae),** Elias Cohen, *Editor-in-Chief*

BIOMEDICAL APPLICATIONS OF THE HORSESHOE CRAB (LIMULIDAE)

Proceedings of a Symposium Held at the
Marine Biological Laboratory,
Woods Hole, Massachusetts, October 1978

Editor-in-Chief
ELIAS COHEN, PHD

Associate Editors
FREDERIK B. BANG, MD
JACK LEVIN, MD
JOHN J. MARCHALONIS, PHD
THOMAS G. PISTOLE, PHD
ROBERT A. PRENDERGAST, MD
CARL SHUSTER, JR, PHD
STANLEY W. WATSON, PHD

ALAN R. LISS, INC. • NEW YORK

Address all Inquiries to the Publisher
Alan R. Liss, Inc., 150 Fifth Avenue, New York, NY 10011

Printed in the United States of America

Library of Congress Cataloging in Publication Data

Main entry under title: Biomedical applications of the horseshoe crab (Limulidae)
(Progress in clinical and biological research; v. 29) Bibliography: p. Includes index.
1. Endotoxin – Analysis – Congresses. 2. Limulus polyphemus – Congresses.
3. Biological assay – Congresses. 4. Lectins – Congresses. 5. Xiphosura – Congresses.
I. Cohen, Elias. II. Bang, Frederik Barry, 1916- III. Series. [DNLM: 1. Horsehoe crabs – Congresses. 2. Limulus test – Congresses. W1 PR668E v. 29/QX463 B615 1978]
QP632.E4B56 616.01'4 79-1748
ISBN 0-8451-0029-7

DEDICATION

Dedicated to Kathryn. Without her forbearance, the Symposium and the Proceedings might not have come into being.

Molly Bang-Campbell

Contents

Dedication v

Acknowledgments xiii

Preface
Elias Cohen xv

I. BIOLOGY AND ECOLOGY OF LIMULUS POLYPHEMUS

Biology of Limulus polyphemus
Carl N. Shuster, Jr. 1

Distribution of the American Horseshoe "Crab," Limulus polyphemus
Carl N. Shuster, Jr. 3

Limulus polyphemus: A Review of the Ecologically Significant Literature
Anne Rudloe 27

Ecology of the Extant Horseshoe Crabs
Koichi Sekiguchi and Koichiro Nakamura 37

Thermoregulatory Behavior and Diel Activity of Limulus polyphemus
William Wallace Reynolds and Martha Elizabeth Casterlin . . . 47

Laboratory Culture of Embryonic and Juvenile Limulus
Kathleen A. French 61

Motility of the Limulus Amebocyte
Peter B. Armstrong 73

The In Vitro Cultivation of Limulus Amebocytes
Frederick C. Pearson and Ellen Woodland 93

Observations on the Potential of Limulus Aquaculture in Israel
Chaim Kropach 103

Limulus polyphemus, a Target of Opportunity
Sidney R. Galler 107

II. DETECTION OF BACTERIAL ENDOTOXIN AND BIOMEDICAL APPLICATIONS OF LIMULUS

A. Biochemistry of the Reaction Between Bacterial Endotoxin and Amebocyte Lysate

Ontogeny and Phylogeny of Response to Gram-Negative Endotoxins Among the Marine Invertebrates
Frederik B. Bang 109

Secretion in Limulus Amebocytes is by Exocytosis
R.L. Ornberg and T.S. Reese. 125

The Reaction Between Bacterial Endotoxin and Amebocyte Lysate
Jack Levin. 131

Studies on Limulus Lysate Coagulating System
Teh-Yung Liu, Robert C. Seid, Jr., Joseph Y. Tai, Su-Mei Liang, Thomas P. Sakmar, and Joan B. Robbins 147

Structural Studies of the Coagulogen of Amebocyte Lysate From Limulus polyphemus
M.W. Mosesson, C. Wolfenstein-Todel, J. Levin, and O. Bertrand. 159

Amino Acid Sequence Studies on Horshoe Crab (Tachypleus tridentatus) Coagulogen and the Mechanism of Gel Formation
T. Takagi, Y. Hokama, T. Morita, S. Iwanaga, S. Nakamura, and M. Niwa 169

Comparative Studies on Hemocytes and Coagulogens of the Asian and the American Horseshoe Crabs
Fumio Shishikura and Koichi Sekiguchi 185

Binding Studies of Radioactive Lipopolysaccharide With Limulus Amebocytes
F.R. Rickles, P.D. Rick, P.B. Armstrong, and J. Levin 203

A New Chromogenic Substrate Method for Assay of Bacterial Endotoxins Using Limulus Hemocyte Lysate
T. Harada, T. Morita, S. Iwanaga, S. Nakamura, and M. Niwa 209

Endotoxin Determination in Viscous Opaque Solutions of Iron Dextran by Limulus Amebocyte Lysate
Stephen L. Gaffin 221

Editor's Summary, Section II, Part A: Biochemistry of the Reaction Between Bacterial Endotoxin and Amebocyte Lysate
Robert A. Prendergast. 229

B. Clinical Applications and Specificity of Limulus Lysate Tests

The Limulus Test: A Status Report
Jack Levin 235

Indications for the Use of the Limulus Lysate Assay in Bacterial Meningitis
Stuart E. Siegel and Ronald Nachum 245

Detection of Endotoxin in Urine: Rapid Diagnosis of Gram-Negative Bacteriuria by the Limulus Assay
James H. Jorgensen 255

The LAL Bead Assay for Endotoxin
Nick S. Harris and Robert Feinstein 265

Association of Endotoxemia and Bacteremia in Burn Patients: A Prospective Study
Albert T. McManus, Robert B. Lindberg, Basil A. Pruitt, Jr., and Arthur D. Mason, Jr. 275

Clinical Utility of the Limulus Test With Blood, CSF, and Synovial Fluid
Ronald J. Elin 279

Correlation of Limulus Amebocyte Lysate Assay With Accepted Endotoxin Assays
Peter A. Tomasulo 293

Endotoxaemia, as Detected by the Limulus Assay, in Fulminant Hepatic Failure and Cirrhosis
S.P. Wilkinson 303

The Limulus Assay for Endotoxemia, as Applied in Gastroenterology
Heinrich Liehr 309

Chemical, Physiochemical and Biological Properties of Bacterial Lipopolysaccharides
C. Galanos, M.A. Freudenberg, O. Lüderitz, E.T. Rietschel, and O. Westphal 321

III. LAL ASSAY OF PHARMACEUTICAL PRODUCTS

Pyrogen Testing of an Intravenous Fat Emulsion
Ruben J. Guzman and Harng-Shen Kuo 333

Stagnation to Anticipation — An Overview of LAL Use in the Pharmaceutical Industry
E.W. Ferry 339

Acceptance of the Limulus Test as an Alternative Pyrogen Test For Radiopharmaceutical and Intrathecal Drugs
James F. Cooper 345

Application of LAL for Detection of Endotoxin in Antibiotic Preparations
Susan J. Harrison, K. Tsuji, and R.M. Enzinger. 353

Limulus Amebocyte Lysate — A Means To Monitor Inactivation of Lipopolysaccharide
Kiyoshi Tsuji and Susan J. Harrison 367

Endotoxin Contamination of Some Commercial Preparations Used in Experimental Research
D. Fumarola and E. Jirillo 379

Applications and Advantages of the Limulus Amebocyte Lysate (LAL) Pyrogen Test for Parenteral Injectable Products
Carmine C. Mascoli and Marlys E. Weary 387

Quantitative Detection of Endotoxin by Nephelometry
John A. Dubczak, Richard Cotter, and Frank R. Dastoli 403

Quantitative Method for Determining Less Than a PG/ML of LPS
Frederica W. Valois. 415

A Turbidimetric Limulus Amebocyte Assay for the Quantitative Determination of Gram-Negative Bacterial Endotoxin
Joseph D. Teller and Kristine M. Kelly 423

Properties of Limulus Amebocyte Lysate and the Turbidimetric Assay for the Quantitative Determination of Gram-Negative Bacterial Endotoxin
R.J.L. Bondar, J.D. Teller, A. Bowanko, and K.M. Kelly 435

Use of the Limulus Amoebocyte Lysate Assay System for Detection of Bacterial Endotoxin in Fluids Associated With Hemodialysis Procedures
Loretta A. Carson, Norman J. Petersen, and Martin S. Favero 453

Characterization of an Endotoxin-Like Substance Recovered From an Intravenous Antibiotic Solution Involving Pyrogenic Reactions
Anita K. Highsmith, Roger L. Anderson, C. Michael West, and Richard E. Dixon 465

Limulus Amebocyte Lysate Assay Used in Evaluating Immunoregulatory Effects of Endotoxin in Type-C RNA Virus Reagents
Robert M. Donahoe and Robert L. Peters. 473

Endotoxin Contamination of Biological Reagents: Detection and Removal of Endotoxin With Limulus Amebocyte Lysate
F.R. Rickles, J. Levin, E. Atkins, and P. Quesenberry 485

Limulus Lysate Testing of Biologics Requirements for the Rabbit Pyrogen Test
H.D. Hochstein and Edward B. Seligmann, Jr. 501

Views on the Reliability of the Limulus Test
Peter J. Weiss 507

IV. BACTERIAL AGGLUTININS

Antimicrobial Defense Mechanisms in Limulus polyphemus
Ronald Nachum 513

Isolation and Characterization of a Bacterial Agglutinin in the Serum of Limulus polyphemus
Kevin J. Gilbride and Thomas G. Pistole. 525

Sites on the Lipopolysaccharide Molecule Reactive With Limulus Agglutinins
Homayoun Rostam-Abadi and Thomas G. Pistole. 537

Bacterial Agglutinins From Limulus polyphemus — An Overview
Thomas G. Pistole. 547

Discussion
Erwin Neter 555

Discussion Commentary
Thomas G. Pistole. 557

V. IDENTIFICATION OF MAMMALIAN CELL RECEPTORS

Some Aspects of the Biomedical Significance of Cell-Surface Sialic Acid and the Haemagglutinin of Limulus polyphemus
G.W.G. Bird 559

Ganglioside-Limulin Interactions
Régine Maget-Dana, Annie-Claude Roche, and Michel Monsigny. . . 567

Reactivity of Limulus polyphemus Hemolymph With D-Glucuronic Acid Containing Glycosubstrates
Peter Vaith, Gerd Uhlenbruck, Werner E.G. Müller, and
Elias Cohen 579

Differentiation of Human Leukemic from Normal Lymphocytes by Purified Limulus Agglutinin
E. Cohen, L.E. Blumenson, M. Pliss, and J. Minowada 589

Discussion to Section V
G.W.G. Bird 601

VI. BIOCHEMISTRY AND BIOPHYSICS OF LIMULUS AGGLUTININS: POTENTIAL APPLICATIONS

Limulin (Limulus polyphemus Lectin): Isolation, Physicochemical Properties, Sugar Specificity, and Mitogenic Activity
Annie-Claude Roche and Michel Monsigny **603**

Functional Implications of the Covalent Structure of Limulin: An Overview
J. Michael Kehoe, Ruth Kaplan, Steven S.-L. Li **617**

Purification and Properties of Lectins From the Japanese Horseshoe Crab, Tachypleus tridentatus
Satoru Shimizu, Mie Ito, Noriko Takahashi, and Makoto Niwa.... **625**

Affinity Chromatography With Agarose-Lectin Derivatives: Separation of Human Glycoproteins and Application to Erythropoietin Purification
J.L. Spivak and M.D. Hollenberg **641**

Interaction of Limulin With Membrane Glycoprotein
Mary A. Fletcher, Timothy M. Lo, and Karen E. Caldwell **655**

Limulus Agglutinins: Past, Present, and Future
John J. Marchalonis and Myron J. Waxdal **665**

VII. PERSPECTIVES

Finale
Frederik B. Bang **677**

Afterword
Lewis Thomas.. **681**

Author Index **685**
Subject Index.. **687**

ACKNOWLEDGEMENTS

The Symposium and Proceedings are possible through the support of:

The Fogarty International Center, through National Institute of Allergy and Infectious Diseases; National Cancer Institute, Division of Research Resources, Bureau of Biologics, FDA and the Fogarty International Center Bethesda, MD (Contract #263-78-C-0266).

Difco Laboratories, Detroit, Michigan

Microbiological Associates, Inc., Walkersville, MD.

Mallinckrodt, Inc., St. Louis, Missouri

Worthington Biochemical Corporation, Freehold, NJ

Seikagaku Kogyo Co., Ltd., Tokyo, Japan

Cutter Laboratories Inc., Berkeley, CA

Travenol Laboratories, Inc., Morton Grove, IL

Ortho Diagnostics Inc., Raritan, NJ

Burroughs Wellcome Co., Research Triangle Park, NC

UpJohn Company, Kalamazoo, MI

Marine Biological Laboratory, Woods Hole, MA

Juliette and Israel Cohen Memorial Fund

Health Research Inc., Roswell Park Division, Buffalo NY

Associates of Cape Cod, Inc., Woods Hole, MA

Dept. of Laboratory Medicine, Roswell Park Memorial Institute, Buffalo, NY

Special acknowledgements are extended to the co-chairman for their encouragement, suggestions and work: Fredrik B. Bang, M.D., Jack Levin, M.D.,

John J. Marchalonis, Ph.D., Robert A. Prendergast, M.D., Thomas G. Pistole, Ph.D., Carl Shuster, Jr., Ph.D., and Stanley W. Watson, Ph.D.

The Marine Biological Laboratory's staff - Homer Smith, Manager and Anne Camille Maher, Public Relations Officer helped before and throughout the meeting. James D. Ebert, Ph.D., Past-President, Director was encouraging and supportive from beginning to end.

Ms. Marie Zembrzycki served as a tireless administrative assistant to the organizer - editor. She and Ms. Grace Lombardo were most superior registration and accommodation secretaries. The Alan R. Liss Co. staff members - Ms. Ann Epner and Mrs. Rita Miranda were invaluable for their helpful instructions and information to all authors of papers.

Elias Cohen, Ph.D., the organizer-editor thanks all of the scientists who came from distant points to present papers or actively participate in discussions that made the Symposium worthwhile and the Proceedings a reality.

PREFACE

An international symposium was held on two current major biomedical applications of the marine arachnoid the Horseshoe Crab, (Horsefoot), *Limulus polyphemus* as well as of related species of *Limulidae*. The meeting concentrated on (a) detection of bacterial endotoxin by *Limulidae* amebocyte lysate, primarily *Limulus* amebocyte lysate (LAL) and (b) the use of agglutinins (lectins) of the hemolymph of *Limulidae* as biochemical "probes" of cellular membrane constituents.

The Symposium held October 7, 8 and 9, 1978, at the Marine Biological Laboratory, Woods Hole, Mass., was opened by remarks by Drs. James D. Ebert and Paul R. Gross, past and current President-Directors, respectively, of that marine biology research center. The tradition of the Laboratory for exemplary basic research was underscored. Many of the voluminous publications of the Laboratory have included investigations of the local denizen of Woods Hole, *Limulus polyphemus*, Horseshoe Crab (Horsefoot).

In a first aquaintance, the author accidently stepped upon the carapace of a large *Limulus*, in a salt marsh on Parsons Island, Chesapeake Bay, MD. In 1950, Drs. Alan and Mabel Boyden and Ralph DeFalco, Rutgers University, demonstrated the avid heteroagglutination of a variety of vertebrate erythrocytes by *Limulus* serum. Those observations led the editor to investigate *Limulus* agglutinins, for "harvest" of tumor cells and leukocytes from whole blood samples of cancer patients with WATNE et al, (1966). In his early work, the editor was unaware that NOGUCHI (1903) had reported non-hemolytic hemagglutination Limulus serum, in historic experiments the Marine Biological Laboratory. MARCHALONIS (1964) had independently corroborated NOGUCHI'S original observations.

Immunologic clinical and basic research was first directed toward identification of mammalian erythrocyte and lymphocyte cell populations, COHEN, et al (1968, 1972, 1974 and 1976).

In the meanwhile, there had been the astounding discovery at Woods Hole by Dr. JACK LEVIN, and F.B. BANG,(1964) that followed the earlier work of Dr. Fredrick B. Bang (1956) that an extract of Limulus amebocyte lysate (LAL) clots in the presence of minute amounts of bacterial endotoxins.

Academic and clinical curiosity prompted the organization of this Symposium; encouraged by a former John Hopkins classmate, who was President-Director of the Marine Biological Laboratory, Dr. James D. Ebert. Some active, distinguished scientists could not participate due to conflicts in work or travel schedules. However, it was considered timely for such a meeting. A willing committee was finally enlisted to help: Drs. Fredrik B. Bang, Jack Levin,Robert Prendergast, Thomas Pistole, Stanley W. Watson, and John J. Marchalonis.

Limulus agglutinin research had been featured in a New York Academy of Sciences meeting, in 1973, on biomedical applications of agglutinins of invertebrate and plant origins, COHEN (1974). However, the amebocyte lysate - bacterial endotoxin gel phenomena, its biochemistry, Horseshoe Crab ecology and conservation, as well as lectin research were embraced by a Japanese conference, published as the Proceedings of the Kasaoka Symposium on the Horseshoe Crab's Contribution to Medical Sciences (1976). The author is indebted to Dr. Makoto Niwa for informing him of the enormous Japanese contribution to Limulidae biology, lysate, amebocyte and agglutinin (lectin) research.

A remarkable private publication is that of the Japanese scientist, Dr. Hiroyuki Nishii, who edited the Monograph About Kabutogami, printed by the Sanyo Printing Co., Okayama, Japan (1975). It describes the paleontology, biology, natural history, art and literature references, ecology and conservation practices, as well as some applications of Japanese and related species of horseshoe crabs.

The history of horseshoe crab research was spearheaded in the 1930's at Woods Hole, by the electrophysiological studies of visual information by H.K. Hartline and his co-workers. A host of workers have studied the hemocyanin respiratory pigment protein of the *Limulus*. Others utilized *Limulus* hemocyanin as an immunologic adjuvant. Carl Shuster, Jr., pioneered American biologic, serologic and ecologic studies of horseshoe crabs, as well as comprehensive morphologic and anatomic studies of the vascular and other organ systems, SHUSTER (1978).

Lectin (agglutinin of plant or invertebrate, origin) research has made biochemical and biophysical advances in recent years due to active workers as SHARON and LIS (1972) and ROCHE and MONSIGNY (1974) and their associates. Most recently, the investigations of SHIMIZU, ITO and NIWA (1977) and associates, led by Professor Niwa, have emphasized the complexity of the interesting lectins of horseshoe crabs, with the requirement for further purification and characterization. Therefore, it was believed that a Symposium and publication in this fascinating area of study was urgent.

The goals of the October 1978 Symposium were to concentrate attention on two of the most recent biomedical applications - endotoxin detection and agglutinin isolation, characterization and applications. Target goals were set, but the chairmen organized and edited Symposium reports into several sections. The format of this Proceedings follows:

The Prologue is a drawing by Ms. Mollie Bang, that symbolized the rapid advancement of knowledge and application of the LAL phenomenon to the biomedical advantage of humans. The Epilogue is a poem by Drs. Jack Levin and Shimon Levit that characterize the past, present and future of Limulidae biomedical research.

Section I — Biology and ecology: Carl Shuster Jr., Chairman - Editor. Distribution, behavior, ecology, laboratory culture, amebocyte biology, potentials of aquaculture of Limulus. Dr. Sidney B. Galler comments on ecological impact and control.

Section II — Biochemistry of bacterial endotoxin and biomedical applications of Limulus. Dr. Robert Prendergast, Chairman - Editor - biochemistry of the reaction between bacterial endotoxin and amebocyte lysate; Dr. Jack Levin, Chairman Editor - clinical applications and specificity of LAL tests. Dr. Chris Galanos and colleagues give a succinct review of the "ABC's"of bacterial endotoxin. A commentary by Dr. Prendergast summarizes the biochemistry of the endotoxin - amebocyte phenomenon.

Section III — Applications in the pharmaceutical field - Dr. Stanley W. Watson, Chairman-Editor - detection of endotoxin in essential biological products, radiopharmaceuticals, as well as devices. Drs. H. Donald Hochstein and Peter Weiss evaluate quality control and reliability of the LAL test.

Section IV — Bacterial agglutinins of Limulus. Dr. Thomas Pistole, Chairman - Editor - The bacterial lectins of Limulus are reported upon with succinct commentary by Dr. Erwin Neter.

Section V — Mammalian Cell Receptors of agglutinins. Dr. Elias Cohen, Chairman - Editor - Receptors of more than erythrocytes and lymphocytes are investigated in the reports presented. Dr. G.W. Bird, evaluates the reports and targets for future goals of Limulidae lectin research.

Section VI — Biochemistry and biophysics of Limulus and related species agglutinins. Dr. John J. Marchalonis, Chairman - Editor - A critical review of the biochemistry and biophysics of the Limulidae agglutinins. Drs. Marchalonis and Waxdal underscore current status and recommend directions of frutiful investigation.

Section VII — The "Finale" Sum-Up by Dr. Frecrik B. Bang, dissects the current problems uncovered by the Symposium and points to where there is a need for future research.

Dr. Lewis Thomas, the "Biology Watcher" was unable to attend the Symposium, but he presents an Afterword, with plaudits for the basic research of Dr. Bang and the other scientists of the Symposium. Skillful observation and serendipity do benefit mankind."

REFERENCES

Bang FB (1956). A bacterial disease of Limulus polyphemus. Bull J Hopkins Hospital 98: p. 325.

Cohen E (1968). Immunologic observations on the agglutinins of the hemolymph of Limulus polyphemus and Birgus latro. Trans NY Acad Sci 20: p. 427.

Cohen E, Minowada J, Pliss M, Pliss L, Blumenson LE (1976). Differentiation of human leukemic from normal lymphocytes by Limulus serum agglutination. Vox Sang 31: p. 117

Cohen E, Roberts SC, Nordling S, Uhlenbruck G (1972). Specificity of Limulus polyphemus agglutinins for

erythrocyte receptor sites common to M and N antigenic determinants. Vox Sang 23: p. 300.
Cohen E, Rozenberg M, Massaro EJ (1974). Agglutinins of Limulus polpyhemus (Horseshoe Crab) and Birgus latro (Coconut Crab). Biomed Perspec Agglut Invert and Plant Origins, Ann NY Acad Sci 234: 28.
Levin J, Bang FB (1964). The role of endotoxin in the extracellular coagulation of Limulus blood. Bull J Hopkins Hosp 115: p. 265.
Marchalonis JJ (1964). Natural hemagglutinin from Limulus polyphemus. Fed Froc 23: p. 1468
Noguchi H, (1903). On the multiplicity of the serum haemagglutinins of cold-blooded animals. Zent Bakt Abt I Orig 34: p. 286.
Roche AC, Monsigny M (1974). Purification and properties of Limulin: A lectin (agglutinin from hemolymph of Limulus polyphemus). Biochim Biophys Acta 371: p. 242.
Sharon N, Lis H (1974). Lectins: Cell agglutinating and sugar-specific proteins. Science 177: p. 949.
Shimizu S, Ito M, Niwa M (1977). Lectins in the hemolymph of Japanese Horseshoe Crab, Tachypleus tridentatus. Biochem Biophys Acta 500: p. 71.
Shuster CN Jr. (1978). The Circulatory System and Blood of the Horseshoe Crab, Limulus polyphemus L.: A Review. US Dept. Energy, Fed Energy Regulatory Comm DOE/FERC/0014: 1-63.

Elias Cohen, Ph.D., Organizer - Editor.

SESSION I: BIOLOGY OF LIMULUS POLYPHEMUS

Carl N. Shuster, Jr.
Office of Regulatory Analysis
Federal Energy Regulatory Commission
Washington, D. C. 20426

This session provides an ecological background to a major consideration of the symposium -- the utility of horseshoe crabs for society. Attention is focused upon selected aspects of the biology and ecology of the American horseshoe crab, Limulus polyphemus (Linnaeus): beginning with the animal in its natural environment, its culture and responses in the laboratory are then considered, with concluding discourses on the amebocytes.

In presenting this information:

- the life cycle, distribution, and major pertinent behavioral patterns (in the field and in the laboratory) are described -- Limulus has anadromous-like spawning migrations; is eurythermal, voluntarily occupying a 15 - 40°C range in a thermal gradient, and exhibiting a low level of thermoregulatory precision with a preference (optimum) between 25 - 30°C; provides an especially good model for studying ecologically-dictated modifications of behavior, and early stages (egg-embryo-larva) are easily maintained and experimented upon in the laboratory.

- all four living species of Limulidae are compared -- the three Asiatic species, Tachypleus tridentatus, Tachypleus gigas, and Carcinoscorpius rotundicauda with Limulus,

- motility and activity of the blood cells are dramatically visualized, and

- basic growth requirements for the in vitro cultivation of Limulus amobebocytes are discussed.

DISTRIBUTION OF THE AMERICAN HORSESHOE "CRAB," *Limulus polyphemus* (L.)

Carl N. Shuster, Jr. 1/
Federal Energy Regulatory Commission
Office of Regulatory Analysis
Washington, D. C. 20426

INTRODUCTION

The American horseshoe "crab," *Limulus polyphemus* (L.), inhabits many stretches of the east coast of North America. It is an estuarine species with anadromous tendencies. Spawning occurs at high tide in late spring on estuarine beaches protected from the force of oceanic waves. Upon hatching and liberation from the beach nests, the young chiefly inhabit nearby intertidal flats. As they grow larger they move further and further away from their natal area into deeper water, many going out onto the continental shelf. Upon maturity, limuli return to the protected beaches to spawn.

Limulus was well-known to American Indians; coastal tribes used the long, serrated telson as a fish-spearhead. The earliest description of this animal was in a report on Virginia in 1588 by Thomas Hariot and it was depicted in one of the drawings of John White, the first English delineator of American natives, flora, and fauna (Quinn, 1955).

I was introduced to *Limulus polyphemus* (L.) by Professor Thurlow C. Nelson of Rutgers University in 1946 (Shuster, 1958b) and have had a continuing interest in the animal ever

---

1/ The views expressed herein are those of the author and do not necessarily represent the views of the Federal Energy Regulatory Commission.

Pages 3—26, Biomedical Applications of the Horseshoe Crab (Limulidae)

since. Part of my studies were at the Woods Hole Oceanographic Institution during the summers of 1949 and 1950 (Shuster, 1950), with the counsel of Dr. Alfred C. Redfield. Drs. Harry A. Charipper and William C. Cole, of New York University and Rutgers University, respectively, also were instrumental in guiding the formative stages of my studies. Much inspiration came from reading an extensive literature, not only on the animal but also on demonstrations of the utility of Limulus for science. If the "oyster is the best-known marine animal" (Clark, 1926), the horseshoe "crab" must be the best-known of all the marine animals utilized in the laboratory (Rudloe, 1971).

DIMENSIONAL ASPECTS OF DISTRIBUTION

Limulus polyphemus (L.) exists along the North American coast, Yucatan to Maine, within many estuaries and out onto adjoining continental shelf areas. In addition to this distribution in space, there is a time factor which is defineable largely in terms of the age of individuals and the season of the year.

Latitudinal Gradient 2/

The general north-south distribution of Limulus has been known for some time. Say (1818) reports the species and a presumed varient (Var. Beta) as common and abundant along the east coast of the United States. Later comments on their distribution were made by Milne-Edwards (1873) and Ives (1891). While no single data bank exists on the distribution of Limulus, by geographical areas and associated abundance, it is certain that each of the major estuaries along the eastern coast of the United States has a Limulus population. That these populations are relatively discrete was determined from a coastal survey during the spawning season in the years 1949 through 1954 and from other reports. Figure 1 and Table 1 summarize the data obtained from these populations.

2/ Based upon portions of Parts I and III of a dissertation in the Department of Biology accepted by the Faculty of the Graduate School of Arts and Science, New York University, in partial fulfillment of the requirements for the degree of Doctor of Philosophy (Shuster, 1958a). Material from subsequent studies is also reported.

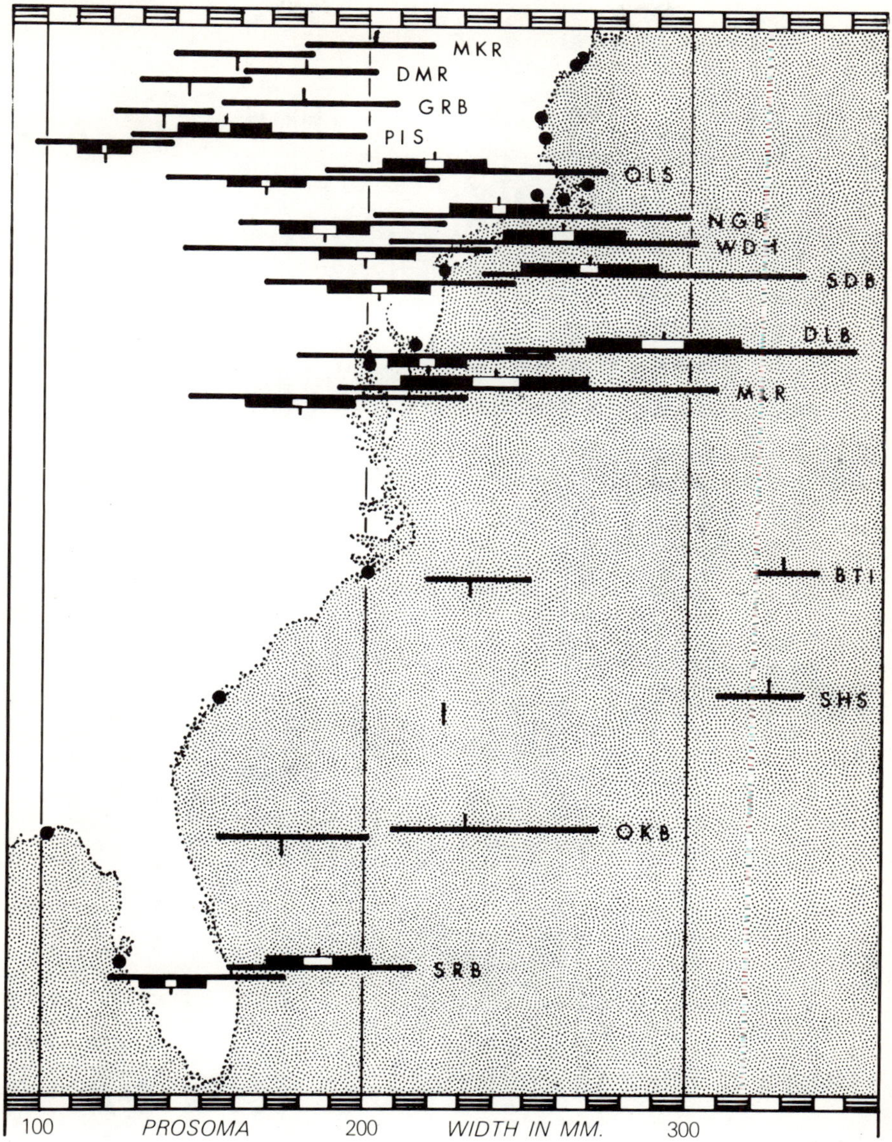

Figure 1. Prosomal widths of adult limuli vary, population to population, north to south (see Table 1): range in width (horizontal line), mean (vertical spike, one standard deviation on either side of the mean (one-half of the bar), two standard errors of the mean on either side of the mean (white portion of each bar) - graphic technique of Hubbs and Hubbs (1953).

Table 1. Adult size of Limulus polyphemus in relation to geographical distribution

Locality Site of Collection	Latitude Date of Collection	CODE	Sex	Prosomal Width (in millimeters) N *	Range	M **	σ ***	$2\sigma_M$ ****
Medomak River, Maine	(44°N)	MKR	M	17	139-182	158	-	-
Meeting House Cove	07/28/53		F	15	180-219	201	-	-
Damariscotta River, Maine	(44°N)	DMR	M	14	129-163	144	-	-
Pleasant Cove	07/28/53		F	11	161-202	180	-	-
Casco Bay, Maine	(44°N)	-	M	4	129-146	140	-	-
Thomas Point	07/30/53		F	5	172-208	191	-	-
Great Bay, New Hampshire	(43°N)	GRB	M	11	121-152	136	-	-
	07/24/53		F	20	154-209	179	-	-
Plum Island Sound, Mass.	(42°30'N)	PIS	*M	608	88-142	118	10	0.8
	07/20/53		M	552	97-140	118	8	0.7
			*F	484	128-200	156	12	1.1
	07/20/53		F	538	126-199	155	14	1.2
Cape Cod, Massachusetts	(41°40'N)	-	M	32	123-212	166	-	-
Barnstable Harbor	June-August 49 & 50		F	13	175-238	214	-	-
Orleans, Massachusetts	(41°40'N)	OLS	M	288	137-222	168	12	1.4
Pleasant Bay	07/10-11, 14/53		F	115	186-272	220	16	3.0
Woods Hole, Massachusetts	(41°30'N)	WDH	**M	99	142-238	199	15	3.0
			**F	99	206-302	260	19	3.8

Narragansett Bay, R.I. Bullocks Cove	(41°20'N)	NGB 06/12/53	M	54	159-224	186	14	3.8
			F	288	201-300	240	15	1.8
Raritan Bay, New Jersey Sandy Hook Bay	(40°25'N)	SDB 05/14/53	M	163	168-246	203	16	2.6
			F	185	235-336	268	21	3.0
Delaware Bay, New Jersey Pierces Point	(39°05'N)	DLB June 53 & 54	M	102	178-258	218	12	2.4
			F	56	243-351	291	24	6.4
Chesapeake Bay, Maryland Miles River	(38°50'N)	MLR June 53 & 54	M	142	142-229	179	17	2.8
			F	62	190-308	239	29	7.4
Beaufort Inlet, N.C. Bird Shoal	(34°40'N)	BTI 06/05/53	M	13	218-251	232		
			F	4	321-340	329	-	-
St. Helena Sound, S. C. off Otter Island	(32°30'N)	SHS June 53	M	1	240	-	-	-
			F	3	308-335	325	-	-
Ochlockonee Bay, Florida Bald Point and Ochlockonee Point	(30°N)	OKB 06/1-2/53	M	24	154-201	174	-	-
			F	17	208-272	231	-	-
Sarasota Bay, Florida and Palma Sola Bay	(27°30'N)	SRB	M	112	121-176	141	10	1.9

*M and *F = based on data collected by Baptist (1953)

**M and **F = from Pomerat (1933).

* N = number of specimens measured.

** M = mean.

*** σ = standard deviation.

**** $2\sigma_M$ = two standard errors of the mean.

A statistical analysis of the prosomal width (Figure 2) of animals from the sixteen localities corroborates field observations -- that Limulus polyphemus tends to form discrete populations and that these populations can be distinguished from each other on the basis of their external anatomy. Four observations concerning these populations emerged from the examination of the morphometric data: (1) the size of the adult animal varies with geographical distribution, (2) the body proportions vary with the size of the animal, (3) these populations can be distinguished from each other morphometrically, and (4) additional examples of sexual dismorphism have been recognized. Only the results of item 1 are described herein.

Adult size and geographical distribution. The data clearly show (Figure 1 and Table 1) that two populations within the Gulf of Mexico and those north of Cape Cod are made up of appreciably smaller animals than are the populations from the Carolinas to Cape Cod (Shuster, 1957a). Indeed, the largest limuli can be found in the estuaries from Georgia to New Jersey. Limulus occurs along the east coast of Florida, in Lake Worth, Lake Boca Raton, and in Biscayne Bay (Voss 1953), but no measurements have been made of these animals. The largest limuli found on the coasts of Yucatan and Campeche may reach 310 millimeters in prosomal width (Villalobos, 1953). But even these southernmost representatives of the species are smaller than those in the middle of their geographical range.

A few populations, such as the limuli from Plum Island Sound, Massachusetts, and the Miles River, Maryland, differ markedly in size from the general trend described above (see Figure 1). It is possible that salinity may be the critical factor delimiting growth in these localities. The salinity in the Miles River varies from 10 to 18 parts per thousand, with an average of about 13 parts per thousand (Pritchard, 1953), as compared to bottom salinites varying from 18.4 to 30.6 parts per thousand in the Ashapoo River and St. Helena Sound, South Carolina (Lunz, 1953).

The Plum Island Sound animals were the smallest adult limuli collected. Since a large number of both males and females was measured in the years 1952 and 1953, it was possible to compare these two samples of the same population to ascertain if there was a significant difference

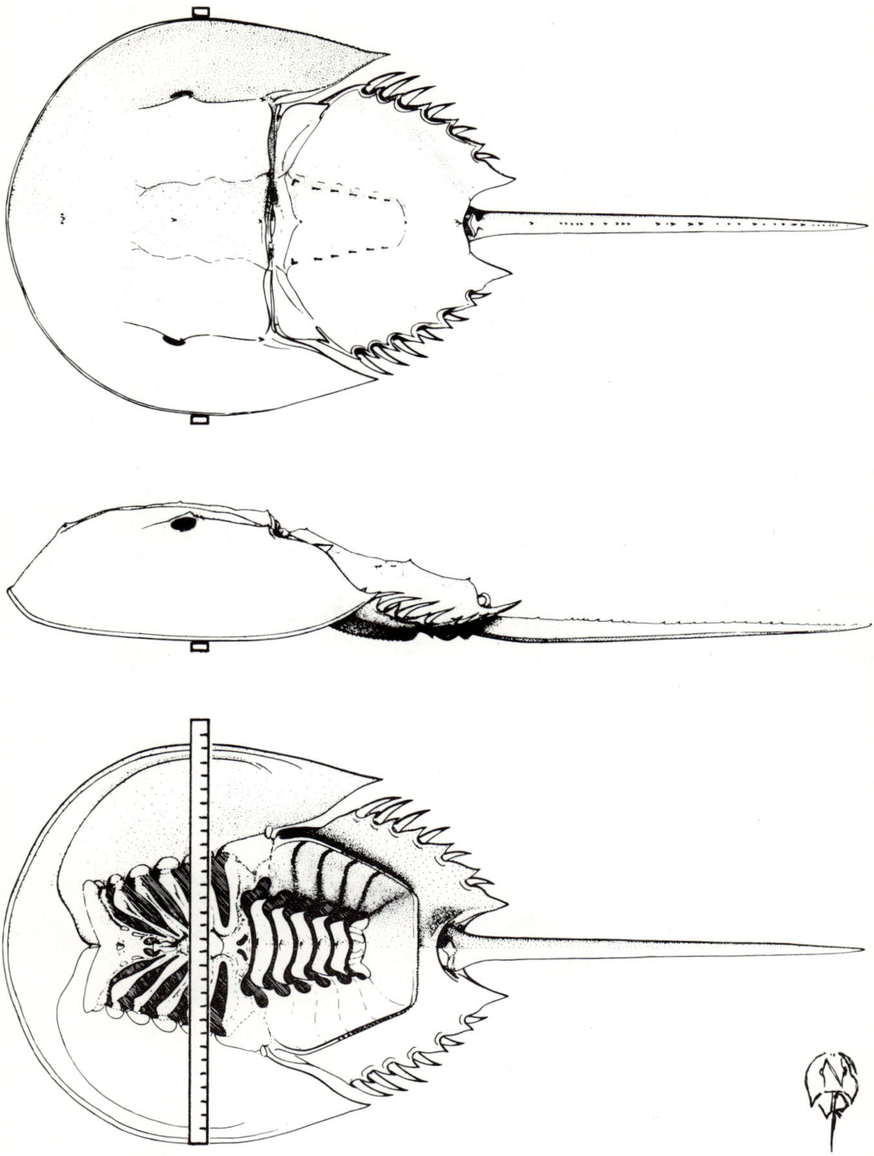

Figure 2. The dorsal, ventral, and lateral aspects of an adult male Limulus polyphemus showing the prosomal width measurement made by a ruler to the nearest millimeter: the ventral appendages are not shown (based on Shuster, 1958a).

between the 1952 and 1953 samples (Table 2). A comparison of these samples of the Plum Island Sound population was then made with a sample from Sandy Hook Bay, New Jersey, to see if there was a significant difference between the two populations (Table 3).

Table 2. Comparison, by the critical ratio method, of the 1952 and 1953 samplings of the adult population of Limulus in Plum Island Sound, Massachusetts.

1952 specimens*	1953 specimens*	Critical ratios**
males	males	0.9
females	females	1.2

*The data on the 1952 sampling were based upon Baptist (1953) and the study by Baptist et al. (1957).

** The critical ratio, $T = \frac{M_1 - M_2}{\sigma_{D_1 - D_2}}$, was determined by dividing the ratio of the difference of the means by $\sigma(M_1 - M_2)$, where $\sigma(M_1 - M_2)$ was computed by the formula

$$\sigma_{D_M} = \sqrt{\frac{\sigma S_1^2}{N_1 - 1} + \frac{\sigma S_2^2}{N_2 - 1}}$$

Table 3. Comparison, by the critical ratio method, of the adult populations of Limulus from Plum Island Sound, Massachusetts, and from Sandy Hook Bay, New Jersey.

Plum Island Sound	Sandy Hook Bay	Critical ratios
1952 males	1953 males	65
1953 males	1953 males	64
1952 females	1953 females	68
1953 females	1953 females	68

Comparison of the 1952 adult males from Plum Island Sound with those collected in 1953 gave a critical ratio of 0.9; the females, 1.2. Since a ratio of less than 3 is not considered significant, it may be assumed that there was no significant difference between the 1952 and 1953 samples. These samples, statistically, could easily be from the same population since a variation of this magnitude can be attributed to random errors of sampling. The statistical evaluation of the data is in agreement with observations in the field. Baptist *et al.* (1957) recovered 41 adult limuli that had been previously tagged in Plum Island Sound. These tagged animals were recovered from the places where they were tagged and other locations up to 19 miles distant (an average of 3 miles), from 2 to 370 days (with an average of 248 days) from the date when they were released. The distances travelled and the time elapsed indicate the range of the population. It, therefore, appears likely that the Plum Island Sound population is a local one.

Comparing the 552 (1953) Plum Island Sound males with the 163 miles from Sandy Hook Bay gave a critical ratio of 64; females, 68 (Table 2). The differences between these populations are so tremendous that there is no chance that it could result from random errors of sampling. The significance of these large critical ratios can best be demonstrated by choosing the much smaller critical ratio of 3.8 for comparison. The actual probability that a variation even of this magnitude (3.8) could be due to the random error of sampling in less than 2 in 1000. There can be little doubt that these populations are morphometrically distinct; geographically, they are also far removed.

Physiological Races

One interpretation of the above morphometric data would be that the horseshoe "crab" is a "temperate zone" creature, since the larger animals are "flanked" by southern and northern populations of smaller-sized animals. Presumably temperature is the chief environmental factor related to the physiological requirements of *Limulus*, particularly to its nutrition and reproduction. This inference is based on the importance of seasonal temperatures as a controlling factor in the geographical distribution of marine organisms (Hutchins, 1947). Further, the distribution of the extant species (Savory, 1935), and the fossil species of Limulidae (Stormer, 1952), from the Mesozoic and Cenozoic seas of

Europe, indicate that the Limulidae "prefer" warmer waters, thriving under "temperate" conditions.

The morphometric information, when related to geographical distribution, suggests that there are physiological races of Limulus. Effects of temperature upon limuli have been reported by Mayer (1914): the thermal death point of specimens from Woods Hole, Massachusetts, was 41° C as compared with 46.25° C at Marquesas Keys, Florida. The resistance to cold was correspondingly different, the Marquesas specimens ceasing to move at 0.8° C, while specimens from Anniaquam, Massachusetts, continued to move at temperatures below -1.1° C when ice crystals were forming in the water, and even survived being frozen into a solid block of ice. Individuals from Florida could survive temperatures of -2.3° C. Maximum activity was reported at temperatures of 41° C (Florida) and 38.1° C (Massachusetts). These findings by Mayer indicate that these animals differ physiologically. The concept of physiological species in estuarine organisms, discussed by Stauber (1950) in his studies on oysters and oyster drills, could apply equally well to Limulus. It would be of interest to conduct further observations and experiments on the comparative physiology of these populations of Limulus.

The population within the Miles River, mentioned earlier because of the smaller size of the adults (when compared with those of Delaware Bay) is further unique (Beaumont and Mansueti, 1955) in that it appears to be the only significant population within Chesapeake Bay. It apparently remains in the area throughout the year. The reasons for the smaller size and unusual distribution are not known.

Thus studies on Limulus polyphemus, throughout its geographic distribution, can be expected to further reveal the discreteness of populations. In addition to adult size, color of the adult carapace, eye pigmentation, serological correspondence among species and protein electrophoresis also demonstrate differences among populations.

Color of the adult carapace, before it is discolored by erosion and the activity of chitinoclastic bateria and attached organisms, may be related to substrate color and water turbidity. Atlantic coast limuli, north of the lighter-colored sands and less-turbid waters of Florida, tend to have dark-colored (browns and greens) carapaces. Limuli along the east and west coasts of Florida are a much lighter brown.

Sokoloff (1978) studied phenotypic variation in the color of the compound eye of Limulus. The resulting data are given in Table 4; employment of the Chi-square test demonstrated that the two populations were clearly different in the distribution of eye pigmentation. Sokoloff (1978) believed that other variations -- body size, the ocelli, spines and the telson -- are also of genetic origin.

Serologically (from data obtained by turbidometric measurements and agar-diffusion records of serum), three of the four extant (Waterman, 1958) species examined (Limulus polyphemus, Tachypleus gigas, and Carcinoscorpius rotundicanda) appear to be con-generic (Shuster, 1962). The existence of a fossil Limulus (coffini) from the Cretaceous of Colorado (Reeside and Harris, 1952) and a presumed tree of the Limulacea (Shuster, 1960, 1971) are evidence of conservative changes in morphometric characteristics.

An analysis of electrophoretic patterns of 24 tissue proteins led to a conclusion that considerable genetic variation exists within and between populations of Limulus polyphemus (Selander et al., 1970). Two populations on the Atlantic coast have fewer polymorphic loci than those from the Gulf of Mexico, but the average proportion of heterozygosity per individual is essentially identical in the two regions (Table 5).

Other Patterns of Distribution

Limulus is a benthic creature. While Limulus is a capable swimmer in quiet water, where it may be seen occasionally swimming, its usual manner of locomotion is "ambling" along the bottom in a "stiff-legged" gait. The swimming aspect has not been observed, in nature, by many persons -- at least as far as publications reveal (Wolff, 1977). In a crowded aquarium, however, young limuli move about a lot and are commonly seen swimming (Barnett, 1955 and Vosatka, 1970). The larvae, from the time they are released from their nests on the beach until undergoing ecdysis, apparently continually try to swim (Pearl, 1904) -- an activity probably favoring their immediate distribution.

Table 4. Comparison of eye color in two northern populations of Limulus polyphemus (from Sokoloff, 1978).

Populations	Black	Variegated	Pigmentless	Total
Cold Spring Harbor *	109	479	64	652
Woods Hole **	37	301	22	360

* A small harbor on the north shore of Long Island, New York.

** From specimens held at the Marine Biological Laboratory, Woods Hole, Massachusetts; locality not given but undoubtedly in the Cape Cod area.

Table 5. Estimates of proportion of loci polymorphic (total loci=25) per population and proportion of loci heterozygous in average individual of Limulus polyphemus (from Selander et al., 1970)

Locality	No. of individuals	No. of loci polymorphic	Proportion of loci polymorphic	Proportion of loci heterozygous per ondividual 1
Atlantic				
Woods Hole	17	5	0.20	0.056
Chincoteague	14	4	0.026	0.058
Gulf of Mexico				
Panacea	19	9	0.36	0.063
Panama City	14	7	0.28	0.050
Unweighted mean	16	6.3	0.25	0.057

1/ Calculated from mean allele frequencies.

Since horseshoe "crabs" are usually found, throughout their range (Yucatan to Maine), wherever the "mud flat" - sandy beach facies of an estuarine habitat exist, their distribution in the Gulf of Mexico is unusual. Within the Gulf, large populations of Limulus are found only along the coasts of Florida and Yucatan. This unusual distribution has not been studied, however. Howard T. Odum (1956) and I have assumed it may in large part be due to the combined lack of semi-diurnal tides and sufficient amplitude of the temperature-moisture regime in which Limulus eggs will hatch.

In most cases, Limulus spends a portion of its life within an estuary and part upon an adjoining continental shelf area. The early life stages are usually spent on the intertidal flats and shallow water areas near their natal beach to which, as adults, they return to mate and spawn. As they grow they wander even farther from the shallow water areas into the deeper estuarine and continental shelf waters. Such a distribution follows a general but typical pattern. These patterns can be aligned, for analytical purposes, with gradients such as salinity and water depth. The ability of Limulus to survive in low salinities was tested by Pearse (1928). Shuster (1957) gave 11 to 30.6ppt salinity as the usual range. Limulus obviously encounters higher salinities out on the continental shelf. Examples from two extremes of bathymetric distribution, intertidal and continental shelf, are dis-

The Bathymetric Pattern

The first example of distribution describes a nearshore concentration which is probably typical of Atlantic coast populations; the rest of the examples are of limuli on the continental shelf.

Observations were made on the distribution (Figure 3) of young limuli on the Cape May shore of Delaware Bay, in front of the New Jersey Oyster Research Laboratory (NJORL), Rutgers -- The State University at Pierces, during the summer of 1951 in conjunction with experiments on shellfish (Shuster, 1958a).

On July 11, twenty young limuli were obtained within one square foot of the tidal flat being examined for

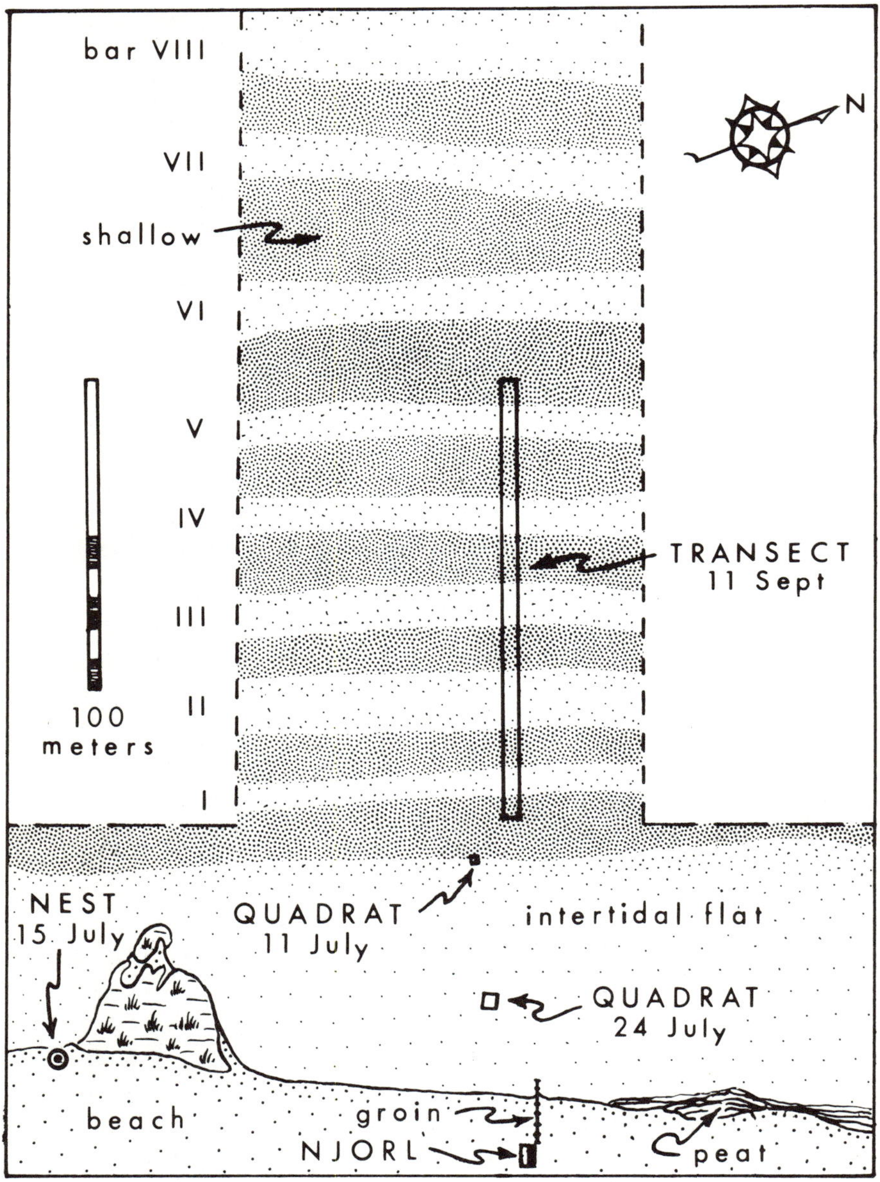

Figure 3. Portion of the inter-tidal flats at the New Jersey Oyster Research Laboratory (Pierces Point, Cape May), showing sand bars exposed at low tide on 11 September 1951. Areas in which limuli were collected in 1951 are indicated by arrows.

juvenile softshell clams Mya arenaria. There were six larvae, eleven first-tailed stage plus two exuvia, and three of the second-tailed stage.

A nest of Limulus, found on July 15, was located seven inches beneath the surface of the beach near the high-water line. The larvae had hatched beneath ground and all (some 3,450) were collected. Of these, one thousand were put in a 10-liter culture jar in recently collected bay water, and kept under constant aeration; temperature varied during the day from 19° to 29° C. Great swimming activity of the larvae was noted during the next three days (July 16-18), but from July 19-20 most of the larvae were "resting" on the bottom of the jar. The larvae were not observed to swim all day on July 21. That evening several exuviae were observed. On July 22, 374 "first-tailed" limuli were counted and removed from the jar to a similar one. On July 23, 568 more were counted and placed in the second jar; July 24, 47 more; July 25 only 1; July 26, 2 more "first-tailed" larvae. By July 29 only three of the 1,000 larvae had not molted.

On July 24 nearly 2,000 young limuli were collected from 16 square feet of tidal flat: 9 larvae; 336 "first-tailed" stage and two exuviae; 950 "second-tailed" stage and 17 exuviae; and 678 "third-tailed" stage. The "second-tailed" stage specimens were put in a culture jar and by July 31, 129 of them had molted. An additional 47 molted by August 1, at which time all specimens were preserved.

The next recorded observation was made on September 11, on a low tide, when it was noted that there were varying numbers of young limuli on the inter-tidal bars which paralleled the shoreline. At the time of the turning of the tide, other chores completed, a collection of these young limuli was made, starting with the furthest offshore bar on which the young were found (Bar 4). All specimens were collected across each bar, following a transect line perpendicular to the shoreline. No attempt was made to collect over equal areas on each bar. However, a one-quart paper milk carton was filled on each successive bar while proceeding towards the beach. No collections were attempted in the sloughs between the bars. The data compiled from this collection are summarized in Table 6. Only three growth stages were found and the distribution of these stages on the bars suggested a pattern of dispersion

Table 6. Early growth stages of Limulus and their distribution on inter-tidal flats at Pierces, Cape May, New Jersey.

Growth Stages	Number of Specimens	Prosomal Width (in MM) Range	Mean
larval*	20	2.8 - 3.8	3.3
Larval**	10	2.9 - 3.3	3.1
II *	21	4.8 - 5.5	5.0
II **	10	4.3 - 4.6	4.4
IV	128	8.5 - 10.5	9.5
V	1634	10.0 - 14.5	12.1
VI	353	14.5 - 19.0	16.6

Inter-tidal Distribution (Figure 3)	Number of Specimens of Each Growth Stage*** and percent Occurrence on Each Bar: 81 of IV	1946 of V	443 of VI
Bar 1	57 (10%)	473 (85%)	30 (5%)
Bar 2	21 (3%)	749 (92%)	28 (5%)
Bar 3	3 (1%)	554 (80%)	140 (20%)
Bar 4	0 (0%)	170 (42%)	225 (58%)

* Larvae collected on 15 July 1951 and the "first-tailed" stage (II) which emerged from these larvae.

** Larvae collected on 26 June 1954 and the "first-tailed" stage (II) which emerged from these larvae.

*** A number of specimens could be measured for body length only. Measurements on these have been added to the number of individuals reported in the top half of this table, to report all of the specimens collected.

of the young stages over the tidal flats. Thus, between July 24 and September 11 the fourth through sixth growth stages appeared and these animals continued to move offshore. Apparently storms of moderate intensity had little effect on this dispersion; several days of winds of moderate strength occurred in the weeks between the two large collections of animals. Presumably, the young limuli burrow into the flats and escape moderate turbulence of the water. It is perhaps significant that there were not any number of these young limuli on any of the bars beyond the fourth bar.

The modes of the three growth stages were clearly defined in the plotted data, but there is an overlap of the size distribution within successive growth stages even in the earliest stages.

Limulus on the Continental Shelf. The presence of Limulus in the inshore areas of the continental shelf has been reported only a few times (viz. Verrill and Smith, 1873; Sumner, Osburn, and Cole, 1911 -- in the vicinity of Vineyard Sound and Woods Hole, Massachusetts, respectively).

Recent records of Limulus polyphemus in north European waters, in an excellent review by Wolff (1977), has stimulated this discussion of the distribution of Limulus on the continental shelf of the Atlantic coast of the United States. The information given by locality from north to south, also identifies the area of the associated estuarine habitats: Plum Island Sound, Cape Cod Bay, Vineyard and Nantucket Sounds, Delaware Bay, Oregon Inlet, and the South Carolina coast.

A large population of Limulus occurs in Massachusetts waters north of Cape Ann. From 1640 "crabs" tagged in 1951-1954 (Baptist, Smith, and Ropes, 1957; Ropes, 1961), more than one fourth of the recoveries were made outside the Sound, mainly on beaches 2-10 miles from where they were released (Wolff, 1977). The data indicate that adult limuli spent a short time in the Sound each year, with peak numbers during June-August followed by an evident out-migration in September.

The next large population, distinquished from that centered in Plum Island Sound by the larger adults, is in the Cape Cod Bay area. There are more than one hundred thousand adults in this area and probably more than a million and a half immature limuli (Turner, 1956). The progressive decrease in the numbers of adults within Barnstable and Wellfleet Harbors during middle and late summer resulted from their migration into the Bay (Shuster, 1950; Turner, 1951).

Limulus occurs in Vineyard Nantucket Sounds (Verrill and Smith, 1873; Sumner, Osburn, and Cole, 1911; Ropes and Martin, 1960). In 1958 Limulus was frequently caught, by jet dredge, in the Horseshoe Shoal-Monomoy Point area and in the area off Edgartown Harbor. It is probable that these areas form the deep water component of the population frequenting the estuaries of the south shore of Cape Cod and the offshore islands.

In May and June, the beaches of Delaware Bay are visited by the largest spawning population of Limulus along the Atlantic coast (Shuster and Botton, 1978). During the winter adult limuli tend to congregate in the deeper waters of the bay and the immediate offshore area. That some of the "crabs" remain in the bay all winter has been shown by their frequent occurrence in oyster dredges (Nelson 1949). Offshore they have been caught by fishermen dredging for blue crabs or trawling for striped bass (Shuster, 1960). In August 1951, an immature male was dredged from a depth of 44 feet at the entrance to Delaware Bay (Haskin 1951). For at least a week in October 1955, adult horseshoe "crabs" were abundant in trawls made in 50 to 60 feet of water off Rehoboth Beach (Booth, 1955).

Surf clam surveys by Rutgers--The State University of New Jersey (Botton 1978), have encountered Limulus off the New Jersey coast. In April 1978, an average of 3 limuli per 5 minute hydraulic clam dredge haul were captured at 19 stations between 3 and 6 miles off Atlantic City. On banks 14 to 15 nautical miles off Atlantic City, large quantities of Limulus, some of which were dried and taken home as souvenirs or for sale, were taken in 1972 during extensive fisheries by Polish fishermen for herring (Wolff, 1977). This summer (Botton, 1978), within 3 miles of the shoreline, hauls containing 30 or more limuli were not unusual. At one station, 2 miles off Stone Harbor, more than 150 limuli were taken.

Limulus polyphemus is taken from time to time in catches made during the National Marine Fisheries Service (NMFS) shellfish surveys on the continental shelf of the east coast of the United (Ropes 1978). In a recent NMFS Albatross IV Scallop Survey (77-08), Limulus was taken at 37 to 66m depths at four locations between 37 and 56 km almost due east of Oregon Inlet, North Carolina. Eleven were mature males, ranging in prosomal width from 19.0 to 21.5cm; three had not lost the atrophied dactylus on the "clasper" legs. All the other limuli (108) appeared to be immature, with prosomal widths from 14.0 to 30.0cm (average of 19.5cm). While the largest of these (27.0 and 30.0cm) may have been mature females, neither they nor any of the other animals had the characteristic eroded spots indicative of mating. In fact, the carapace of all the limuli (119) was glossy and had no attached fouling organisms. Further information on this and other collections will be reported by Ropes at a later date.

Since 1953, Bears Bluff Laboratories and the S. C. Division of Marine Resources have conducted trawl sampling at regular stations throughout the estuarine and near shore marine area of coastal South Carolina. Data collected includes numerous records of Limulus polyphemus, both from estuarine and near shore marine waters. Locations where this species has been collected include the Murrells Inlet, Bulls Bay, Charleston Harbor, North Edisto River, St. Helena Sound, Port Royal Sound, and Calibogue Sound estuaries. Offshore marine locations where collections have been made include off Morris Island, Kiawah Island and Turtle Island. These data are available to interested investigators at the South Carolina Marine Center in Charleston (Bearden 1978).

STIMULUS TO SHOREWARD MIGRATION

Despite its polarized light sensitivity (Waterman, 1954, 1955, 1959) and its highly light optimized eye (Levi-Setti 1975), (Levi-Setti, Park, and Winston 1975), there is no evidence that eye aids the animal in navigation or in responding to seasonal changes in light intensity. But, if Limulus does use these laboratory-confirmed characteristics of the compound eye, these could be the mechanisms for sensing seasonal patterns in light in deep water and navigating back to their "natal" estuary.

Occurrence Outside Natural Range

Recent records of *Limulus polyphemus* in north European waters, including a review of all previous records outside their natural range, are given in an excellent review by Wolff (1977).

REFERENCES

Anon (1974). Solar energy - the *Limulus* connection. Univ Chicago Div Phy Sci Repts 21(2):1-2

Baptist JP (1953). Personal communication (US Fish & Wildlife Service, Newburyport, Mass).

Baptist JP Smith OR Ropes JW (1957). Migrations of the horseshoe crab, *Limulus polyphemus* in Plum Island Sound, Massachusetts. US Dept Int FWS Spec Sci Rept--Fisheries No 220:1-15.

Barnett L Editorial Staff of Life (1955). The World We Live In. Time Inc (New York):127 (photograph of swimming limuli).

Bearden CM (1978). Personal communication (Marine Resources Div SC, Wildlife & Marine Resources Dept, Charleston).

Beaumont B Mansueti R (1955). Distribution of horseshoe crabs in Chesapeake Bay mystifies experts. MD Tidewater News 12(4):1, 3-4.

Booth RA (1955). Personal communication (Univ of Delaware Marine Labs, Lewes).

Botton ML (1978). Personal communication (NY Oyster Research Lab Rutgers - The State Univ of NJ).

Clark AH (1926). The Smithsonian Institution, its function and its future. Science 63 (1963):147-157.

Haskin HH (1951). Personal communication (NJ Oyster Research Lab Rutgers - The State Univ of NJ).

Hubbs CL Hubbs C (1953). An improved graphical analysis and comparison of series of examples. System Zool 2(2):49-56, 92.

Hutchins LW (1947). The basis for temperature zonation in geographical distribution. Ecol Monogr 17(3):325-335.

Ives JE (1891). Crustacea from the northern coast of Yucatan, the harbor of Vera Cruz, the West coast of Florida and the Bermuda Islands. Proc Acad Nat Sci Phila 43:176-200.

Levi-Setti R (1975). Solar Energy Collection. US Patent No 3,899,672:11 pages.

Levi-Setti R Park PA Winston R (1975). The corneal cones of Limulus as optimized light collectors. Nature 253 (5487):115-116.

Lunz GR (1953). Personal communication (Bears Bluff Laboratories, State of SC, Wadmalaw Island).

Mayer AG (1914). I. The effects of temperature upon tropical marine animals. Papers Tortugas Lab Carnegie Inst Publ 183, 6:1-24.

Milne-Edwards A (1873). Etudes sur les Xiphosures et les Crustaces de la region Mexicane. Ann Sci Nat 17(5):1-67 & 12 pls.

Nelson TC (1949). Personal communication (NY Oyster Research Lab Rutgers - The State Univ of NJ).

Odum HT (1956). Personal communication (Institute of Marine Science, Univ Texas).

Pearl R (1904). On the behavior and reactions of Limulus in early stages of its development. J Comp Neur & Psych 14:138-164.

Pearse AS (1928). On the ability of certain marine invertebrates to live in diluted sea water. Biol Bull 54(5):405-409.

Pomerat CM (1933). Mating in Limulus polyphemus. Biol Bull 64(2):243-252.

Pritchard DW (1953). Personal communication (Chesapeake Bay Inst Annapolis MD).

Quinn DB (1955). The Roanoke Voyages, 1584-1590. Documents to Illustrate the English Voyages to North America Under the Patent Granted to Walter Raleigh in 1584. Hakluyt Soc 2nd Ser No CIV Vols 1 & 2:1003 pp.

Reeside JB Jr Harris DV (1952). A Cretaceous horseshoe crab from Colorado. J Wash Acad Scis, 42(6):174-178.

Ropes JW (1961). Longevity of the horseshoe crab, Limulus polyphemus (L). Trans Amer Fish Soc 90:79-80.

Ropes JW (1978). Personal communication (Northeast Fisheries Ctr Nat Marine Fish Serv NOAA/USDC, Woods Hole, MA).

Ropes JW Martin CE (1960). The abundance and distribution of hard clams in Nantucket Sound, Massachusetts, 1958. US Dept Int FWS Spec Sci Rept -- Fisheries No. 354:1-12.

Rudloe J (1971). The Erotic Ocean: A Handbook for Beach-comers-Chapter 23. Limulus polyphemus, The Horseshoe Crab. World Publishing: 263-267.

Savory TH (1953). The Archnida. Edward Arnold & Co (London): xi+218 p.

Say T (1818). An account of the Crustacca of the United States. J Acad Nat Sci Phila Pt II 1(5):433-436.

Selander RK Yang SY Lewontin RC Johnson WE (1970). Genetic variation in the horseshoe crab (Limulus polyphemus), a phylogentic "relic." Evolution 24(2): 402-414.

Shuster CN Jr (1950). Observations on the natural history of the American horseshoe crab, Limulus polyphemus. 3rd Rept on Invest of Methods of Improving the Shellfish Resources of Mass, Woods Hole Oceanographic Inst:18-23.

Shuster CN Jr (1953). Odyssey of the horseshoe crab. Audubon Mag 55(4):162-163, 167.

Shuster CN Jr (1954). A horseshoe "crab" grows up. Ward's Nat Sci Bull 28(1):1, 3-6.

Shuster CN Jr (1957a). Xiphosura (with especial reference to Limulus polyphemus. Geol Soc Amer Mem 67 Vol 1:1171-1174.

Shuster CN Jr (1957b). Horseshoe "crabs." Delaware Conservationist 1(3):15.

Shuster CN Jr (1958a). On morphometric and serological relationships within the Limulidae, with particular reference to Limulus polyphemus. Diss Abst 18(2):371-372.

Shuster CN Jr (1958b). "Study these" -- the story of the horseshoe "crab." The Staff Reporter (Wilmington Public Schools, DE) 10(8):4-5.

Shuster CN Jr (1960). Horseshoe "crabs" -- in former years, during the month of May, these animals dominated Delaware Bay shores. Estuarine Bull (Univ DE) 5(2):1-9.

Shuster CN Jr (1962). Serological correspondence among horseshoe "crabs" (Limululidae). Zoologica 47(1):1-8+ 1 pl.

Shuster CN Jr (1971). Xiphosurida. In Encyclopedia of Science and Technology. McGraw-Hill Book Co Inc.

Shuster CN Jr (1978). The Circulatory System and Blood of the Horseshoe Crab, Limulus polyphemus L.: A Review. US Dept Energy, Fed Energy Regulatory Comm, DOE/FERC/ 0014: 1-63.

Shuster CN Jr Horrell HC (1966). Limulas exoskeleton as a teaching aid. Turtox News 44(1):40-41.

Shuster CN Jr Botton M (1978). An Estimate of the 1977 spawning population of the Horseshoe "Crab" Limulus polyphemus in Delaware Bay (manuscript).

Sokoloff A (1978). Observations on populations of the horseshoe crab Limulus (= Xiphosura) polyphemus. Res Population Ecol 19(2):222-236.

Sumner FP Osburn RC Cole LJ (1911). A biological survey of the waters of Woods Hole and vicinity. Bull Bur Fisher 31 (Part I):1-442; (Part II):545-794.

Stauber LA (1950). The problem of physiological species with special reference to oysters and oyster drills. Ecol 31(1): 109-118.

Stormer L (1952). Phylogeny and taxonomy of fossil horseshoe crabs. J Paleontol 26(4):630-640.

Turner HJ Jr (1951). The horseshoe crab and the soft clam industry. 4th Rept on Invest of the Shellfisheries of Mass, Woods Hole Oceanogr Inst:5-7.

Turner HJ Jr (1956). The Shellfish Predators of Massachusetts. Mass Dept of Nat Res:1-20.

Verill AE Smith SI (1873). Report upon the invertebrate animals of Vineyard Sound and the adjacent waters with an account of the physical characters of the region. Rept of the US Fish Comm 1871-72:295-778 + pls I-XXXVIII.

Villalobos A (1953). Personal communication (Institute of Biology, Mexico City, Mexico).

Vosatka ED (1970). Observations on the swimming, righting, and burrowing movements of young horseshoe crabs, Limulus polyphemus. Ohio J Sci 70(5):276-283

Voss GL (1953). Personal communication (Marine Lab Univ of Miami Coral Gables, FL).

Waterman TH (1950). A light polarization analyzer in the compound eye of Limulus. Science, 111(2880):252-254.

Waterman TH (1954a). Directional sensitivity of single ommatidia in the compound eye of Limulus. Proc Nat Acad Scis 40(4):252-257.

Waterman TH (1954b). Polarized light and angle of stimulus incidence in the compound eye of Limulus. Proc Nat Acad Sci 40(4):258-262.

Waterman TH (1955). Polarized light and animal navigation. Scientific Amer:88-94.

Waterman TH (1958). On the doubtful validity of Tachypleus hoeveni Pocock, an Indonesian horseshoe crab (Xiphosura). Postilla Yale Peabody Mus Nat Hist No. 36: 1-17.

Waterman TH (1959). Animal navigation in the sea. Gunma J Med Scis 8(3):243-262.

Wolff T (1977). The horseshoe crab (Limulus polyphemus) in north European waters. Vidensk Meddr dansk naturh Foren 140:39-52.

LIMULUS POLYPHEMUS: A REVIEW OF THE ECOLOGICALLY SIGNIFICANT LITERATURE

ANNE RUDLOE
Department of Biological Science
Florida State University
Tallahassee, Florida 32306

Despite its abundance, phylogenetic importance and significance as a major neuro-physiological preparation, remarkably little has been known of the ecology and behavior of the horseshoe crab, Limulus polyphemus.

The life cycle in essence is as follows: sexually mature adults deposit eggs in the sand of a low energy beach through the spring and summer at the high water mark on a spring high tide. The eggs hatch in approximately 5 weeks in the northern Gulf of Mexico, producing trilobite larvae. The larvae remain in the sand for several weeks and are released into the water column on an appropriate high tide. They then molt through a series of juvenile instars, morphologically similar to adults, and are found on the intertidal sand flat adjacent to the breeding beach. As they grow, they move lower in the intertidal, becoming subtidal by the time they mature. Adults are found to depths of approximately 30 meters, but major concentrations are found in depths of about 5.6 meters on sand bottom. They remain offshore for the duration of the life cycle, returning to the beach only to breed. The species ranges from Maine to Yucatan on the Atlantic and Gulf Coasts of North America.

The most comprehensive work to date on the general biology of Limulus polyphemus is that of Shuster (1950, 1952, 1953, 1954 a,b) who provides extensive information on the external anatomy and morphology, comparative serology and natural history of Limulus. His morphological and serological work (1962) sheds light on the relationships

Pages 27–35, Biomedical Applications of the Horseshoe Crab (Limulidae)

of the genera of horseshoe crabs.

The horseshoe crab, is an especially good model for studying ecologically dictated modifications of behavior. It moves during the course of the life cycle from the high supralittoral through the intertidal and finally into the subtidal environment. The demands imposed by each environment change, requiring adoptive changes in biological responses, most particularly in behavior. All phases are accessible to field as well as laboratory observations.

This species is also valuable as a subject for ecological study for a variety of additional factors.

Due to *Limulus'* position as a "living fossil", eventual comparisons with ecologically similar crustacea could conceivably reveal something of the course of behavioral evolution, both in terms of more "primitive" and more "advanced" means of accomplishing common behavioral objectives, and in terms of alternate sensory capabilities. *Limulus polyphemus* is one of the few surviving species of the Xiphosura, the sword tails, the horseshoe crabs. A widespread and diverse group in the Paleozoic, today only 4 species in 3 genera survive. In addition to *Limulus*, which ranges from Maine to Yucatan on the North American coast, *Tachypleus tridentatus* and *T. gigas* range from Japan through Southeast Asia. The 4th species, *Carcinoscorpius rotundicaudata* is a freshwater form from Malaysia. The majority of the species in the fossil record occurred in brackish to freshwater environments. Forms nearly identical to *Limulus* date from the Triassic some 230 million years ago, and very similar horseshoe crabs appear in the Devonian, some 400 million years ago.

The presently accepted taxonomic organization of the Xiphosuridae was established by Pocock (1902) who recognized five species in three genera. The phylogenetic history and fossil taxonomy of the Xiphosuran line is presented by Stormer (1952) who traces a continuous line of development from the Silurian, some 425 million years ago. More recent palentological work on this group has included publications by Fisher (1975, 1977) on the biomechanics and locomotion of *Lumulus* and *Mesolimulus* as well as work by Eldredge (1970) on burrowing behavior.

The lateral compound eye of the horseshoe crab is "one of the best understood examples of sensory receptor function" (Adolph, 1971) at the electrophysiological level, as a result of almost 40 years of research using *Limulus* preparations.

There is a vast literature concerning the neurophysiology of the *Limulus* compound eye. Hartline (1927) first introduced the preparation, which has since received an enormous amount of attention (Wolbarsht & Yeandle, 1967) from physiologists. The advantages it offers include the size and constancy of the optic nerve fibers and electrical responses that parallel data from more complex eyes. Recently Barlow et al (1977) have reported a circadian rhythm in optic fiber responses that parallels rhythms observed in intact animals. And yet the visual system is not at all well known in terms of its relevance to the intact animal.

Nor have the reams of work on the visual neurophysiology and biochemistry of *Limulus* been paralleled by an equivalent quantity of research concerning the physiology, behavior, or ecology of the horseshoe crab.

Scattered observations on various aspects of the natural history of *Limulus* have appeared occasionally. In recent years these have included reports of hermaphroditism, estimates of the life span, descriptions of burrowing and swimming movements, and tagging studies employing Petersen tags (Baptist, 1953; Ropes, 1961; Eldredge, 1970; Vosatka, 1970; Baptist, et. al., 1957).

In addition to these older publications, several physiological studies have appeared in the last few years. Crabtree and Page (1974) and Thompson and Page (1975) have described external oxygen receptors on the book gills. Studies of osmoregulation have included McMannus (1969), Robertson (1970) and Hannan and Evans (1973), while respiratory physiology has been examined by Hammen and Lum (1964), Johansen and Peterson (1975), and Magnum (1976). Eagles (1973) has identified the lateral spines as mechanoreceptors. Frankel (1960) has described upper temperature limits, while Bursey (1977) has described the histological response of *Limulus* to injury. Brown and Kanouse (1973) provide a study of fertilization of *Limulus* eggs, and Hummun (1976) describes the interaction of beach neurofauna with developing Limulus eggs. Larval molting physiology has been studied by Jelga (1972) and Jelga, Costlow and Alspaugh (1972). Selander et. al. (1970) have examined Limulus for genetic variation.

The limited work on learning in invertebrates has included some research on Limulus, (Lahue, 1973), as studies of conditioned and unconditioned reflexes have appeared (Smith and Baker, 1960; Wasserman and Patton, 1970; Wasserman, 1973; Wasserman, in press).

Research on the sensory capabilities of Limulus has been substantial using neurophysiological approaches. Work by Waterman (1950; 1954a, b) has demonstrated the sensitivity of the lateral eye to polarized light. Wald and Krainan (1963) extended this work with electroretinograms of the median eye of young Limulus and found they are sensitive to ultraviolet light. Spectral sensitivities of the lateral, ventral and median eyes were measured (Millecchia and Mauro, 1969a, b; Nolte and Brown, 1969, 1970, 1972), while attempts to measure the animals' ability to discriminate visual patterns have been inconclusive (Campenhausen, 1967). Adolph (1971) recorded optic nerve spikes underwater and found continuous optic nerve activity under dim light in turbid water. Lall and Chapman (1973), working with intact animals in nature, found positive responses to full spectrum sunlight mediated through the lateral eyes and dorsal ocelli. Positive responses to near ultraviolet were shown to be mediated by the dorsal ocellus.

Work on the non-visual sensory systems of Limulus has included descriptions of chemoreceptors associated with respiration, responses to chemical, thermal, tactile and proprioceptive stimuli and descriptions of articular proprioceptors of the walking legs (Barber, 1951, 1956, 1958, 1960; Waterman and Travis, 1953; Pringle, 1956; Barber and Segel, 1960; Hayes and Barber, 1967; Ward, 1969). Eagles (1973) described the anatomy and innervation of the tailspine system relative to the behavioral function of the telson, and Wyse (1971) describes the receptors of Limulus chelae. Most of the literature, however, treats receptor physiology rather than exploring the functional significance of the receptor to intact horseshoe crabs. Nevertheless, these studies establish a broad spectrum of possible modes of reception for guidance stimuli.

The emerging importance of Limulus to biomedical research, as the source of Limulus lysate, requires more complete knowledge of the biology of the species, so that it can be wisely managed as a natural resource. This biochemical preparation from horseshoe crab blood provides a rapid and accurate assay for endotoxemia or gram negative sepsis (Levin et. al., 1970). There has developed in recent years a substantial harvest of horseshoe crabs for this purpose. Such harvesting has most often occurred on breeding beaches. If the resource is not to be depleted, we must have a better knowledge of Limulus populations and growth rates.

Limulus polyphemus, a Paleozoic relic, is behaviorally a relatively simple creature.

Yet it has succeeded evolutionarily. The eggs are protected from most potential predators, buried in the highest intertidal, and nearly all survive to hatching. The trilobite larva is a stage specialized for dispersal with an initial "swimming frenzy" and positive phototaxis that facilitates release from the nest. The juveniles are found in an environment that, due to its physiological rigor at low tide, permits few potential predators or competitors access to them, and their burial after low tide protects them from the pelagic invaders of the rising tide (Rudloe, in press). The adults orient precisely to wave surge and other orientational stimuli as they move to and from the breeding beaches (Rudloe and Herrnkind, 1976). Even as the continents have drifted, the Tethys has given way to the Atlantic and sea levels have fluctuated over hundreds of meters, there has evidently always been - somewhere - a sandy beach, a sandy somewhat estuarine tide flat next to it, and sand bottoms offshore. A horseshoe crab has needed little else. Our current interest in Limulus as a biomedical resource must not endanger the elegantly adapted species and bring its long story to an end.

LITERATURE CITED

Adolph AR (1971). Recording of optic nerve spikes underwater from freely moving horseshoe crabs. Vision Research 11: 979.

Baptist J (1953). Record of a hermaphroditic horseshoe crab, Limulus polyphemus. Biol Bull 64:243.

Baptist J et al (1957). Migrations of the horseshoe crab, Limulus polyphemus, in Plum Island Sound, Mass. US Fish and Wildlife Ser Sp Sci Rep Fish 220.

Barber S (1951). Contact chemoreception in Limulus. Anat Rec 124:561.

Barber S (1956). Chemoreception and proprioception in Limulus. J Exp Biol 131:51.

Barber S (1958). Properties of Limulus articular proprioceptors. Anat Rec 132.

Barber S (1960). Structure and properties of Limulus articular proprioceptors. J Exp Zool 143:283.

Barber S, Segel M (1960). Structure of Limulus articular proprioceptors. Anat Rec 137:336.

Barlow RB, Balanowski SJ, Brachman ML (1977). Efferent optic nerve fibers mediate circadian rhythms in the Limulus eye. Science 197:86.

Brown GG, Knouse JR (1973). Effects of sperm concentrations, sperm aging, and other variables on fertilization in the horseshoe crab, Limulus polyphemus. Biol Bull 144(3): 462.

Bursey CR (1977). Histological response to injury in the horseshoe crab, Limulus polyphemus. Am J Zool 55:458.

Campenhausen, C (1967). The ability of Limulus to see visual patterns. J Exp Biol 46:557.

Crabtree RL, Page CH (1974). Oxygen-sensitive elements in the book gills of Limulus polyphemus. J Exp Biol 60(3):631.

Eagles DA (1973). Lateral spine mechanoreceptors in Limulus polyphemus. Comp Biochem Physiol A Comp Physiol 44(2):557.

Eldredge N (1970). Observations on burrowing behavior in Limulus polyphemus (Chelicerata, Merostomata) with implications on the functional anatomy of trilobites. Am Mus Novitates No. 2436.

Fisher DC (1975). Swimming & burrowing in Limulus and mesolimulus. Fossils & Strata 4:281.

Fisher DC (1977). Functional significance of spines in the Pennsylvanian horseshoe crab, Euproops danae. Paleontology 3:175.

Frankel G (1960). Lethal high temperatures for 3 marine invertebrates, Limulus polyphemus, Littorina littorea & Pasurus longicarpus. Oikos 11:171.

Hammon CS, Lum SC (1964). Carbon dioxide fixation in marine invertebrates. Quantitative relations (Bdelloura candida, Lumulus polyphemus, Heterakis gallinae). Nature Zol (4917)416.

Hannan, JV, Evans DH (1973). Water permeability in some euryhaline decapods and Limulus polyphemus. Comp Biochem Physiol A Physiol 44:1199.
Hummun WD et al (1976). Meio-Fauna-macrofauna interactions. I. Sand beach meiofauna affected by maturing Lumulus eggs. Chesapeake Sci 17:292.
Jelga TC (1972). Development and molting physiology of horseshoe crab larvae. Am Zool 12:724.
Jelga TC, Costlow JD, Alspaugh J (1972). Effects of ecdysones and some synthetic analogs on horseshoe crab larvae. Gen & Comp Endocrin 19:159.
Johansen K, Peterson JA (1975). Respiratory adaptations in Limulus polyphemus. In Vernberg J (ed): "Physiol Ecol Estuarine Organisms", University of South Carolina Press, Columbia, SC.
Lahue R (1973). Learning in chelicerata. In Corning WC, Dyal JA, Williams (eds): "Invertebrate Learning", Plenum Press, New York.
Lall A, Chapman R (1973). Phototaxis in Limulus under natural conditions: evidence for reception of near-ultraviolet light in the median dorsal ocellus. J Exp Biol 58:213.
Levin J, Poore TE, Zauber NP, Oser RS (1970). Detection of endotoxin in the blood of patients with sepsis due to gram-negative bacteria. New Engl J Med 283:1313.
Magnum CP (1976). The ionic environment of hemocyanin in Limulus polyphemus. Biol Bull 150:453.
McMannus AMP (1969). Osmotic relations in the horseshoe crab Limulus polyphemus. Am Mid Nat 81:569.
Millecchia R, Mauro A (1969). The ventral photoreceptor cells of Limulus. II. The basic photoresponse. J Gen Physiol 54:310.
Millecchia R, Mauro A (1969). The ventral photoreceptor cells of Limulus. III. A voltage clamp study. J Gen Physiol 54:331.
Nolte J, Brown J (1969). The spectral sensitivities of single cells in the median ocellus of Limulus. J Gen Physiol 54:636.
Nolte J, Brown J (1970). The spectral sensitivities of single receptor cells in the lateral, median, and ventral eyes of normal and white eyed Limulus. J Gen Physiol 55:787.
Nolte J, Brown J (1972). Ultraviolet induced sensitivity to visible light in ultraviolet receptors of Limulus. J Gen Physiol 59:186.

Pocock R (1902). The taxonomy of recent species of Limulus. Ann Mag Nat Hist 9:256.
Pringle J (1956). Proprioception in Limulus. J Exp Biol 33:658.
Robertson JD (1970). Osmotic and ionic regulation in the horseshoe crab Limulus polyphemus (Linnaeus). Biol Bull 141(3):157.
Ropes J (1961). Longevity of the horseshoe crab Limulus polyphemus. Trans Am Fish Soc 90:79.
Rudloe AD, Herrnkind WF (1976). Orientation of Limulus polyphemus in the vicinity of breeding beaches. Mar Beh Physiol 4:75.
Selander RK, Yang SY, Lewontin RC (1970). Genetic variation in the horseshoe crab (Limulus polyphemus), a phylogenetic relic. Evol 24(2): 402.
Shuster C (1950). Observations on the natural history of the American horseshoe crab, Limulus polyphemus. Woods Hole Ocean Inst Contr 564:18.
Shuster C (1953). Odyssey of the horseshoe crab. Audubon Mag 55:162.
Shuster C (1954). A method for recording measurements of certain mollusks, arthropods, and fishes. Prog Fish Colt 16:39.
Shuster C (1954). On morphometric and serological relationships within the Limulidae, with particular reference to Limulus polyphemus. unpubl PhD dissertation, New York University.
Shuster C (1962). Serological correspondence among horseshoe crabs. Zoologica 47:1.
Smith J, Baker H (1960). Conditioning in the horseshoe crab. J of Comp and Physiol 53:279.
Stormer L (1952). Phylogeny and taxonomy of fossil horseshoe crabs. J Paleont 26:630.
Thompson C, Page CH (1975). Nervous control of respiration: oxygen-sensitive elements in the prosoma of Limulus polyphemus. J Exp Biol 62:545.
Vosatka E (1970). Observations on the swimming, righting, and burrowing movements of young horseshoe crabs, Limulus polyphemus. Ohio J of Sci 70:276.
Wald G, Krainin J (1963). The median eye of Limulus: an ultraviolet receptor. Proc Nat Aca Sci 50:1011.
Ward D (1969). Leg extension in Limulus. Biol Bull 136:288.
Wasserman G (1973). Unconditioned response to light in Limulus: mediation by lateral, median, and ventral eye. Loci Vision Research 13:95.

Wasserman G (in press). Unconditioned response to light in *Limulus*: role of time of day, day of testing, and intertrial interval. Vision Res.

Wasserman G, Patton D (1970). *Limulus* visual threshold obtained from light-elicited unconditioned tail movements. J Comp and Physiol Psych 73:111.

Waterman T (1950). A light polarization analyzer in the compound eye of *Limulus*. Science 111:252.

Waterman T (1954). Relative growth and the compound eye in *Xiphosura*. J Morphology 95:125.

Waterman T (1954). Polarized light and angle of stimulus incidence in the compound eye of *Limulus*. Proc Nat Aca Sci 40:258.

Waterman T, Travis D (1953). Respiratory reflexes and the flabellum of *Limulus*. J Cell and Comp An atomy 41:261.

Wolbarsht M, Yeandle S (1967). Visual processes in the *Limulus* eye. A Rev Physiol 29:513.

Wyse GA (1971). Receptor organization and function in *Limulus* chelae. Z Vergl Physiol 73:249.

ECOLOGY OF THE EXTANT HORSESHOE CRABS

Koichi SEKIGUCHI* and Koichiro NAKAMURA**

* Institute of Biological Sciences, The University of Tsukuba, Ibaraki, Japan and ** Biological Laboratory, Kunitachi High School, Tokyo, Japan

The horseshoe crabs had lived already in Mesozoic sea in almost the same forms as they are now, and therefore, they are sometimes called living fossils. It has been knew since early times that they occur in the coast of North America, Japan, Thailand and India. However, not many biological studies have been done.

We have been working for many years on mainly Tachypleus tridentatus, one of the four extant species of horseshoe crabs, and recently have collected all 4 species together. We are now doing comparative studies of them in morphology, physiology and embryology. We also have had chances to study the ecology of them in Japan, America, India and Thailand. In this paper, the extant horseshoe crabs are compared from the ecological point of view.

MATERIALS

Horseshoe crabs: Limulus polyphemus, Tachypleus tridentatus, Tachypleus gigas, and Carcinoscorpius rotundicauda were used.

RESULTS AND DISCUSSION

Morphological characteristics: Figure 1 shows the distinct characteristics in morphology in the 4 species of horseshoe crabs. Except the shapes of the genital operculum and the cross section of the telson, all the characteristics relate to the copulating behavior. The size of the body shows a variation even in one species (Sekiguchi, et al. 1976), but

Pages 37–45, Biomedical Applications of the Horseshoe Crab (Limulidae)

generally, L. polyphemus and T. tridentatus are large and T. gigas and C. rotundicauda are small.

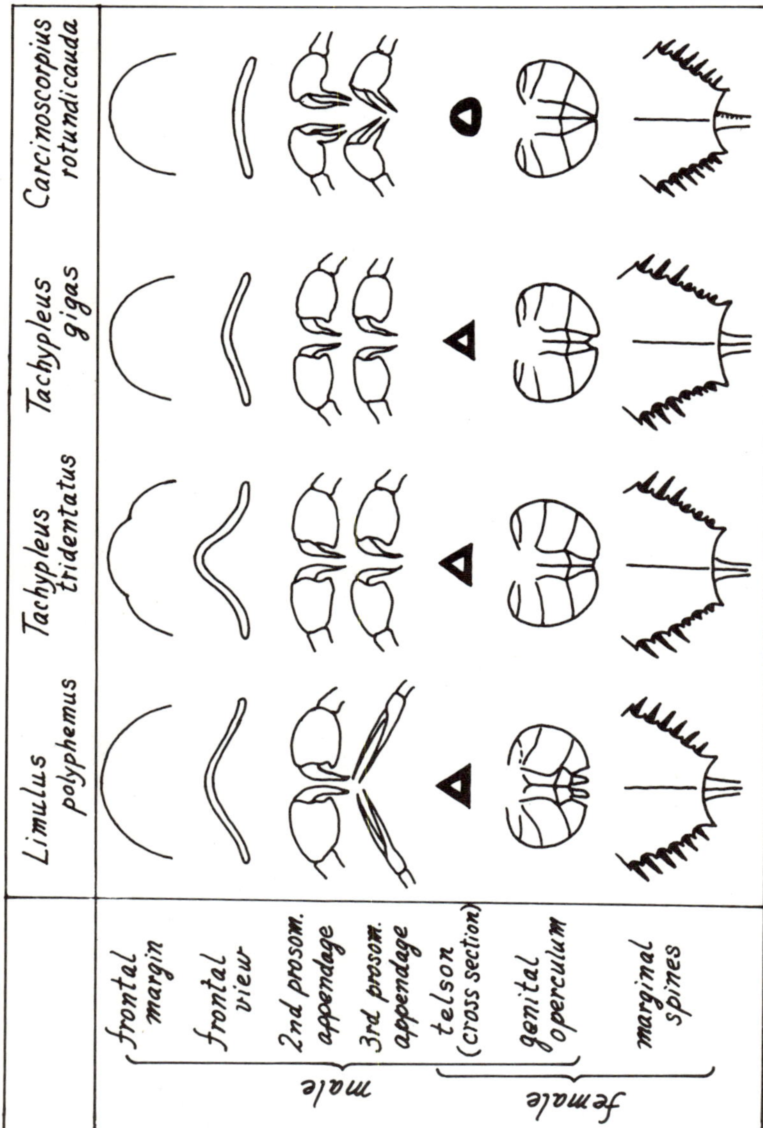

Fig. 1 Distinct morphological characteristics observed in the extant four species of the horseshoe crabs.

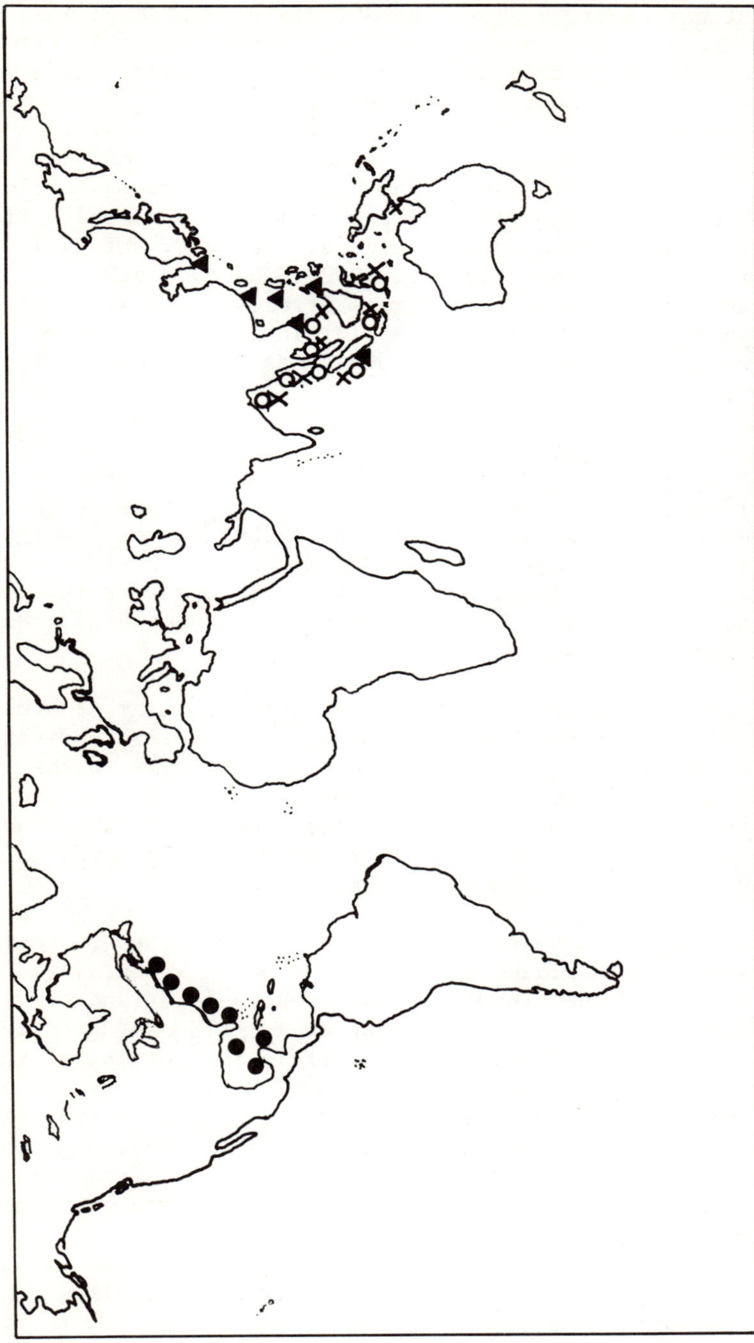

Fig. 2 Distribution of the extant 4 species of the horseshoe crabs. L. polyphemus: ●, T. tridentatus: △, T. gigas: ×, C. rotundicauda: ○.

Distribution: Figure 2 shows the distribution of each species. L. polyphemus occurs in the Atlantic coast of North America, widely distributing from Maine to Yucatan (Shuster, 1957). The other 3 species occur in the Pacific Ocean. T. tridentatus occurs in western and southern Japan, Taiwan, Philippines, and north Borneo (Pocock, 1902). Recently a specimen was recorded from Sumatora (Tsuchiya, K., personal communication). T. gigas and C. rotundicauda live in almost the same district. They were found on the coast of India, Thailand, Malaysia, Philippines, and Borneo (Pocock, 1902: Smedley, 1929). T. gigas was recorded also from Torres Straits (Pocock, 1902). L. polyphemus and T. tridentatus distribute from north to south, while T. gigas and C. rotundicauda distribute from east to west.

There is a question whether T. tridentatus shares the habitat with T. gigas or C. rotundicauda. It was recorded that T. tridentatus were found with T. gigas in northern Borneo (Pocock, 1902), but no record with C. rotundicauda. We showed the western limit of the distribution for T. gigas and C. rotundicauda was the coast of India, but that of T. tridentatus is not clear, although there is a report that this species was collected in Vietnam (Waterman, 1953). More exploration is necessary for the further details.

Habitat: All 4 species live in the sea of a calm by or a mouth of a river with the bottom of muddy sand. C. rotundicauda are often found not only in the sea water but also in the brackish water of the upper stream of a river. We found C. rotundicauda in a river about 4 km from the sea in Thailand. Annandale (1909) recorded that he found C. rotundicauda in a river 90 miles from the sea in India.

Breeding season: In L. polyphemus, the breeding season begins in May in southern places and lasts for about a month. In northern places, the breeding season begins later and terminates in July. The breeding season of T. tridentatus is from middle of July to August in Japan. It is interesting to know whether it begins earlier in southern places, such as Taiwan or Philippines. The breeding season of T. gigas is from end of April to August in Thailand. It is fairly long, but the peak is from June to July. C. rotundicauda lays eggs all round a year in the Gulf of Siam, Thailand. These facts are summarized in Table 1.

Breeding habits: A male of a horseshoe crab mates his partner by holding the female's opisthosoma by his claspers. In T.

	L. polyphemus	T. tridentatus	T. gigas	C. rotundicauda
Breeding season	May(south) to July (northe)	Middle July to August (Japan)	End of April to August (Thailand)	All round a year(Thailand)
Place to spawn	In the sand of the beach between high and low tide lines at the major tide. About 15 cm deep.			In the mud of sea shore or river bed. 2-5 cm deep.
Number of males to females	much more males than females	1:1	1:1	1:1

Table 1. Characteristics related to breeding in 4 species of the horseshe crabs.

tridentatus, all the adults found in the breeding season are couples. Therefore, the ratio of males to females is 1:1. It is very difficult to find solitary males or females in this species. L. polyphemus has a different habit. In breeding season, males come to the shore and wait for the females' arrival. When the females arrive, the males make couples (Shuster, 1953). But the number of the males is much larger than that of the females, many solitary males are found in nature. T. gigas or C. rotundicauda are found in couples. But sometimes, we can see solitary males of T. gigas or C. rotundicauda in Thailand. This is because the people eat the eggs of these horseshoe crabs there, and only females are caught and sold in the market. The males are backed to the sea.

The coupled animals are hard to separate. We have kept a couple of T. tridentatus in a laboratory. They did not separate until the next year.

At the spawning time, the couple come to the beach with a high tide at a major tide and lay eggs in the sand. A mature female carries many eggs (Table 2), but does not lay all the eggs at once. The number of the eggs laid at one place, or a nest, varies depending on species as shown in Table 2. When a female lays eggs, she lays eggs in one nest and moves twward the shore a little where she lays another egg-mass. Sometimes, several nests were made by one female in one night.

	L. polyphemus	T. tridentatus	T. gigas	C. rotundicauda
Number of eggs carried by a female	ca. 90,000	ca. 20,000	ca. 8,000	ca. 10,000
Number of eggs per a nest	ca. 900	200 - 500	ca. 400	80 - 150
Diameter of an egg (mm)	1.6 - 1.8	3.0	4.0	2.0

Table 2. Number of eggs and their size in 4 species of the horseshoe crabs.

To know whether one couple come to the shore only once in one major tide, we marked the individual of *T. tridentatus* with a float connected to the animal by a thread, so that we could observe and identify the same couple moving under the water. We found that a couple went back to the deep water after the spawnings at one high tide, and came to the shore again with the next high tide (Sekiguchi, et al. Unpublished data).

Though *T. gigas* and *C. rotundicauda* distribute in the same district, the former lays eggs in the sea water, while the latter lays eggs both in the sea and the brackish water. It seems that *C. rotundicauda* is adapted to a wide range of salinity. The nests of *C. rotundicauda* are found nearer to the surface of the ground than others (Table 1).

At the spawning time of *T. tridentatus*, we can observe air bubbles comming up from the water. The bubbles make a ring on the surface of the water. We can count how many couples of *T. tridentatus* are under the water by counting the number of the bubble rings.

Embryonic development: We have succeeded in the artificial fertilization in all 4 species of the horseshoe crabs. The details of the embryonic development in *T. tridentatus* was described elsewhere (Sekiguchi, 1973). Under the same condition, *L. polyphemus* showed the fastest growth. As shown in

Table 3, it took only 15 days from fertilization to hatching for an embryo of L. polyphemus, while it took 45 days in T. tridentatus. An hatched larva of L. polyphemus molted six times during the year it hatched (Shuster, 1950), while the larva of T. tridentatus showed no molting until the next spring in Japan. T. gigas and C. rotundicauda were intermediate as shown in Table 3.

	Days from fertilization to hatching	Number of moltings in the hatched year
L. polyphemus	15	6
T. tridentatus	45	0
T. gigas	32 - 35	2 - 3
C. rotundicauda	32 - 35	2 - 3

Table 3. Time for the embryonic development and the number of the moltings during the hatched year in 4 species of the horseshoe crabs.

Although there are studies by Shuster (1954) and by Goto and Hattori (1929), the relation between the age and the size of horseshoe crabs is not well known. From our embryological studies, it seems that T. tridentatus took more time to become an adult than L. polyphemus.

All the horseshoe crabs lay eggs in the muddy sand between the high and low tide lines of the major tide. Therefore, the fertilized eggs of the horseshoe crabs are mostly over the sea level. It is necessary for the egg to keep the water inside. The inner egg membrane is tough and effectively keeps the water. During the development, the outer eggshell is punctured by the swelling of the egg. The reason of this swelling seems to be that the egg takes water from the environment. It is interesting to know what mechanism the inner egg membrane has to take water inside but not to let it out.

In conclusion, from our ecological studies, C. rotundicauda differs from others in the breeding season and in the habitat, or the adaptiveness to lower salinity. In the ratio of mature males to females, L. polyphemus seems unique. L.

polyphemus is also unique in that it grows faster than any other species. However, we think that these 4 species of the extant horseshoe crabs are mostly similar in ecology.

The horseshoe crabs are phylogenically interesting animals in Phylum Arthropoda. It is especially interesting that only one species L. polyphemus live in the Atlantic Ocean widely distributed, while 3 species live in the Pacific Ocean and their living areas are overlapped. Furthermore, their distributing area is almost same as that of L. polyphemus in distance. It is a question why L. polyphemus had not differentiated into more than one species in spite of its broad spreading.

Recently, the plasma of L. polyphemus has been known as a useful biomedical agent for "Limulus test". The plasma of T. tridentatus is also effective as that of L. polyphemus, and therefore, many of them have been killed for collecting the plasma. However, we think these animals are very interesting and important for the biological sciences as well as useful in the medical sciences. They will be extinct soon if we do not try to protect them. Fortunately, we can take about 1/3 of the whole plasma without killing them. This system, a "blood donation" system in the horseshoe crabs has advantages to both the animals and the human beings. In Japan, the Association for Conservation of Horseshoe Crabs was established in this year. We are campaigning to protect the horseshoe crabs. We think it is important for all the scientists working on the horseshoe crabs in the world to cooperate for the conservation of this small, harmless, useful and valuable animals for the progress in the biological and the medical sciences.

ACKNOWLEDGMENT

We are deeply indebted to the cooperation of Prof. John D. Costlow of Duke University Marine Laboratory, North Carolina, U.S.A., Prof. Smarn Srythunya of Srinakarinwirot University, Bangsean, Thailand, and Mr. T. K. Sen of Zoological Survey of India, Calcutta, India.

REFERENCES

Annandale N (1909). The habits of Indian king crabs. Rec Ind Mus 3:294.

Goto S, Hattori O (1929). Notes of the spawning habits and growth stages of the Japanese King Crabs. X Cong Intl Zool

Pocock RI (1902). The taxonomy of recent species of Limulus. Ann Mag Nat His 9:256.

Sekiguchi K (1973). A normal plate of the development of the Japanese horseshoe crab, Tachypleus tridentatus. Sci Rep Tokyo Kyoiku Daigaku Sec B 15:153.

Sekiguchi K, Nakamura K, Sen TK, Sugita H (1976). Morphological variation and distribution of a horseshoe crab, Tachypleus gigas, from the Bay of Bengal and the Gulf of Siam. Proc Jap Soc Syst Zool 12:13.

Sekiguchi K, Nishiwaki S, Makioka T, Srithunya S, Machjajib S, Nakamura K, Yamasaki T (1977). A study on the egg-laying habits of the horseshoe crabs, Tachypleus gigas and Carcinoscorpius rotundicauda in Chonburi area of Thailand. Proc Jap Soc Syst Zool 13:39.

Shuster Jr CN (1950). Observations on the natural history of the American horseshoe crab, Limulus polyphemus. 3rd Rep Invest Meth Improv Shellfish Resources Massachusetts 564:18.

Shuster Jr CN (1953). Odyssey of the horseshoe crab. Zoolog 47:1.

Shuster Jr CN (1954). A horseshoe "crab" grows up. Ward's Nat Sci Bull 28:3.

Shuster Jr CN (1957). Xiphosura (with especial reference to Limulus polyphemus). Geol Soc Amer 67:1171.

Smedley N (1929). Malaysian King-Crab. Bull Raffles Mus 2:73.

Waterman TH (1953). Xiphosura from Xuong-Ha. Amer Sci 41:292.

THERMOREGULATORY BEHAVIOR AND DIEL ACTIVITY OF *LIMULUS POLYPHEMUS*

William Wallace Reynolds and
Martha Elizabeth Casterlin
Department of Biology
The Pennsylvania State University
Wilkes-Barre, Pennsylvania 18708 U.S.A.

The primitive chelicerate arthropod *Limulus polyphemus* occurs along the Gulf of Mexico and Atlantic coasts of North America from Yucatan to Maine (Shuster 1957) and Nova Scotia (Robertson 1970). *Limulus* is tolerant of a wide range of environmental extremes, including hypoxia (Johansen and Petersen 1975), salinities ranging from full sea water down to 5% sea water (1.7 ppt)(Pearse 1928; Robertson 1970), and temperatures ranging from below 0°C to over 40°C (Mayer 1914; Fraenkel 1960). In spite of its eurythermality, the northernmost and southernmost populations are smaller in maximum size than those in the center of the species' geographic range, suggesting that temperature plays a significant role in delimiting the range (Shuster 1957). While *Limulus* shows "marked adaptability to extremes of temperatures" (Mayer 1914), Shuster (1957) notes that "presumably temperature is the chief environmental factor related to the physiological requirements of *Limulus*," and further observes that "Limulidae 'prefer' the warmer water, thriving under 'temperate' conditions." Individual *Limulus* exhibit seasonal migrations that appear related to temperature. Adults migrate shoreward from deeper waters for spawning in the spring and early summer, earlier in southern waters and later farther north (Shuster 1954, 1957). As winter approaches, there is a "great exodus to their winter habitat in deeper

Pages 47–59, Biomedical Applications of the Horseshoe Crab (Limulidae)

waters" (Shuster 1953), apparently to avoid cold winter temperatures; and Shuster (1953) further notes that "when fall approaches the small horseshoe crabs stay on the mud flats, burying themselves to escape the cold water and ice." Field observations indicate that *Limulus* also tend to burrow into the sand when stranded on the beach at low tide during the day during spring and summer, in this case apparently to avoid high temperatures. Thus burrowing, as well as onshore-offshore migrations, appear to constitute thermoregulatory behavior in this species.

Another environmental factor of importance in the behavioral ecology of *Limulus* is light intensity and its variations with the day-night cycle. Experiments on photic orientation of *Limulus* have variously reported it to be photopositive (attracted to light)(Loeb 1893; Cole 1923, 1924; Wolf and Zerrahn-Wolf 1937; von Campenhausen 1967; Lall and Chapman 1973) or photonegative (avoiding light) (Loeb 1893; Northrop and Loeb 1923; Cole 1924), at various stages in their life cycle. Newly hatched *Limulus* larvae apparently are photopositive (Loeb 1893; Northrop and Loeb 1923), but become photonegative within less than 24 hours under laboratory conditions (Cole 1923, 1924; Sokoloff 1978). Townes (1938) reported that young limuli, 5-9 mm long, swim near the surface at night. Apparently the adults become somewhat photopositive during the spawning season when they move into shallow water and onto the beach and are therefore most readily available for experimental work (Lall and Chapman 1973); however, Shuster (1954) notes that "the maximum numbers of spawning animals collect during the hours of subdued light of an evening, full moon, or dawn tide." Both the lateral eyes and the dorsal ocelli participate in the photic responses (Lall and Chapman 1973).

In addition to direct phototactic responses to light, it is known that the photoperiod (light-dark cycle) often entrains (synchronizes) endogenous circadian rhythms in many organisms, and such rhythms are often expressed in terms of locomotor activity cycles as well as physiological rhythms. Circadian rhythms have been described for the sensitivity and neural activity of the *Limulus* eye

(Barlow et al. 1976a, 1976b). Barlow et al. (1976a), measuring electroretinograms and optic nerve responses under constant darkness, found that "when *Limulus* is kept in constant darkness, the lateral eye produces larger ERG and optic nerve responses during nighttime than during daytime. These responses to brief flashes of constant light intensity increase in amplitude at dusk, remain high during the night, decrease at dawn, and remain low during the day. The level of spontaneous activity recorded from a single optic nerve fiber is inversely related to the amplitude of the light-evoked responses. Cutting the lateral optic nerve abolishes these circadian changes in neural activity. Fibers in the proximal stump of the cut lateral optic nerve give synchronous bursts of impulses during nighttime and little or no activity during daytime." Such circadian changes in sensitivity of the eye should result in corresponding changes in photic behavior, and would presumably be related to diel activity patterns.

Apparently there has been a complete lack of experimental data on behavioral thermoregulation (preferred and avoided temperatures) and diel activity of *Limulus*, so we undertook an investigation using our electronic shuttlebox (Ichthyotron--described by Reynolds et al. 1976, and by Reynolds 1977a) which permits an animal to control water temperatures, and thereby its own body temperature, by shuttling between two chambers differing slightly in temperature. The direction of movement determines whether heating or cooling is initiated in both chambers; thus the temperature continually changes either upward or downward in accordance with the animal's movements. Movements of the animal are monitored by a pair of photocells as it interrupts narrowly focussed white-light beams during the course of its movements between the "warmer" and "cooler" chambers, so that information on the diel organization of locomotor activity of the animal is continuously monitored and recorded alongside the temperature. A wide variety of aquatic organisms have been tested in this way, including amphibian larvae (Casterlin and Reynolds 1977a, 1978a), teleost fishes (Reynolds and Casterlin 1976a, 1976b, 1977a,

1977b, 1978a, 1978b), holostean fish (Reynolds et al. 1978a), sharks (Reynolds and Casterlin 1978c; Casterlin and Reynolds 1979), lamprey ammocoete larvae (Reynolds and Casterlin 1979a), and decapod crustaceans (Casterlin and Reynolds 1977b). In general, although some minimum level of activity is necessary for thermoregulation, there is little relationship between temperature and activity level in this device (Reynolds and Casterlin 1976b; Reynolds 1977a; Reynolds et al. 1978b).

We obtained ten juvenile *Limulus*, 20-50 mm in prosomal width, from the Gulf Specimen Co. in Panacea, Florida (northeastern Gulf of Mexico). We tested each of these individually for 3-day periods in the Ichthyotron during the month of May, using a natural photoperiod and window light. The animals were held and tested in artificial sea salts at about 30 ppt salinity. Shelters were provided, but there was no substrate in which they could burrow; nor could they crawl out of the water. The animals were not fed during the tests, to prevent feeding activity or time of feeding from interfering with the results. Activity was quantified as numbers of light-beam interruptions per hour. Water-temperature data from the second and third day were used, allowing the initial 24 hours for gravitation to the thermal "final preferendum" (Fry 1947; Richards et al. 1977; Reynolds and Casterlin 1979b).

The distribution of voluntarily occupied (preferred) temperatures did not differ significantly between day and night (i.e,, there was no diel rhythm of preferred temperature evident, such as occurs in some species--cf. Reynolds et al. 1978b), so the pooled data were combined into a single histogram (Fig. 1) to show the relative amounts of time spent at each temperature (1°C class intervals). The range of preferred temperatures spanned 25°C, from 15 to 40°C, out of a potentially available range of 0 to 50°C. The broad span of occupied temperatures is indicative of a lesser degree of thermoregulatory precision than is typically exhibited by other taxa tested (see above), which generally occupy a range of 10°C or less. This

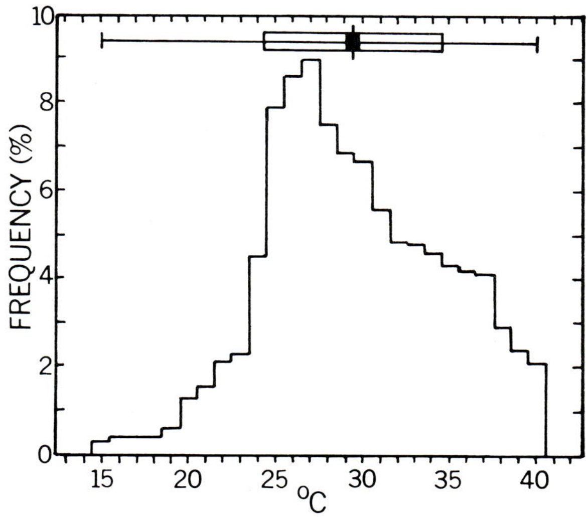

FIG. 1. Relative frequency distribution (%) of occupied temperatures from day/night pooled data of ten juvenile *Limulus* thermoregulating individually in an electronic shuttlebox for 48 hr. Above is a modified Dice-Leraas diagram showing range, mean, ± 2 standard errors (shaded area), and ± 1 standard deviation (unshaded bar).

wide range of preference may be related to the wide range of thermal tolerance (from less than 0°C to more than 40°C) in this very eurythermal species. The final thermal preferendum can be characterized by any of a number of measures of central tendency (Reynolds and Casterlin 1976a; Reynolds 1977b), which differ from one another owing to the skewness of the distribution (Fig. 1). The midpoint of the range was 27.5°C; the modal preferendum was 27°C; the median was 29°C; and the mean was 29.4°C ± 0.19 (standard error), or ± 5.27 (standard deviation). The skewness (Pearson's coefficient = (mean - mode)/ standard deviation) was +0.46 (cf. Reynolds and Casterlin 1976a).

The diel locomotor activity pattern (Fig. 2) was markedly nocturnal, with mean hourly nocturnal activity (7.8%/hr) four times that exhibited during

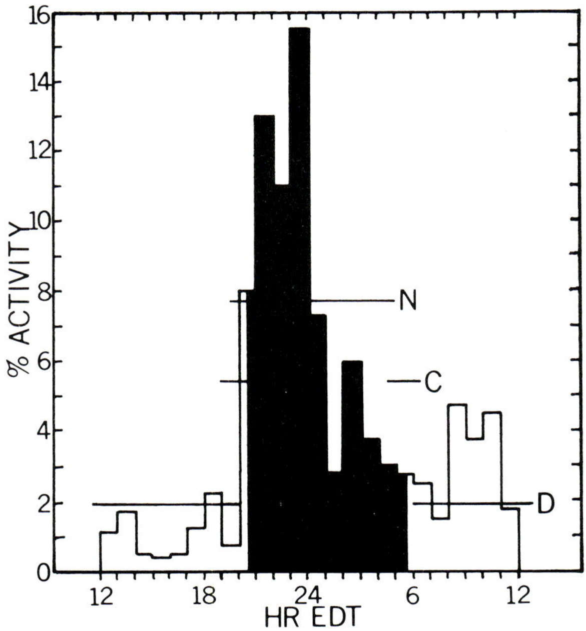

FIG. 2. Locomotor activity, expressed as hourly percentages of 24-hr total, showing diel pattern relative to photoperiod (EDT = Eastern Daylight Time; shaded portion represents night). Horizontal lines show nocturnal (N), crepuscular (C) and diurnal (D) mean hourly activity. Data was pooled for ten juvenile *Limulus* tested individually in an electronic shuttlebox for 3-day periods with natural lighting.

the day (1.9%/hr). Crepuscular activity (dawn and dusk average) was intermediate (5.4%/hr). Locomotor activity increased rapidly at dusk (Fig. 2) to about the nocturnal average, increased to even higher levels during the early hours of darkness, then declined gradually toward diurnal levels at dawn; there was no secondary dawn peak such as is characteristic of a bimodal crepuscular pattern. Although the abrupt increase in activity at dusk might be interpreted as a direct (photokinetic) response to the change in light intensity, activity levels changed considerably (Fig. 2) during the night while light intensity did not, and decreased to diurnal levels before light intensity began to increase at dawn. This, along with the evidence of Barlow et al. (1976a, 1976b) for endogenous circadian rhythmicity in the optic nerve, suggests that the activity pattern represented in Fig. 2 is controlled at least in part by an internal rhythm. Such a rhythm would presumably be entrained by photoperiod, so the diel pattern would be related indirectly, if not directly, to the day/night cycle. The pattern does not indicate complete inactivity during the day (Fig. 2), so it is not surprising to observe some diurnal activity in the field, although we have usually found them motionless and usually partially buried on the beach at low tide during the day.

The diel activity pattern is not expressed in terms of thermoregulatory performance in this species --i.e., there is no rhythm of preferred temperature. This is not surprising in light of results obtained previously with other organisms. For example, the bluegill sunfish (*Lepomis macrochirus*) is more than twice as active during the day as at night, yet thermoregulates a bit more precisely at night (Reynolds and Casterlin 1976a), with no significant change in mean preferred temperature. The diurnal pattern of the bluegill is essentially the same whether or not the fish is thermoregulating or held at constant high or low temperatures, or even when the fish controls the light switch (Reynolds and Casterlin 1976b), indicating the endogenous nature of the rhythm and demonstrating that "forced" thermoregulatory movements in the Ichthyotron do not significantly affect the diel activity pattern (a low level of activity during the inactive period suffices for thermoregulation). Also, results obtained in the Ichthyotron agree well with those obtained for the same species by other investigators using different methodologies (Reynolds and Casterlin 1976a, 1976b, 1978a). Goldfish (*Carassius auratus*) exhibit independent rhythms of locomotor activity and of preferred temperature, which display no apparent relationship to one another (Reynolds et al. 1978b).

The low level of thermoregulatory precision evident in Fig. 1 might be a reflection of the primitive phylogenetic position of *Limulus* (Shuster 1971), and/or of its wide range of thermal tolerance (Mayer, 1914). The goldfish exhibits a similarly eurythermal tolerance range (0 - 41°C; Fry 1947), and its mean thermal preferendum is similar (28°C; Reynolds et al. 1978b), but its thermoregulatory precision is much greater, as it voluntarily occupies a range of temperatures in the Ichthyotron not exceeding about 10°C, compared to 25°C for *Limulus*.

The final thermal preferendum, as measured by the gravitational method in our tests, is independent of prior thermal acclimation state (Fry 1947; Reynolds 1977c; Richards et al. 1977; Reynolds and Casterlin 1979b). Several nonthermal factors can affect preferred temperatures within a few °C, including ontogenetic stage (Reynolds and Casterlin 1978d, 1979b), season or photoperiod (Richards et al. 1977; Reynolds and Casterlin 1979b), and bacterial infection (Reynolds et al. 1976b, 1978c, 1978d; Casterlin and Reynolds 1977a, 1977b; Covert and Reynolds 1977). There typically is little or no difference in thermal preference (final preferenda) among different geographic regions within a species' range (Reynolds and Casterlin 1976a). Differences in thermal final preferendum values as determined independently by different investigators using various methodologies, and perhaps as affected by the variables enumerated above and others, typically agree within a range of a few °C (Richards et al. 1977). Since the various measures of central tendency for the distribution of preferred temperatures displayed by *Limulus* (Fig. 1) vary from 27 to 29.4°C, it seems reasonable to suggest that the preferred temperature of *Limulus* as a species falls within a range of about 25 to 30°C.

A further question arises as to the extent to which a thermal preferendum can be considered a thermal optimum (Reynolds 1977b). It does seem reasonable to infer that, for a behavioral mechanism to have evolved which leads an animal to prefer some temperatures and to avoid others, there must be some physiological and/or ecological benefits which result from such behavior (Reynolds 1977b). It is not unusual for the preferendum of a species to lie near the upper end of the range of normally experienced temperatures in nature (or even exceeding those in the northern part of the geographic range of some species), and often the preferendum lies within a few degrees of the upper lethal temperature (Reynolds and Thomson 1974; Reynolds and Casterlin 1976a, 1979b), even exceeding the ultimate upper incipient lethal temperature during brief excursions into the resistance zone. Preferenda increase still further during a "behavioral fever," an increase in preferred temperature in response to bacterial infection (Reynolds et al. 1976b, 1978c, 1978d; Casterlin and Reynolds 1977a, 1977b; Covert and Reynolds 1977). This behaviorally mediated febrile phenomenon has been found to enhance survival from infection (Covert and Reynolds 1977), possibly by speeding the movement of granulocytes to the site of infection (Bernheim et al. 1978) via thermally-induced increases in blood flow (Reynolds 1977d; Reynolds and Casterlin 1978e). The upper lethal temperature (40°C or more--Mayer 1914) of *Limulus* far exceeds its preferendum, a situation similar to that in the goldfish (Reynolds et al. 1978b, 1978d; Covert and Reynolds 1977) but atypical of most other species (Reynolds and Casterlin 1979b). While water temperatures approaching 40°C may infrequently be encountered by *Limulus* except in very shallow estuarine waters in the southern portions of its range, it can be exposed to very high substrate and body temperatures due to solar radiation when stranded on the beach during the day (Fraenkel 1960), leading to burrowing behavior.

In view of the significant effects typically exerted by temperature and by circadian rhythms on physiological processes, experiments with *Limulus* should take these important factors into account. Physiological processes are likely to be optimum in *Limulus* within a temperature range of about 25-30°C (responses to infection may be optimized at a slightly higher febrile temperature). The time of day at which experiments are conducted can also significantly influence the results of experiments where circadian rhythmicity occurs, often even in isolated tissues.

Barlow RB Jr, Bolanowski SJ Jr, Brackman ML (1976a). Efferent mediated circadian changes in the neural activity of the *Limulus* lateral eye. Biol Bull 151:400-401.

Barlow RB Jr, Chamberlain SC, Kaplan E (1976b). Efferent inputs and serotonin enhance the sensitivity of the *Limulus* lateral eye. Biol Bull 153:414.

Bernheim HA, Bodel PT, Askenase PW, Atkins E. (1978). Effects of fever on host defense mechanisms after infection in the lizard *Dipsosaurus dorsalis*. Brit J Exp Pathol 59:76-84.

Casterlin ME, Reynolds WW (1977a). Behavioral fever in anuran amphibian larvae. Life Sci 20:593-596.

Casterlin ME, Reynolds WW (1977b). Behavioral fever in crayfish. Hydrobiologia 56:99-101.

Casterlin ME, Reynolds WW (1978). Behavioral thermoregulation in *Rana pipiens* tadpoles. J Thermal Biol

Casterlin ME, Reynolds WW (1979). Diel activity patterns of the smooth dogfish shark, *Mustelus canis*. Bull Mar Sci 29:

Cole WH (1923). Circus movements of *Limulus* and the tropism theory. J Gen Physiol 5:417-426.

Cole WH (1924). The effect of laboratory age upon the phototropic reactions of *Limulus*. J Gen Physiol 6:295-297.

Covert JB, Reynolds WW (1977). Survival value of fever in fish. Nature 267:43-45.

Fraenkel G (1960). Lethal high temperatures for three marine invertebrates: *Limulus polyphemus*, *Littorina littorea* and *Pagurus longicarpus*. Oikos 11:171-182.

Fry FEJ (1947). Effects of the environment on animal activity. Univ Toronto Stud Biol Ser 55, Publ Ontario Fish Res Lab 68:1-62.

Johansen K, Petersen JA (1975). Respiratory adaptations in *Limulus polyphemus* (L.). In Vernberg FJ (ed): "Physiological Ecology of Estuarine Organisms," Columbia: Univ South Carolina Press, pp 129-145.

Lall A, Chapman R (1973). Phototaxis in *Limulus* under natural conditions: evidence for reception of near-ultraviolet light in the median dorsal ocellus. J Exp Biol 58:213-244.

Loeb J (1893). Uber kunstliche Umwandlung positiv heliotropischer Tiere in negativ heliotropische und umgekehrt. Pflug Arch Ges Physiol 53:81-107.
Mayer AG (1914). The effects of temperature upon tropical marine animals. Papers Tortugas Lab 6: 1-24.
Northrop JH, Loeb J (1923). The photochemical basis of animal heliotropism. J Gen Physiol 5:581-595.
Pearse AS (1928). On the ability of certain marine invertebrates to live in diluted sea water. Biol Bull 54:405-409.
Reynolds WW (1977a). Fish orientation behavior: an electronic device for studying simultaneous responses to two variables. J Fish Res Board Can 34:300-304.
Reynolds WW (1977b). Temperature as a proximate factor in orientation behavior. J Fish Res Board Can 34:734-739.
Reynolds WW (1977c). The final preferendum of fishes: shuttling behaviour and acclimation overshoot. Hydrobiologia 57:123-124.
Reynolds WW (1977d). Thermal equilibration rates in relation to heartbeat and ventilatory frequencies in largemouth blackbass, *Micropterus salmoides*. Comp Biochem Physiol 56A:195-201.
Reynolds WW, Casterlin ME (1976a). Thermal preferenda and behavioral thermoregulation in three centrarchid fishes. In Esch GW, McFarlane RW (eds): "Thermal Ecology II," Springfield, Va.: U.S. Natl. Technical Information Service, pp 185-190.
Reynolds WW, Casterlin ME (1976b). Locomotor activity rhythms in the bluegill sunfish, *Lepomis macrochirus*. Am Midl Nat 96:221-225.
Reynolds WW, Casterlin ME (1977a). Temperature preferences of four fish species in an electronic thermoregulatory shuttlebox. Progr Fish-Cult 39: 123-125.
Reynolds WW, Casterlin ME (1977b). Diel activity in the yellow bullhead. Progr Fish-Cult 39:132-133.
Reynolds WW, Casterlin ME (1978a). Behavioral thermoregulation and diel activity in white sucker (*Catostomus commersoni*). Comp Biochem Physiol 59A: 261-262.
Reynolds WW, Casterlin ME (1978b). Behavioral thermoregulation in the rock bass *Ambloplites rupestris*. Comp Biochem Physiol 60A:263-264.

Reynolds WW, Casterlin ME (1978c). Thermoregulatory behavior in the smooth dogfish shark, *Mustelus canis*. Fed Proc 37:427.

Reynolds WW, Casterlin ME (1978d). Ontogenetic change in preferred temperature and diel activity of the yellow bullhead, *Ictalurus natalis*. Comp Biochem Physiol 59A:409-411.

Reynolds WW, Casterlin ME (1978e). Estimation of cardiac output and stroke volume from thermal equilibration and heartbeat rates in fish. Hydrobiologia 57:49-52.

Reynolds WW, Casterlin ME (1979a). Behavioral thermoregulation by ammocoete larvae of the sea lamprey (*Petromyzon marinus*) in an electronic shuttlebox. Hydrobiologia

Reynolds WW, Casterlin ME (1979b). Behavioral thermoregulation and the "final preferendum" paradigm. Am Zool

Reynolds WW, Casterlin ME, Covert JB (1976b). Behavioural fever in teleost fishes. Nature 259: 41-42.

Reynolds WW, Casterlin ME, Covert JB (1978c). Febrile responses of bluegill (*Lepomis macrochirus*) to bacterial pyrogens. J Thermal Biol

Reynolds WW, Casterlin ME, Matthey JK, Millington ST, Ostrowski AC (1978b). Diel patterns of preferred temperature and locomotor activity in the goldfish *Carassius auratus*. Comp Biochem Physiol 59A:225-227.

Reynolds WW, Casterlin ME, Millington ST (1978a). Circadian rhythm of preferred temperature in the bowfin *Amia calva*, a primitive holostean fish. Comp Biochem Physiol 60A:107-109.

Reynolds WW, Covert JB, Casterlin ME (1978d). Febrile responses of goldfish (*Carassius auratus*) to *Aeromonas hydrophila* and to *Escherichia coli* endotoxin. J Fish Diseases

Reynolds WW, McCauley RW, Casterlin ME, Crawshaw LI (1976a). Body temperatures of behaviorally thermoregulating largemouth blackbass (*Micropterus salmoides*). Comp Biochem Physiol 54A:461-463.

Reynolds WW, Thomson DA (1974). Responses of young Gulf grunion, *Leuresthes sardina*, to gradients of temperature, light, turbulence and oxygen. Copeia 1974:747-758.

Richards FP, Reynolds WW, McCauley RW, Crawshaw LI, Coutant CC, Gift JJ (1977). Temperature preference studies in environmental impact assessments: an overview with procedural recommendations. J Fish Res Board Can 34:728-761.
Robertson JD (1970). Osmotic and ionic regulation in the horseshoe crab, *Limulus polyphemus* (Linnaeus). Biol Bull 138:157-183.
Shuster CN Jr (1953). Odyssey of the horseshoe crab. Audubon Magazine 55:162-163.
Shuster CN Jr (1954). A horseshoe "crab" grows up. Ward's Natural Science Bull 28:1-6.
Shuster CN Jr (1957). Xiphosura (with especial reference to *Limulus polyphemus*). In Hedgpeth JW (ed): "A Treatise on Marine Ecology and Palaeoecology," Vol. 1, Geol. Soc. Amer. Mem. 67:1171-1174.
Shuster CN Jr (1971). Xiphosurida. In "Encyclopedia of Science and Technology," New York: McGraw-Hill Book Co., Inc.
Sokoloff A (1978). Observations on populations of the horseshoe crab *Limulus (=Xiphosura) polyphemus*. Res Pop Ecol 19:222-236.
Townes HK Jr (1938). Ecological studies on the Long Island marine invertebrates of importance as fish food or bait. A Biological Survey of the Salt Waters of Long Island, Part II. NY State Cons. Dept. Salt-water Survey No. 14.
von Campenhausen C (1967). The ability of *Limulus* to see visual patterns. J Exp Biol 46:557-570.
Wolf E, Zerrahn-Wolf G (1937). Reactions of *Limulus* to illuminated fields of different areas and flicker frequency. J Gen Physiol 20:767-776.

LABORATORY CULTURE OF EMBRYONIC AND JUVENILE LIMULUS

Kathleen A. French
Laboratory of Sensory Physiology and
Boston University Marine Program
Marine Biological Laboratories
Woods Hole, MA 02543

Watching _Limulus_ develop is a venerable passtime. Biologists have for more than a century observed and described the process. In 1870, the Reverend Dr. S. Lockwood published a somewhat fanciful account of his careful observations, and for the next forty years A. S. Packard (1870, 1872, 1873, 1875, 1885), J. S. Kingsley (1890, 1892, 1893), and W. Patten (1912) produced papers describing the microanatomy and gross morphology of fixed specimens of embryonic and larval horseshoe crabs. The major motivation of most of these men was to place _Limulus_ in its proper phylogenetic niche. They arrived at various answers, usually placing the animal in the Crustacea or the Arachnida. Patten (1912), however, based on embryological data, argued that the Xiphosura were the direct ancestors of the vertebrates.

Now biologists have again taken up the study of the development of the horseshoe crab. Their interest extends from classical embryological questions (Scholl, 1977) to the physiology of the organism as it changes during ontogeny. Live healthy animals at all stages are therefore required, and as a result, _Limulus_ must be cultured in the laboratory.

METHODS AND RESULTS

Fertilization

Fertilized _Limulus_ eggs can successfully be obtained in two ways. During the spawning season, which occurs primarily in late May and early June on Cape Cod and peaks around the

Pages 61–71, Biomedical Applications of the Horseshoe Crab (Limulidae)

full moon, nests can be located on favorite spawning beaches in the morning following a spring tide. The eggs are deposited near the high tide line, so prior to the next high tide, it is not difficult to find the nests and dig for eggs.

Outside of the normal spawning period--or from mid-April until mid-August or later in Woods Hole--viable gametes can be obtained from adults, and the eggs fertilized _in vitro_. Gentle pressure anterior to the genital flap is usually sufficient to release mature ova or sperm. Alternative methods for obtaining gametes (Shrank, et al., 1967; Brown and Knouse, 1973) may also be used. Pooling eggs or sperm from more than one animal has no detectable effect on development. The process of fertilization and the effects of several variables upon it have been investigated (Shoger and Bishop, 1967; Shoger and Brown, 1970; Brown and Humphreys, 1971; Brown and Knouse, 1973; Mowbray and Brown, 1974).

Embryonic Development

The fertilized eggs are maintained at room temperature in sea water, which is changed every two or three days. About 60-70% of eggs fertilized _in vitro_ begin to develop and reach at least the limb bud stage. The fraction is somewhat larger

Table 1. Time from fertilization to first hatch in the first and last batches of eggs fertilized _in vitro_ for two breeding seasons

	Fertiization date	Days to first hatch
		1976
First batch	3 April*	45
Last batch	28 June**	25
		1978
First batch	12 April*	43
Last batch	13 June**	28

*Approximate temperature during incubation was 15° C.
**Approximate temperature during incubation was 23° C.

in eggs collected in the field. The rate of development varies greatly among eggs in a single clutch and is strongly temperature dependent (Table 1).

Although Patten (1912) described 15 embryonic stages, far fewer can reliably be recognized in the unstained, living animal. Limulus appears to have the same molting pattern within the egg as does Tachypleus (Sekiguchi, 1970), and the molts offer convenient milestones for the demarcation of stages. Tachypleus undergoes four embryonic molts, the first of which is only a partial molt. Some time after the first molt, the heavy, opaque chorion usually splits, leaving the embryo enclosed by a filmy, transparent membrane (the deutovum). This event occurs approximately concurrently with the second embryonic molt. In Limulus, the chorion ruptures at about the same stage in development, and depending on the timing, the change in form following three molts can be discerned. I have used these to indicate developmental stage. (See Figure 1 and Table 2.)

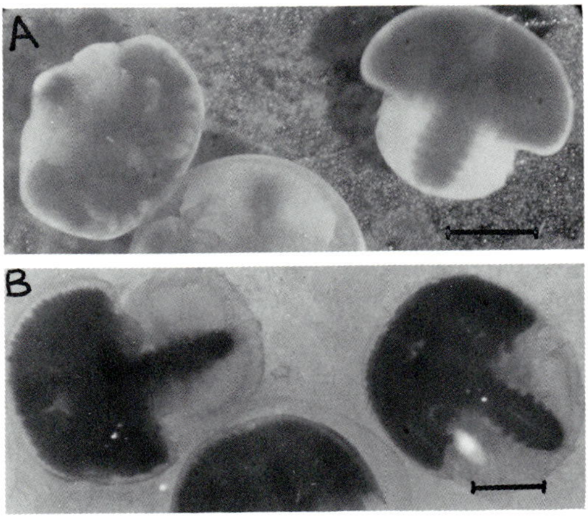

Figure 1. Dorsal (right) and ventral views of Limulus embryos with the deutovum removed. Animal at the center is in intact egg. Calibration bars are 1 mm. A. Third intermolt. B. Fourth intermolt.

Table 2. Developmental stages of *Limulus* following the shedding of the chorion

Stage	Average Size (Mm)	Some Distinguishing Characters
Second intermolt	1.8	Body nearly spherical; legs visible; opisthosome suggested but not well defined; some slow and sporadic movement of legs.
Third intermolt	2.5	Body flatter; opisthosome well defined; gill flaps immobile; eyes appear; very active movement of legs.
Fourth intermolt	3.1	Edge of prosomium thin and sharply defined; gill flaps and body joint mobile; hairs on body surface; legs and body very active.
Trilobite larva	3.1	Free swimming; spends most of the time ventral side up; yolk still plentiful; no telson.
First juvenile tailed stage	4.5	Quite translucent; short but distinct telson
Second juvenile tailed stage	6.8	Telson is longer compared to the rest of the body; some brown pigment apparent in carapace .

Larval and Juvenile Stages

When the trilobite larvae emerge from the egg membrane, they still contain some yolk and do not seem to need to eat. They swim actively, usually by flapping their gills while lying ventral side up. In my laboratory, they are transferred at this time to filtered running sea water (filter pore size of 50 u or less). The flow rate can be quite low. The animals are most successfully kept in mesh-sided boxes with autoclaved sand in the bottom. Animals raised without sand eventually become covered with long filamentous epiphytes. A density of one animal per cm^2 or less seems optimum, but higher densities may be satisfactory.

The larvae may molt as soon as three weeks after emerging from the egg. In this molt they acquire a telson, and following the molt they spend more time walking with their dorsal side up. They also begin to burrow in the sand.

First tailed stage juveniles appear to be devoid of yolk and begin to eat. They will accept a variety of foods: bits of Mytilus, small pieces of Nereis,frozen brine shrimp, the alga Enteromorpha, marine nematodes, and small whole polychaetes. Because the animals eat small living polychaetes most readily, a project to culture Capitella as a food species has been started.

A few animals have completed the second molt after hatching by the end of the summer season in the culture situation, but most of the juveniles enter the winter in the first tailed stage. In comparison, several Limulus that were collected in grab smaples from Falmouth Harbor on 3 October of this year were all first tailed stage juveniles, based on size. Apparently development is not severely retarded by the laboratory diet and conditions.

Diseases

Larval and juvenile Limulus kept in dishes without sand and/or in crowded conditions develop an unidentified filamentous growth on the gill plates and the legs. In an experiment to determine an effective treatment of the disease, the following agents were used singly or in combination: autoclaved sand, exposure to a 0.1% solution of malachite green in sea water for 2 to 5 minutes, and an antibiotic solution of 60 mg/l penicillin and 50 mg/l streptomycin in filtered sea water (Capo, 1978; Fisher, et al., 1978). Two groups of 48 infected animals were exposed to each treatment. Half of each group were larvae and half were juveniles. The animals were kept in finger bowls at room temperature, with the water changed weekly. Cast molts and dead animals were counted periodically. The results after 21 days of treatment are shown in Table 3.

Table 3. The effect of several treatments on the mortality from epiphytic disease, after 21 days, in groups of 24 larvae and 24 juveniles. Each value is the average for two groups.

Treatment	Trilobites Dead	Juveniles Dead	Trilobites Molted
No treatment	2.5	10	5.5
Malachite green-2 minutes	3	8.5	4
Malachite green-5 minutes	0.5	6	6
Antibiotic	2	10	5.5
Sand	1.5	3.5	4.5
Malachite green-3 minutes + sand	0.5	2.5	5.5
Malachite green 3 minutes + sand + antibiotic	0.5	1.5	5

Apparently, adding sand to the culture produces the greatest reduction in mortality. Observation suggests that the most effective cure for the condition is molting. (Cast molts often had heavily coated, brown gills.) In sand, the newly molted animals seem to remain free from reinfection, but in the dishes without sand, few animals were free of epiphytes, even though several trilobites had molted. It is interesting to note that the infection seems to have little effect on the organisms' ability to molt, because the process occurred in lightly- and heavily-infected animals.

Temperature effects

The sizeable effect of temperature on the rate of development can be used in the laboratory. Lockwood (1870) first reported that the embryos withstand very cold temperatures: "Hoping to continue observations upon the growth of my interesting little family the ensuing year the jars were carefully put away. Little regard, however, was paid to temperature, which, on several occasions, went down to the freezing point. On the 3d of May, 1870, I emptied the jars to see how my charge was getting on, when lo, not one of last year's hatching was alive! but wonderful to say at least a dozen little fellows, all hatched this spring and all alive, had taken their place. . .Just think of it--these eggs are within two weeks only of being a year old!"

Eggs refrigerated from June, 1977, to March, 1978, appearred to develop normally when brought to room temperature, but trilobite larvae and first tailed stage juveniles died in the refrigerator. It should be noted that these animals were not frozen at any time. Their tolerance for low temperatures and developmental arrest implies that early developmental stages can be available for study throughout the year.

Heating fertilized eggs to increase the rate of development was tried with limited success. Holding eggs at 24° C in April speeded development but produced large numbers of anomalies. The practice was therefore abandoned.

APPLICATIONS

Circulating blood in larval and juvenile <u>Limulus</u> was reported by Packard (1872), and he ascertained the pattern of circulation by watching the movements of cells carried by the moving plasma. Patten (1912, page 231) mentioned and included drawings of "ameoboid fiber cells" found in blood channels and of "blood corpuscles of the usual type." It is difficult, however, to compare his drawings with the blood cells that can be observed in fresh preparations with Nomarski optics (Figure 2). The cell shown is identical in appearance with cells in the blood stream of later embryos and larvae and appears to be of the same type as amoebocytes from adult animals (e.g., Armstrong, 1977).

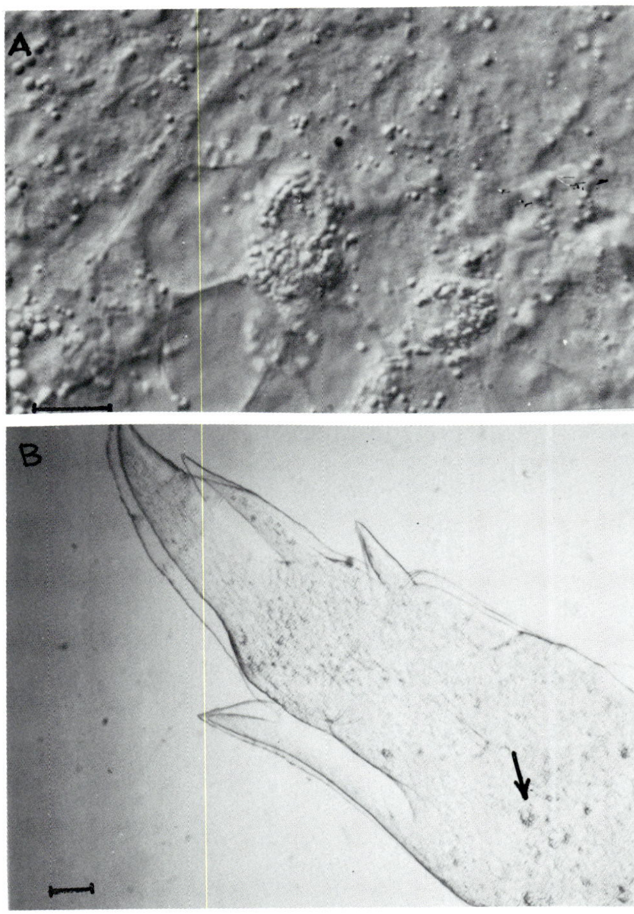

Figure 2. A. Granular cell in the last walking leg of a third intermolt Limulus embryo. Calibration bar is 20 u. B. Leg in which cell above was seen. A granular cell may be seen faintly at the arrow. Calibration bar is 50 u.

These cells can readily be found in the legs of third intermolt embryos. They have not been seen in the legs of second intermolt animals, but they may simply be rarer, or they may be found in other portions of the animals.

When a third intermolt embryo is squashed on a slide, granular cells can be seen surrounded by amorphous yolk. Over the course of minutes, they send out processes and begin amoeboid movement (Figure 3).

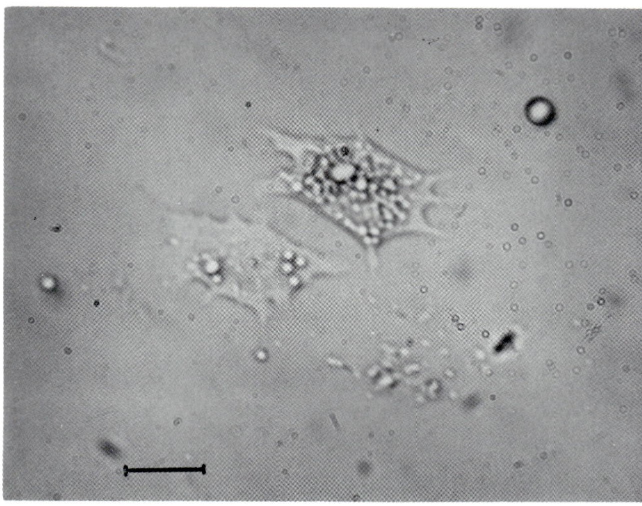

Figure 3. Granular cells and yolk particles several minutes after a third intermolt embryo was squashed on a slide. Calibration bar is 20 u.

The study of embryonic and larval Limulus blood cells could add to our understanding of the cells in all Limulus for several reasons. The tissue that serves as a source of the cells might be discovered in young animals that must be producing many cells during their relatively rapid molt schedule. Furthermore, although Yamamichi and Sekiguchi (1974) successfully cultured Tachypleus explants and whole embryos removed from the deutovum, under normal conditions the deutovum appears to protect the embryo from infection. Even though the membrane may become heavily covered by filamentous growth, the embryo inside is clean. What would be the effect upon the developing amoebocytes of contact with microbes at the third or fourth embryonic intermolt stages? When do the cells acquire their adult characteristics and biochemical properties? These questions might be answered using cultured embryos.

A second system that is well-studied in the adult horseshoe crab and that may be profitably investigated in young animals is the photorecptive organs. The lateral, ventral, and median eyes become pigmented at approximately the same time early in the third intermolt period and increase in size with each molt thereafter. These, too, are an experimentally accessible preparation that may allow us to expand our understanding of the physiology fo the animal.

Because Limulus can be cultured relatively easily, its physiology and biochemistry can be studied experimentally from the very earliest developmental stages, offering the possibility of more clearly understanding the adult animal and general physiological principles.

I wish sincerely to thank Daniel Gibson, Thomas Capo, and Dr. Frederick Bang for helpful and stimulating discussions and the Rowland Foundation for support through a grant to the Laboratory of Sensory Physiology.

REFERENCES

Armstrong PB (1977). Interaction of the motile blood cells of the horseshoe crab, Limulus. Exp Cell Res 107:127.

Brown GG, Humphreys WJ (1971). Sperm-egg interactions of Limulus polyphemus with scanning electron microscopy. J Cell Biol 51:904.

Brown GG, Knouse JR (1973). Effects of sperm concentration, sperm aging, and other variables on fertilization in the horseshoe crab, Limulus polyphemus L. Biol Bull 144:462.

Capo T (1978). Personal communication.

Fisher WS, Nilson EH, Steenbergen JF, Lightner DV (1978). Aquaculture 14:115.

Kingsley JS (1890). The ontogeny of Limulus. Amer Nat 24:678.

Kingsley JS (1892). The embryology of Limulus. J Morph 7:35.

Kingsley JS (1893). The embryology of Limulus-Part II. J Morph 8:195.

Lockwood S (1870). The horse foot crab. Amer Nat 4:257.

Mowbray RC, Brown GG (1974). Fertilization studies with egg sections of the horseshoe crab, Limulus polyphemus. Biol Reprod 10:62.

Packard AS (1870). The embryology of Limulus polyphemus. Amer Nat 4:498.

Packard AS (1872). The development of Limulus polyphemus L. Mem Boston Soc Nat Hist 2:155.

Packard AS (1873). Farther observations of the embryology of Limulus, with notes on its affinities. Amer Nat 7:675.

Packard AS (1875). On the development of the nervous system in Limulus. Amer Nat 9:422.

Packard AS (1885). On the embryology of Limulus polyphemus. Amer Nat 19:722.

Patten W (1912). "The Evolution of the Vertebrates and Their Kin." Philadelphia: P. Blakiston's Son and Co.

Scholl G (1977). Beitrage zur Embryonalentwicklung von Limulus polyphemus L. Zoomorph 86:99.

Schrank WW, Shoger RL, Schechtman LM, Bishop DW (1967). Electrically induced spawning in the male and female horseshoe crab, Limulus polyphemus. Biol Bull 133:453.

Sekiguchi K (1970). On the embryonic moultings of the Japanese horse-shoe crab, Tachypleus tridentatus. Sci Rep Tokyo Kyoidu Daigaku, Sec B 14:121.

Shoger RL, Bishop DW (1967) Sperm activation and fertilization in Limulus polyphemus. Biol Bull 133:485.

Shoger RL, Brown GG (1970). Ultrastructural study of sperm-egg interactions of the horseshoe crab Limulus polyphemus L. J Submicr Cytol 2:167.

Yamamichi Y, Sekiguchi K (1974). Embryo and organ cultures of the horseshoe crab, tachypleus tridentatus. Dev Growth Diff 4:295.

MOTILITY OF THE *LIMULUS* AMEBOCYTE

Peter B. Armstrong

Department of Zoology
University of California
Davis, CA 95616, U.S.A.
and
Marine Biological Laboratory
Woods Hole, MA 02543, U.S.A.

INTRODUCTION

Almost all animals phylogenetically more advanced than the flatworms possess blood cells capable of active, pseudopod-directed motility (Andrew, 1965). Although the range of functions served by motile blood cells varies among different animal species, hemostasis and defense against invading pathogens are nearly universal. The pseudopodial activities of motile blood cells are important in both processes. During hemostasis, pseudopodial contraction serves to produce retraction of the hemostatic plug (Lüscher and Bettex-Galland, 1971; Pollard et al., 1977; Shepro et al., 1969). In cellular defense functions, pseudopodial activity is essential for phagocytosis (Allison et al., 1971; Reaven & Axline, 1973; Stossel, 1977) and active cell migration is involved in diapedesis and migration of inflammatory cells to sites of bacterial infection (Marchesi, 1970). The present report presents results from studies on the details of the active pseudopod-directed motility of the blood cells (amebocytes) of the horseshoe crab *Limulus* and the motile behavior of populations of these cells. Cells were studied both in culture, during which time the cells migrated on glass microscope coverslips, and *in vivo*, in gill leaflets isolated from young animals.

Limulus is almost unique among those animals whose blood cells have been studied in having but a single cell type in the coelomic fluid, the granular amebocyte (Dumont

Pages 73–92, Biomedical Applications of the Horseshoe Crab (Limulidae)

et al, 1966; Levin & Bang, 1964; Loeb, 1902). In the intact animal, the circulating cells are nonadhesive and do not protrude pseudopods (Fig. 1).

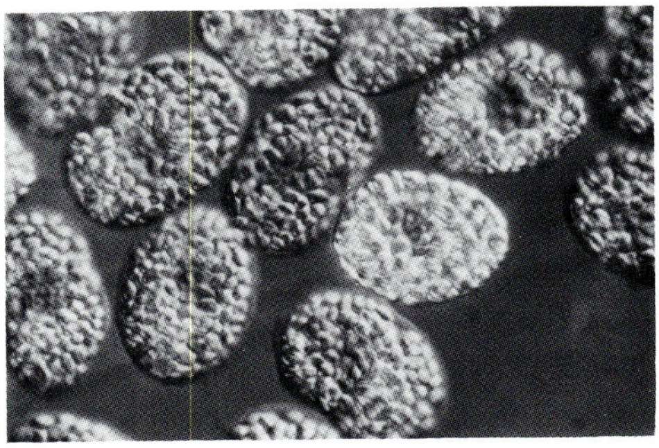

Fig. 1. Limulus coelomocytes treated with N-ethyl maleimide to prevent transformation. The cells are oval, nonadhesive and immobile. Cell morphology and behavior change dramatically upon the transformation that ordinarily follows removal of the cells from the animal. x990.

The unactivated state of circulating cells can be observed by bleeding and washing at 4°C followed by immediate aldehyde fixation, by bleeding into propranolol (Murer et al., 1975) or N-ethyl maleimide (Solum, 1970), both of which stabilize the cells, or by observing cells in the intact animal as they circulate in the capillaries of the abdominal spines. In all cases, the circulating cells are comprised of a homogeneous population of oval, disc-shaped granular cells (Fig. 1). The disc shape apparently is maintained by a marginal band of microtubules (Cohen et al., 1977; Dumont et al., 1966).

Following extravasation or trauma, the blood cells transform to become adhesive and actively motile. Nothing is known of the nature of the triggers for transformation or the nature of the changes that lead to the acquisition of adhesiveness and motility. Changes presumably occur in the structure of the plasma membrane and the cell surface to lead to adhesion and in the organization of the contractile and cytoskeletal proteins to lead to changes in cell shape and to motility.

METHODS

In studies of cellular motility *in vitro*, mature animals were bled by cardiac puncture into embryo dishes. Throughout these studies, all glassware was rendered endotoxin-free by incubation at 180°C for 4 hours. The blood cells settled to the bottom of the dish to form a layer 1-2 mm thick. The cells of this layer adhered to one another to make a coherent mat of tissue ("amebocyte tissue"). Fragments of amebocyte tissue 2-3 mm on a side were dissected out and were transplanted to viewing chambers constructed by sealing #1-1/2 glass coverslips to both sides of a 1.6 cm diameter hole cut through a 0.75 mm thickness sheet of aluminum (Fig. 2). Gel-free *Limulus* plasma or sterile, endotoxin-free 3% NaCl buffered with 2 mM $NaHCO_3$ served as the culture medium. Cell behavior and morphology were indistinguishable in the two media.

Cell motility was studied *in vivo* in gill leaflets isolated from juvenile animals 25-40 mm in width. Isolated gill leaflets were placed on a microscope slide in a few drops of artificial sea water and a coverslip was placed over the preparation. The walls of the gill leaflets of young animals were thin and almost perfectly transparent and allowed the use of a high numerical aperture microscope condenser and oil immersion objectives. A variable fraction of the blood cells contained within the gill leaflet underwent activation (due, apparently, to the trauma attendant with dissection of the gill leaflet) and migrated about over the inner surfaces of the gill leaflet and over the vertical pillars that prevent the upper and lower gill surfaces from coming together. The amebocytes remained actively motile for several hours if the coverslip was sealed to prevent dessication.

Motility of the amebocytes was studied with time-lapse microcinematography and real-time observation using phase contrast and Nomarski differential interference contrast optics. The details of locomotion were studied with oil immersion objectives (Planapo 63/1.4 phase contrast and Plan 100/1.25 with Nomarski optics) and the collision behavior was studied principally with dry objectives (Neofluar 16/0.32 phase contrast and Plan 40/0.4 with Nomarski optics). Time lapse film records were analyzed with an L/W analyst projector. All observations were conducted at room temperature. Using these procedures, it has been possible to arrive at a detailed understanding of the motile behavior of the Limulus blood cell both in culture and in vivo.

Fig. 2. Adult horseshoe crabs were bled into 5 ml embryo dishes (left). The cells settled to the bottom and aggregated into amebocyte tissue. Squares of amebocyte tissue were dissected free and explanted to double cover-slip observation chambers (right). Cells were viewed as they emigrated from the explants onto the microscope coverglasses.

CELLULAR MOTILITY IN VITRO

Within 1/4 to 1 hour after explantation of the amebocyte tissue to a chamber, the first cells began emigrating from the explant onto the microscope coverglasses that formed the roof and floor of the observation chamber. By 2-4 hours, the coverglasses were populated with large numbers of motile cells that had migrated out from the explants (Fig. 3a). As they emerged from the explants, the amebocytes were compact and unflattened and had limited areas of adhesion to the substratum (Fig. 3b). Their cytoplasm contained abundant oval granules. These are the granules that contain the clottable protein system.

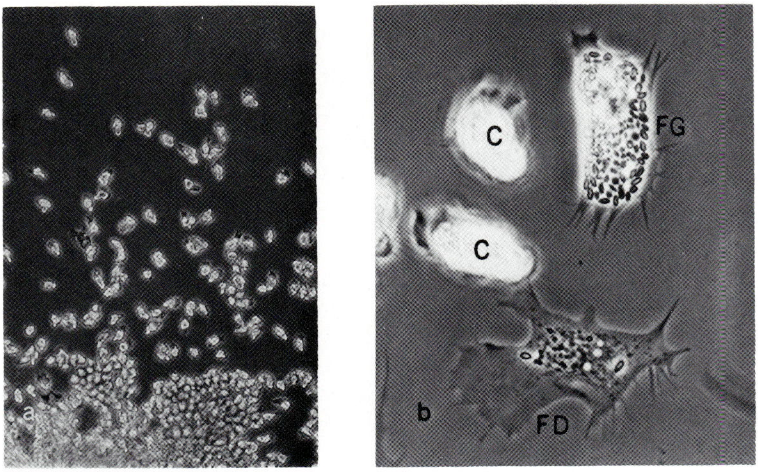

Fig. 3a. Migration of Limulus amebocytes from an explant of amebocyte tissue onto a microscope coverglass. The explant is at the bottom margin of the photograph and the short axis of the photograph lies parallel to the margin of the explant. Taken 60 min after the culture was established.

Fig. 3b. Limulus amebocytes that have migrated onto the coverglass of the culture chamber. Two contracted (C), one flattened granular (FG) and one flattened degranulated (FD) cell are visible.

After the cells had been in contact with the glass for a period of time, partial flattening occurred. The flattened granular cells were still actively motile, though less so than the compact cells (Fig. 3b). Flattening was reversible as long as the cells retained their cytoplasmic granules. Spontaneous degranulation occurred in culture. Degranulated cells survived for long periods of time in culture but were only slightly motile or completely immobile.

Motility of both compact and flattened granular cells involved the protrusion of hyaline pseudopods which were devoid of cytoplasmic granules. The absence of granule and the hyaline appearances of the pseudopods made it possible to distinguish them easily from the nonmotile portions of the cell. When studied with Nomarski optics and a 100/1.25 microscope objective, the focal depth is very narrow (about 0.3 μm) and critical focusing on the glass-medium interface revealed that three modes of pseudopodial protrusion were shown by cells in culture. Pseudopods may be protruded in contact with the substratum, may be extended into the medium from the anterio-dorsal surface of the cell with no initial contact with the substratum, with the pseudopod subsequently being lowered to the substratum or the distal end of the pseudopod may be elongated continually off of the substratum while the more proximal portions of the pseudopod are continually being brought down to contact the substrate. In the latter two cases, the ventral surface of the pseudopod remained out of focus when the microscope was focused at the glass-medium interface until contact was established with the glass. In all cases, cell movement was rapid. In the absence of contact with other cells, a given cell showed sustained polarity of movement, changing direction only infrequently.

Following protrusion of a pseudopod, it became invaded by the cytoplasmic granules and the trailing portion of the cell moved forward. The posterior end of the cell often had the appearance of a contracted knot or uropod. As the cell moves, thin filamentous processes ("retraction fibers") were drawn out between the uropod and residual sites of cell-substrate adhesion.

In many respects, the locomotion of the Limulus amebocyte resembles that of vertebrate granulocytes. In both cases hyaline pseudopods are protruded, often initially out of contact with the substratum (Armstrong, 1977, Armstrong & Lackie, 1975), later to be invaded by granular endoplasm (Archer & Hirsch, 1963; Armstrong & Lackie, 1975; De Bruyn, 1944, 1946; Lewis, 1927; Ramsey, 1974; Robineaux, 1964). In both cells the posterior end of the cell is contracted into a uropod with trailing retraction fibers.

The cytoplasmic flow which produces forward movement of the posterior end of the cell might be produced by contraction at the posterior end of the cell (either in a circumferential or a longitudinal direction) or might be generated by an interaction of individual granules or of the endoplasm with microtubules or microfilaments in the endoplasm. Although the contracted cell frequently shows a uropod-like structure at its trailing end, which might be proposed as a site of circumferential contraction involved in granule flow, flattened granular cells lack such a structure yet still show cytoplasmic streaming. In flattened cells, where the flow of individual granules can be observed more easily than in contracted cells, individual granules display considerable independence of movement. At a given instant, adjacent granules often moved at different speeds and in different directions. These observations suggest that an internal pressure gradient produced by contraction of the posterior end of the cell is not the sole source of granule movement. The observations do not exclude the possibility that in the more rapidly moving contracted cell, circumferential or longitudinal contraction of the cortical cytoplasm may contribute to cytoplasmic flow. I have observed that motility is unaffected in the presence of 100 μg/ml of colcemid, suggesting that intact microtubules are not required for cytoplasmic flow (Armstrong, in preparation).

CELLULAR MOTILITY *IN VIVO*

The morphology and active motility of the blood cell was observed *in situ* in isolated gill leaflets maintained on a microscope slide. The amebocytes in the gill that transformed following isolation displayed the full range of morphological states that were present in culture (*e.g.* the compact, motile state, the flattened, granular state, and the flattened, degranulated state). In the isolated gill leaflets, the transformed granular cells were actively motile, migrating over the inner surfaces of the walls of the gill leaflets and over the vertical pillars that prevent upper and lower gill surfaces from coming together. Cell motility was morphologically identical in the isolated gill leaflet to that observed in culture, being accomplished by the protrusion of hyaline pseudopods and the flow of granular cytoplasm into the pseudopods.

The comparison of motility *in vivo* with that observed in culture is interesting because it demonstrates that the motility seen in culture is an accurate reflection of the behavioral capabilities of these cells in their natural environments. Observation of cell motility *in vivo* has been successful in only a few systems. Certain embryonic tissues, including intact teleost (Armstrong, 1978; Trinkaus, 1973), and sea urchin (Gustafson & Wolpert, 1967) embryos, the chick embryo cornea (Bard & Hay, 1975), and the frog tadpole tailfin (Billings-Gagliardi, 1977; Clark, 1912; Clark et al., 1936, Radice, 1977; Speidel, 1932, 1933) have proven thin and transparent enough for the study of cell motility in vivo. In adult organisms, the test cells of the tunicate (Izzard, 1974), and the motile blood cells in the wing veins of the cockroach (Arnold, 1961) have been studied. Only in the study of Bard & Hay (1975) have detailed comparisons been made of cell motility in culture and in vivo. They reported that the motility of corneal mesenchyme cells differed in the two circumstances. Cells which showed broad lamellipods when migrating on glass exhibited thin pseudopods in the intact cornea. Ruffling activity occurred in vitro but was absent in vivo. In contrast, the Limulus amebocyte was indistinguishable in culture and in vivo.

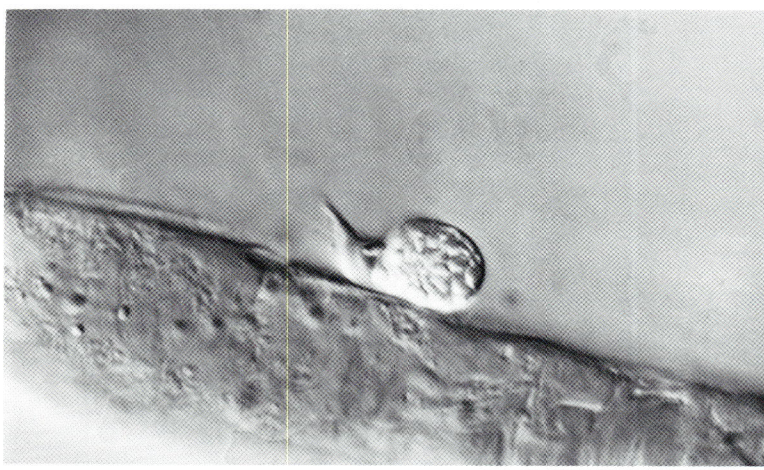

Fig. 4. Side view of a Limulus amebocyte migrating in an isolated gill leaflet. The cell is moving from right to left. The hyaline pseudopod is in the process of being protruded into the plasma-filled space of the gill leaflet, prior to being lowered to contact the substratum.

The amebocytes migrating in isolated gill leaflets could be viewed not only from above but also from the side when they moved over the pillars that keep the lumen of the gill leaflet open. Observation of cells from the side were especially useful in studying the events of pseudopodial protrusion. Thick myaline pseudopods were protruded from the anterior-dorsal surface and subsequently were lowered to contact the substratum only when extension was complete (Fig. 4). The ability of cells to protrude the pseudopods initially into fluid filled space without contact with the substratum suggests that the forces that extend the pseudopod are not those of passive spreading onto local areas of high energy surface as has been suggested by Carter (1967, 1970) since the amebocyte pseudopod is not in contact with a surface during extension.

MOTILE BEHAVIOR OF AMEBOCYTE POPULATIONS IN VITRO

The motile behavior of populations of cells is strongly influenced by contact interactions between cells. Contact paralysis, defined as the inhibition of pseudopodial activity over those portions of a cell's surface that lie in contact with other cells, represents one such mode of cell-cell interaction. Contact paralysis has been studied most extensively for vertebrate fibroblasts in culture (Fig. 5) (Abercrombie, 1970; Abercrombie & Ambrose, 1958; Armstrong & Armstrong, 1973; Armstrong & Lackie, 1975; Harris, 1973; Martz & Steinberg, 1972; Trinkaus et al., 1971) and may contribute to the monolayering shown by cultured fibroblasts with the inhibition of pseudopodial extension at sites of cell contact preventing the overlapping of cells (Abercrombie, 1970).

Although studies of contact paralysis have, in recent years, been limited almost exclusively to vertebrate tissue cells, the concept was first formulated by Leo Loeb (1920, 1921) from studies on the motile blood cells of Limulus. Following collision between cells migrating in culture, most amebocytes exhibited contact paralysis in that they almost never protruded new pseudopods from those portions of their surfaces that directly contacted other cells (Fig. 6) (Armstrong, 1977). In only one case (out of 89 collisions analyzed from the film records) did one cell actively push a second cell out of the way. Other than this one case, the consequences of a failure of contact paralysis (i.e., a cell using its pseudopod to push off from other cells, to push other cells out of the way, or to indent the surfaces of

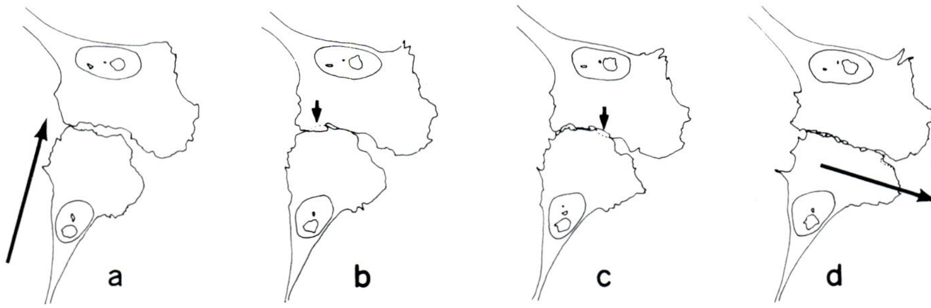

Fig. 5. Contact paralysis of pseudopodial activity following collision between two chick heart fibroblasts migrating in monolayer tissue culture is shown in this series of tracings of selected frames from a time-lapse film. The fibroblast is roughly triangular in shape, with the apex of the triangle being the tail and the leading pseudopod occupying the curved, scalloped portion of the cell's margin. The arrow in the first frame (a) indicates the initial direction of motility of the lower cell. Following collision, the pseudopods of both cells are inhibited from further expansion along the margin of contact, limiting overlapping (arrows in b and c). This results in a change in direction of movement of the lower cell, indicated by the arrow in the last frame (d). It can be proposed that the paralysis of pseudopods at sites of cell contact inhibits cell movement in coherent tissues in vivo where cells are in contact with other cells on all sides.

other cells) were not observed. The area of the cell surface subject to contact paralysis was strictly limited to the immediate regions of contact. Although the Limulus amebocyte showed contact paralysis of the pseudopods in the sense that further extension of an already existing pseudopod was halted upon collision with another cell and new pseudopods were not produced by those portions of a cell that were in contact with other cells, cell behavior following collision was quite different from that shown by vertebrate fibroblasts. In vertebrate fibroblasts, overlapping rarely occurs but in the Limulus amebocyte, it frequently did (Table 1). Contracted amebocytes overlapped contracted cells, flattened granular cells (Fig. 7) and

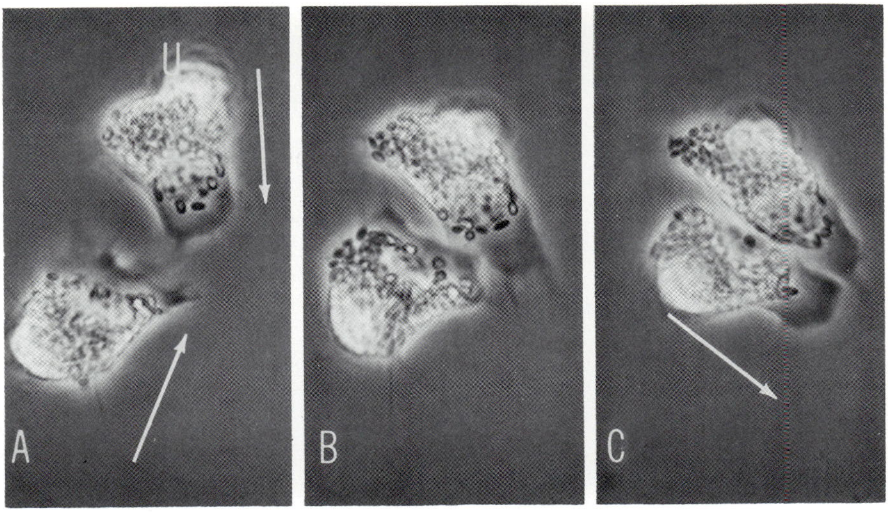

Fig. 6. Sequential photographs showing contact paralysis of the pseudopods. The cells which collide in (a) and (b) have come into close contact along the margin of initial collision (c). Between the time of initial contact (a) and (c), both cells have changed the direction of movement as the result of collision with both moving in parallel off to the right. There is no sign of pseudopodial activity along the mutual interface. The arrows in (a) and (c) indicate, respectively, the initial and final directions of cell movement, U, uropod.

flattened degranulated cells in over 40% of the collisions. Contracted cells moved freely over the dorsal surfaces of cells which they overlapped.

The marked differences in overlapping behavior of amebocytes and vertebrate tissue cells may find its explanation in the manner in which pseudopods are protruded in the different cells. Vertebrate tissue cells extend their pseudopods in close apposition to the surface, usually with only a small rim of the anterior margin raised above the substratum (Heath & Dunn, 1978; Izzard & Lockner, 1976). When contact occurs, the extent of overlapping is limited. In contrast, the <u>Limulus</u> amebocyte often extend the entire pseudopod well above the substratum, lowering it to contact the substratum only after extension is complete. It is suggested that the ability to protrude pseudopods into the

Table 1. Collision analysis of the behavior of contracted Limulus amoebocytes

The behavior of cells that had emigrated onto microscope coverglasses from ameobocyte tissue explants following cell-cell collision was determined by careful frame-by-frame analysis of time-lapse film records. Each collision in the films was recorded and scored.

Identity of colliding cells	Consequences of collision				Total no. of collisions observed
	Extent of overlapping			Partial underlapping	
	None	Partial	Complete		
Contracted-contracted	25	17[a]	38[a]	9	89[b]
Contracted-flattened	3	2[c]	7[c]	0	12

[a] At least one of the two cells involved in the collision overlapped the other either partially or completely.

[b] These 89 collisions between contracted amoebocytes were also analyzed in regard to the degree of failure of contact paralysis (which occurred in only one case).

[c] The contracted amoebocyte always overlapped the flattened cell, never vice versa.

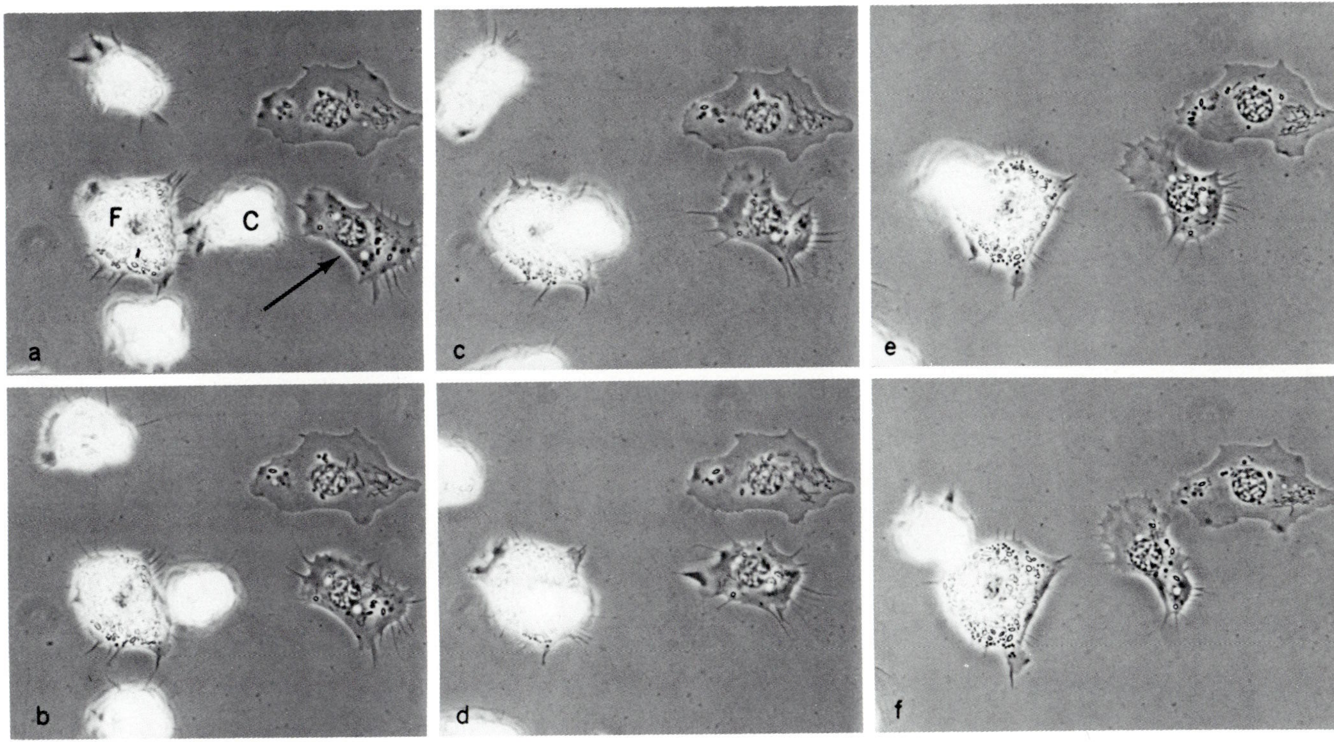

Fig. 7 Sequential photographs showing the overlapping of a flattened granular amebocyte (f) by a contracted amebocyte (C). The contracted cell moves directly onto (c,d) and across (e,f) the dorsal surface of the flattened cell. Note the movement of one of the two flattened degranulated cells (arrow, (a)) during this sequence. x636.

medium permits the amebocyte to extend pseudopods over neighboring cells initially without contacting those cells and thus without the intervention of contact paralysis. When such a pseudopod is lowered, it contacts the dorsal surface of the overlapped cell and one cell can then move bodily onto the second, even though further extension of the pseudopod is inhibited by contact paralysis.

PHAGOCYTOSIS *IN VITRO*

In most higher animals, phagocytosis represents an important aspect of the cellular defense functions of the motile blood cells (Salt, 1970; Zweifach, et al., 1973). Phagocytosis represents another aspect of cellular motility, involving as it does the contractile activities of cellular actomyosin (Allison et al., 1971; Stossel, 1977). Although the extracellular clotting system has received considerable attention in its defense role against gram negative organisms (Bang, 1956; Levin, 1976; Levin & Bang, 1964, 1968; Solum, 1970), less attention has been paid to the phagocytic capabilities of the *Limulus* amebocyte. The literature is divided on the issue of whether the cells are even capable of phagocytosis with claims that the cells are phagocytic (Ruediger & Davis, 1907; Stagner & Redmond, 1975) being contrasted with reports that the cells are incapable of phagocytosis (Bang, 1956; Levin & Bang, 1968; Loeb, 1902; Shirodkar et al., 1960).

In reinvestigating the problem of amebocyte phagocytosis, cells were exposed either to *E. coli* (both live and heat-killed) or carbonyl iron particles. Cells were brought into contact with the particles either by bleeding directly into a particle suspension, followed by explantation of amebocyte tissue fragments to observation chambers in the usual manner, or by coating the top coverslip of the observation chamber with particles and allowing cells to encounter particles as they emigrated from explants of amebocyte tissue.

Although several attempts were made, the cells appeared unable to phagocytose *E. coli* under these conditions. In the presence of bacteria, the cells degranulated, flattened and became immobile. Massive amounts of extracellular gel were formed. Cells were able to phagocytose carbonyl iron particles (Fig. 8), especially when these had previously been rendered endotoxin-free by heating. Migrating cells carried the particles about with them as they moved over the microscope coverglass of the culture chamber. The particles flowed about in the cytoplasm with the cytoplasmic granules. That the particles were truly within the cells rather than being carried on their surfaces was established by critical focusing of the microscope using a Plan 100/1.25 objective, 10x oculars, a green interference filter and Nomarski differential interference contrast optics. Using the formula for depth of focus of the microscope (Piller, 1977), it was calculated that under these conditions, the depth of focus was 0.29 μm. The cells were as much as 12 μm deep and the particles were about 1 μm or less in size.

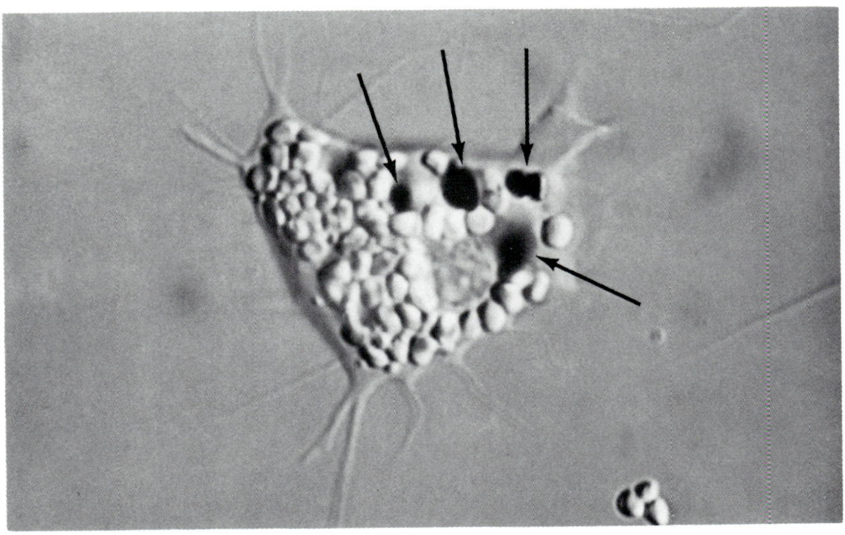

Fig. 8. In this preparation, a horseshoe crab was bled into an embryo dish containing a sterile, endotoxin-free suspension of carbonyl iron particles. Fragments of amebocyte tissue were then explanted to observation chambers and cells containing phagocytosed particles migrated out onto the microscope coverglass. The particles moved with the cells, flowing with the cytoplasmic granules. The localization of the particles within the cytoplasm (rather than adherent to the cell surface) was established by critical focusing of the microscope which revealed that particles were located between dorsal and ventral surfaces of the cell. Arrows indicate carbonyl iron particles. x2165.

Using critical focusing, it was possible to demonstrate unequivocally that the particles lay within the cytoplasm. The amebocytes are, thus, capable of phagocytosis and seem able to express this ability best under endotoxin-free conditions. Endotoxin causes Limulus blood cells to degranulate, flatten and become immobile. Such cells show little movement of cell processes. Endotoxin also, as noted above, stimulates the formation of a voluminous extracellular gel. Both processes probably interfere with phagocytosis: immobilization by preventing the formation of phagocytic vacuoles and the extracellular gel by preventing direct contact between particles and cells. It is suggested that Limulus blood cells handle foreign particles in two ways. The extracellular clotting system deals with gram negative organisms and phagocytosis deals with other types of foreign particles.

PROSPECTS FOR FUTURE STUDY

The study of cellular motility is currently an area of intensive investigation. The capacity for motility is nearly ubiquitous in animal tissue cells and is important in a variety of phenomena. Most of the current investigation of motility in animal tissue cells has centered on the vertebrate fibroblast maintained in culture. As noted earlier, motility of this cell is characterized by an absence of cytoplasmic flow. The motility of the Limulus amebocyte and vertebrate leukocyte is morphologically quite different, probably reflecting significant differences in the organization of the contractile and cytoskeletal elements. A careful fine-structural and immunofluorescence analysis of the organization of the cytoskeletal elements in cells that show extensive cytoplasmic streaming should prove illuminating in elucidating the mechanisms of their motility.

A second area of investigation for which the Limulus amebocyte is ideally suited is the biochemical investigation of motility. The ability to obtain amebocytes in large quantities as homogeneous populations by the simple expedient of bleeding an animal is a very attractive feature for their use in biochemical studies. Of interest will be not only the identities and characteristics of the contractile proteins (actin and myosin) but the identities of the associated proteins that control the state of polymerization of actin and myosin and their enzymatic and contractile activities. Presumably these latter proteins play an important role in maintaining the inactivated cell in an immobile condition and alter their association with actin and myosin when the cell is activated and begins extending pseudopods and engaging in locomotion.

ACKNOWLEDGEMENTS

This study was supported in part by Cancer Research Funds of the University of California and NSF Grant #PCM 77-18950. Figures 1, 3, 6, 7 and Table I are from Armstrong (1977) Exp. Cell Res. 107:127 reprinted by permission of Academic Press, Inc. Figure 5 is from Armstrong and Lackie (1975) J. Cell Biol. 65:439, reprinted by permission of The Rockefeller University Press.

REFERENCE LIST

Armstrong M (1976). Contact inhibition in tissue culture. In vitro 6:128.

Abercrombie M, Ambrose EJ (1958). Interference microscope studies of cell contacts in tissue culture. Exp Cell Res 15:332.

Allison AC, Davies P, dePetris S (1971). Role of contractile microfilaments in macrophage movement and morphogenesis. Nature, New Biol 232:153.

Andrew W (1965). "Comparative Hematology." London: Grune & Stratton.

Archer GT, Hirsch JG (1963). Motion picture studies on degranulation of horse eosinophils during phagocytosis. J Exp Med 118:287.

Armstrong PB (1977). Interaction of the motile blood cells of the horseshoe crab, Limulus. Studies on contact paralysis of pseudopodial activity and cellular overlapping in vitro. Exp Cell Res 107:127.

Armstrong PB (1978). Time-lapse microcinematographic studies of cell motility in the embryonic yolk sac of Fundulus heteroclitus. Biol Bull (Woods Hole) 155.

Armstrong PB, Armstrong MT (1973). Are cells in solid tissues immobile? Mesonephric mesenchyme in vitro. Devel Biol 35:187.

Armstrong PB, Lackie J (1975). Studies on intercellular invasion in vitro using rabbit peritoneal neutrophil granulocytes (PMN's) I. Role of contact inhibition of locomotion. J Cell Biol 65:439.
Arnold JW (1961). Further observations on amoeboid haemocytes in Blaberus giganteus (L) (Orthoptera: Blattidae). Canad J Zool 39:755.
Bang FB (1956). A bacterial disease of Limulus polyphemus. Bull Johns Hopkins Hosp 98:325.
Bard JBL, Hay ED (1975). Behavior of fibroblasts from the developing avian cornea morphology and movement in situ and in vitro. J Cell Biol 67:400.
Billings-Gagliardi S (1977). Mode of locomotion of Schwann cells migrating in vivo. Amer J Anat 150:73.
Carter SB (1967). Haptotaxis and the mechanism of cell motility. Nature 213:256.
Carter SB (1970). Cell movement and cell spreading: a passive or an active process? Nature 225:858.
Clark ER (1912). Further observations on living growing lymphatics: their relation to the mesenchyme cells. Amer J Anat 13:351.
Clark ER, Clark EL, Rex RO (1936). Observations on polymorphonuclear leucocytes in the living animal. Amer J Anat 59:123.
Cohen WD, Nemhauser I, Jaeger R (1977). Rapid visualization of the marginal band system in blood cells of marine species. Biol Bull (Woods Hole) 153:420.
DeBruyn PPH (1944). Locomotion of blood cells in tissue culture. Anat Res 89:43.
DeBruyn PPH (1946). The ameboid movement of the mammalian leukocyte in tissue culture. Anat Rec 95:177.
Dumont JH, Anderson E, Winner G (1966). Some cytologic characteristics of the hemocytes of Limulus during clotting. J Morphol 119:181.
Gustafson T, Wolpert L (1967). Cellular movement and contact in sea urchin morphogenesis. Biol Revs 42:442.
Harris AK (1973). Cell surface movements related to cell locomotion. In "Ciba Foundation Symposium on Locomotion of Tissue Cells," Amsterdam: Elsevier, p 3.
Heath JP, Dunn GA (1978). Cell to substratum contacts of chick fibroblasts and their relation to the microfilament system. A correlated interference-reflexion and high-voltage electron-microscope study. J Cell Sci 29:197.
Izzard CS (1974). Contractile filopodia and in vivo cell movement in the tunic of the ascidian, Botryllus schlosseri. J Cell Sci 15:513.

Izzard CS, Lochner LR (1976). Cell-to-substrate contacts in living fibroblasts: an interference reflection study with an evaluation of the technique. J Cell Sci 21:129.
Levin J (1976). Blood coagulation in the horseshoe crab (Limulus polyphemus): a model for mammalian coagulation and hemostasis. In "Animal Models of Thrombosis and Hemorrhagic Disease" (DHEW Publication No (NIH) 76-982), Washington DC, p 87.
Levin J, Bang FB (1964). The role of endotoxin in the extracellular coagulation of Limulus blood. Bull Johns Hopkins Hospital 115:265.
Levin J, Bang FB (1968). Clottable protein in Limulus: its localization and kinetics of its coagulation by endotoxin. Thrombosis Diath Haemorrh 19:186.
Lewis WH (1927). Migration of neutrophilic leucocytes. Arch exp Zellforsch 4:442.
Loeb L (1902). On the blood lymph cells and inflammatory processes of Limulus. J Med Res 7:145.
Loeb L (1920). The movements of the amoebocytes, and the experimental production of amoebocyte (cell fibrin) tissue. Washington University Studies 8:3.
Loeb L (1921). Amoeboid movement, tissue formation and consistency of protoplasm. Amer J Physiol 56:140.
Lüscher, EF, Bettex-Galland M (1971). Thrombosthenin. Johnson SA (ed): "The Circulating Platelet," New York: Academic Press, p. 225.
Marchesi VT (1970). Mechanisms of blood cell migration across blood vessel walls. In Gordon AS (ed): "Regulation of Hematoporesis," New York: Appleton-Century Crofts, vol II, p 943.
Martz E, Steinberg MS (1972). Contact inhibition of what? An analytical review. J Cell Physiol 81:25.
Mürer EH, Levin J, Holme R (1975). Isolation and studies of the granules of the amebocytes of Limulus polyphenus, the horseshoe crab. J Cell Physiol 86:533.
Piller H (1977). Depth of focus (focal depth). In "Microscope Photometry," Berlin: Verlag, p 15.
Pollard TD, Fugiwara K, Hardin R, Weiss G (1977). Contractile proteins in platelet activation and contraction. Ann NY Acad Sci 283:218.
Radice GP (1977). Active and passive cellular movements during epithelial wound closure in vivo. J Cell Biol 75: 411a.
Ramsey WS (1974). Leucocyte locomotion and chemotaxis. In Sorkin E (ed): "Chemotaxis, its Biology and Biochemistry," Basel: S. Karger, p 179.

Reaven EP, Axline SG (1973). Subplasmalemmal microfilaments and microtubules in resting and phagocytizing macrophages. J Cell Biol 59:12.

Robineaux R (1964). Movements of cells involved in inflammation and immunity. In Allen RD, Kaniya N (eds): "Primitive Motile Systems in Cell Biology," New York: Academic Press, p 351.

Ruediger GF, Davis DJ (1907). Phagocytosis and opsonins in lower animals. J Infect Dis 4:333.

Shepro D, Belamarich FA, Merk FB, Chao FC (1969). Changes in thrombocyte ultrastructure during clot retraction. J. Cell Sci 4:763.

Shirodkar MV, Warwick A, Bang FB (1960). The in vitro reaction of Limulus amebocytes to bacteria. Biol Bull 118:324.

Solum NO (1970). Some characteristics of the clottable protein of Limulus polyphenus blood cells. Thrombos Diathes Haemorrh 23:170.

Speidel CC (1932). The movements of individual sheath cells and nerve sprouts correlated with process of myelin-sheath formation in amphibian larvae. J Exp Zool 61:279.

Speidel CC (1933). Activities of growth cones, sheath cells and myelin segments as revealed by prolonged observation of individual nerve fibers in frog tadpoles. Am J Anat 52:1.

Stagner JI, Redmond JR (1975). The immunological mechanisms of the horseshoe crab, Limulus polypheus. Marine Fish Rev 37:11.

Stossel TP (1977). Motile functions of phagocytic effector cells. In Cooper MD, Dayton DH (eds): "Development of Host Defenses," New York: Raven Press, p 187.

Trinkaus JP (1973). Surface activity and locomotion of Fundulus deep cells during blastula and gastrula stages. Devel Biol 30:68.

Trinkaus JP, Betchaku T, Krulakowski LS (1971). Focal inhibition of ruffling during contact inhibition of cell motion. Exptl Cell Res 64:291.

Zweifach BW, Grant L, McCluskey RT (1973). "The Inflammatory Process, 2nd Ed," New York: Academic Press.

THE *IN VITRO* CULTIVATION OF *LIMULUS* AMEBOCYTES

Frederick C. Pearson and Ellen Woodland
Rhode Island College
Providence, Rhode Island 01908

INTRODUCTION

Interest in the laboratory cultivation of *Limulus* cells and tissues is not new. As early as 1959, when cell culture was still emerging from its infancy, Sanborn and workers reported on the successful *in vitro* cultivation of a number of cell types from *Limulus* tissues. These included primary explants of ovary, hepatopancreas, nerve, leg and cardiac muscle all of which were cultivated by use of a simple hanging drop method. Although the medium employed in this study was not named, it contained inorganic salts, multiple sugars and organic acids all of which are consistent with contemporary cell culture media. Even though cells in culture require an elaborate list of essential amino acids, there was no indication that such growth factors were included in this medium. However, five to ten percent sterile *Limulus* serum was incorporated in the growth medium and may have masked the need for exogenous amino acids. Cells were maintained up to thirty days, but after ten days vacuolation and granulation become pronounced.

The general morphology of cells cultivated from *Limulus* primary explants are not unlike those encountered in explants of vertebrate tissue. Cells which emerge from the margins of the explant develop into either compact, rounded cells indicative of epithelial morphology or long spindle shaped cells indicative of fibroblasts. *Limulus* explants of muscle and nerve predominantly generate fibroblasts and other tissues tend to produce cells with more compact morphology.

Pages 93–102, Biomedical Applications of the Horseshoe Crab (Limulidae)

Ovary, hepatopancreas, cardiac tissue and amebocytes have been cultivated in our laboratory and maintained for several months in polystyrene tissue culture flasks employing frequent media changes.*

When using clotted blood of Limulus as an explant, Sanborn was able to maintain amebocytes for three months. However, no cell migration from explant to medium was noted since the solid nature of the clot impeded cell movement.

In 1962, Wolff reported being able to maintain cultures of hepatopancreas, heart muscle, gonad, intestine and nervous tissue for up to 37 days.

Later work conducted by Yamamici on embryo and organ culture of the Japanese horseshoe crab, Tachypleus tridentatus, demonstrated that rhythmic beating of cardiac explants continued for up to twelve days in calf serum agar, medium 199 agar and calf serum agar 199. When these media were used without the addition of agar, contraction of cardiac tissue was observed in culture for over one month. Migration of fibroblasts from cardiac tissue was most pronounced when medium 199 was used and supplemented with 10% calf serum. We did not find medium 199 supplemented with calf serum to be an effective growth medium for Limulus amebocytes, however, Yamamici did not attempt to cultivate amebocytes during this study.

The following paper presents basic growth requirements for amebocytes cultivated in vitro and suggests the potential use of such a system for the standardized production of Limulus lysate.

Although the production and application of the Limulus Amebocyte Lysate test has become more standardized in recent years, experience still supports the notion that significant variation occurs in lysate produced by various manufacturers and that lot-to-lot variation is exhibited by individual manufacturers. Despite the use of a standard endotoxin, a reference lysate and the widespread use of the clot endpoint, the problem of lysate variability remains. The successful cultivation of Limulus amebocytes in the laboratory suggests the possibility of producing a lysate that is

* Mrs. Lois Patterson, Biology Department, Rhode Island College

not subject to the variability of the natural environment. Such a standardized production would most probably reduce batch variability and seasonal variability of lysate. Bleeding of horseshoe crabs might be deleted and thus the collection, transportation and possible depletion of the horseshoe crab population could be circumvented. Although the U. S. horseshoe crab population appears to be stable and in good numbers currently, other creatures which have once crowded this earth no longer exist due to overutilization, as well as diminished and deterioriating habitat. While the American horseshoe crab has coped with twentieth century habitat and survived its challenge from the fertilizer, fishing and medical industries, its Japanese counterpart has not been so fortunate. Niwa concluded that the Japanese horseshoe crab is an "endangered species and seemingly on its way to extinction due to reclamations and contamination". Fortunately, loss of habitat and pollution has not been as pronounced along the majority of the East Coast as it has in Japan. Although Limulus appears to be miraculously resilient to being bled for the collection of amebocytes, the impact of bleeding activity on the horseshoe crab population has not been clearly documented in the literature. Proposed rules which were released in the Federal Register on August 11, 1978 underscore the potential problem by requesting manufacturers of LAL to guarantee that the production of LAL will not have an adverse impact on existing crab populations and horseshoe crabs will be returned alive to their natural environment after a single collection of their blood.

RESULTS

A number of common cell culture media were selected for their potential use in the cultivation of Limulus amebocytes. Minimal Essential Medium (MEM), Leibovitz (L-15), Medium 199 and Grace's Insect Medium (GIM) were used. Grace's medium was selected for its established use in the culture of invertebrate cells. Although the other selections are used primarily for the cultivation of primate cell lines, the work of both Yamamici and Sanborn suggested their applicability in the cultivation of Limulus cells. As can be seen in Figure 1, the growth promoting ability of Grace's Insect Medium is superior to Minimal Essential Medium. Since animal cells in culture generally require a protein supplement for optimal growth, early cultures of amebocytes were grown in various concentrations of fetal calf serum (FCS). Serum supplements of 5, 10, 15 and 20% infrequently improved growth over unsupplemented medium and were generally inhibitory to amebocytes.

Although amebocytes cultivated in GIM seemed to grow relatively well, appropriate levels of electrolytes and a suitable protein supplement were needed. Electrolyte determinations conducted on horseshoe crab serum and GIM revealed significantly higher concentrations of both sodium and chloride ions in horseshoe crab serum (Table 1). Increased amounts of potassium and calcium were also noted in serum relative to levels in GIM. The medium was rendered more isotonic with horseshoe crab serum by the addition of appropriate levels of sodium, potassium and chloride and the effect on growth can be seen in Figure 2. Not only did the addition of electrolytes increase cell yield but when cells were viewed by inverted phase contrast microscopy, they appeared less vacuolated and more elliptical. Amebocytes cultivated in GIM alone exhibited a large number of vacuoles and frequent pleomorphism.

Serum protein electrophoresis of horseshoe crab serum demonstrated approximately 6.0 gm% total protein with a major peak of albumin constituting 67% of the total protein. Due to the high protein content of amebocyte environment in vivo coupled with the general need of exogenous protein for cells grown in vitro, an appropriate protein supplement was sought. Albumin did not improve growth and most albumins proved inhibitory. Lactalbumin hydrolysate failed to increase cell growth and yeastolate proved strongly inhibitory. Additional protein supplements commonly used in cell culture such as horse serum were also found to be inhibitory to amebocytes (Table 3). Homologous serum was the only protein supplement found that significantly improved cell yield (Table 2). Horseshoe crab serum was prepared by collecting crab blood in 500 ml pyrogen free bottles, allowing clot formation to take place at 4°C overnight, subsequently harvesting the dark blue supernatant and clarifying by centrifugation. Serum was filter sterilized prior to use and routinely employed in cultures at 10%.

The serum of donor horseshoe crabs ranges from 6.89 to 7.51 and averages 7.28. Since the pH of horseshoe crab serum is alkaline and the pH of Grace's medium is acidic, culture medium was routinely adjusted to pH 7.2 - 7.4. The effect of pH on the growth of amebocytes can be seen in Figure 4.

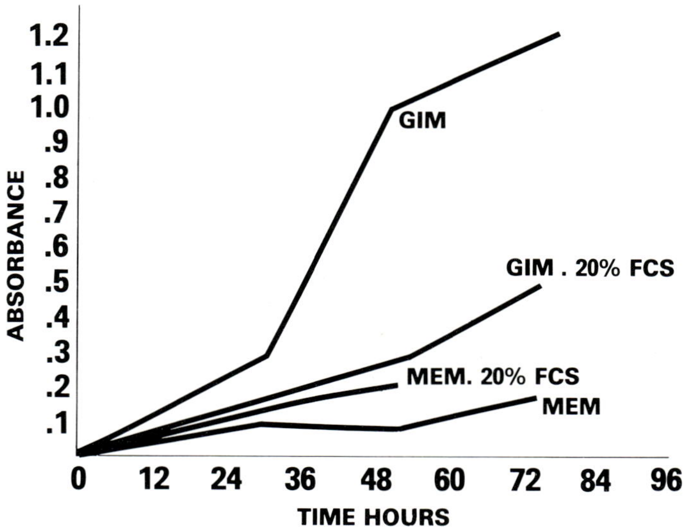

Fig. 1. Effect of FCS on the growth of amebocyte.

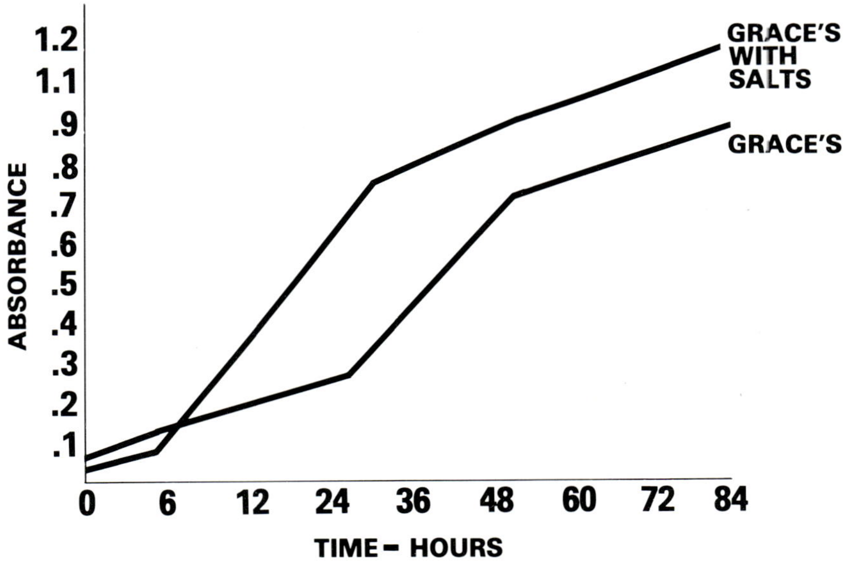

Fig. 2. Effect of added electrolytes on amebocyte growth.

TABLE 1.

	HUMAN SERA	LIMULUS SERA	GRACES MEDIUM
NA +	140*	400	12
K+	3.5	11.6	0.05
CA++	5	16	13
CL –	106	588	3.6
OSMOLARITY		1007	270

* MILLIEQUIVALENTS PER LITER = MEQ/L

TABLE 2.
EFFECT OF HOMOLOGOUS SERUM ON AMOEBOCYTE GROWTH

T (HR)	G.I.M.	G.I.M. 10% SERUM
0	.06*	.03
5	.22	.22
25	.80	.91
48	1.10	1.30

* OPTICAL DENSITY

TABLE 3.
EFFECT OF ADDED PROTEINS ON AMEBOCYTE GROWTH

1. G.I.M.
2. G.I.M. + 5% ALBUMIN
3. G.I.M. + 5% LAH
4. G.I.M. + 5% YEASTOLATE

T(HR)	1	2	3	4
0	0.25*	.015	0.15	0.15
18	1.0	.34	.82	-
44	1.2	.75	1.15	.02

* OPTICAL DENSITY

Cells cultivated at temperatures between 12-24°C (room temperature) produced the greatest cell yields. Amebocytes grown at 37 were severely inhibited and cell yields obtained at 4°C were less than those at room temperature (Figure 5).

Amebocytes grew equally well in the presence or absence of light.

Various nutritional components including inorganic salts, carbon sources, amino acids and vitamins incorporated in the medium may be seen in Tables 4, 5 and 6.

Attempts were made to isolate active Limulus amebocyte lysate from cells cultivated in modified Grace's Insect Medium. The method of Jorgensen and workers was used. (Table 7). Cell crops were harvested in pyrogen free, siliconized, 250 ml polypropylene centrifuge bottles containing 125 ml of 0.125% N-ethylmalimide in 3% sodium chloride at pH 7.4. Cells were centrifuged at 50X g for 10 minutes, washed twice in 3% sodium chloride and then lysed in distilled water 1:3 at 4°C for 18-24 hours. Cells were then vortexed for one minute and centrifuged at 100 x g for 10 minutes to clarify the cell lysate. The supernatant was decanted and stored at ⁻70°C.

Lysate preparations were tested for potency by using graded endotoxin solutions ranging from 0.1 ng to 1000 ng per ml. Since solid clots were not formed by lysate preparations, a graded endpoint was used ranging from 0 (negative) to +3 (a clot that would run when the tube was tilted 180). Intermediate values were based on degree of opacity, viscosity and production of floccules. To date, lysate preparations from amebocytes grown in vitro have given only slight positive results (+1 reaction) in the presence of 1000 ng of endotoxin.

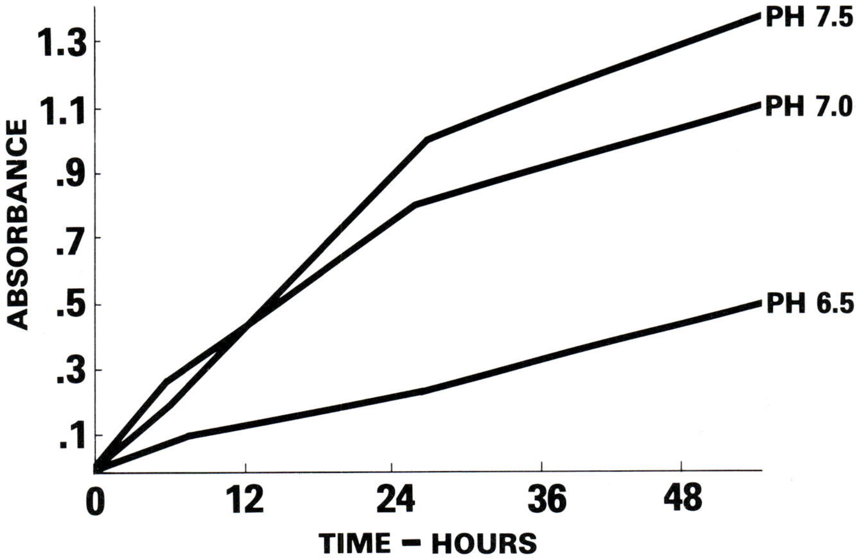

Fig. 3. Effect of pH on amebocyte growth.

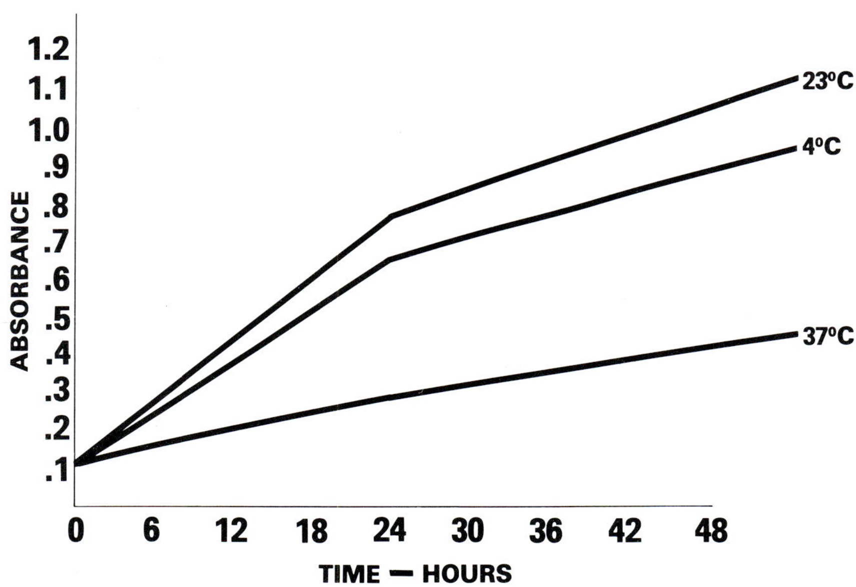

Fig. 4. Effect of temperature on amebocyte growth.

TABLE 4.
MODIFIED GRACES INSECT MEDIUM

INORGANIC SALTS	MG/L
$CaCl_2$	780.00
KCL	4130.00
NaCL	1000.00
$MgCl_2 \cdot 6H_2O$	2280.00
$MgSO_4 \cdot 7H_2O$	2780.00
$NaHCO_3$	350.00
$NaH_2PO_4 \cdot H_2O$	1013.00
CARBON SOURCES	
ALPHA KETOGLUTARIC	370.00
MALIC ACID	55.00
GLUCOSE	2000.00
SUCROSE AND FRUCTOSE DELETED	

TABLE 5.
MODIFIED GRACES INSECT MEDIUM

AMINO ACID	MG/L
L—ALANINE	225.00
L—ARGININE HCL	700.00
L—ASPARAGINE	350.00
L—ASPARTIC ACID	350.00
L—CYSTINE	100.00
L—GLUTAMIC ACID	600.00
GLYCINE	650.00
L—HISTIDINE	2500.00
L—ISOLEUCINE	50.00
L—LEUCINE	75.00
L—LYSINE HCL	625.00
L—METHIONINE	50.00
L—PHENYLALANINE	150.00
L—PROLINE	350.00
DL—SERINE	1100.00
L—THREONINE	175.00
L—TRYPTOPHAN	100.00
L—TYROSINE	50.00
L—VALINE	100.00

TABLE 6.
MODIFIED GRACES INSECT MEDIUM

VITAMINS	MG/L
BIOTIN	0.01
D—CA PANTOTHENATE	0.02
CHOLINE CHLORIDE	0.20
FOLIC ACID	0.02
I—INOSITOL	0.02
NIACIN	0.02
PARA-AMINOBENZOIC ACID	0.02
PYRIDOXINE HCL	0.02
RIBOFLAVIN	0.02
THIAMINE HCL	0.02

TABLE 7.
PREPARATION OF AMEBOCYTE LYSATE

125ML CELLS

125ML 0.125% N—ETHYLMALEIMIDE
IN 3% NACL PH 7.4

CENTRIFUGE 50XG 10'

WASH 2X 3% NACL

LYSE CELLS 1:3 DISTILLED WATER
AT 4°C 18—24 HRS.

CENTRIFUGE 150XG 10'

STORE LYSATE — 70°C

REFERENCE

Dailey ME Fremont Smith F and Carroll MP (1931) Relative composition of sea water and the blood of Limulus polyphemus. J Biol Chem 93:17

Niwa M and Ulagori O (1975) Biochemistry of horseshoe crab amebocyte. Title not translated.

Sanborn R Haskell A and Fisher F (1959) Cells of Limulus polyphemus in tissue culture. Biol Bull 117:399

United States Federal Register (1978) 43:35731-35734

Wolff EM (1962) Culture organotipique de tissues de jeunes Limulus (Xiphosura polyphemus). Bull Soc Zool France 87: 120-126

Yamamichi Y and Sekiguchi K (1974) Embryo and organ cultures of the horseshoe crab, Tachypleus tridentatus.

OBSERVATIONS ON THE POTENTIAL OF LIMULUS AQUACULTURE IN ISRAEL

Chaim Kropach

Israel Oceanographic and Limnological Research Ltd.

P. O. Box 1212, Elat, Israel

Work on *Limulus polyphemus* was begun by the author and Steve Gaffin as a joint-project under the sponsorship of the Israel Oceanographic and Limnological Research (I. O. L. R.) and the Technion, Haifa, in the summer of 1977. The help and encouragement of Dr. Jacob Fine in initiating this project is appreciated and gratefully acknowledged.

RESEARCH FACILITIES, MATERIAL AND METHODS

In order to get the *Limulus* program underway, and to insure a source of material for biochemical work, a colony of horseshoe crabs was established in the mariculture laboratory of I. O. L. R. in Elat. This facility is located 7 km south of Elat, on the eastern border of the Sinai desert. It is provided with sea-water pumps, a running-water system, holding tanks, and a food-preparation unit.

The land along the Gulf-of-Elat coast is arid and hot, temperatures ranging from 12 °C (and occasionally less) in the winter to over 40 °C in the summer. Water temperatures in the Gulf are relatively constant, ranging from a minimum of 20°C in the winter to 28 °C in the summer, with little bathymetric difference at any given time (Kropach, pers. obs.) The local winter is short (approximately mid-December-March) and mild. Radiation is intense, especially in the summer months, and evaporation rate high (about 3.5 m/yr.) Sea water in the Gulf is nutrient-poor, oxygen-rich and of high salinity, averaging 40.6 ‰.

Pages 103–106, Biomedical Applications of the Horseshoe Crab (Limulidae)

Horseshoe crabs were obtained from Gulf Specimen Co., Panacea, Florida. Most of them were kept in an especially-built circular pond of 5 m diameter, containing 20 m^3 sea water, over a sandy bottom. Other animals were kept in fibre-glass or PVC containers whose floors remained uncovered.

The floor of the circular pond served as a filter to insure cleanliness of the sand (Kropach, in prep.). Thus, it provided a breeding substrate for the adults and permitted proper aeration for the eggs and for larval and post-larval limuli. Water flowed continuously through the pond and all other containers. The pond was partitioned into sections permitting differential feeding of the animals.

The animals were fed natural food (fish, clams, shrimp, and squid) as well as processed food. Larvae were collected by passing outflow water through a cloth mesh, and were placed in various containers of varying substrates and food.

RESULTS

On a whole Limulus had become well acclimated in the pond and seemed to be in good condition in all the other containers where they had been placed. The intense radiation in the summer appeared to cause restlessness in the animals. This problem was solved by providing a cover of polyethylene sheets or palm-tree fronds over the various tanks and the pond.

In the pond, the horseshoe crabs remained most of the time under the sand. Young animals (1½-4"), when provided with a sand-substrate also chose to remain under sand most of the time.

The animals fed well on all diets given to them. Mortality throughout the year did not appear to be differential and was not restricted to a given diet-group. Examination of dead females from various compartments did not show variations in gonadal development of individuals which had been kept on different diets. All had developed ova as early as December and were potential breeders regardless of nutritional regimen.

There was no measurable growth in adult Limulus, and none had molted. The young animals (1½-4"), kept on natural food and on sand-substrate, have gone through molts and have grown. Measurements of five linear dimensions on the exoskeleton have consistently shown growth of 25-35 % in body size after molt. Weight increment was 50-150 %, depending, probably, on the interval-length between successive molts. This varied from 3 months to a whole year. Some animals have not molted and have shown no growth or weight increase at all.

Reproductive-type behavior was exhibited by the animals throughout the year. Larvae appeared from the first week of March to the end of August. Larval growth during the first six months was from 2.5 mm to 15 mm carapace width.

The main limiting factors to larval survival which have been recognized were algal growth on carapace and gills and predation by amphipods. Adult mortality was mainly due to disease affecting the animals after a year in the laboratory.

DISCUSSION

The long geological record of the horseshoe crabs, their living in a particularly demanding and multi-faceted habitat in the subtidal zone, and their distribution along a highly-pronounced climatological gradient on the Atlantic seaboard of N. America make these animals "generalists" and as such potentially good colonizers. Therefore, the idea of acclimating these organisms to an area outside of their natural distribution seemed feasible. This was proven so in our study.

Growth in Limulus takes place through successive molts, which do not occur after the animals have reached maturity. Therefore, there was no measurable morphometric growth in the adult horseshoe crabs in this study, nor was there any gain in weight. The presence of gravid females, however, and the spawning activity indicates good health and body maintenance. Future work will determine the effect of various diets on adult animal maintenance, especially with regard to amebocyte production and cellular content. Similarly, raising the young on various types of food will help determine the most efficient diet for rapid growth. The most important problem that will have to be solved is disease control.

The question whether commercial farming of Limulus should be established is, of course, not only biological. As aquaculture problems are being ironed-out, economic factors such as cost of pond construction and maintenance, the running of a sea-water system, space, food-supply, etc. would have to be taken into consideration.

CONCLUSIONS

1. The one-year study on Limulus polyphemus in the laboratory in Elat bears out theoretical thinking on the organism's colonizing potential.
2. Limulus aquaculture is a feasible operation.
3. Experimental work on Limulus aquaculture should continue in areas outside the species distribution.
4. Work should focus on improving diet and disease control.

Limulus Polyphemus, a Target of Opportunity

by Sidney R. Galler, Deputy Assistant
Secretary for Environmental Affairs
Department of Commerce

Washington, D. C. 20230

The distinguished scientist and philosopher, Sir Ritchie Calder, once asserted that science produces knowledge not wisdom; wisdom is gained from knowledge tempered with judgment. His observation is an appropriate introduction to my perspective on this symposium, particularly, this session concerned with the biology and ecology of Limulus polyphemus.

Succinctly stated, there is a direct correlation between the discovery of the utility of an organism and the rate of its economic exploitation. This symposium affords a unique opportunity to alert the participants and the general scientific community to the risks of disseminating information about the biomedical application of the horseshoe crab without, at the same time, recognizing the parallel responsibility for preparing a practical plan for its conservation and protection.

Our Japanese colleague has already made it clear that the Asiatic species have become endangered in Japan and that, consequently, a society for the protection of the species has been established. At this time when L. Polyphemus is becoming of increasing economic interest as an ingredient in chicken feed and as conchbait, we should be mindful of the fact that large sectors of its habitat are becoming increasingly

Pages 107–108, Biomedical Applications of the Horseshoe Crab (Limulidae)

degraded through man's activities. Thus it is important that we begin to sensitize the public to the importance of instituting appropriate measures for the protection and preservation of this species. Indeed, the emerging biomedical importance of the horseshoe crab justifies taking such measures.

In this connection, we should not be lulled into believing that the cleaning up of a few habitats; *e.g.*, the Delaware Bay, while most welcome, will automatically ensure the survival of this animal. That would be a self delusion that could lead to a weakening of our attempts to husband a species whose survival in the long term is increasingly at risk.

In conclusion, I hope that one of the key results of this symposium will be the inauguration of a cooperative program of practical measures designed to prevent *Limulus Polyphemus* from going the way of the passenger pigeon and the dodo bird. What a pity it would be were we to allow one of the oldest species to be eradicated by one of the youngest species on earth!

ONTOGENY AND PHYLOGENY OF RESPONSE TO GRAM-NEGATIVE ENDOTOXINS AMONG THE MARINE INVERTEBRATES

Frederik B. Bang

The Johns Hopkins University School of Hygiene and Public Health, Department of Pathobiology, 615 North Wolfe Street, Baltimore, Maryland

If we return to the biological approach to disease so clearly enunciated by Metchnikov (1892) when he spoke of a new branch of biology, that is, pathology, three things in his work seem particularly important to me. They are:

1. Study of natural diseases of animals, especially invertebrates. These are the experiments of nature.

2. The use of transparent animals, so that you can observe in the living animal what is happening. This was most beautifully illustrated in the studies of the fungus disease of Daphnia, water fleas, in the tank of the Zoological Garden in Paris, where a natural epidemic was under way (Metschnikoff, 1884).

3. Use of both the developing larvae and the adult to analyze disease phenomena (ontogeny and phylogeny of disease).

In order to follow these three principles as they apply to Limulus, I must of course go beyond the approach first published in 1956 (Bang, 1956). Thus during the past two summers, as I began again to look at the effect of endotoxin on Limulus, I realized how limited my original approach had been. Several things now stand out:

a. There is a wide variation in susceptibility to endotoxin among the marine arthropods.

b. There are beautiful windows into the pathological and

Pages 109–123, Biomedical Application of the Horseshoe Crab (Limulidae) physiological processes in Limulus. These are the transparent spines of the abdomen (opisthoma) which are movable and easily transilluminated. Circulation can be readily observed in these.

c. There are disease processes active in Limulus that we have not even thought about, and which need detailed explanation.

d. Limulus needs, and probably has, the capacity to react to endotoxin as soon as it hatches from the egg.

My interest in the ontogeny of the immune response started last January when Kathy French showed me the beautiful larvae in the embryo, and the tiny hatched animals, which - like so many embryos - are highly transparent. Inside these larvae are clearly apparent circulating granular cells which are identical in appearance to those of the adult Limulus. These I will continue to call amebocytes. They have been found as early as the trilobite state (Fig. 1), and form typical clots when the tips of the limbs or

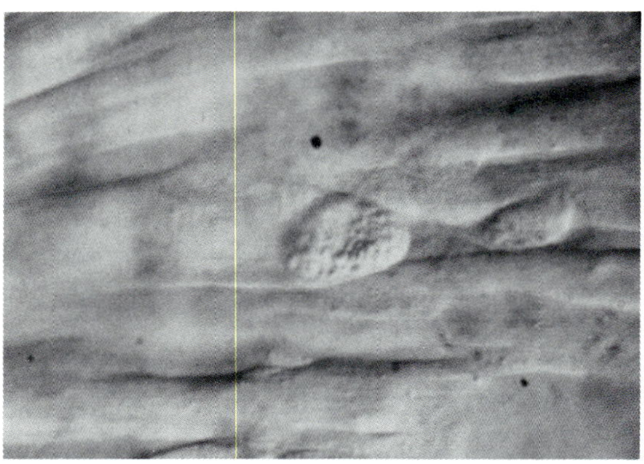

Fig. 1: Amebocyte within muscle tissue in living trilobite stage larva. (Courtesy: Kathy French - Niemarski optics.)

the tail is cut (Fig. 2). Whether these cells respond to the presence of endotoxin is not clear as yet (Fig. 3).

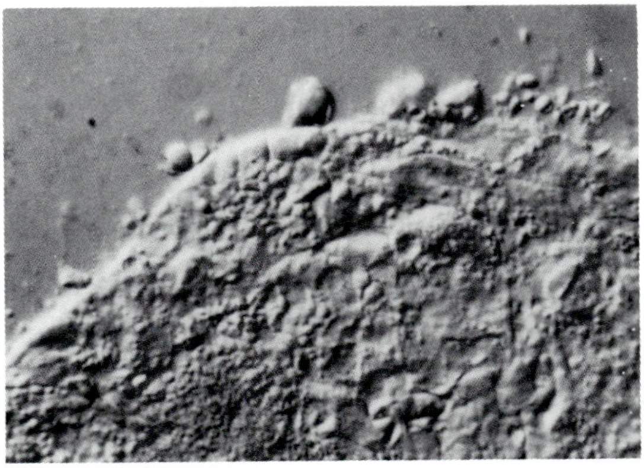

Fig. 2: Clot produced by cutting tail of first stage *Limulus* larva. (Courtesy: Kathy French - Niemarski optics.)

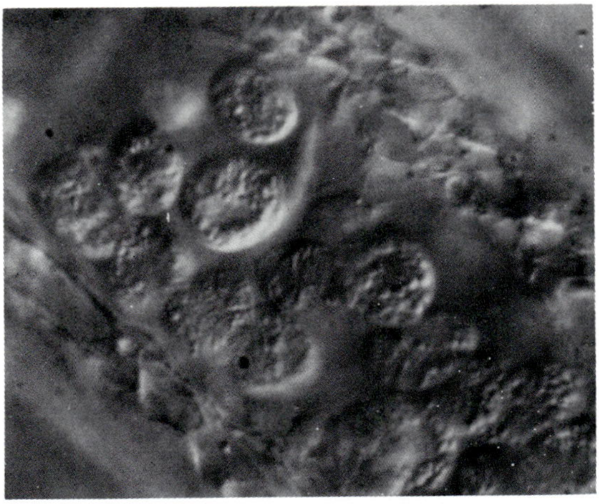

Fig. 3: Clump of amebocytes within channel in tip of foot following endotoxin injection. (Courtesy: Kathy French - Niemarski optics.)

Cells were found collected together in areas between the muscle fibers, but we could not tell whether they were abnormally clumped or not.

It may be useful at this stage to look again at the types of change that are found in the adult animal. Here are the marked change in cell shape (Fig. 4), sticking of

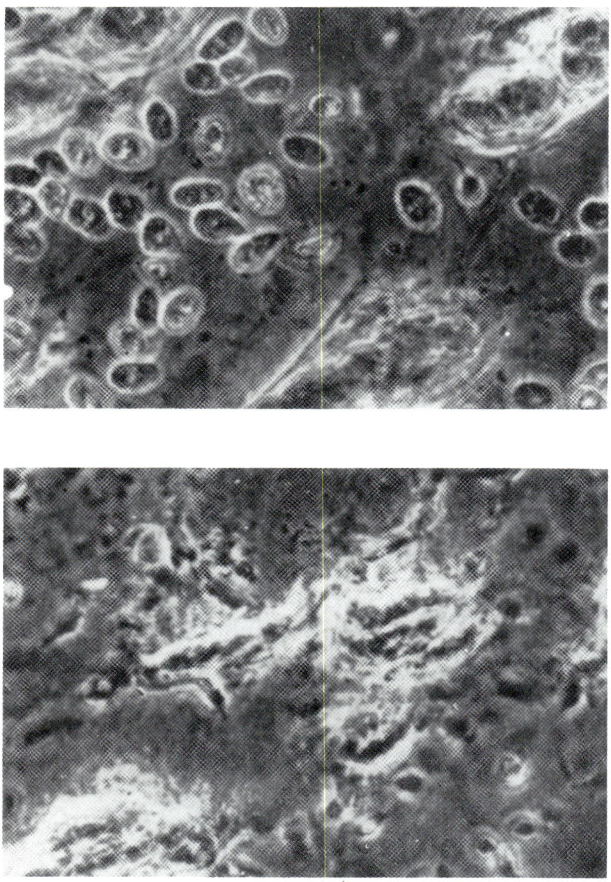

Fig. 4 (a)

Fig. 4 (b)

Fig. 4: Change in shape of amebocytes as seen in book gills of normal (a) and endotoxin-injected (b) Limulus. (From F.B. Bang, Bull. Johns Hopkins Hospital 98: 325-351, 1956.)

cells to the sides of the vascular channels, and finally the change in the fluid pre-gel to gel, as seen with the electron microscope (Fig. 5). Studies on larvae for these kinds of changes are needed.

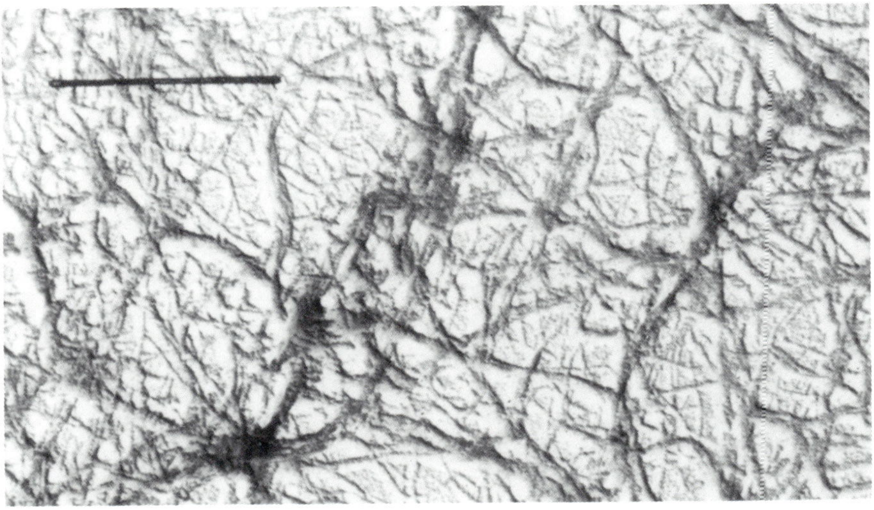

Fig. 5: Electron microscopic view of gelation of Limulus coagulen produced by endotoxin. (From F.B. Bang, Bull. Johns Hopkins Hospital 98:325-351, 1956.)

One of the Metchnikovian ideals is to be able to see what is going on in the living animal. Despite years of studying Limulus, I was ignorant of how beautiful an animal Limulus is in this regard. Not only can one readily see the circulation of blood in the young embryos, but as I looked first at the tail and later at the lateral spines on the abdominal segment, even in animals as large as 10 cm in length, one can see clearly blood flow in and out of each spine. The afferent channel can be seen in the central portion of the spine, and here individual cells rushing through to the very tip are apparent. Return through efferent channels is slower and the channels are more superficial and circuitous. In larger animals, the vessels carry a stream of cells and are usually several times the width of an individual amebocyte.

Thus clumping is readily observed (Table 1).

Table 1

Blood circulation in living *Limulus*

1.	Specific channels
2.	Clot formation after cut
3.	Control of circulation
4.	Spontaneous thrombus
5.	Bacterial and ciliate invasion of clot
6.	Carmine particles
7.	Endotoxin effect

If the tip of the spine is cut with scissors, there follows within a few seconds a flow of cells out into the seawater. Very soon these cells start to stick to each other at the edge of the opening in the chitin, and within ten seconds a loose clot covers the surface, retraction of the clot follows. Not only do the extruded cells come in contact with the masses of bacteria that are growing on the surface of the chitin, but frequently the animal passes the tip of a leg or swimmeret directly over the cut end. This tip usually has a mass of bacteria growing on it. Thus contact with gram-negative endotoxin is immediate and maintained. The clot extends a short distance into the spine, and circulation may be seen continuing behind the spine.

During the process of clot formation at the cut tip, there is at first a slowing of blood flow to the cut spine, and at the same time there seems to be a more rapid flow initiated in the neighboring spines. This was seen repeatedly, and leads to the question of nervous control of circulation. Patten's monumental study (Patten and Redenbaugh, 1899-1900) of the anatomy of *Limulus* does show nerves extending into the spines, and yet their function is, as far as I know, undetermined. Muscle action within the spines might well control flow in and out of spines.

From our standpoint, the effect of endotoxin is of greatest interest. Injection of crude seawater had no effect on the circulation. However, relatively large amounts of endotoxin, 100 μg, did cause cessation of circulation. This set in within one to two minutes of the injection, was not accompanied by visible clot formation. A series of small 1.5 cm to 4.5 cm long *Limuli* were given repeated injections of endotoxin at three to four day intervals. When circulation stopped, the animal did not seem to be significantly affected, but perhaps some showed slowing of movement. Most of them recovered their circulation within a day or two and returned to full circulation, with again a stoppage of circulation after the second or third inoculation. We did not determine whether repeated injection caused increased mortality. If circulation did not return within two days, death of the animal occurred within the next day. Larger adult animals responded to this same dose (200 μg) of endotoxin with sticking of cells and slight degranulation (Table 2).

Table 2

Effects of recurrent (4-day intervals) injections of endotoxin on circulation of blood in *Limulus*

Limulus length (cm)	Dose of endotoxin (μg)	Effects
1.4	100	circulation stopped; recovery.
1.8	40,10	circulation stopped; recovery. death on 4th day after 2nd injection.
2.5	200,200	circulation stopped; recovery. death on 4th day after 2nd injection.
4.5	100,100,100	recovery 1 day after 1st and 2nd injection. death on 3rd day after 3rd injection.
5.5	50,100,30	recovery after each injection.
8.0	DDT + 100	death - ? DTT.
8.0	100	recovery.

(continued)

Table 2 (continued)

Limulus length (cm)	Dose of endotoxin (μg)	Effects
30.0	200	sticky cells in gill leaflets.
30.0	200	sticky cells in gill leaflets.

We have not determined whether the primary action took place on the heart or the clotting system. In embryos, the heart beat could be observed and was stopped at the same time as the peripheral flow was stopped.

Any view of the phylogeny of response of animals to endotoxin must take into account the tremendous bias that has been exercised in the collection of data on the effect of gram-negative endotoxin. A huge amount of data has been accumulated for mammals (Morrison and Ulevitch, 1978), less for birds, and a small amount of data on coldblooded vertebrates. Even fish can be shown to develop a behavioral "fever" when inoculated with endotoxin (Reynolds, Covert and Casterlin, 1978). The exact role of endotoxin in human disease is still to be settled. However, its role in the cause of fever is but one aspect. It may cause massive vascular changes resulting in the Shwartzman phenomenon. This was a guiding principle in the original work on the disease in *Limulus*. It is also a well known stimulant of B lymphocytes and has both direct and indirect effect on the macrophages (Katz, 1977).

Much less data is available for the invertebrates, and most of this in turn deals with the arthropods. Gram-negative bacilli kill oyster spat (Tubiash, Chanley and Leifson, 1965), but this action is probably not by way of endotoxin, since boiled bacteria did not have an effect on the growth of the spat (Brown and Losee, 1978). We have not seen any effect of the injection of large amounts of bacteria into *Mya* (undiluted suspensions of agar plates) and have seen clumping in *Asterias* only when cloudy suspensions of bacteria are injected into the coelomic cavity. In unpublished work, we (B.G. and F.B. Bang) have shown that commercial preparations of endotoxin seem to prepare the urn cells of *Sipunculus* for a greater response to bacterial suspensions.

Among the arthropods, the work of McKay et al. (McKay and Jenkin, 1970; McKay, Jenkin and Tyson, 1973) on phagocytosis in crayfish is particularly interesting. This work, as well as other studies, now needs to be repeated since Block et al. (Block, Georgopoulos, Mayer and Drews, 1978) have shown that ubiquinone-8, a possible contaminant of endotoxin preparations, does increase phagocytosis by mouse macrophages, and inoculation 24 hours before infection protects against *E. coli*.

In the past, we have emphasized the need for marine arthropods to be protected against the gram-negative bacteria to which they are constantly exposed in the estuarine waters. However, all the aquatic arthropods have a living exoskeleton, on the surface of which masses of bacteria grow. Thus they constantly have a special problem in keeping gram-negative bacteria out. Surface growth of bacteria (ZoBell, 1946) may then be much more important than the bacteria in the surrounding water: a wound of the exoskeleton leads to immediate exposure to masses of bacteria.

Bacterial diseases of marine arthropods are common, and their pathogenesis is beginning to be understood (Bang, 1970). Endotoxin kills crustaceans of several kinds: *Carcinus* is susceptible to amounts of endotoxin above 100 μg (Bang, unpublished). Within a few hours, the animal ceases movement, becomes limp and examination of the transparent epipodites shows sticking and clumping of the cells, often with ballooning and disintegration. Lower doses are not without effect. In recent work at the Station Biologique in France, we have found that a ciliate parasite (Bang, Audouin and Leglise, 1972) of *Cancer* (Bang, 1970), a large crab, may be transferred to *Carcinus* readily *if* sublethal doses of endotoxin are administered just before the parasite inoculation (Table 3) (Bang, Audouin and Leglise, 1972). Without endotoxin, the *Carcinus* develops a very irregular infection, despite the fact that the parasite survives in *Carcinus* and *Cancer* by eating the host's amebocytes.

Some of the interaction between amebocytes and gram-negative bacteria has been studied in tissue culture (Shirodkar, Warwick and Bang, 1960). With a virulent *vibrio* there is cell destruction and clot formation, but an avirulent organism was overcome without massive cellular changes.

Table 3

Effect of previous injection of endotoxin on susceptibility of *Carcinus* to the ciliate, *Anophrys*

Dose of endotoxin (μg)	Effect of endotoxin on *Carcinus*	Susceptibility of survivors at 3 days
50	3/4 dead	1/1 infected
25	0/6 dead	4/6 infected
seawater	0/6 dead	0/6 infected
no injection	0/6 dead	0/6 infected

Limulus is not the only arthropod which will develop a gel in reaction to endotoxin. The parasitic crustacean *Sacculina* has a sac-like body hanging from the genital area of the carapace. Here it substitutes for a mass of eggs and can continually be bumped by contact of the host crab *Carcinus* with the stones among which it lives. *Sacculina* coelomic fluid is free of cells and gels immediately on contact with endotoxin (Barker and Bang, 1966).

Susceptibility as measured by death of *Limulus* is much below what one might expect from the known extreme sensitivity of the amebocyte lysate. This 1000-fold difference means of course that many factors come into play in the intact animal.

Thus it should not be too much of a surprise that some marine arthropods are much more sensitive. *Emerita*, the little mole shrimp, which lives in the sand at the beach where the waves constantly hit the shore, and which feeds on the debris which rushes back with the receding waves (Efford, 1966), is the most sensitive animal (as measured in terms of death) that we have studied. Crude seawater itself, when injected into the thorax, kills within one to two days. Larger amounts of endotoxin injected paralyze the animals within an hour or so (Table 4). The amebocytes may be observed stuck to the walls of their channels in the swimmerets which are readily cut off the abdomen. Can this extreme sensitivity be a protective mechanism whereby the animal protects itself against the constant barrage of pebbles with which it is bombarded by the receding waves?

Table 4

Susceptibility of *Emerita* (mole shrimp) to endotoxin

Dose of endotoxin (μg)	Number injected	Dead at 2 days	Mortality %
50	2	2	100
10	2	2	100
5	10	10	100
3	6	6	100
0.3	8	8	100
0.03	8	2	25
artificial seawater	21	0	0
natural seawater	22	17	72

Thus there is a very large variation in the susceptibility of marine arthropods to endotoxin, and we really have no idea of the general mode of action in these animals. Does it always work via the vascular system, or does it affect other tissues such as the heart? Are other arthropods just as good a source of a pre-gel (coagulen), as Levin and I originally called this extracellular clot in *Limulus* (Levin and Bang, 1968)?

Other diseases of Limulus. One of the most remarkable features of the horseshoe crab, or as our Japanese colleagues call it, the "kabuto-gani" (kabuto = helmet, gani = crab), is its book of gills (Fig. 6), in which one may see the cells intact, freely circulating and without adherence in the normal animal. These gills are for exchange of oxygen and presumably discharge of carbondioxide from the circulating hemocyanin (Prosser, 1973). Thus they act as lungs, and destruction of the gills should eventually lead to death. Among the animals kept in the Supply Department

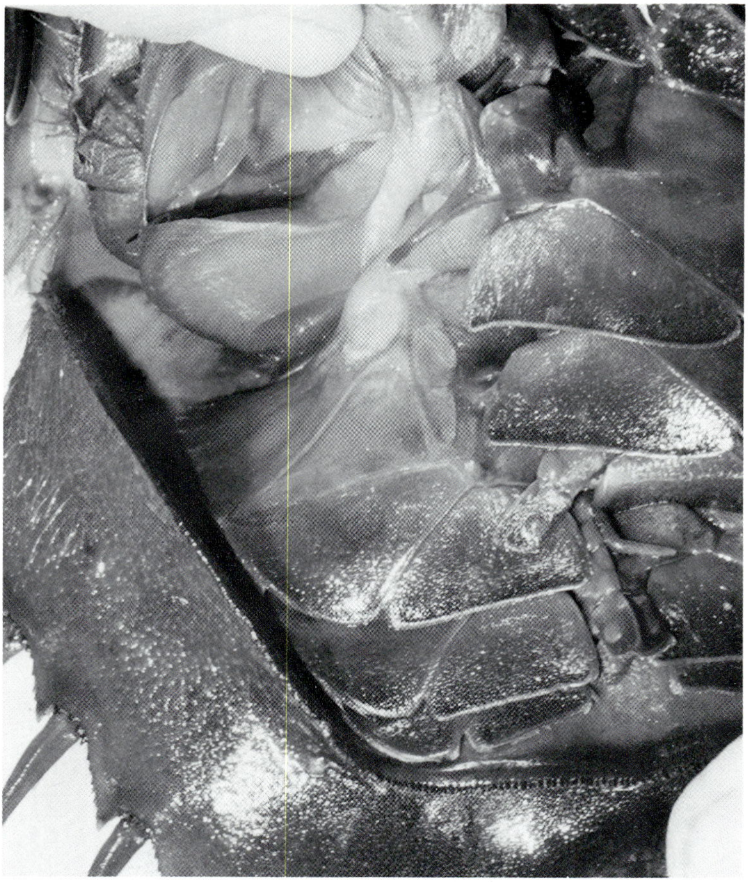

Fig. 6: Normal book gills of Limulus.

at the Marine Biological Laboratory in Woods Hole, not infrequently are found ones that are weakened or just dead, and on the surface of their gill leaflets may be found thick layers of bacteria. This presumably blocks oxygen transport, and one might jokingly say that Limulus had died of "external pneumonia." Others are found with greatly dilated leaflets, so that this two-layered structure has become a bag of blood in which the partially oxidized hemocyanin may be clearly seen (Fig. 7). This might be called localized

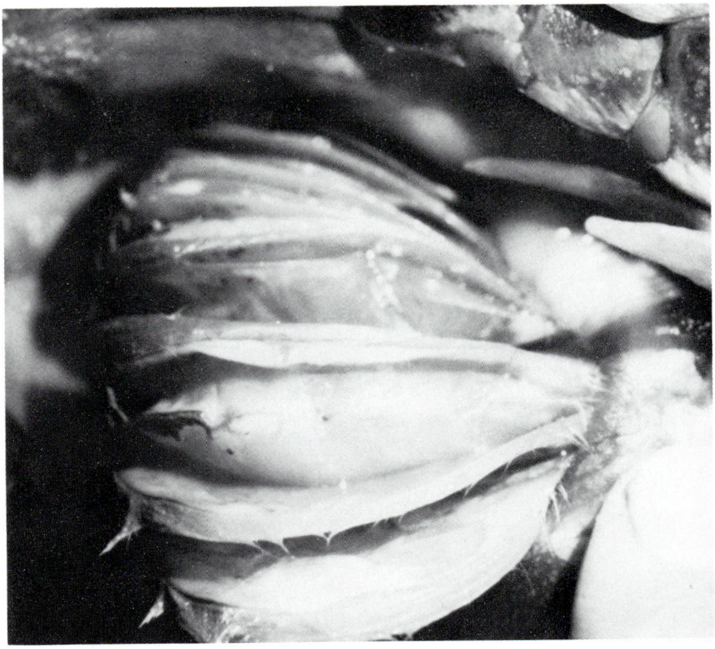

Fig. 7: Spontaneous ballooning disease of Limulus gill.

pulmonary edema. Finally, others are found with great areas of disintegrated and destroyed leaflets, even with barnacles and crepidula stuck next to the eroded areas (Fig. 8). This we might call "phthisis." All joking aside, we need to study these processes, determine what their causes are, and investigate Limulus for other spontaneous diseases. What about virus diseases like those of the blue crab (Johnson, 1977). Surely this ancient animal has more to teach us than merely a stereotyped reaction to endotoxin!

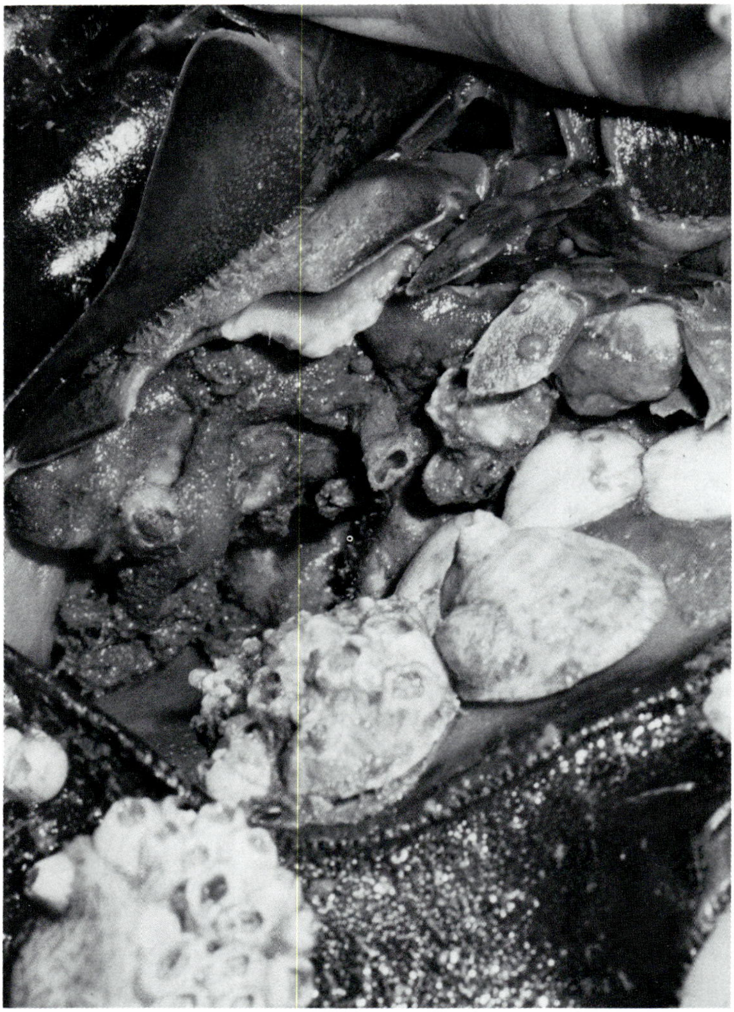

Fig. 8: Destruction of book gills by unknown process and replacement by crepidula and barnacles.

References

Bang FB (1956). A bacterial disease of Limulus polyphemus. Bull Johns Hopkins Hosp 98:325.

Bang FB (1970). Disease mechanisms in crustacean and marine arthropods. In Snieszko SF (ed): "Diseases of Crustaceans," chapter 2 in "A Symposium on Diseases of Fishes and Shellfishes," Washington, D.C.: American Fisheries Society Special Publication No. 5, pp. 383-404.

Bang FB. Unpublished data from work at the Station Biologique de Roscoff, France.

Bang FB, Audouin J, Leglise M (1972). Ciliate infection of the blood of the edible crab, Cancer pagurus, in holding tanks in Brittany, France. J Invert Pathol 20:226.

Barker WH, Jr, Bang FB (1966). The effect of infection by gram-negative bacteria, and their endotoxins, on the blood clotting mechanism of the crustacean Sacculina carcini, a parasite of the crab Carcinus maenas. J Invert Pathol 8: 88.

Block LH, Georgopoulos A, Mayer P, Drews J (1978). Nonspecific resistance to bacterial infections. Enhancement by ubiquinone-8. J Exp Med 148:1228.

Brown C, Losee E (1978). Observations on natural and induced epizootics of vibriosis in Crassostrea virginica larvae. J Invert Pathol 31:41.

Efford IA (1966). Feeding in the sand crab, Emerita analoga (Stimpson) (Decapoda, Anomura). Crustaceana 10:167.

Johnson PT (1977). A virus disease of the blue crab, Callinectes sapidus. Histopathology and differential diagnosis. J Invert Pathol 29:201.

Katz DH (1977). Lymphocyte differentiation, recognition and regulation. In Dixon FJ, Jr, Kunkel HG (eds): "Immunology: An International Series of Monographs and Treatises," New York: Academic Press (749 pp.).

Levin J, Bang FB (1968). Clottable protein in Limulus: Its localization and kinetics of its coagulation by endotoxin. Thromb Diathes Haemorrh (Stuttg) 19:186.

McKay D, Jenkin CR (1970). Immunity in invertebrates. Correlation of the phagocytic activity of haemocytes with resistance to infection in the crayfish (Parachaeraps bicarinatus). Aust J Exp Biol Med Sci 48:609.

McKay D, Jenkin CR, Tyson CJ (1973). Effect of endotoxin on resistance of the freshwater crayfish (Parachaeraps bicarinatus) to infection. J Inf Dis 128:S165.

Metchnikov E (1892). "Leçons sur la Pathologie Comparée de l'Inflammation." Paris: Masson. Translation: "Lectures

SECRETION IN LIMULUS AMEBOCYTES IS BY EXOCYTOSIS

Dr. R.L. Ornberg and Dr. T.S. Reese

NINCDS, National Institutes of Health

Bethesda, Maryland 20014

Previous studies of hemolymph coagulation in the horseshoe crab, Limulus polyphemus conclude that the circulating blood cells, the amebocytes, secrete the blood coagulation system in response to bacterial endotoxins. Minute quantities of endotoxin cause amebocytes to aggregate and degranulate concomitant to clot formation (Bang, 1955; Levin and Bang, 1964). All of the clotting proteins released by the amebocytes are contained in numerous large granules filling the cytoplasm (Murer et al, 1975). The cellular process of degranulation, however, whereby the clotting proteins are secreted is still obscure. Bang reported an increase in free granules in the lymph of endotoxin injected crabs implying that stimulated amebocytes simply rupture (Band, 1955). Dumont et al examined the ultrastructural changes of stimulated amebocytes and suggested that the granule membrane fused with the plasmalemma during release (Dumont et al, 1968).

In this paper we have examined the morphology of secretion in Limulus amebocytes. Instead of conventional chemical fixation however, we have quickly frozen secreting amebocytes to capture transient morphological events. Our results ultimately concur with those of Dumont et al in that granule contents are released as a result of granule membrane fusion, i.e. release occurs by granule exocytosis. In addition we find that more internal granules release by fusing with membranes of previously released peripheral granules.

Pages 125–130, Biomedical Applications of the Horseshoe Crab (Limulidae)

MATERIALS AND METHODS

Horseshoe crabs (6"-8" carapace) were obtained from the Supply Department of the Marine Biological Laboratory in Woods Hole, Massachusetts. Lyopholized lipopolysaccharide from E. coli B12 was purchased from Difco Laboratories, Inc. and dissolved in amebocyte saline. This saline was formulated from the ionic composition of Limulus blood and contained 4:54 mM NaCl, 10 mM KCl, 10 mM $CaCl_2$, 32 mM $MgCl_2$, 14 mM $MgSO_4$, 20 mM HEPES (N-2-Hydroxyethylpiperazine-N-2 ethane-sulfonic Acid) buffer at pH 7.8.

The apparatus and method for quick freezing biological tissue for morphologic studies have been described (Heuser et al, 1978). In simple terms, tissue for study is frozen upon contact with a copper block cooled to 4.2°K with liquid helium. For these experiments a drop of freshly drawn blood containing amebocytes was placed on the specimen stage of the freezing apparatus and frozen when the stage hit the precooled copper freezing block. The amebocytes were stimulated upon contact with a piece of endotoxin soaked filter paper which served as a support for the blood drop on the specimen stage. The final concentration of endotoxin in the blood drop was approximately 4 micrograms per milliliter. Amebocytes withdraw by cardiac puncture were placed on the filter paper within 10-15 seconds after withdrawal and were frozen at 30-60 seconds thereafter.

Frozen amebocytes were prepared for conventional thin section electronmicroscopy by a freeze-substitution process (Feder and Sidman, 1958). During this process, the quick-frozen amebocytes are warmed from -196°C to room temperature in a solution of 5% osmium tetroxide in acetone. The acetone melts at about -95°C and dissolves away the tissue ice while the osmium tetroxide fixes the tissue at very low temperature. After substitution, amebocytes were stained en bloc with 5% uranyl acetate in methanol for 10-12 hours and embedded in Araldite epoxy resin. Thin sections (400-600 Å) were further stained with uranyl acetate followed by lead citrate and examined in an AEI 6B electronmicroscope at 60 kV.

RESULTS

Light microscope studies have shown that degranulation in

amebocytes stimulated with endotoxin follows a rapid time course, being essentially complete in 60-90 seconds. While this explosive secretion and associated morphology may elude fixation and capture by slow chemical fixatives such as aldehydes, it is ideally suited for quick-freeze fixation. Electrical measurements of ice formation and thus freezing rates indicate that the first 20 μm of tissue hitting the freezing block is frozen in approximately two milliseconds with the apparatus used in these experiments (Heuser et al, 1978). Thus with the appropriate timing of secretory events and freezing, there is an excellent opportunity of seeing a secretory event in progress. Such an event has been captured in Figure 1. This micrograph illustrates a secretory granule (G_1) releasing contents into a vacuole created by previously released granules. In addition, a neighboring granule (G_2) has fused and opened into granule G_1. During release the

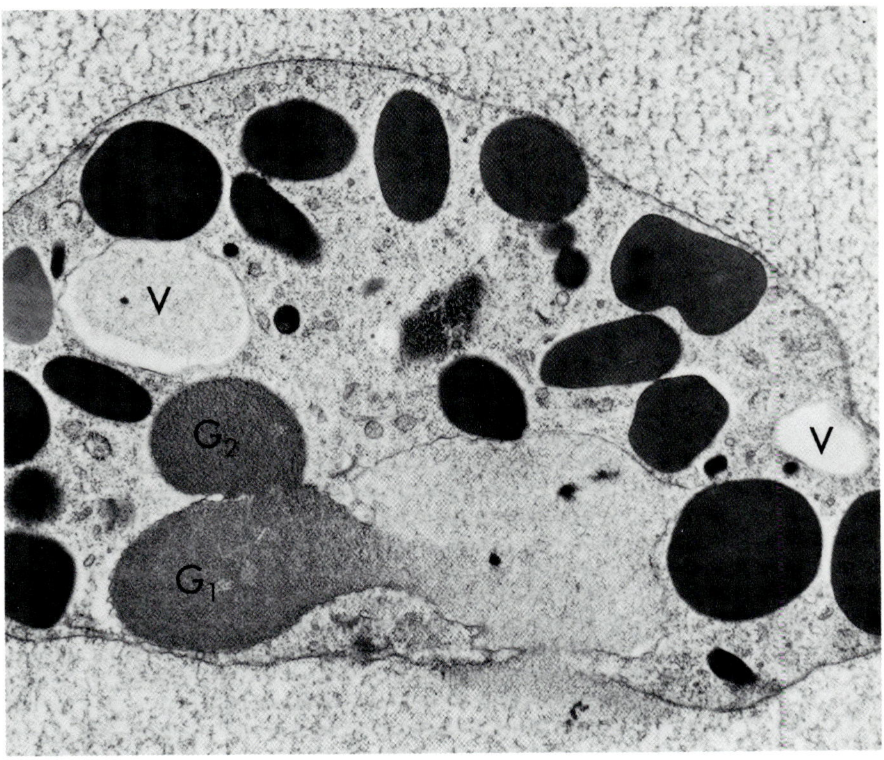

Figure 1 Amebocyte stimulated with endotoxin for 30 seconds before quick freezing. See text for explanation. X 125,000.

granule contents transform from an amorphous electron opaque material to a fiberous less dense material. The quick-freeze method has captured a gradient of this material spewing from granule, G_1. Eventually, the granule is devoid of contents except for diffuse flocculent material and the granual can resemble a vacuole, V. These changes in granule content, occur only after release has started and a granule opening is formed. The vacuoles have been examined in serial sections of stimulated cells and are always open to the extracellular spaces.

It is clear from Figure 1 that the clottable proteins in the granule are released from the amebocyte by granule exocytosis. Exocytosis refers to a sequence of morphological events occurring during the release of membrane bound products from some secretory cells. These events include the approach of the granule to the plasmalemma, the fusion of granule and plasma membranes, the formation of an opening for the dispersal of contents, and the incorporation of the granule membrane into the plasma membrane. Though only the latter exocytotic stages are shown in Figure 1, it is not unreasonable to assume that the membranes interacted at some fusion site during an earlier stage. The fact that two granules have fused with one releasing into a neighbor granule is an example of compound exocytosis.

DISCUSSION

Our results confirm the suggestions of Dumont et al in that the granule membrane fuses with the plasmalemma. However the sequences of events concerning the changes in granular contents following stimulation are different. Dumont et al describe similar structural changes in the granule contents to those reported here but imply that they occur before the granule has opened. We find that stimulated amebocytes show no change in the structure of granule content before fusion and that the transformation of contents from amorphous to fibrous form occurs after the granule opening has formed. This transformation is most probably due to the solvation of the proteineous contents by the extracellular fluid.

The morphology of exocytosis has been studied in detail in a variety of cells which share functional similarities with amebocytes, i.e. mammalian platelets and most cells. Striking similarities between mammalian platelets and Limulus amebocytes have been well documented (Levin and Bang, 1968). However the morphology of exocytosis with two systems is considerably different. In platelets the secretory granules move centrally to fuse with and release into a subsurface canalicular system which, since it is continuous with extracellular space, serves as a conduit for the secretory products (White, 1970; Benhke, 1967). In amebocytes a canalicular system does not exist and the granules do not converge to a central point in the cell. On the contrary, amebocyte secretory granules move toward the plasmalemma with outer granules fusing with the plasmalemma followed by sequential exocytosis of more internal granules. This amebocyte release reaction resumbles secretion in most cells (Rohlich et al, 1971).

The mainstream of studies of exocytosis is currently centered around the early morphological events and underlying physiological mechanisms in the secretory reaction. Most secretory cells presently used for these studies require a number of isolation procedures which may or may not effect secretory activity. In contrast, Limulus amebocytes require no isolation other than removal from the animal since they are the only circulating cell in the horseshoe crab and are present in quantity, 2 x 10^6 cell per milliliter of blood (Levin and Bang, 1968). Experiments using amebocytes as our model system to examine the early stages of exocytosis are in progress.

REFERENCES

Bang F (1956). A bacterial disease of Limulus polyphemus. Bull Johns Hopkins Hosp 98:325.
Behnke O (1967). Electron microscopic observations on the membrane system of the rat blood platelet. Anat Rec 58:121.
Dumont JN, Anderson E, Winner G (1966). Some cytologic characteristics of the hemocytes of Limulus during clotting. J Morphol 119:181.
Feder N, Sidman RL (1958). Methods and principles of fixation by freeze substitution. J Biophys Biochem Cytol 4:593
Heuser JE, Reese TS, Dennis MJ, Jan Y, Jan L, Evan, L (1978). Exocytosis of synaptic vesicles captured by quick freezing and correlated with quantal transmitter release. J Cell Biol in press.

Levin J, Bang F (1964). A description of cellular coagulation in the Limulus. Bull Johns Hopkins Hosp 115:337.
Levin J, Bang F (1968). Clottable protein in Limulus: its localization and kinetics of its coagulation by endotoxin. Thrombos Diathes Haemorrh 19:186.
Murer EH, Levin J, Holme R (1975). Isolation and studies of the granules of the amebocytes of Limulus polyphemus, the horseshoe crab. J Cellular Physiol 86:533.
Rohlich P, Anderson P, Uvnas B (1971). Election microscope observations on compound 48/80-induced degranulation in rat mast cells. J Cell Biol 51:465.
White JG (1968). Fine structural alterations induced in platelets by adenosine diphosphate. Blood 31:604.
White JG (1970). A search for the platelet secretory pathway using electron dense tracers. Am J Pathol 58:31.

THE REACTION BETWEEN BACTERIAL ENDOTOXIN AND AMEBOCYTE LYSATE

Jack Levin, M.D.

Department of Medicine, The Johns Hopkins University School of Medicine and Hospital
Baltimore, Maryland

INTRODUCTION

The amebocyte is the only type of circulating cell in the blood of Limulus polyphemus, the horseshoe crab[1]. It is a nucleated cell, the cytoplasm of which is packed with granules. When Limulus blood is shed, there is rapid aggregation of the amebocytes. Aggregation is associated with marked changes in cell shape, flattening with an apparent increase in cell diameter, development of long processes that extend from cell to cell, and striking degranulation (Levin and Bang, 1964; Dumont, Anderson, and Winner, 1966; Levin, 1967).

Although the important role of amebocytes in the coagulation of the blood of Limulus has been recognized for many years (Howell, 1885; Loeb, 1904; Maluf, 1939; Copley, 1952), it was not until the development of new techniques that it was unequivocally shown that amebocytes contain all elements of the coagulation system of Limulus blood (Levin and Bang, 1964a; Levin and Bang, 1968). Not only the clottable protein but all other factors required for coagulation are contained within amebocytes. Cell-free plasma does not clot and is not required for coagulation, since extracts of washed amebocytes (amebocyte lysate) are gelled following incubation with bacterial endotoxins (Levin and Bang, 1964a;

[1]The term horseshoe crab is a misnomer in the zoological sense because Limulus is not a crustacean, but a member of the class Merostomata (subclass, Xiphosura).

Pages 131–146, Biomedical Applications of the Horseshoe Crab (Limulidae)

Levin and Bang, 1968).

SOME ASPECTS OF THE REACTION BETWEEN BACTERIAL ENDOTOXINS AND AMEBOCYTE LYSATE

Amebocyte lysate, prepared from washed amebocytes that have been disrupted by freezing and thawing or by exposure to distilled water (Levin and Bang, 1968), gels following incubation with bacterial endotoxin (Fig 1). The rate of the reaction is dependent upon the concentration of endotoxin (Levin and Bang, 1968; Levin, Tomasulo, and Oser, 1970). The reaction can be measured easily by recording alteration in optical density (Fig 2) or increase in light scattering (Fig 3). The rate of the reaction is not affected by the concentration of protein in the preparation of amebocyte lysate (Fig 4). As little as 0.0001 µg/ml of E. coli endotoxin (Difco Laboratories, Detroit, Michigan) in saline can be detected reproducibly with amebocyte lysate although, at low concentrations of endotoxin, a solid gel may not form.

Figure 1. Amebocyte lysate (Limulus polyphemus). Left: Before addition of endotoxin, amebocyte lysate is a clear, colorless liquid. Right: Amebocyte lysate following addition of endotoxin. A solid gel has formed.

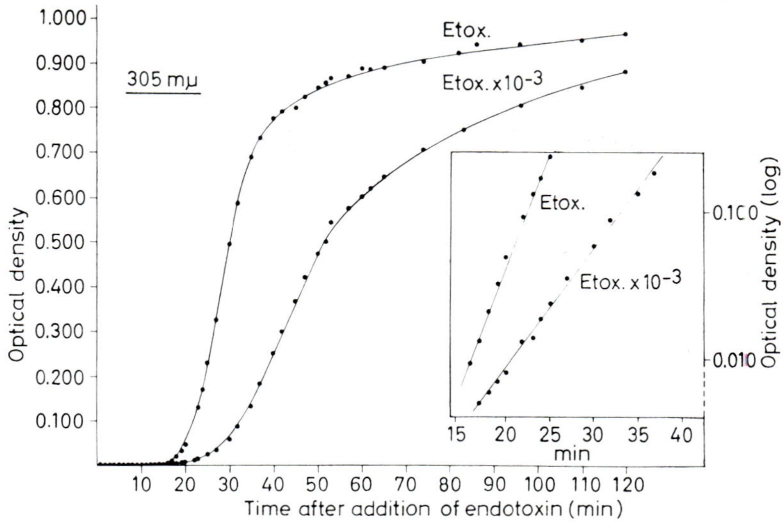

Figure 2. Increase in optical density following the addition of different concentrations of endotoxin to amebocyte lysate. 0.1 ml of Vibrio endotoxin was added to 0.9 ml samples of the same lysate. The relative concentrations of endotoxin (Etox.) are indicated. Protein concentration was constant. The maximal rate of increase in optical density following the addition of endotoxin to amebocyte lysate was greater with a higher concentration of endotoxin (panel on right). (from Levin and Bang, 1968; with permission of F.K. Schattauer Verlag).

The early stages of the reaction also can be detected by continuous and sensitive measurement of alterations of pH (Gaffin, 1977) or use of a synthetic substrate (Nakamura, et al, 1977) for the procoagulant enzyme in amebocyte lysate (vide infra). It is noteworthy that each of the techniques has demonstrated a lag period between the addition of endotoxin to amebocyte lysate and the detection of change in some characteristic of the incubation mixture.

Based on their initial observations, Levin and Bang (1968) postulated that an enzymatic system mediates the conversion, by endotoxin, of the clottable protein (coagulogen) into a gel (coagulin). This has been borne out by subsequent investigations. Young, Levin, and Prendergast (1972) carried out the first studies directed at characterizing the

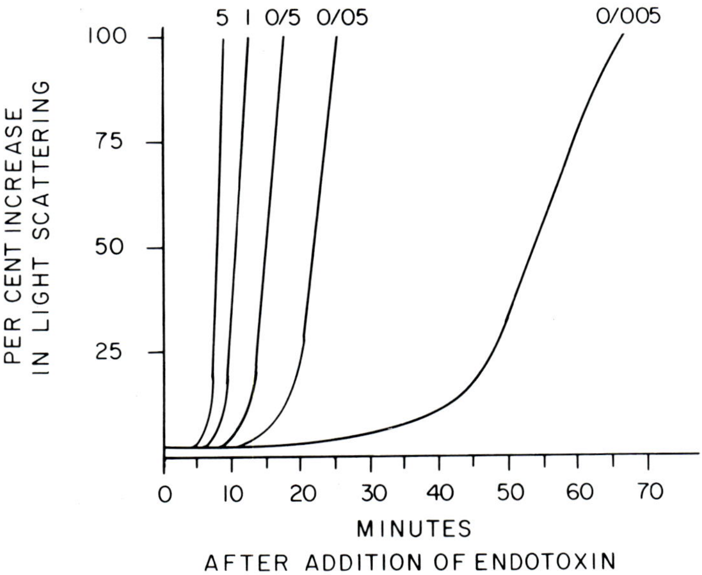

Figure 3. Increase in light scattering after addition of endotoxin to amebocyte lysate. The record is a direct tracing of continuous recordings of increase in light scattering. Each reaction mixture consisted of 0.9 ml of lysate and 0.1 ml of endotoxin. The final endotoxin concentration in each reaction mixture (μg/ml) is shown along the top. 100% represented the maximum increase in light scattering observed 24 hrs after the addition of E. coli endotoxin (5 μg/ml) to amebocyte lysate. (from Levin, Tomasulo, and Oser, 1970; with permission of C.V. Mosby Co.)

components of amebocyte lysate and their role in the reaction with bacterial endotoxin. Fractionation of amebocyte lysate by Sephadex column chromatography, in the absence of bacterial endotoxin, demonstrated two major peaks. The peak which appeared in the void volume of either Sephadex G50 or G75 was heat labile; the second and major retained fraction was heat stabile, and constituted approximately 50% of the total protein in amebocyte lysate.

Their studies demonstrated that the heat labile component, with a molecular weight of greater than 75,000, was

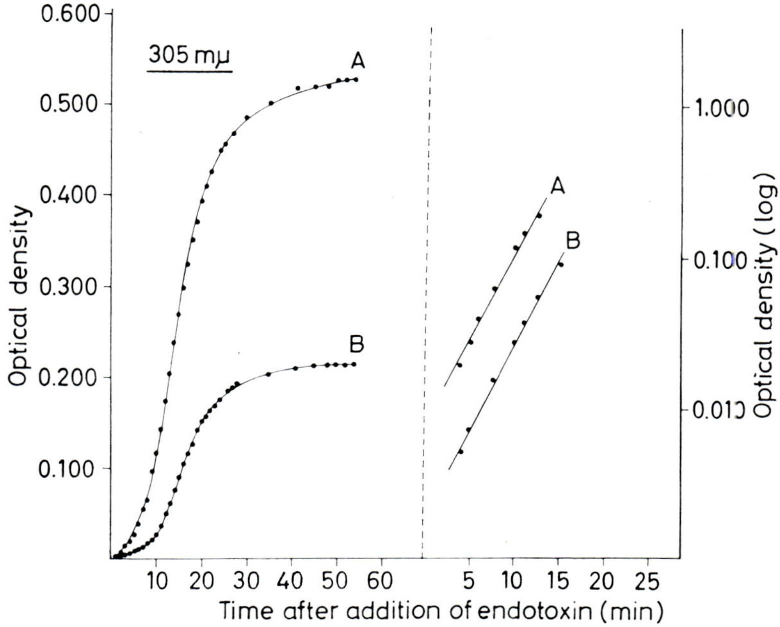

Figure 4. Increase in optical density following the addition of endotoxin to different concentrations of amebocyte lysate. 0.1 ml of E. coli endotoxin (50 μg/ml) was added to 0.9 ml samples of amebocyte lysate. The concentration of protein in sample A was 2.5 times greater than that in sample B. The final optical density of sample A was 2.5 times greater than sample B (panel on left). The maximal rates of increase in optical density, when the concentration of endotoxin was constant and the concentration of clottable protein varied, were similar (panel on right). (from Levin and Bang, 1968; with permission of F.K. Schattauer Verlag).

activated by endotoxin and accelerated the rate of the reaction between endotoxin and the heat stabile fraction, which contained the clottable protein (Fig 5). As had been shown previously, the concentration of the clottable protein did not alter the rate of the reaction (Fig 6). Preincubation of the enzymatic fraction with endotoxin accelerated the rate of the reaction which subsequently occurred when the clottable protein was added to the incubation

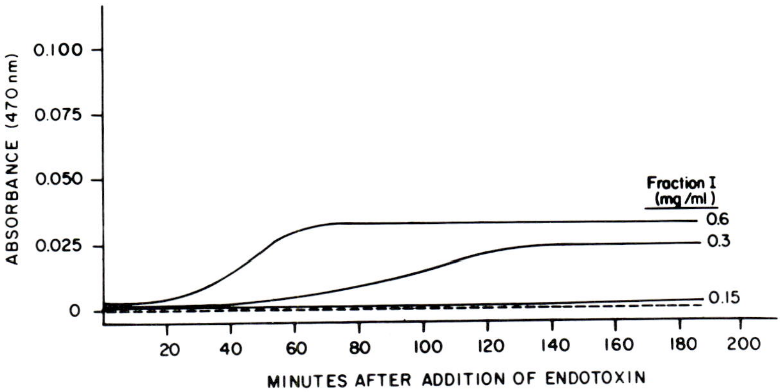

Figure 5. The effect of various concentrations of fraction I (enzymatic fraction) on the reaction of fractions I and II (clottable protein) with endotoxin. The reaction mixtures consisted of 0.01 ml of fraction I, in the concentrations indicated, plus 0.1 ml of fraction II (5 mg/ml) and 0.1 ml of endotoxin (2.5 μg/ml). The control, indicated by the dotted line, was obtained by the addition of sodium chloride (0.9%) instead of endotoxin to a reaction mixture which contained 5 mg/ml of fraction II and 0.6 mg/ml of fraction I. The records are direct tracings of continuous recordings of increase in optical density. The rate of increase in optical density was proportional to the concentration of fraction I. (from Young, Levin, and Prendergast, 1972; with permission of Rockefeller Universitv Press).

mixture (Fig 7). These workers also demonstrated that the rate of the reaction was temperature and pH dependent (Young, Levin, and Prendergast, 1972). Subsequent investigations by other laboratories have confirmed these findings (Sullivan and Watson, 1975; Niwa, Hiramatsu, and Waguri, 1975; Sullivan, Valois, and Watson, 1976), and it is apparent that the clottable protein constitutes approximately 50% of amebocyte lysate (Solum, 1970; Solum, 1973; Nakamura, et al, 1976a). Studies carried out by Japanese investigators have utilized Tachypleus tridentatus, not Limulus polyphemus. However, it appears that the lysates prepared from the amebocytes of these animals are similar and, for the purposes of this paper, the data have been used interchangeably. It is possible that certain specific details, such as the amino

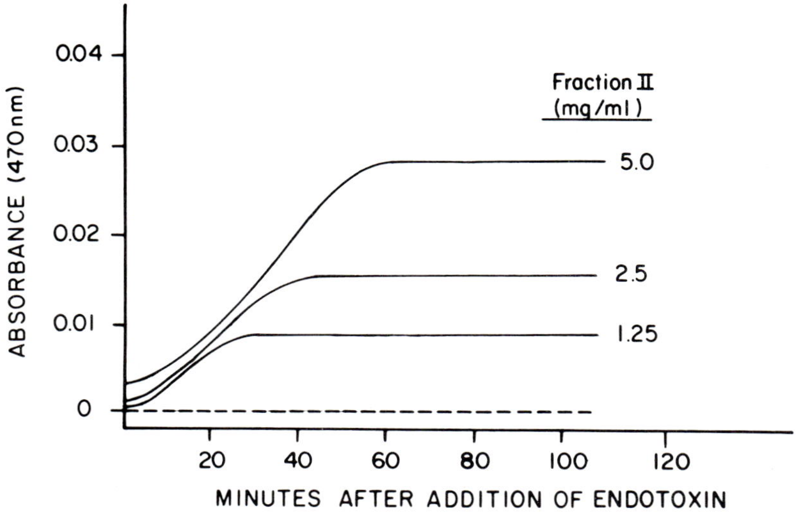

Figure 6. The effect of various concentrations of fraction II on the reaction of fractions I (enzymatic fraction) and II (clottable protein) with endotoxin. The reaction mixtures consisted of 0.1 ml of fraction II, in the concentrations indicated, plus 0.01 ml of fraction I (6 mg/ml) and 0.1 ml of endotoxin (2.5 μg/ml). The control, indicated by the dotted line, was obtained by the addition of sodium chloride (0.9%) instead of endotoxin to a reaction mixture which contained 0.6 mg/ml of fraction I and 5 mg/ml of fraction II. The records are direct tracings of continuous recordings of increase in optical density. The rate of increase in optical density was independent of the concentration of fraction II, but the maximum optical density was proportional to the concentration of fraction II. (from Young, Levin, and Prendergast, 1972; with permission of Rockefeller University Press).

acid sequences of the enzyme and clottable protein or the site of cleavage of the clottable protein differ.

Young, Levin, and Prendergast (1972) demonstrated that the reaction between endotoxin and the enzymatic fraction could be inhibited by diisopropyl fluorophosphate, para-chloromercuribenzoate, or para-chloromercuriphenyl sulfonate, observations compatible with an enzyme in which sulfhydryl

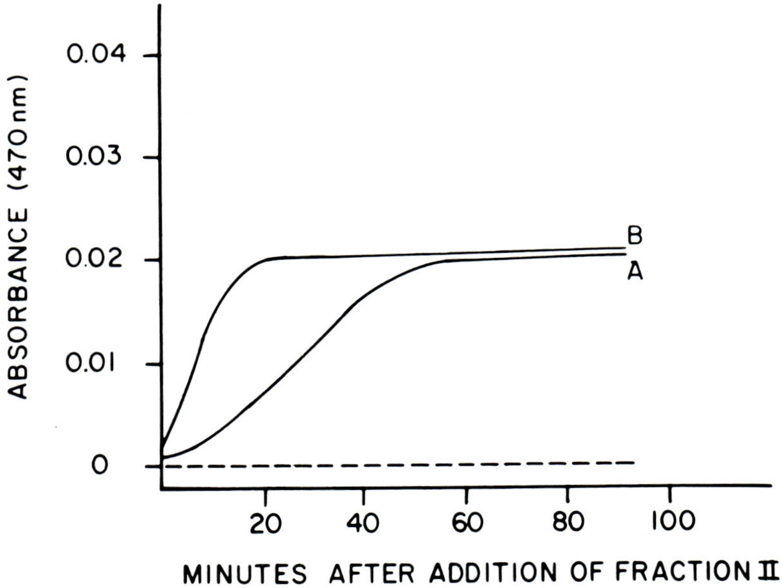

Figure 7. The effect of pre-incubation of fraction I and endotoxin on the reaction of fractions I and II with endotoxin. Endotoxin (0.1 ml, 5 μg/ml) was incubated with fraction I (0.01 ml, 6 mg/ml) at 37° C. Fraction II (0.1 ml, 5 mg/ml) was added either immediately (A) or 60 min later (B). The control, indicated by the dotted line, was obtained by the addition of sodium chloride (0.1 ml, 0.9%) to fractions I and II. The records are direct tracings of continuous recordings of increase in optical density. Preincubation of fraction I with endotoxin increased the subsequent rate of increase in optical density when fraction II was added. (from Young, Levin, and Prendergast, 1972; with permission of Rockefeller University Press).

or hydroxyl groups are important. Other studies also have indicated that the enzyme is a serine protease (Niwa, Hiramatsu, and Waguri, 1975; Tai and Liu, 1977). It has been reported not to hydrolyze TAME (Sullivan and Watson, 1975). EDTA (10^{-3} M), in buffered solutions, did not appear to inhibit the reaction (Young, Levin, Prendergast, 1972), suggesting that the previous observation that EDTA was inhibitory was secondary to low pH (Levin and Bang, 1968).

However, other workers have reported that EDTA is inhibitory (Niwa, Hiramatsu, and Waguri, 1975) and that Ca^{2+} and Mg^{2+} accelerate the reaction (Niwa, Hiramatsu, and Waguri, 1975; Tai and Liu, 1977). Ca^{2+}, per se, does not cause gelation (Fumarola, et al, 1975).

Further insight into the nature of the enzyme with which endotoxin reacts has been provided by the observation of Nakamura, et al (1977) that the enzyme, activated by endotoxin, hydrolyzes BZ-Ile-Glu-Gly-Arg-p-nitroanilide, which is also a substrate for the mammalian blood coagulation factor, Xa; the latter is also a serine proteinase. It is noteworthy that the enzyme from both Tachypleus and Limulus showed greatest activity for this anilide substrate, and that there was a correlation between endotoxin concentration and amidase activity. Some studies have reported that preparations of thrombin gel amebocyte lysate, but that this effect is blocked by heparin, anti-thrombin III, or hirudin (Solum, 1970; Fumarola, et al, 1975). Nevertheless, it was concluded that thrombin was capable of causing gelation through a mechanism other than that by which it causes coagulation of fibrinogen. It would appear more likely that the preparations of thrombin were contaminated with endotoxin, a possibility considered by Solum (1970).

Tai and Liu (1977) have reported the purification of the procoagulant enzyme that is activated by endotoxin. The molecular weight is at least 150,000 and the protein appears to consist of a single peptide chain. It contains γ-carboxyglutamic acid and apparently requires both Ca^{2+} and endotoxin for its activation (Tai and Liu, 1977). Sullivan and Watson (1975) reported the isolation of the enzyme apparently after activation by endotoxin; in their study, the molecular weight was 84,000, and two identical subunits (43,000 molecular weight) were observed after SDS gel electrophoresis. The enzyme, which produces limited proteolysis and gelation of the clottable protein (coagulogen) of Tachypleus, has been reported to cleave Arg-Thr and Arg-Gly linkages (Nakamura, et al, 1976). However, Tai, et al (1977) reported that an Arg-Lys bond was cleaved during this reaction in Limulus. Regardless, it is interesting that the high activity this enzyme demonstrates against a synthetic substrate involves cleavage of an Arg linkage (Nakamura, et al, 1977).

THE CLOTTABLE PROTEIN (COAGULOGEN)

The clottable protein obtained from the amebocytes of the two genera of horseshoe crabs studied has a molecular weight with a mean range of approximately 23,000-27,000 for Limulus polyphemus (Solum, 1970; Young, Levin, and Prendergast, 1972; Gaffin, 1976; Tai, et al, 1977), while the molecular weight for Tachypleus tridentatus appears to be less (approximately 16,000-18,000) (Nakamura, et al, 1976a; Shishikura, Chiba, and Sekiguchi, 1977). Nevertheless, all reports indicate that gelation is associated with limited proteolysis of the coagulogen; and to date, have concluded that the coagulogen is composed of a single polypeptide chain. The peptide, containing 28 amino acids, that is released during coagulation from the coagulogen of Tachypleus, appears homologous to primate fibrinopeptide B (Nakamura, et al, 1976b). A 45 amino acid peptide has been reported to be released from the coagulogen of Limulus (Tai, et al, 1977).

The gel that forms following addition of adequate concentrations of bacterial endotoxin is stable for many weeks, if kept in a moist environment. However, it is mechanically fragile (Levin and Bang, 1968; Solum, 1970); and will not re-form once disrupted. Once formed, the gel does not lyse spontaneously (Levin and Bang, 1968), nor after the addition of plasmin (Fumarola, 1975). The gel does not retract. The gel is insoluble in 8 M urea or 2-mercaptoethanol (Solum, 1970; Young, Levin, and Prendergast, 1972; Gaffin, 1976); Triton X-100, or 8 M urea <u>and</u> 0.1 M 2-mercaptoethanol (Gaffin, 1976); and also in 8 M guanidine (Levin, unpublished observation). However, some authors have indicated it is soluble in 6 M urea (Fumarola, et al, 1975; Tai, et al, 1977). The reason for these disparate observations is unclear, but may depend upon the duration of time between gelation and incubation of the gel with urea (Gaffin, 1978, personal communication). A combination of SDS and 2-mercaptoethanol does solubilize the gel (Gaffin, 1976), but SDS alone does not. Therefore, Gaffin (1976) concluded that both hydrophobic bonding and disulfide bond formation were present within the gel (or clot).

DISCUSSION

The reaction between lysates of amebocytes of the horseshoe crab, Limulus polyphemus, and bacterial endotoxin provides the basis for the most sensitive, currently available in vitro assay for endotoxin. Existing data indicate that endotoxin does not react directly with the clottable protein (coagulogen) in amebocyte lysate, but with apparently only one enzyme, activation of which results in limited proteolysis of the clottable protein. Polymerization then occurs and a gel is produced. Polymerization continues after gelation has occurred, the latter taking place when the increase in light scattering reaches approximately 50% of the maximum finally achieved (Levin, unpublished observations). The rate of the reaction is dependent upon both the concentration of endotoxin and the procoagulant enzyme (Levin and Bang, 1968; Young, Levin, and Prendergast, 1972). As is the case for other biological activities of endotoxin, the reaction between endotoxin and amebocyte lysate depends upon the presence of Lipid A (Levin, Tomasulo, and Oser, 1970; Niwa, Hiramatsu, and Waguri, 1975).

It appears that the procoagulant enzyme is normally in the zymogen form, but that during its activation by endotoxin, the mechanism of which remains unknown, the enzyme apparently is cleaved (i.e., Tai and Liu (1977) obtained a molecular weight of 150,000 for the native molecule, whereas Sullivan and Watson (1975) reported a molecular weight of 84,000 after exposure of lysate to endotoxin). Many zymogens are cleaved, during their activation, at Arg-Val, Arg-Ile, or Lys-Ile linkages (Neurath and Walsh, 1976). It will be interesting to learn whether the zymogen of amebocyte lysate is similar, despite the many evolutionary differences between the horseshoe crab and mammals. The suggestion by Neurath and Walsh (1977) that zymogens are rarely found in primitive animal forms may not be correct, if one considers Limulus or Tachypleus as being in that category.

A lag period occurs after addition of endotoxin either to whole amebocyte lysate or to partially purified fractions of lysate, before alterations in the incubation mixture can be detected. However, if the procoagulant enzyme and endotoxin are pre-incubated, subsequent addition of the clottable protein to the mixture results in immediate evidence of a reaction (Young, Levin, and Prendergast, 1972). These observations parallel the pattern of coagulation of mammalian plasma that follows either addition of a thromboplastin or of pre-formed thrombin, respectively.

Available data indicate that Arg-Gly, Arg-Thr, and/or Arg-Lys linkages may be broken during the reaction between the activated enzyme and the clottable protein (Nakamura, et al, 1976; Tai, et al, 1977), in a manner at least partially homologous to the reaction between thrombin and fibrinogen, in which Arg-Gly links are broken with release of fibrinopeptides A and B (Doolittle, 1976). Nakamura, et al (1976b) have reported homology between the peptide released from the clottable protein of Tachypleus and primate fibrinopeptide B. Another similarity appears to be the presence of γ-carboxyglutamic acid in both the procoagulant enzyme of the horseshoe crab and in the vitamin-K dependent coagulation factors in mammals (Fernlund, et al, 1975). In addition, both the procoagulant enzyme of the horseshoe crab and many mammalian procoagulant factors are serine proteases, and are inhibited by diisopropyl fluorophosphate. Ca^{2+} also is apparently required by both systems. However, despite these intriguing similarities, the term 'invertebrate fibrinogen' is not appropriate, since there are clearly many differences between the clottable proteins of the horseshoe crab and other invertebrates and mammalian fibrinogens (Doolittle, 1970), including the low molecular weight of the coagulogen of the horseshoe crab and different subunit structures.

In marked contrast to mammalian coagulation mechanisms (which are based on a series of factors present in plasma), the coagulation mechanism of the horseshoe crab is localized entirely intracellularly and only appears in the plasma following its release from amebocytes. Furthermore, Murer, Levin, and Holme (1975) have shown that the granules, which are so prominent morphologically and which disappear during the aggregation of amebocytes, are the cellular sites that contain the components of the coagulation system of Limulus. However, a study of Tachypleus indicated that the clottable protein was in the cytoplasmic matrix of the majority of hemocytes, while only 15% of the cells demonstrated the clottable protein in cytoplasmic granules (Shishikura, Chiba, and Sekiguchi, 1977).

Overall, despite the great development of hemostasis during evolution (Quick, 1967), there remain striking homologies and similarities between the 'simple' coagulation mechanism of Limulus (Fig 8) and mammalian coagulation. Although coagulation has become extracellular and diversified in higher animals, the activation of blood coagulation in mammals by endotoxin may be derived, in part, from this

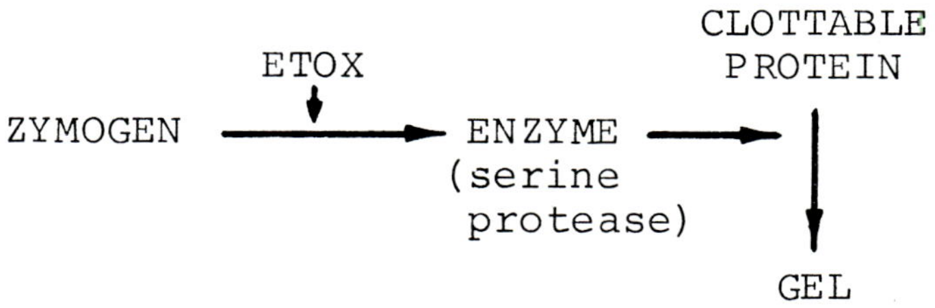

Figure 8. Coagulation of amebocyte lysate by endotoxin. Endotoxin (ETOX) activates an enzyme, which then causes limited proteolysis of the clottable protein. Following polymerization, a gel is produced.

more primitive mechanism which serves also to control gram-negative infection in the horseshoe crab (Levin, 1976). An understanding of blood coagulation in the horseshoe crab may shed light on some of the biochemical mechanisms involved in mammalian coagulation, by providing an example of a primitive system from which human coagulation has developed (Levin and Bang, 1968; Levin, 1976). Further isolation and characterization of the enzyme system in amebocyte lysate which is activated by endotoxin may lead not only to a better assay for endotoxin but also to a greater insight into other biologic effects of endotoxin.

ACKNOWLEDGMENT

Supported in part by Research Grant HL 01601 from the National Heart, Lung, and Blood Institute, National Institutes of Health, Bethesda, Maryland.

Request for Reprints

Jack Levin, M.D., The Johns Hopkins Hospital, Baltimore, Maryland 21205 USA.

REFERENCES

Copley AL (1952). The rheology of blood. A survey. J Colloid Sci 7:323.
Doolittle RF (1970). Evolution of fibrinogen molecules. Thromb Diath Haemmorh 39(suppl):25.
Doolittle RF (1976). The evolution of vertebrate fibrinogen. Fed Proc 35:2145.
Dumont JN, Anderson E, Winner G (1966). Some cytologic characteristics of the hemocytes of Limulus during clotting. J Morphol 119:181.
Fernlund P, Stenflo J, Roepstorff P, Thomsen J (1975). Vitamin K and the biosynthesis of prothrombin. J Biol Chem 250:6125.
Fumarola D, Pasquetto N, Telesforo P, Donati MB (1975). Studies on the clotting mechanism of Limulus polyphemus. Thromb Res 7:401.
Gaffin SL (1976). The clotting of the lysed white cells of Limulus induced by endotoxin--I: Preparation and characterization of clot-forming proteins. Biorheology 13:273.
Gaffin SL (1977). Absorption of protons during the clotting of Limulus lysate induced by endotoxin. FEBS Lett 84:275.
Howell WH (1885). Observations upon the chemical composition and coagulation of the blood of Limulus polyphemus, Callinectes hastatus, and Cucumaria sp. Johns Hopkins Univ Circ 5:4.
Levin J, Bang FB (1964). A description of cellular coagulation in the Limulus. Bull Johns Hopkins Hosp 115:337.
Levin J, Bang FB (1964a). The role of endotoxin in the extracellular coagulation of Limulus blood. Bull Johns Hopkins Hosp 115:265.
Levin J (1967). Blood coagulation and endotoxin in invertebrates. Fed Proc 26:1707.
Levin J, Bang FB (1968). Clottable protein in Limulus: its localization and kinetics of its coagulation by endotoxin. Thromb Diath Haemorrh 19:186.
Levin J, Tomasulo PA, Oser RS (1970). Detection of endotoxin in human blood and demonstration of an inhibitor. J Lab Clin Med 75:903.
Levin J (1976). Blood coagulation in the horseshoe crab (Limulus polyphemus): a model for mammalian coagulation and hemostasis. In "Animal Models of Thrombosis and Hemorrhagic Diseases", (Proceedings of a Symposium of the National Academy of Sciences), Washington: US Department of Health, Education, and Welfare, DHEW Publication No (NIH) 76-982, p 87.

Loeb L (1904). On the spontaneous agglutination of blood cells of arthropods. Univ Pa Med Bull 16:441.

Maluf NSR (1939). The blood of arthropods. Q Rev Biol 14:149.

Mürer EH, Levin J, Holme R (1975). Isolation and studies of the granules of the amebocytes of Limulus polyphemus. J Cell Physiol 86:533.

Nakamura S, Takagi T, Iwanaga S, Niwa M, Takahashi K (1976). A clottable protein (coagulogen) of horseshoe crab hemocytes. Structural change of its polypeptide chain during gel formation. J Biochem 80:649.

Nakamura S, Iwanaga S, Harada T, Niwa M (1976a). A clottable protein (coagulogen) from amebocyte lysate of Japanese horseshoe crab (Tachypleus tridentatus). Its isolation and biochemical properties. J Biochem 80:1011.

Nakamura S, Takagi T, Iwanaga S, Niwa M, Takahashi K (1976b). Amino acid sequence studies on the fragments produced from horseshoe crab coagulogen during gel formation: homologies with primate fibrinopeptide B. Biochem Biophys Res Commun 72:902.

Nakamura S, Morita T, Iwanaga S, Niwa M, Takahashi K (1977). A sensitive substrate for the clotting enzyme in horseshoe crab hemocytes. J Biochem 81:1567.

Neurath H, Walsh KA (1976). Role of proteolytic enzymes in biological regulation (a review). Proc Natl Acad Sci USA 73:3825.

Neurath H, Walsh KA (1978). The role of proteases in physiological regulation: an overview. In Magnusson S, et al (eds): "Regulatory Proteolytic Enzymes and Their Inhibitors", New York:Pergamon Press, Vol 47 Symposium A6, p 1.

Niwa M, Hiramatsu T, Waguri O (1975). Quantitative aspects of the gelation reaction of horseshoe crab amoebocyte and bacterial endotoxin. Jpn J Med Sci Biol 28:98.

Quick AJ (1967). Hemostasis as an evolutionary development. Thromb Diath Haemorrh 18:1.

Shishikura F, Chiba J, Sekiguchi K (1977). Two types of hemocytes in localization of clottable protein in Japanese horseshoe crab, Tachypleus tridentatus. J Exp Zool 201:303.

Solum NO (1970). Some characteristics of the clottable protein of Limulus polyphemus blood cells. Thromb Diath Haemorrh 23:170.

Solum NO (1973). The coagulogen of Limulus polyphemus hemocytes. A comparison of the clotted and non-clotted forms of the molecule. Thromb Res 2:55.

Sullivan JD Jr, Watson SW (1975). Purification and properties of the clotting enzyme from Limulus lysate. Biochem Biophys Res Commun 66:848.

Sullivan JD Jr, Valois FW, Watson SW (1976). Endotoxins: the Limulus amoebocyte lysate system. In Bernheimer AW (ed): "Mechanisms in Bacterial Toxinology", Somerset, NJ: John Wiley and Sons, Inc, p 218.

Tai JY, Liu T-Y (1977). Studies on Limulus amoebocyte lysate. Isolation of pro-clotting enzyme. J Biol Chem 252:2178.

Tai JY, Seid RC Jr, Huhn RD, Liu T-Y (1977). Studies on Limulus amoebocyte lysate II. Purification of the coagulogen and the mechanism of clotting. J Biol Chem 252:4773.

Young NS, Levin J, Prendergast RA (1972). An invertebrate coagulation system activated by endotoxin: evidence for enzymatic mediation. J Clin Invest 51:1790.

Studies on *Limulus* Lysate Coagulating System

Teh-Yung Liu, Robert C. Seid, Jr., Joseph Y. Tai,
Su-Mei Liang, Thomas P. Sakmar and Joan B. Robbins

Division of Bacterial Products
Bureau of Biologics
Food and Drug Administration
8800 Rockville Pike
Bethesda, Maryland 20014

In 1964, Levin and Bang discovered that gram-negative bacterial endotoxin could rapidly induce gelation of *Limulus* amoebocyte lysate (1). Because of this unique property and its extreme sensitivity, *Limulus* lysate has been used to detect endotoxin in a variety of biological fluids. For the past three years, our laboratory has been engaged in the study of the *Limulus* lysate coagulating system. Our study is directed toward the understanding of the chemical and the physiological principles underlying the clotting processes. We have purified a proclotting enzyme (2), a coagulogen (3), and more recently, an active enzyme (4) and an endotoxin binding protein (5) from the *Limulus* amoebocytes.

The proclotting enzyme was purified by gel filtration of lysate on Sephadex G-75, followed by two steps of ion-exchange chromatography on DEAE-cellulose and DEAE-Sephadex. The purified proenzyme has a molecular weight of about 150,000 and requires both calcium and endotoxin for exhibiting maximum activity.

The active clotting enzyme(s) from the *Limulus* amoebocyte lysate has been purified to apparent homogeniety by gel filtration on BioGel A-5m (Fig. 1) followed by affinity chromatography using an endotoxin coated hydroxylapatite (Fig. 2). The *Limulus* enzyme(s) was found to be stabilized by the detergent Tween-20. Without the detergent, enzymatic activity decreased rapidly. Two active forms of the clotting enzyme were demonstrated to exist in the *Limulus* amoebocyte lysate. The enzymatic material existing as higher molecular weight aggregates was purified to give a single

Pages 147—158, Biomedical Applications of the Horseshoe Crab (Limulidae)

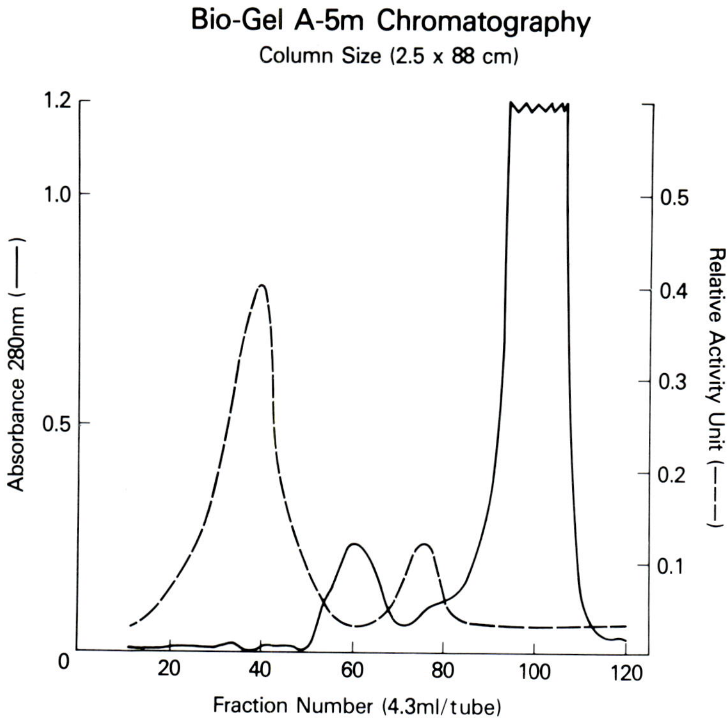

Figure 1. Gel permeation chromatography of *Limulus* clotting enzyme extract from lysate clot. The arrows indicate elution positions of markers for void volume (V_o) and total volume (V_t). Dashed line and solid bar represent amidase and clotting activities, respectively.

protein band on sodium dodecyl sulfate-polyacrylamide gel electrophoresis with a molecular weight of 79,000. The enzymatic material eluting in the lower molecular weight region of the gel-permeation column yielded a protein with a molecular weight of about 40,000 (Fig. 3). The amino acid compositions of the two proteins were very similar indicating a "monomer"-"dimer" relationship but attempts to interconvert the 79,000 protein into the 40,000 protein have not been successful. Both enzymes have the same calcium and endotoxin requirement for exhibiting maximal clotting activity and share the same specificity toward synthetic substrates.

For the isolation of coagulogen, the amoebocytes were lysed and extracted with 30% acetic acid. This step removes

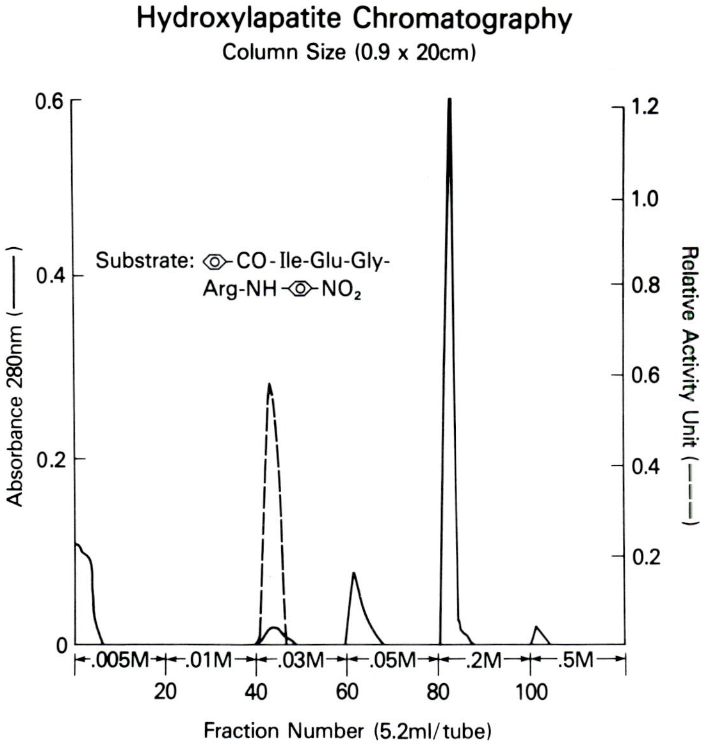

Figure 2. Affinity chromatography on endotoxin coated hydroxylapatite resin. *A*, chromatogram of *Peak I* material obtained from BioGel A-5m filtration, Figure 1. *B*, chromatogram of *Peak II* material. A stepwise increase in potassium phosphate was used to elute the clotting enzyme. Dashed line and solid bar indicate amidase and clotting activities, respectively.

contaminating proteins as insoluble precipitate and preserves the coagulogen as an intact protein. The clear acid-extract was placed on a Sephadex G-100 column in 30% acetic acid to yield five fractions (Fig. 4) Only fraction II gave a positive clot when the purified *Limulus* enzyme was added.

The purity of the protein in fraction II was ascertained by end-group analyses as well as by SDS-gel electrophoresis (Fig. 5). On SDS-gel electrophoresis both the native coagulogen and the reduced carboxymethylated coagulogen

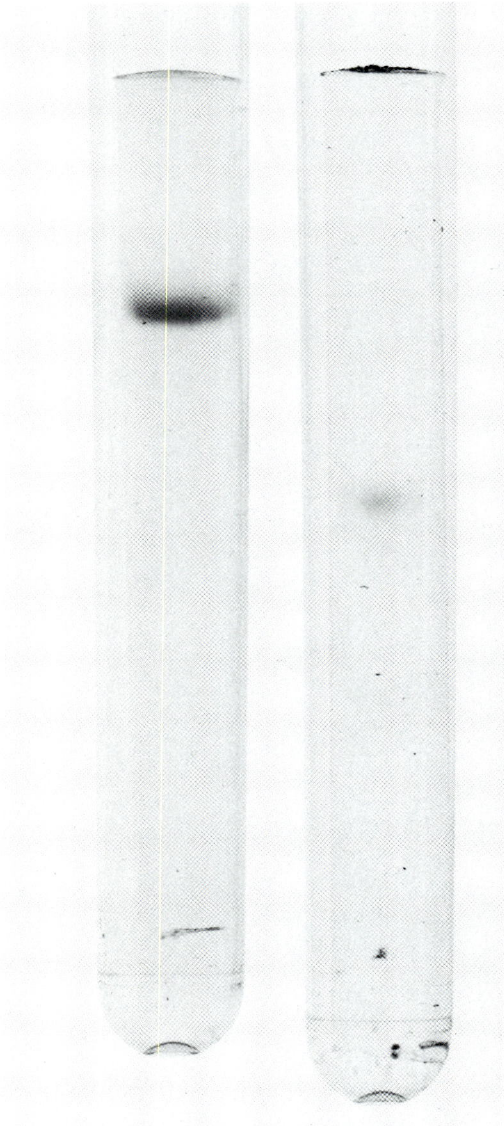

Figure 3. Acrylamide gel electrophoresis in 0.1% sodium dodecyl sulfate, 0.1 mM dithiothreitol, 0.1M pH 7.2 phosphate buffer. *Gel 1* represent 0.03M clotting enzyme fraction, Fig. 2*A*, obtained from affinity chromatography; *Gel 2* represents 0.03M enzyme fraction of Figure 2*B*.

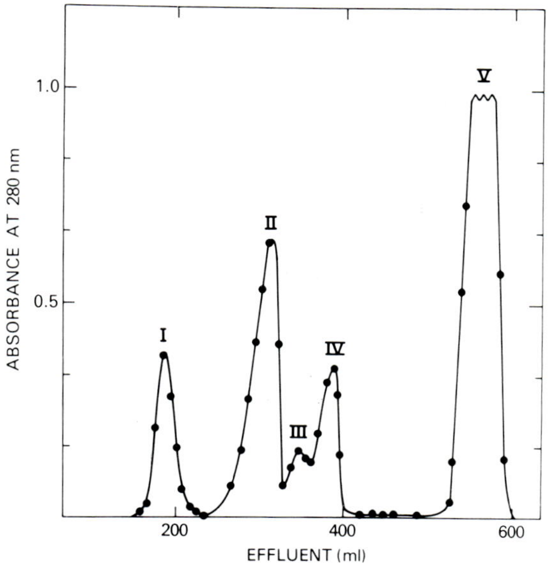

Figure 4. Gel filtration of 30% acetic acid extract of *Limulus* lysate on Sephadex G-100. The column (2 x 150 cm) was equilibrated and eluted with 30% acetic acid. A_{280}, absorbance of eluate at 280 nm.

migrated to similar positions, suggesting that the coagulogen is a single polypeptide chain. The coagulogen consists of about 220 amino acids, corresponding to a molecular weight of 24,500 which agrees with the result obtained from SDS-gel electrophoresis. The protein is devoid of methionine and has a high half-cystines content. There are a total of 18 half-cystines, all of which are involved in disulfide bond formation since free SH groups were not found by either titration with 5,5'-dithiobis(2-nitrobenzoic acid) (DTNB) or alkylation with iodoacetamide in 6M guanidine-hydrocloride. No carbohydrate was found in the protein.

Amino terminal analyses showed glycine as the sole N-terminal residue. Carboxypeptidase A digestion of the reduced carboxymethylated protein indicated serine as its C-terminal residue.

Figure 6 shows the isolation of the coagulin and the soluble C-peptide. Purified coagulogen was incubated with clotting enzyme in the presence of calcium and endotoxin.

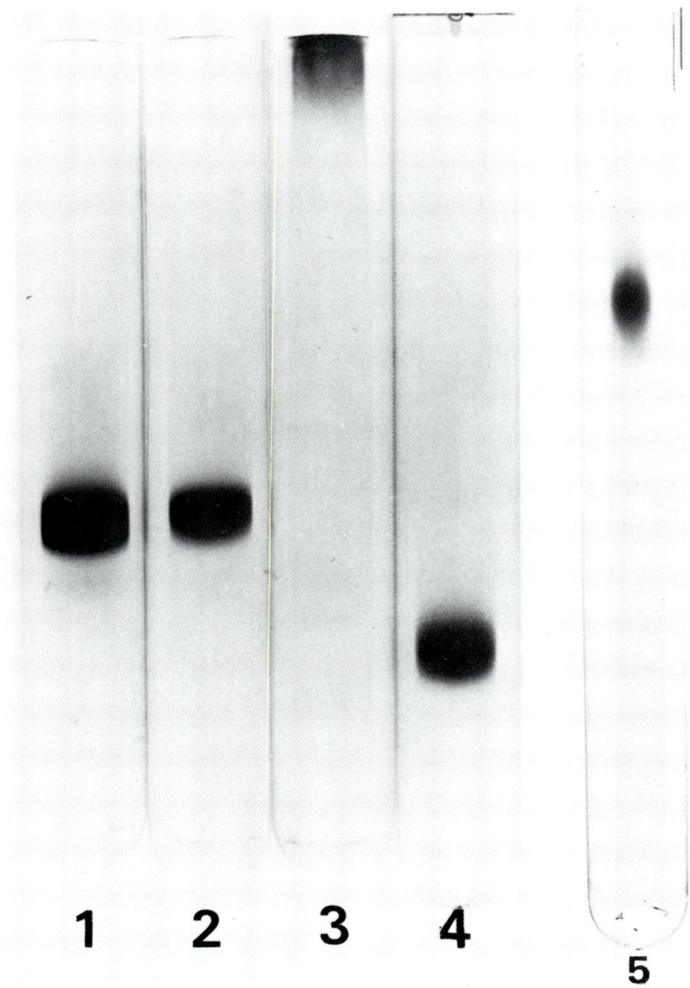

Figure 5. Acrylamide gel (7%) electrophoresis in 0.1% sodium dodecyl sulfate, 0.1M, pH 7.0, phosphate buffer of coagulogen, and the insoluble coagulin. *Gel 1*, coagulogen from *Peak II* in Fig. 4; *Gel 2*, reduced and carboxymethylated coagulogen; *Gel 3*, insoluble coagulin; *Gel 4*, reduced and carboxymethylated insoluble coagulin; *Gel 5*, 6M urea-acrylamide gel electrophoresis of C-peptide in 0.1% sodium dodecyl sulfate, 0.1M, pH 7.2, phosphate buffer. The acrylamide concentration is 12.5% and the ratio of acrylamide to methylenebisacrylamide is 10:1.

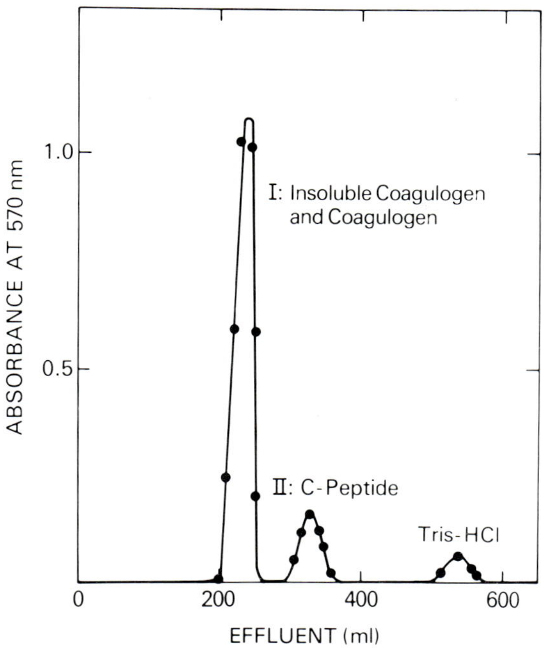

Figure 6. Gel filtration of the clotted coagulogen induced by the *Limulus* clotting enzyme through Sephadex G-50 in 30% acetic acid. Eluents were analyzed by ninhydrin after alkaline hydroylsis.

The resulting gel was dissolved in 30% acetic acid and placed on Sephadex G-50 column. Two fractions were obtained. The fraction at the void volume is a mixture of intact coagulogen and the coagulin or the N-peptide. The second fraction contained a small pure peptide of 45 amino acids termed the C-peptide. The first fraction was further purified on a Sephadex G-100 column in 30% acetic acid to yield the pure N-peptide (Fig. 7).

The amino acid composition of the N- and C- peptides is shown in Table I. The C-peptide is water soluble and consists of about 45 amino acids. It retains more than 50% of the charged amino acid residues, Lys, Arg, Asx and Gly from the parent coagulogen. The N-peptide is insoluble in aqueous solution. It retains the 9 disulfide bonds together with most of the hydrophobic amino acids.

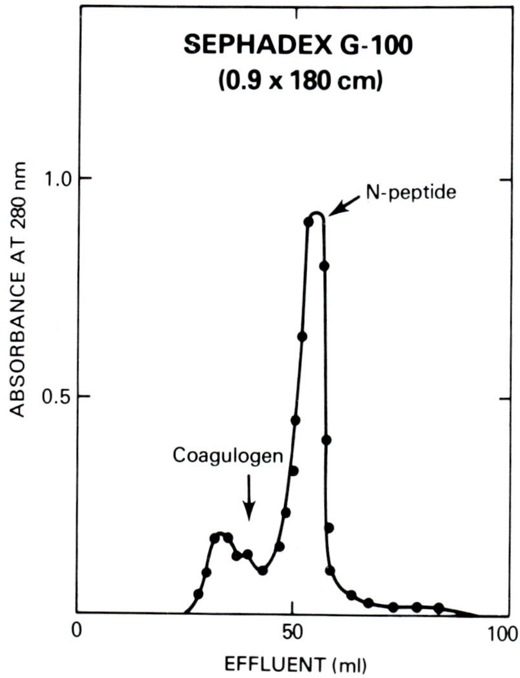

Figure 7. Gel filtration of *Fraction I* from Figure 6 through Sephadex G-100 in 30% acetic acid. A_{280}, absorbance of eluate at 280 nm.

The gelation of coagulogen can also be induced by trypsin to yield an identical N-peptide. However, the corresponding C-peptide is further hydrolyzed to several smaller peptides.

Some characteristics of the N- and C-peptides are summarized in Figure 8. The N-terminal sequence of the N-peptide was determined to be Gly-Asx-Pro which is identical to that of the intact coagulogen. The C-peptide gave the sequence Lys-Ile-Val, which is different from the coagulogen. However, its C-terminal serine is identical to the parent coagulogen. Thus, it is apparent that a single peptide bond of Arg-Lys is hydrolyzed by the clotting enzyme. Trypsin cleaves the same peptide bond. The N-peptides interact among themselves in a non-covalent fashion to form the insoluble clot.

TABLE I

AMINO ACID COMPOSITION OF LIMULUS COAGULOGEN, N-PEPTIDE AND C-PEPTIDE[a]

Amino Acid	Coagulogen	N-peptide		C-peptide CE-induced
		CE-induced	Trypsin-induced	
Lysine	11.9 (12)	6.0 (6)	6.0 (6)	6.0 (6)
Histidine	4.9 (5)	4.6 (5)	4.8 (5)	0.0
Arginine	16.8 (17)	13.8 (14)	14.1 (14)	3.4 (3)
Aspartic Acid	14.8 (15)	9.7 (10)	10.1 (10)	5.0 (5)
Threonine[b]	15.2 (15)	11.9 (12)	11.9 (12)	3.1 (3)
Serine[b]	13.7 (14)	10.8 (11)	10.9 (11)	3.1 (3)
Glutamic Acid	29.0 (29)	19.3 (19)	19.2 (19)	10.0 (10)
Proline	12.0 (12)	11.8 (12)	12.1 (12)	0.0
Glycine	16.6 (17)	14.0 (14)	14.3 (14)	1.8 (2)
Alanine	9.1 (9)	6.0 (6)	6.0 (6)	3.3 (3)
Half-Cystine[c]	18.2 (18)	17.8 (18)	17.9 (18)	0.0
Valine[d]	16.7 (17)	12.1 (12)	12.1 (12)	4.4 (5)
Methionine	0.0	0.0	0.0	0.0
Isoleucine[d]	7.7 (8)	2.6 (3)	2.5 (3)	5.2 (5)
Leucine	7.1 (7)	7.2 (7)	7.0 (7)	0.0
Tyrosine	5.6 (6)	5.9 (6)	6.1 (6)	0.0
Phenylalanine	15.0 (15)	14.7 (15)	14.9 (15)	0.0
Tryptophan	0.9 (1)	1.0 (1)	0.9 (1)	0.0
Total Residues	217.0	171.0	171.0	45.0

a. Average of duplicate analysis of 22, 48, and 72 hr. hydrolysates.

b. The values of threonine and serine were corrected for destruction during hydrolysis.

c. Half-cystine determined as S-sulfocystine is listed. Half-cystine determined as cysteic acid (Moore, S. (1963) *J. Biol. Chem.* *238*, 235-237) gave a value of 18± 0.2.

d. The 72 hr. hydrolysate values for isoleucine and valine are listed.

e. The numbers given in the parentheses are nearest integer values.

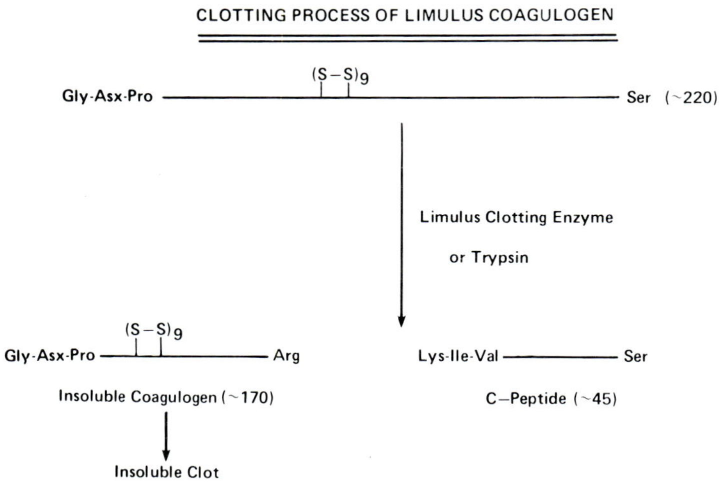

Figure 8. Schematic representation of the clotting process of *Limulus* coagulogen.

The specificity of the clotting enzyme was examined with the aid of a number of synthetic substrates (Table II). Benzoyl-Ile-Glu-(γ-OCH_3)-Gly-Arg-p-nitroanilide (-pNA) was hydrolyzed 30 times faster than Benzoyl-Phe-Val-Arg-pNA, whereas Benzoyl-Arg-pNA and Benzyloxycarbonyl-Gly-Pro-Arg-pNA were not hydrolyzed. More work is necessary to establish the clotting enzyme specificity which differs significantly from the specificity of trypsin.

The existence of a cross-linking enzyme similar to that found in the mammalian coagulation system was investigated in collaboration with Dr. S.I. Chung of the National Institute of Dental Research, NIH (6).

Transglutaminase activity was demonstrated in the freshly prepared *Limulus* lysate by its ability to incorporate primary amines into the acetylated β-chain of oxidized insulin, casein, and other lysate proteins. The enzyme is activated maximally by Ca^{2+}, minimally by Mn^{2+}, Sr^{2+}, and totally inhibited by sulfhydryl reagents, such as iodoacetamide, mercury benzoate, and S-methyl-methyl sulfonate. The inhibitors to serine proteases showed no effect.

Finally, we would like to touch upon the possible physiological role of the *Limulus* clotting system. Minute amounts of endotoxin cause cellular degranulation and lysis,

TABLE II

RELATIVE SPECIFICITY OF LIMULUS CLOTTING ENZYME

Peptide Substrate	Relative Rate
1. ⟨O⟩–CO–Ile–Glu–(γ–OCH_3)–Gly–Arg–NH–⟨O⟩–NO_2	100%
2. ⟨O⟩–CO–Phe–Val–Arg–NH–⟨O⟩–NO_2	2.8%
3. ⟨O⟩–CH_2OCO–Gly–Pro–Arg–NH–⟨O⟩–NO_2	~0%
4. ⟨O⟩–CO–(DL)–Arg–NH–⟨O⟩–NO_2	~0%

followed by the formation of an extracellular gel. No enzymatic activity related to clotting occurs within the amoebocyte. All factors which eventually contribute to a clot must be expelled into the blood plasma when the cells degranulate or lyse. Activation of a zymogen by calcium and endotoxin is an early stage in the clotting process, in spite of the fact that endotoxin cannot enter the cell intact. Consequently, *in vivo*, endotoxin must mediate the degranulation and lysis of amoebocytes before the activation of zymogen occurs. We propose the existence of endotoxin receptor site(s) on the outer membrane of *Limulus* amoebocyte. A possible physiological advantage of such a system may be to localize and immobilize minute quantities of endotoxin in *Limulus* blood without the occurrence of massive intravascular coagulation. Receptor site bound endotoxin may stimulate the release of clotting zymogen which has an affinity for this endotoxin. The invading endotoxin is quickly engulfed by a clot.

A protein with a molecular weight of about 80,000 has been isolated and purified from amoebocyte membranes (Fig. 9). This protein binds with endotoxin and enhances the *Limulus* lysate coagulation process as effectively as about a 100-fold excess of endotoxin. It is possible that the "endotoxin-receptor protein" plays a role in the *Limulus* coagulating system.

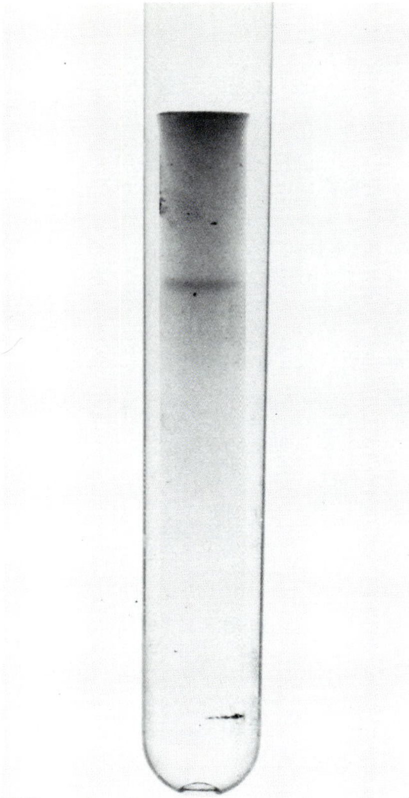

Figure 9. SDS-gel electrophoresis of the endotoxin membrane receptor of *Limulus* amoebocyte. The acrylamide concentration is 7.5% and the ratio of acrylamide to methylenebisacrylamide is 39:1.

REFERENCES

1. Levin, J., and Bang, F.B. (1964) *Bull. John Hopkins Hosp. 115*, 337-345
2. Tai, J.Y., and Liu, T.-Y. (1977) *J. Biol. Chem. 252*, 2178-2181
3. Tai, J.Y., Seid, R.C., Jr., Huhn, R.D., and Liu, T.-Y. (1977) *J. Biol. Chem. 252*, 4773-4776
4. Seid, R.C.,Jr., and Liu, T.-Y. (1978) Manuscript submitted for publication.
5. Sakmar, T.P., Liang, S.-M., and Liu, T.-Y. (1978) Manuscript submitted for publication.
6. Chung, S.I., Seid, R.C., Jr., and Liu, T.-Y. (1977) *Thrombos. Haemostas. 38*, 182

STRUCTURAL STUDIES OF THE COAGULOGEN OF AMEBOCYTE LYSATE FROM LIMULUS POLYPHEMUS*

M.W. Mosesson[a], C. Wolfenstein-Todel[a], J. Levin[b], and O. Bertrand[c]
a-SUNY, Downstate Med. Ctr., B'klyn, N.Y. 11203; b-The Johns Hopkins Univ. Sch. Med., Balt., Md. 21205; c-Hôpital Beaujon, 92118 Clichy, France.

INTRODUCTION

Circulating in the hemolymph of the horseshoe crab (Limulus polyphemus) is a single type of cell, designated the amebocyte. When stimulated by gram-negative bacteria or endotoxin, this cell undergoes degranulation (2) and a gel is produced from a soluble precursor protein (coagulogen) contained within the amebocyte granules (3-5). Amebocyte aggregation and degranulation can be inhibited if the amebocytes are collected in the presence of N-ethyl maleimide (4). Subsequently, a lysate that contains the coagulogen can be prepared from such cell preparations. This protein can be converted from a soluble form to an insoluble gel (coagulin) following the addition of bacterial endotoxin to the lysate (4). The conversion to a gel is mediated by a proteolytic enzyme in the lysate, the activation of which from the zymogen form is triggered by endotoxin (4,6,7).

Several studies relating to the structure of the coagulogen have been reported. On the basis of gel filtration experiments, Young et al. (6) estimated its M_r to be approximately 27,000. Na dodecyl sulfate gel electrophoretic analyses by Solum (8) led to the conclusion that each molecule consists of a single chain having a M_r of 23,100. Tai et al. (9) drew similar conclusions and have reported the M_r of the chain to be 24,500. Gaffin (10) found the molecu-

*Some of these experiments were carried out at the Marine Biological Laboratory, Woods Hole, Mass. Certain of these results will be reported elsewhere as a journal article (1).

Pages 159–168, Biomedical Applications of the Horseshoe Crab (Limulidae)

lar weight of the protein to be 21,000. Structural studies of the corresponding protein from Tachypleus tridentatus (11-13), the horseshoe crab of a closely related genus, are of interest since dodecyl sulfate gel electrophoretic analyses of unreduced samples containing this coagulogen also indicated a M_r of about 23,000. The coagulogen from this species, like that from Limulus, also is believed to consist of a single chain.

This report concerns recent studies (1) that were carried out to characterize the structure of the coagulogen from Limulus polyphemus. The results suggest a molecular structure comprised of two identical chains linked by noncovalent forces.

MATERIALS AND METHODS

Amebocyte lysate was prepared by the disruption of washed amebocytes in sterile, pyrogen-free distilled water, as described previously (4,6). This extract of whole cells has been shown to contain the entire blood coagulation system of Limulus (3,4). Purified coagulogen was prepared from fresh lysate that had not been frozen.

For certain control experiments, amebocyte preparations also were lysed in H_2O containing diiosopropyl phosphofluoridate (1 mM, final concentration) (Schwarz/Mann, Orangeburg, N.Y.); Kunitz' pancreatic trypsin inhibitor (Trasylol, FBA Pharmaceuticals, New York, N.Y., 1000 units/ml, final concentration); or 9 M urea.

Na dodecyl sulfate polyacrylamide gel electrophoresis (14) was performed in gels in which the final acrylamide concentration was 9 or 12%. For disulfide bond reduction, β-mercaptoethanol (1%, v/v final concentration) or dithiothreitol (14 mM, final concentration) was added. Gels were stained with Coomassie brilliant blue. The polypeptide markers used have been summarized elsewhere (1).

Acidic polyacrylamide gel electrophoresis (10% acrylamide, pH 2.7) was carried out in gels containing 2 M urea and 5% (v/v) acetic acid, as described by Brummel and Montgomery (15) and were stored as previously described (1).

Coagulogen was purified by a two-step procedure involving first a minor modification (1) of the "acidification" procedure described by Solum (8) followed by preparative gel electrophoresis under the acidic conditions stated above (15). Details have been provided elsewhere (1).

Coagulogen prepared according to Tai et al. (9) was used in certain comparative experiments. This material was a generous gift from Dr. T-Y. Liu, Bureau of Biologics, FDA, Bethesda, Md.

Molecular exclusion chromatography was carried out on Sephadex columns that had been equilibrated with one of the following solutions: 0.1 M NaCl, 0.01 M sodium phosphate buffer, pH 7, containing 0.02% Na azide; 8 M urea in 5% acetic acid solution; 8 M urea in 0.1 M Tris HCl buffer, pH 8.2. Details are reported elsewhere (1).

RESULTS

Purified coagulogen at a concentration of 1 mg/ml, in phosphate buffered saline, pH 7.2, formed a solid gel after the addition of a source of converting enzyme. This source consisted of the supernatant obtained by adding endotoxin, at a final concentration of 0.5 μg/ml, to whole lysate and then removing the resultant gel. Judging from the absence of coagulogen in the supernatant obtained following gelation (as demonstrated by dodecyl sulfate gel electrophoresis), all of the coagulogen had been clotted.

Analytical acidic gel electrophoresis demonstrated that highly purified coagulogen preparations migrated as a single cathodal band (Fig.1) and occupied the same position as did the coagulogen band from unmodified starting lysate or from lysate subfractions obtained after step 1. Final step 2 coagulogen also migrated as a single component in dodecyl sulfate-containing gels (Fig.2) and the migration rate was the same as that of coagulogen in the starting material. After sample reduction with β-mercaptoethanol or dithiothreitol, the coagulogen still migrated as a single band in dodecyl sulfate gels but the band was now in a more anodal position, corresponding to that of the reduced coagulogen band in lysate. Furthermore, there were no differences between the dodecyl sulfate gel electrophoretic characteristics of reduced coagulogen in lysate prepared from cells lysed by the standard technique and those of coagulogen prepared by lysing cells in the presence of diisopropyl phosphofluoridate, Kunitz' pancreatic trypsin inhibitor, or 9 M urea solutions. Reduced S-carboxymethylated coagulogen (16) also occupied a more anodal position than did unreduced material but its migration rate was somewhat lower than that of reduced material.

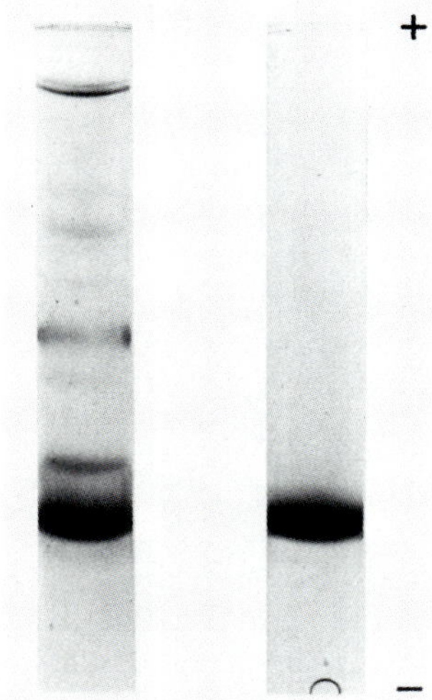

FIGURE 1

Acidic gel electrophoresis (pH 2.7); supernatant solution of amebocyte lysate after the Solum "acidification" procedure (left); final purified material obtained by preparative electrophoresis (right). The anode and cathode are indicated on the right. Reproduced from ref.1 with permission.

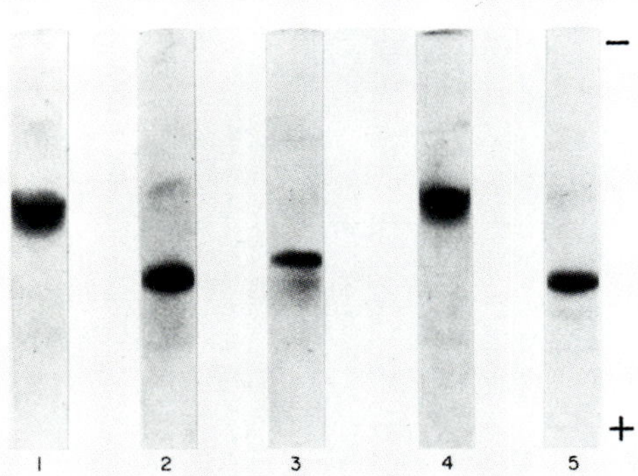

FIGURE 2

Na dodecyl sulfate gel electrophoresis (9% acrylamide) comparing our step 2 coagulogen with coagulogen prepared as described in ref.9. The gels are numbered as follows: 1, Step 2 coagulogen, unreduced; 2, same + dithiothreitol; 3, same, reduced and S-carboxymethylated; 4, ref.9 coagulogen, unreduced; 5, same + dithiothreitol. Reproduced from ref.1 with permission.

We also have compared our coagulogen preparation in this electrophoretic system with a sample that had been prepared as described in ref.9 (Fig.2). This sample was indistinguishable from ours with respect to the anodal migration rate of the coagulogen band both before and after reduction with dithiothreitol.

Dodecyl sulfate gel electrophoretic analyses of unreduced coagulogen yielded values for the apparent M_r ranging from 21,000 to 30,000. Following disulfide bridge reduction, the coagulogen band occupied a more anodal position corresponding to M_r 11,000 to 17,000. Despite the variation in apparent M_r from one run to another, a constant feature displayed by this material was that the ratio of the M_r of unreduced:reduced coagulogen in any given run was about two. For purposes of considering this information in relation to other data, we have calculated the mean apparent M_r of unreduced coagulogen from seven separate analyses to be 23,500 ± 2,500; that of reduced material from the same runs was 12,400 ± 1,500 (Fig.3). Assuming these values, the apparent M_r of reduced S-carboxymethylated coagulogen is approximately 1,600 greater than that of reduced coagulogen (i.e., 14,000).

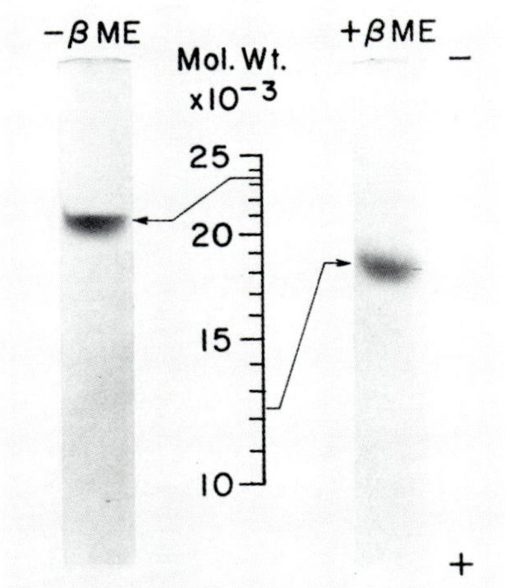

FIGURE 3

Na dodecyl sulfate gel electrophoreiss (12% acrylamide) of purified coagulogen before (left) and after (right) reduction of disulfide bridges. βME=β-mercaptoethanol. The mean apparent molecular weight of each band is indicated. The anode and cathode are indicated on the right.

The effect of another dissociating solvent, viz., urea, on the apparent molecular weight was explored by means of gel sieving chromatography. The coagulogen peak of unreduced coagulogen (two preparations of purified coagulogen, and one preparation of step 1 coagulogen) on Sephadex G-75 in phosphate buffer, pH 7, emerged in a position corresponding to M_r 19,000 ± 1,000 (Fig.4). Similar analyses of unreduced material on Sephadex G-100 in 8 M urea, 5% acetic acid solution yielded a single coagulogen peak corresponding in position to M_r 11,500 ± 1,000. These results suggested that the chains of the coagulogen molecule could be dissociated in the presence of 8 M urea without disulfide bridge reduction. Analogous results were obtained when unreduced coagulogen was chromatographed in 8 M urea buffered at pH 8.2. However, under these conditions, only 85% of the material emerged in the major peak corresponding to M_r 11,500. The rest was distributed between two other peaks, one of which corresponded to M_r 19,000 (~5%), and the other of which emerged in the void volume (~10%).

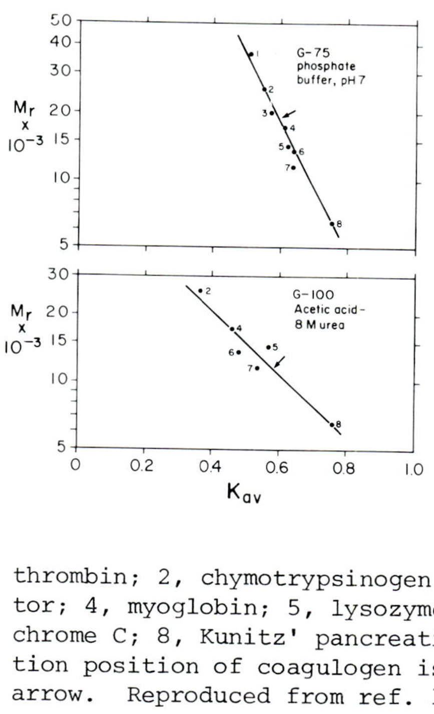

FIGURE 4

Estimation of the molecular size of coagulogen by gel filtration. The partition coefficient (Kav) is plotted against the log of the molecular weight (M_r). The upper portion of the figure presents data from chromatography on a column of Sephadex G-75 (1.5 X 90 cm) equilibrated and developed with 0.1 M NaCl, 0.01 M sodium phosphate buffer, pH 7.0, containing 0.02% Na azide. The lower portion of the figure presents data from chromatography on a column of Sephadex G-100 (1.5 X 90 cm) equilibrated and developed with 8 M urea, 5% acetic acid. The following proteins were used for calibration: 1, human thrombin; 2, chymotrypsinogen A; 3, soybean trypsin inhibitor; 4, myoglobin; 5, lysozyme; 6, ribonuclease; 7, cytochrome C; 8, Kunitz' pancreatic trypsin inhibitor. The elution position of coagulogen is indicated in each case by the arrow. Reproduced from ref. 1 with permission.

DISCUSSION

Most reported estimates of the size of the Limulus polyphemus amebocyte coagulogen indicate that the molecular weight lies between 21,000 and 27,000 (6,8-10). Our present experiments indicate a size in the range of 19,000 to 23,500 or more depending upon the method used for molecular weight estimation. More importantly, in each measurement system, a manipulation that could be expected to induce subunit dissociation resulted in reduction of the apparent molecular weight to approximately one half. These observations are summarized below.

Our dodecyl sulfate gel electrophoretic analyses indicated a size of approximately 23,500 ± 2,500 for unreduced samples and 12,400 ± 1,500 following disulfide bridge reduction. Owing to its high positive charge, as evidenced by its rapid cathodal migration rate in acidic-urea gels (Fig.1, ref.8), molecular weight analysis of the coagulogen by dodecyl sulfate gel electrophoresis could, in principle, result in over-estimation of the true size (14). These unusual charge characteristics probably also contributed to the variability in apparent M_r that we have observed. Nevertheless, these results are also consistent with a molecular structure consisting of two chains with equal masses. The finding of a single NH_2-terminal sequence for the first 6 residues (1) suggests that these chains would be identical.

Further evidence consistent with a molecular structure comprised of two chains was obtained from gel sieving experiments (Fig.4). When gel sieving chromatography was carried out at neutral pH, an apparent M_r of 19,000 ± 1,000 was found. Chromatography of the same material in solutions containing 8 M urea indicated that the M_r had fallen to 11,500 ± 1,000. Thus, the results not only suggest a two-chain structure for the native coagulogen molecule, but also indicate that the bonds joining the chains of each molecule are not covalent. It is pertinent to emphasize, however, that disulfide bridges evidently play an indirect role in such interchain bridging, since treatment with reducing agents resulted in halving of the apparent molecular weight in the presence of Na dodecyl sulfate (Fig.3).

We have examined the reports of other investigators who concluded that each coagulogen molecule consists of a single chain. Solum (8) found that reduced coagulogen migrated considerably more slowly than did unreduced material in the presence of Na dodecyl sulfate. We cannot explain why his observations are opposite from ours.

Tai et al. (9) found that reduced, S-carboxymethylated coagulogen had nearly the same electrophoretic migration rate as that which was unreduced. Our results differ from theirs in that in our hands, both their preparation and ours underwent a decreast in M_r of approximately 50% after disulfide bridge reduction (Fig.2). We also have observed that S-carboxymethylated material migrates somewhat less anodally than does material that has been reduced but not alkylated. This tendency might have contributed to the migration rate that Tai et al. (9) observed and to their conclusion that the coagulogen molecule consists of a single chain.

Tai et al. (8) also believed that their compositional analyses were consistent with a single chain structure of M_r 24,500 since a minimum M_r in that range could be calculated from the tryptophan content. By contrast, we found virtually no tryptophan by two methods of analysis (1), one of which (viz. amino acid analysis after methanesulfonic acid hydrolysis, ref.17) was the procedure they employed.

Comparison of our results with structural studies (10-12) of the coagulogen from Tachypleus tridentatus indicates that these proteins are similar in several respects. Both coagulogens are positively charged. When subjected to dodecyl sulfate gel electrophoresis, both proteins have an apparent M_r in the range of 23,000 before disulfide reduction; both exhibit an apparent fall in molecular size (i.e., increased migration rate) following reduction of disulfide bridges. There are similarities in the amino acid composition, particularly with respect to their relatively high contents of cystine and glutamic acid (or glutamine) and the absence of methionine. Comparison of the NH_2-terminal sequence indicates that three of the first six residues of the Limulus coagulogen chain are identical with those of the Tachypleus coagulogen (1,13). [The report by Takagi (this meeting) of the amino acid sequence of the Tachypleus protein indicates that its chain consists of 175 residues. Thus, if the two coagulogens are as similar as we suspect them to be, our estimations of the size of the reduced Limulus peptide will prove to be low.] There also appear to be important similarities in the sequence of proteolytic activation of the molecule; like Nakamura et al. (11), we observed that endotoxin-mediated conversion to coagulin involves formation of an intermediate derivative chain plus a major ter-inal fragment (1). On the other hand, Tai et al. (9) have proposed a proteolytic activation sequence for Limulus coagulogen that is different from the one proposed for Tachypleus (11). Thus, however tempting it is to propose that these two species' molecules have similar structures, it is clear that in order to reach that conclusion, further substantiation of the structure of both types of coagulogen will be necessary.

SUMMARY

These studies concern structural characterization of the soluble clottable protein (coagulogen) from the amebocytes of the horseshoe crab (Limulus polyphemus). Upon electrophoresis under acidic conditions (unreduced) or in dodecyl sulfate-containing gels (reduced or unreduced), purified coagulogen migrated as a single band and occupied the same position as did the corresponding coagulogen band from the starting material. Dodecyl sulfate gel electrophoretic analyses of unreduced samples indicated an apparent molecular weight ranging from 21,000 to 30,000 (mean 23,500 ± 2,500) whereas reduced samples had an apparent molecular weight ranging from 11,000 to 17,000 (mean 12,400 ± 1,500). Gel sieving chromatography of coagulogen in phosphate buffer, pH 7, yielded an apparent molecular weight value of 19,000 ± 1,000. The same type of analysis in solutions containing 8 M urea yielded a value of 11,500 ± 1,000. These findings are consistent with a molecular structure comprised of two identical chains joined by non-covalent forces.

ACKNOWLEDGMENTS

These investigations were supported by a Josiah Macy Jr. Faculty Scholar Award and by Research Grants HL-01601 and HL-17419 from the National Heart, Lung, and Blood Institute, Bethesda, Maryland. We thank Mrs. Lori Hubert for her invaluable secretarial services.

REFERENCES

1. Mosesson MW, Wolfenstein-Todel C, Levin J, Bertrand O (1978). Characterization of amebocyte coagulogen from the horseshoe crab (Limulus polyphemus). Thromb Res (In press).
2. Levin J, Bang FB (1964). A description of cellular coagulation in the Limulus. Bull Johns Hopk Hosp 115:337.
3. Levin J, Bang FB (1964). The role of endotoxin in the extracellular coagulation of Limulus blood. Bull Johns Hopk Hosp 115:265.

4. Levin J, Bang FB (1968). Clottable protein in Limulus: Its localization and kinetics of its coagulation by endotoxin. Thromb Diath Haemorrh 19:186.
5. Mürer EH, Levin J, Holme R (1975). Isolation and studies of the granules of the amebocytes of Limulus polyphemus, the horseshoe crab. J Cell Physiol 86:533.
6. Young NS, Levin J, Prendergast RA (1972). An invertebrate coagulation system activated by endotoxin: Evidence for enzymatic mediation. J Clin Invest 51:1790.
7. Tai JY, Liu T-Y (1977). Studies on Limulus amebocyte lysate. Isolation of pro-clotting enzyme. J Biol Chem 252: 2178.
8. Solum NO (1973). The coagulogen of Limulus polyphemus hemocytes. A comparison of the clotted and non-clotted forms of the molecule. Thromb Res 2:55.
9. Tai JY, Seid RC Jr, Huhn RD, Liu T-Y (1977). Studies on Limulus amoebocyte lysate II. Purification of the coagulogen and the mechanism of clotting. J Biol Chem 252: 4773.
10. Gaffin SL (1976). The clotting of the lysed white cells of Limulus induced by endotoxin I. Preparation and characterization of clot-forming proteins. Biorheology 13: 273.
11. Nakamura S, Takagi T, Iwanaga S, Niwa M, Takahashi K (1976). A clottable protein (coagulogen) of horseshoe crab hemocytes. Structural change of its polypeptide chain during gel formation. J Biochem 80:649.
12. Nakamura S, Iwanaga S, Harada T, Niwa M (1976). A clottable protein (coagulogen) from amebocyte lysate of Japanese horseshoe crab (Tachypleus tridentatus). Its isolation and biochemical properties. J Biochem 80:1011.
13. Nakamura S, Takagi T, Iwanaga S, Niwa M, Takahashi K (1976). Amino acid sequence studies on the fragments produced from horseshoe crab coagulogen during gel formation: homologies with primate fibrinopeptide B. Biochem Biophys Res Comm 72:902.
14. Weber K, Osborn M (1975). In Neurath H, Hill RH (eds): "The Proteins," Academic Press, NY, 3rd ed., vol.1, p179.
15. Brummel MS, Montgomery R (1970). Acrylamide gel electrophoresis of the S-sulfo derivatives of fibrinogen. Anal Biochem 33:28.
16. Konigsberg W (1972). In Hirs CHW, Timasheff SH (eds): "Methods in Enzymology", Academic Press, NY, vol.25, p185.
17. Simpson RJ, Neuberger MR, Liu T-Y (1976). Complete amino acid analysis of proteins from a single hydrolysate. J Biol Chem 251:1936.

AMINO ACID SEQUENCE STUDIES ON HORSESHOE CRAB (*Tachypleus tridentatus*) COAGULOGEN AND THE MECHANISM OF GEL FORMATION

T.Takagi[1],Y.Hokama,T.Morita,S.Iwanaga,S.Nakamura[2], and M.Niwa[3]

Institute for Protein Research,Osaka University,
[1] Biological Institute,Tohoku University,[2]Primate Research Institute,Kyoto University,[3] Osaka City Universtity Medical School.

The amoebocyte in horseshoe crab hemolymph participates in hemostasis and in the defence against invading bacteria. The amoebocyte contains a clottable protein,coagulogen,and a zymogen or zymogens which react with coagulogen to produce a gel after activation by bacterial endotoxin. The gelation test using the amoebocyte lysate has been widely employed as a very sensitive assay for endotoxin. The coagulation of horseshoe crab amoebocyte lysate seems to be very similar to mammalian blood coagulation,which involves sequential activation of several proenzymes. It is interesting to study the molecular events of the coagulation in the horseshoe crab both from the standpoint of evolution and to understand the complicated vertebrate blood coagulation system.

PREPARATION OF AMOEBOCYTE LYSATE

Hemolymph of *Tachypleus tridentatus* (adult,both sexes) was collected by cardiac puncture. Fifty to 150 ml per individual(1-3 kg) was drawn into a bottle which contained 3 % NaCl containing 0.1 M caffeine amounting to one tenth of the original bled volume of hemolymph. Amoebocytes were obtained by centrifuging pooled hemolymph at 2,500 rpm for 10 min and washed twice with 3 % NaCl. The cells were dispersed in one tenth of the original volume of a lysate buffer consisting of 0.05 M Tris-HCl,0.001 M $CaCl_2$ and 0.15 M NaCl,pH 7.2, and the suspension was frozen and thawed. The thawed cells were homogenized by a mechanical blender and centrifuged at 2,500 rpm for 15 min. The sediment,which still contained unbroken amoebocyte,was homogenized again with one thirties or more

Pages 169—184, Biomedical Applications of the Horseshoe Crab (Limulidae)

of the original bled volume of lysate buffer and centrifuged. The combined supernate was spun again at 4° at 5,000 rpm for 30 min to remove cell debris. The lysate thus obtained contained 15-20 mg of protein per ml.

ISOLATION OF *Tachypleus* COAGULOGEN

Two different methods were used to isolate coagulogen. Method I : Ten ml of lysate was gel-filtered on a pyrogen-free Sephadex G-75 column equilibrated with pyrogen-free saline and using pyrogen-free glasswares. The elution profile

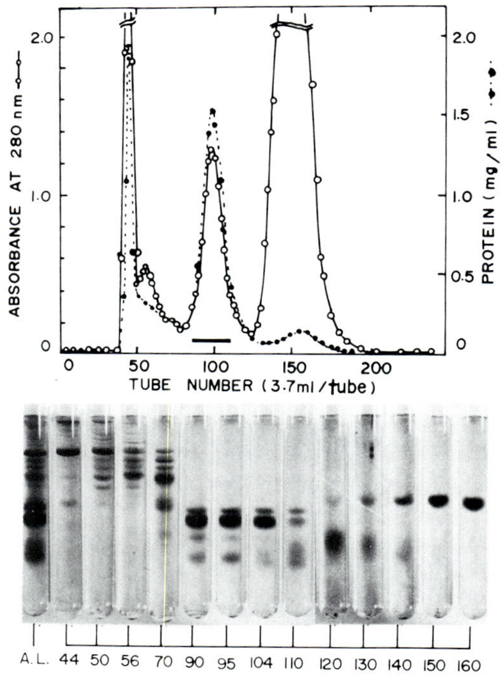

Fig 1. Gel-filtration of *Tachypleus* lysate by Sepadex G-75 column(Method I). The lysate (198 mg protein) was passed through a column(2.9x100 cm) equilibrated with pyrogen-free saline. Fraction size=3.7 ml.Flow rate=40 ml/h.Coagulogen fractions indicated by a solid bar were combined. SDS-gel electrophoretic patterns in the presence of 2-mercaptoethanol of fractions are shown under the chromatogram. A.L.=whole lysate. (Nakamura et al. 1976a)

and electrophoretic patterns are shown in Fig 1. The first peak contained a potent clotting enzyme. A coagulogen (detected by clot formation with trypsin) was found in the second peak, indicated by a solid bar.These fraction were concentrated by adding ammonium sulfate to give 40 % saturation. After dialysis it was further purified by CM-32 column chromatography. Through these procedures,40 mg of coagulogen was obtained.

Mehtod II : Thirteen ml of the lysate was first acidified with 1 ml of 1.0 N HCl to inactivate the clotting enzyme(s). After standing for 15 min,it was neutralized by adding 1 ml of 1.0 N NaOH. The precipitate was removed and the supernate was gel-filtered on a Sephadex G-75 column (2.6 x 100 cm)

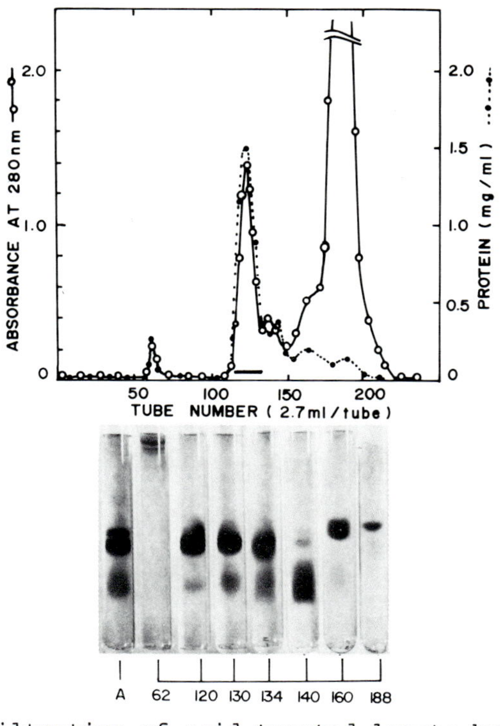

Fig 2. Gel filtration of acid-treated lysate by a Sephadex-G75 column(Method II). The sample (150 mg Protein) was passed through a column(2.6x100 cm) equilibrated with non-sterilized saline. Fraction size=2.7 ml,Flow rate= 37 ml/h. Coagulogen fractions(solid bar) were combined.SDS-gel electrophoresis patterns in the presence of 2-mercaptoeyhanol of fractions are shown under the chromatogram.A=acid-treated lysate. (Nakamura et al. 1976a)

equilibrated with non-sterilized saline. The elution profile and SDS-gel electrophoretic patterns are shown in Fig 2. The fractions indicated by a solid bar were combined and concentrated by 40 % of ammonium sulfate saturation. It was gel-filtered again on a Sephadex G-50 column. Through these procedures 50 mg of coagulogen was obtained from 13 ml of lysate. Method II was employed for large scale preparations.

PURITY AND PROPERTIES OF ISOLATED COAGULOGEN

The purified preparation obtained by either Method I or II gave a single protein band on gel electrophoresis at pH 3.2 in the presence of 6 M urea(a sample prepared by Method II is shown in Fig 3). On gel electrophoresis at pH 8.3,coagulogen did not penetrate the gel and was retained at the top of the gel. SDS-gel electrophoresis showed a single band in the presence or absence of 2-mercatoethanol,suggesting that it consists of a single polypeptide chain(Fig 3). The molecular weight of the reduced coagulogen determined from the relative mobility was 19,500 ± 1,000. The isoelectric point estimated

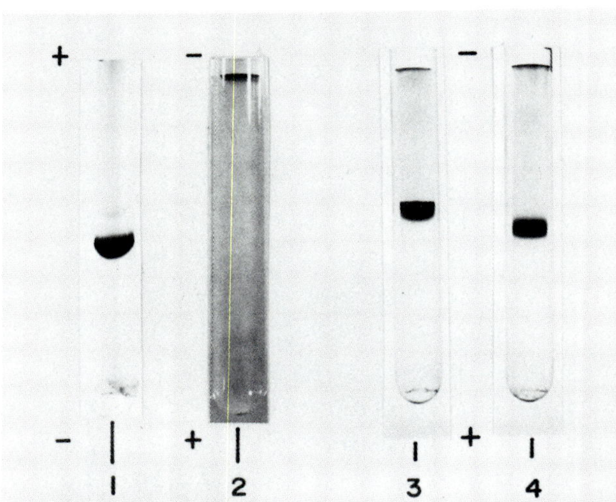

Fig 3. Polyacrylamide gel electrophoreses of purified coagulogen(Method II). (1) at pH 3.2 in the presence of 6 M urea. (2) at pH 8.3. (3) SDS-gel in the absence of 2-mercaptoethanol (4) SDS-gel in the presence of 2-mercaptoethanol. (Nakamura et al. 1976a)

by gel electrofocusing was 10.0. The amino-terminal residue was found only alanine by dansyl method and subsequent sequence analysis by dansyl-Edman method gave Ala-Asx-Thr. On the other hand, carboxyl-terminal residue was found phenylalanine by hydrazinolysis and recovered 0.56 mole per mole coagulogen. The C-terminal phenylalanine was also confirmed by tritium labeling method. These results shows that *T.tridentatus* coagulogen consists of a single polypeptide chain. The amino acid composition is shown in Table 1. No methionine was found. Coagulogen appears to have relatively high contents of glutamic acid,CM-cysteine,phenylalanine and valine. No free sulfhydryl group was titrated. Eight disulfide bridges can be expected and probably account for the high stability of this protein.

GEL FORMATION

Clotting enzyme was prepared from the pooled fractions obtained from the first peak in Fig 1. Ten ml of the fraction was mixed with 0.1 ml of 0.5 M Tris-HCl buffer,pH 7.5,containing

Table 1. Amino Acid Composition of *T.tridentatus* Coagulogen.

Amino acid	Residues per mole
Aspartic acid	10.6 (10)
Threonine[a]	10.7 (10)
Serine[a]	11.4 (11)
Glutamic acid	20.0 (20)
Proline	10.5 (10)
Glycine	13.9 (14)
Alanine	8.0 (8)
CM-cysteine	15.2 (16)
Valine	15.4 (16)
Isoleucine	5.5 (6)
Leucine	5.9 (6)
Tyrosine	5.6 (5)
Phenylalanine	13.4 (13)
Lysine	7.5 (7)
Histidine	5.3 (5)
Arginine	16.6 (16)
Tryptophan [b]	1.3 (1)
Total	176.8(175)

Value of 24 h hydrolysis.Values in parentheses were obtained by sequence analysis. a:Extraporated.b: Hydrolysed with 4 M methane sulfonic acid.

0.125 M $MgSO_4$ and 0.1 ml of *Salmonella minnesota* R595 endotoxin (5 µg/ml.) After incubation at 37° for 90 min ,it was centrifuged and the supernatant was purified by DEAE-cellulose chromatography.The reaction system for a gel formation was as follows: A mixture containing 100 µl of coagulogen solution (1.6 mg/ml),40 µl of 50 mM Tris-HCl buffer,pH 7.8,containing 50 mM $MgSO_4$,and 40 µl of clotting enzyme was incubated at 37°. At indicated times,15 µl alliquotes were subjected to SDS-gel electrophoreses and for the estimation gel protein,10 µl of 50 mM diisopropyl fluorophosphate(DFP) was added to terminate the reaction. Gels were washed and estimated protein concentrations. The results of SDS-gel electrophoreses of reduced or non-reduced samples are shown in Fig 4. In the non-reduced ones,

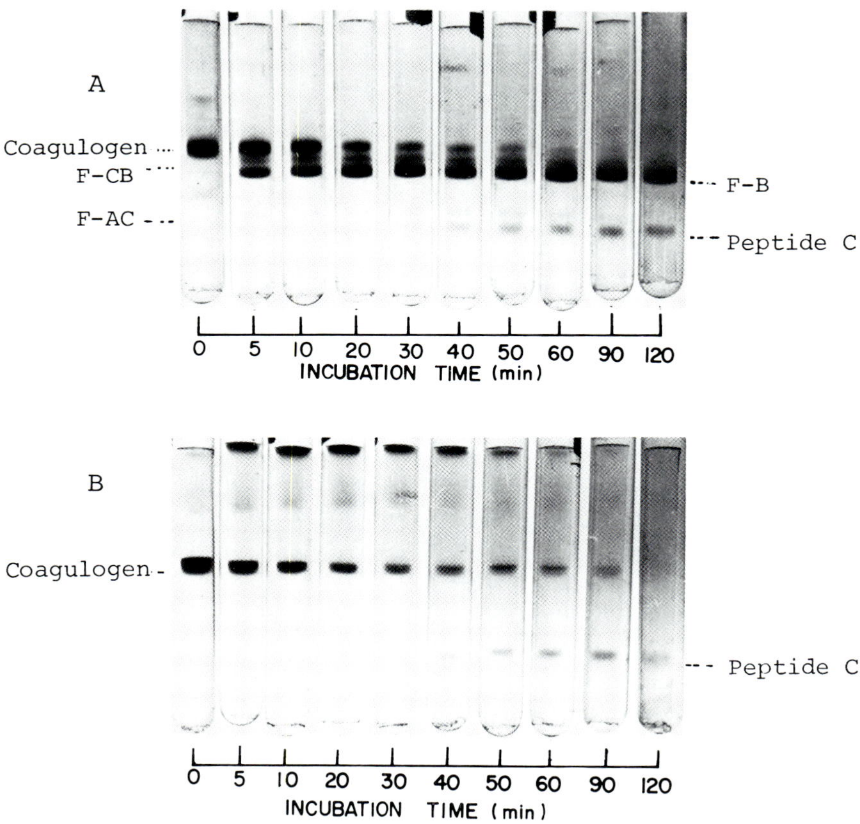

Fig 4. SDS-gel electrophoreses of the reaction products of coagulogen with clotting enzyme. A ; reduced samples. B ; non-reduced samples.

coagulogen showed a single band at 0 time. After 5 min, clotted protein appeared at the top of the gel. After that, the coagulogen band gradually disappeared and a new band, peptide C, appeared. At longer incubation times only peptide C could be seen and clotted protein was no longer seen even on the top of the gels. These results suggest that the clotting enzyme hydrolysed coagulogen and released peptide C, whereas the remaining molecule polymerized. In the reduced samples, coagulogen again showed a single band at 0 time. After 5 min, however, new intermediate bands could be observed. One was disignated as F-CB and another as F-AC. Both appeared within 5 min, increased rapidly, and then disappeared gradually. Finally only two protein bands were observed. One was B-chain (F-B in Fig B) and the other was peptide C.

To estimate the relative amount of these fragments, the bands on the gels were scanned with a densitometer and the amounts of gel protein formed were measured separately. The results are shown in Fig 5. The amount of F-AC increased in the initial stage of gel formation and its decrease seemed

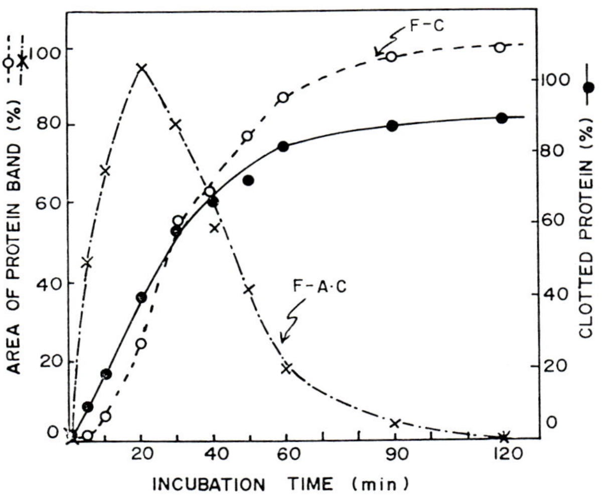

Fig 5. Densitometric analyses of the fragments obtained from coagulogen during proteolysis and estimation of gel protein accumulated in the reaction mixture. The stained gels in Fig 4. were scanned with a densitometer and the relative intensities of the bands were calculated, taking the maximum intensities of each fragments as 100 %. Gel protein was estimated by the method of Miller.

to be roughly proportional to the increase in the amount of peptide C. The amount of gel protein also increased in parallel with release of peptide C.

In the next experiment,the reaction was stopped at an intermediate stage by addition of DFP. The reaction mixture was reduced and alkylated and applied on a Sephadex G-50 column. The elution pattern is shown in Fig 6.

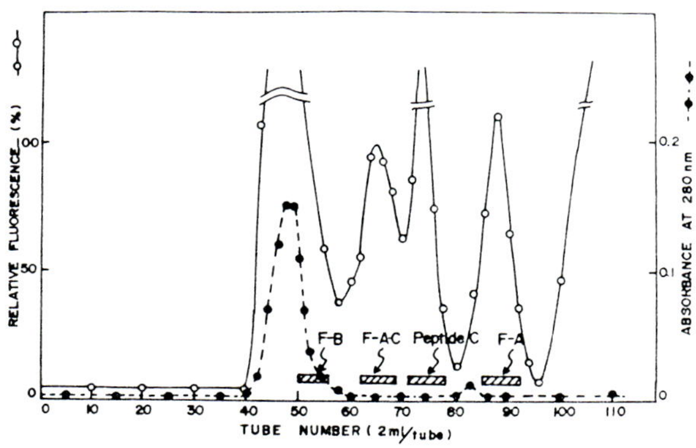

Fig 6. Gel filtration of the peptide fragments produced from coagulogen during gel formation. The reaction mixture was reduced and alkylated, and passed through a Sephadex G-50 column. F-B=B-chain, F-A=A-chain.

We pooled each of the peaks and determined their amino acid compositions as well as their amino- and carboxyl-terminal residues. The results are given in Table 2. A-chain(F-A in Fig 6) consisted of 18 residues with amino-terminal alanine an carboxyl-terminal arginine and subsequent N-terminal sequence was determined to be Ala-Asx-Thr,which is the same as that of intact coagulogen. Peptide C was 28 residues and its N-terminus and C-terminus were threonine and arginine, respectively. Fragment F-AC was estimated to contain 46 residues which agrees with the sum of total residues in A-chain and Peptide C. The large fragment B-chain (F-B) consisted of 129 residues with N-terminal glycine and its C-terminal phenylalanine was identical with that of intact coagulogen. From these results,it was concluded that coagulogen consists of three polypeptide segments,A-chain peptide C and B-chain, and has a total 175 amino acid residues, as shown in Fig 7.

Table 2. Amino Acid Compositions and N- and C-terminal Residues of Polypeptide Fragments from *Tachypleus tridentatus* Coagulogen

Amino Acid	A-chain	Peptide C	F-AC	B-chain
Aspartic acid	3.1 (3)	1.1 (1)	4.1	6.5 (6)
Threonine	1.0 (1)	2.9 (3)	4.1	6.9 (7)[a]
Serine	0	1.9 (2)	2.1	9.1 (9)[a]
Glutamic acid	1.1 (1)	6.1 (6)	6.8	13.4 (13)
Proline	1.9 (2)	0	2.0	8.0 (8)
Glycine	2.1 (2)	2.1 (2)	4.0	10.4 (10)
Alanine	2.0 (2)	3.0 (3)	4.8	3.3 (3)
CM-cysteine	2.1 (2)	0	1.4	14.8 (14)
Valine	1.1 (1)	3.9 (4)	4.7	10.2 (11)
Isoleucine	1.0 (1)	2.8 (3)	3.9	2.0 (2)
Leucine	2.1 (2)	0	2.1	4.0 (4)
Tyrosine	0	0	0	5.0 (5)
Phenylalanine	0	0	0	13.1 (13)
Lysine	0	2.8 (3)	3.0	4.3 (4)
Histidine	0	0	0	5.2 (5)
Arginine	1.1 (1)	0.9 (1)	1.9	13.8 (14)
Tryptophan	0	0	0	1.3 (1)[b]
Total	18	28		129
N-terminal	Ala	Thr	Ala	Gly
C-terminal	Arg	Arg	Arg	Phe

Figures represent residues per mole calculated from values for 24 h hydrolysate. Values in parenthes were obtained from sequence analyses. a : Extrapolated value. b : Hydrolysed with 4 M methanesulfonic acid.

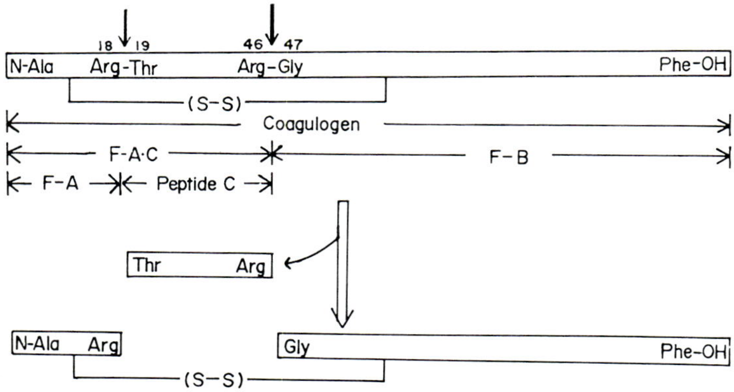

Fig 7. Alignment of the peptide segments and structural change of *Tachypleus tridentatus* coagulogen during gel formation. The arrows indicate the site cleaved by clotting enzyme. (Nakamura et al. 1977b).

PRIMARY STRUCTURE OF *Tachypleus tridentatus* COAGULOGEN

For sequence analyses,each polypeptide segment was prepared in large scale as following. Purified coagulogen (200 mg) was dissolved in a buffer (0.05 M Tris-HCl containing 0.1 M NaCl,pH 7.5) and appropriate amount of clotting enzyme was added to clot within 10 min. It was incubated for 3 h and gel was washed with the buffer and centrifuged. The supernate contained peptide C. The gel was suspended in 6 M guanidine-hydrochloride-0.5 M Tris containing 10 mM EDTA,pH 8.5. It was reduced by dithiothreitol (10 mM) and alkylated with iodo-acetic acid (20 mM). Then it was washed with 70 % aqueous acetone for three times. The supernate contained S-CM-A-chain and the precipitate was S-CM-B-chain. Amino acid sequence of peptide C was established by the direct Edman method and analyses of tryptic peptides and thermolytic peptides (Nakamura et al. 1976c). A-chain was also sequenced all 18 residues by mannual Edman method and confirmed by its thermolytic peptides.(Nakamura et al. 1976c). The results are shown in Table 3. From the results of amino acid sequence of A-chain and peptide C, we found several things. First, the C-terminal octapeptide sequence of these peptides exhibit great homology, suggesting that a specific oligopeptide sequence immediately preceeding the bond to be cleaved is required for the clotting enzyme to split the Arg-Thr and Arg-Gly linkages so as to initiate gel formation. In this respect,the specificity of

Table 3. Amino Acid Sequence of A-chain(F-A) and Peptide C. (Nakamura et al. 1976c).

FRAGMENTS	COMPLETE OR PARTIAL AMINO ACID SEQUENCE
	1 5 10 15
F-A	H-ALA-ASP-THR-ASN-ALA-PRO-ILE-CYS-LEU-CYS-ASP-GLU-PRO-GLY-VAL-LEU-
	18
	GLY-ARG-OH
	1 5 10
PEPTIDE C	H-THR-GLN-ILE-VAL-THR-THR-GLU-ILE-LYS-ASP-LYS-ILE-GLU-LYS
	← T-I → ← T-III →
	ALA-VAL-GLU-ALA-VAL-ALA-GLN-GLU-SER-GLY-VAL-SER-GLY-ARG-OH
	← T-II →
	1 102
F-B	H-GLY-PHE-(SER)-ILE-PHE-----------------------------------PHE-OH

horseshoe crab clotting enzyme seems to resemble that of vertebrate thrombin. This is supported by the facts that bovine α-thrombin is able to clot the purified coagulogen.(Takagi, Morita,Miyata and Iwanaga, unpublished data). The most interesting finding,however,is a remarkable sequence homology of peptide C and A-chain with primate fibrinopeptide B. Especially, the last eight C-terminal residues of peptide C are very similar to those of Rhesus monkey fibrinipeptide(Wooding and Doolittle,1972),as shown in Fig 8.

Tachypleus tridentatus Coagulogen	
A-chain	Asp-Glu-Pro-Gly-Val-Leu-Gly-Arg
Peptide C	Gln-Glu-Ser-Gly-Val-Ser-Gly-Arg
Rhesus monkey F-B[a]	Glu-Glu-Ser-Pro-Phe-Ser-Gly-Arg
Human fibrinopeptide B[b]	Glu-Gly-Gly-Phe-Phe-Ser-Ala-Arg

Fig 8. Sequence homology among *Tachypleus tridentatus* coagulogen fragments and primate fibrinopeptides. a : Wooding and Doolittle,1972, b : Blombäck et al. 1966.

These facts suggest that horseshoe crab coagulogen and primate fibrinogen are derived from a common ancester or that the coagulogen is a prototype of primate fibrinogen. By these findings we have expected that B-chain of coagulogen might have sequence homology with either subunits, α-, β- or γ-chain of fibrinogen. The amino acid sequence of B-chain was determined by direct Edman method and analyses of its tryptic,α-chymotryptic,thermolytic and *Staphylococcal* protease digested peptides, as shown in Fig 9. The sequence of 129 residues of B-chain was completed. The sequence homology, with fibrins, that is what we expected, was found in only N-terminal several residues, as shown in Fig 10,the other part was quite different from fibrinogen.

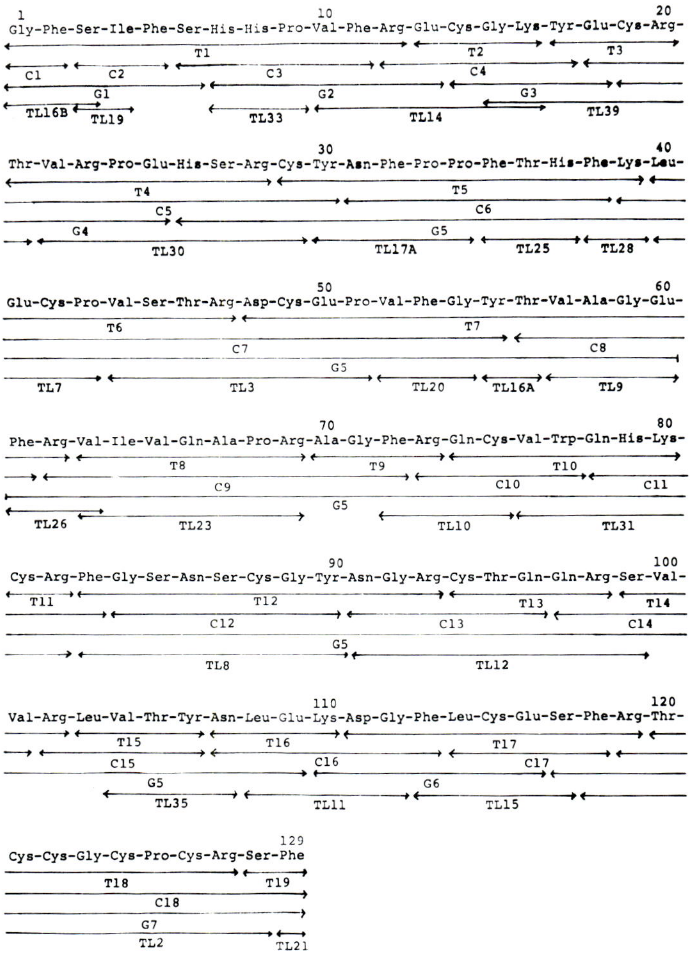

Fig 9. The amino acid sequence of B-chain. T=trypsin,C=chymotrypsin, G=*Staphylococcal* protease,TL=thermolysin. Details on amino acid sequence of B-chain will appear elsewhere (Takagi,Hokama,Morita,Iwanaga and Niwa,in preparation).

Human fibrin α-chain[a]	Gly-Pro-Arg-Val-Val-Glu-Arg---
Human fibrin β-chain[a]	Gly-His-Arg-Pro-Leu-Asp-Lys---
Chicken fibrin α-chain[b]	Gly-Pro-Arg-Ile-Leu-Glu-Asn---
Chicken fibrin β-chain[b]	Gly-His-Arg-Pro-Leu-Asp-Lys---
Tachypleus tridentatus Coagulin B-chain	Gly-Phe-Ser-Ile-Phe-Thr-His---

Fig 10. Comparison of B-chain with α- and β-chains of human and chicken fibrins. a : Iwanaga et al.1969. b : Murano et al. 1977.

Although the sequence homology of B-chain with vertebrate fibrinogen was not found,we could accomplish the sequence of coagulogen. Fig 11 shows the complete sequence of *Tachypleus tridentatus* coagulogen. Coagulogen has 16 cysteines,especially

1 Ala-Asp-Thr-Asn-Ala-Pro-Ile-Cys-Leu-Cys(10)-Asp-Glu-Pro-Gly-Val-Leu-Gly-Arg-↓Thr-Gln(20)-

Ile-Val-Thr-Thr-Glu-Ile-Lys-Asp-Lys-Ile(30)-Glu-Lys-Ala-Val-Glu-Ala-Val-Ala-Gln-Glu(40)-

Ser-Gly-Val-Ser-Gly-Arg-↓Gly-Phe-Ser-Ile(50)-Phe-Ser-His-His-Pro-Val-Phe-Arg-Glu-Cys(60)-

Gly-Lys-Tyr-Glu-Cys-Arg-Thr-Val-Arg-Pro(70)-Glu-His-Ser-Arg-Cys-Tyr-Asn-Phe-Pro-Pro(80)-

Phe-Thr-His-Phe-Lys-Leu-Glu-Cys-Pro-Val(90)-Ser-Thr-Arg-Asp-Cys-Glu-Pro-Val-Phe-Gly(100)-

Tyr-Thr-Val-Ala-Gly-Glu-Phe-Arg-Val-Ile(110)-Val-Gln-Ala-Pro-Arg-Ala-Gly-Phe-Arg-Gln(120)-

Cys-Val-Trp-Gln-His-Lys-Cys-Arg-Phe-Gly(130)-Ser-Asn-Ser-Cys-Gly-Tyr-Asn-Gly-Arg-Cys(140)-

Thr-Gln-Gln-Arg-Ser-Val-Val-Arg-Leu-Val(150)-Thr-Tyr-Asn-Leu-Glu-Lys-Asp-Gly-Phe-Leu(160)-

Cys-Glu-Ser-Phe-Arg-Thr-Cys-Cys-Gly-Cys(170)-Pro-Cys-Arg-Ser-Phe(175)

Fig 11. The complete amino acid sequence of *Tachypleus tridentatus* coagulogen.

5 residues are concentrated in C-terminal region, and all of them are joined with disulfide bonds. The positions of these disulfide bonds are now under investigation.

Our results on the structure of *Tachypleus tridentatus* coagulogen is quite different from the results of *Limulus* coagulogen obtained by Tai et al. Although complete amino acid sequence of *Limulus* coagulogen has not yet been determined, they showed that amino-terminal sequence was Gly-Asx-Pro (*Tachypleus* ; Ala-Asp-Thr), carboxyl-terminal was Ser (*Tachypleus*;Phe), peptide C was liberated from C-terminal (*Tachypleus* ; from near N-terminal), peptide C was 45 residues(*Tachypleus* ; 28), amino-terminal sequence of peptide C was Lys-Ile-Val (*Tachypleus* ; Thr-Gln-Ile) and the bond cleaved by clotting enzyme was Arg-Lys (*Tachypleus* ; Arg-Thr and Arg-Gly). These differences also reflected on their immunochemical properties (Shishikura and Sekiguchi,1977, Niwa,1977). The anti-serum to *Tachypleus* coagulogen did not form any prepitin line against *Limulus* coagulogen, whereas *Tachypleus gigas* and *Carcinoscorpius rotundicauda* coagulogens did cross-react with *Tachypleus tridentatus* coagulogen. These results may indicate the close phylogenic relationship among the Asian three species of horseshoe crab and may provide an interesting problem in molecular evolution.

EFFECTS OF OTHER PROTEASES ON COAGULOGEN

Trypsin has been known to make a gel with *Limulus* coagulogen(Solum,1970).It acts on the same bonds as *Tachypleus* clotting enzyme and also on the bonds in peptide C. So that, during the gel formation, small fragments of peptide C were released. Trypsin did not digested the gelled coagulin further even more than 24 h incubation. α-thrombin, blood coagulation Factor Xa and plasmin also act on coagulogen and make a gel. The bonds cleaved by these enzymes seem to be same as those by clotting enzyme judging from analyses of SDS-gel electrophreses. Other than these trypsin-like proteases, α-chymotrypsin and thermolysin were also effective on a gel formation with coagulogen. By SDS-gel electrophoreses of clotted protein, coagulogen was digested in three fragments by these enzymes. Alanine, serine and lysine were found as amino-terminal residues in chymotryptic gel protein. So that α-chymotrypsin probably splitted Phe(48)-Ser(49) and Phe(84)-Lys(85) bonds Fig 11). The amino-terminal residues of thermolytic gel protein were found to be alanine, valine and isoleucine. We do not yet know where these residue were derived from. From these results, the bonds, Arg(18)-Thr(19) and Arg(46)-Gly(47), cleaved by *Tachypleus* clotting enzyme were also susceptible to trypsin and trypsin-like enzymes, and as a result formed a gel. However, cleavage of these two bonds were not specific to the gel formation.

ACKNOWLEDGEMENTS

We thank Miss K.Fukunishi for her assistance in amino acid analyses. The authors are also indebted to Dr.M.Yoshida, Marine Laboratory of Okayama University and Dr.H.Nishii, Kasaoka.Okayama Pref. for the support in collection of horseshoe crabs. Helpful advices through this study by Dr. K.Takahashi, Primate Research Institute, Kyoto University are greatly appreciated.

REFERENCES

Blombäck B, Blombäck M, Edman P, Hessel B (1966). Human fibrinopeptides. Isolation, characterization and structure. Biochim Biophys Acta 115:371

Iwanaga S, Wallen P, Grondahl NJ, Henschen A, Blomback B (1969) On the primary structure of human fibrinogen, Isolation and characterization of N-terminal fragments from plasmic digests. Eur J Biochem 8:189.

Miller GL (1959). Protein determination for large numbers of samples. Anal Chem 31:964

Murano G, Waltz D, Williams L, Pindyck J, Mosseson MW (1977) Primary structure of the amino terminal regions of chicken fibrin chain. Thromb Res 11:1.

Nakamura S, Iwanaga S, Harada T, Niwa M (1976a). A clottable protein (coagulogen) from amoebocyte lysate of Japanese horseshoe crab(*Tachypleus tridentatus*). J Biochem 80:1011

Nakamura S, Takagi T, Iwanaga S, Niwa M, Takahashi T (1976b) A clottable protein (coagulogen) of horseshoe crab hemocytes. Structural change of its polypepetide chain during gel formation. J Biochem 80:649

Nakamura S, Takagi T, Iwanaga S, Niwa M, Takahashi T(1976c) Amino acid sequence studies on the fragments produced from horseshoe crab coagulogen during gel formation. Homologies with primate fibrinopeptide B. Biochem Biophys Res Comm 72:906

Niwa M (1977) Host defence in invertebrates: With special

reference to the coagulation system in horseshoe crabs. Proc 48th Annual Meeting Zool Soc Japan 86:551(in Japanese)

Shishikura F,Sekiguchi K (1977) Comparative studies on coagulogens from the four species of the horseshoe crab. ibid 86:539(in Japanese)

Solum NO (1970) Some characteristics of the clottable protein of *Limulus polyphemus* blood cells. Thromb Diathes Haemorrh 23:170

Tai JY,Seid RC Jr,Huhn RD,Liu T-Y (1977) Studies on Limulus amoebocyte lysate II. Purification of the coagulogen and the mechanism of clotting. J Biol Chem 252:4773

Wooding GL,Doolittle RF (1972) Primate fibrinopeptides: Evolutional significance. J Human Evol 1:553

COMPARATIVE STUDIES ON HEMOCYTES AND COAGULOGENS OF THE ASIAN AND THE AMERICAN HORSESHOE CRABS

Fumio SHISHIKURA and Koichi SEKIGUCHI

Institute of Biological Sciences,
the University of Tsukuba
Sakuramura, Ibaraki 300-31, Japan

A blood clotting mechanism has been found in the American horseshoe crab, *Limulus polyphemus* (Bang, 1956), which has a single type of hemocytes called amoebocytes with homogenous granules (Dumont *et al.*, 1966; Levin and Bang, 1968). Recently Mürer *et al.* (1975) have prepared the granules by sucrose gradient centrifugation and demonstrated the presence of all the clotting factors in the granule extract obtained by freezing and thawing.

The clotting mechanism has also studied with the Japanese species, *Tachypleus tridentatus* (Kobayashi and Yamamoto, 1974; Niwa *et al.*, 1974). The classification of hemocytes in the Tachypleus blood, however, has not been established. Only a simple discription of the hemocytes was reported previously (Ohuye, 1938), suggesting two kinds of granules in a hemocyte. Therefore, the clotting system proposed in the Limulus hemocytes that all the clotting factors are present in one kind of granules in a single type of hemocytes may not be applicable to the Tachypleus hemocytes. These criteria lead us to study the different characteristics between these animals.

The present communication describes, first, the classification of the Tachypleus hemocytes, and second, the localization of coagulogen in the hemocytes by a fluorescent antibody technique, and last, the biochemical and immunological characteristics of coagulogens prepared from the four species of horseshoe crabs, *Tachypleus tridentatus*, *Tachypleus gigas*, *Carcinoscorpius rotundicauda*, and *Limulus polyphemus*.

Pages 185–201, Biomedical Applications of the Horseshoe Crab (Limulidae)

L. polyphemus was found to be very different from the other three species in some characteristics studied, which suggested the phylogenetic relationship among these animals.

MATERIALS AND METHODS

Horseshoe crabs

Tachypleus tridentatus, a Japanese horseshoe crab, was collected at Kasaoka, Okayama prefecture and Imari, Saga prefecture in Japan. *Tachypleus gigas* and *Carcinoscorpius rotundicauda*, two Southeast Asian species, were collected at the Gulf of Siam, Thailand and transported to Japan. *Limulus polyphemus*, the American species, was obtained from Duke University Marine Laboratory, Beufort, North Carolina, and Marine Biological Laboratory, Woods Hole, Massachusetts. These animals were maintained in running sea water at Shimoda Marine Research Center of the University of Tsukuba.

Preparation of Blood Smears

To avoid the initiation of clotting, the animals were precooled for 12 hr at 4 °C in sea water. All glassware were baked at 200 °C for 5 hr. Hemolymph of the horseshoe crab was withdrawn directly into a sterile disposable syringe (2 ml) by inserting an 18 gauge needle at the joint between the thoracic and abdominal segments. The joint area had been cleaned with 100 % ethylalcohol in advance. The smears were fixed for about 20 min at 4 °C with vapord formalin or osmic acid in a chamber.

Cytological and Histochemical Stainings

Wright's eosin methylenblue solution (Merk Chemical Co.) was used according to the procedures of Sano (1970). Histochemical stainings for crabohydrates were carried out essentially according to the methods described by Culling (1974). The methods include Schiff reaction, periodic acid-Schiff (PAS) reaction, alcian blue (pH 1.0 and 2.5)-PAS reaction, Hale's colloidal iron, methylation-Hale's colloidal iron, methylation-saponification-Hale's colloidal iron, and Sudan black B. Dimedon-PAS reaction was performed by the method of Sano (1970). Toluidine blue O staining was carried out at pH 8.5 which were adjusted with 0.05 M citric acid and 0.25 M sodium phosphate (dibasic) solution.

Preparation of Hemocyte Lysate

Blood of the horseshoe crabs was obtained by sterile disposable syringes (30 ml) with an 18 gauge needle inserting the same joint area as above. The syringes with the blood inside were left standing with the needle upward for about 20 hr in a cold room at 4 °C, and the supernatant (cell-free plasma) was pushed out from the syringes. The hemocyte was left in them. The lysis of hemocytes was performed by suspending the cells in sterile distilled water and by subsequent homogenizing in an ice bath. Hemocyanin and cell debris were promptly removed by centrifuging the homogenate at 100,000 x g for 5 hr at 4 °C (Marchalonis and Elelman, 1968). The supernatant (hemocyte lysate) was used for the preparation of coagulogen and clotting enzyme.

Preparation of Coagulogen and Clotting Enzyme

Coagulogen and clotting enzyme were prepared by means of DEAE Sephadex A-50 column chromatography, Sephadex G-75 column chromatography, and finally preparative polyacrylamide gel electrophoresis in acid or in alkaline buffer. Immediately after the centrifugation of lysed hemocytes, the hemocyte lysate was applied on a DEAE Sephadex A-50 column (2.5 cm x 20 cm) which was equilibrated with 0.02 M Tris-HCl buffer, pH 8.0 (Sullivan and Watson, 1975), and eluted with Tris-HCl buffer to get a crude coagulogen fraction. The fraction was dialyzed for two days against distilled water, and then lyophilized. The lyophilized crude coagulogen was fractionated by Sephadex G-75 column (2.0 cm x 65 cm) chromatography with the same buffer. The final procedure of purification of coagulogen was carried out by polyacrylamide gel electrophoresis in acid buffer (Brummel and Montgomery, 1970) using preparative gels (2.0 cm x 10 cm). The purity of the coagulogen was tested by a sodium dodecylsulfate SDS gel electrophoresis after the method essentially similar to those described by Weber and Osborn (1969). To identify the coagulogen in the fractions a 0.2 ml of the clotting enzyme oftained from *T. tridentatus* or the same volume of trypsin (Sigma Chemical Co.) was added.

On the other hand, when the absorbance at 280 nm of the eluted fraction from the DEAE Sephadex A-50 column decreased under 0.1 Tris-HCl buffer containing 0.3 M NaCl was applied to the column. A crude clotting enzyme fraction could be obtained and the fractionation of crude clotting enzyme was carried out by polyacrylamide gel electrophoresis in alkaline buffer (Davis, 1964) using preparative gels (2.0 cm x 10 cm) at 4 °C with a low current. On the process of purification of the clotting enzyme, heat-treated Pre-gel (Teikoku Hormone MFG, Co., Japan), in which the clotting enzyme was completely inactivated, was used as a substrate for the enzyme (for 60 min at 80 °C; Sullivan and Watson, 1975).

Clot formation by the clotting enzyme or trypsin is usually detected by measureing the turbidity of the reacted mixture (Young *et al.*, 1972) or by decanting (Levin and Bang, 1968). Observation of the gel formed on a slide glass was carried out by a light microscope under dark field illumination.

Immunological Techniques

Rabbit anti-coagulogen serum was obtained by a single injection of the coagulogen (50 μg) purified from the Pregel emulsified with complete Freund's adjuvant (Difco Laboratories). Anti-coagulogen of *L. polyphemus* serum was prepared in the same way. The animals were bled five weeks after priming and the sera were prepared. The immunological characteristics of these sera were examined by Ouchterlony's double diffusion test (Ouchterlony, 1953). Agar plates were prepared from 0.6 % Bactoagar (Difco Laboratories)-0.6 % Agarose (Nakarai, Co., Tokyo) in 0.01 M sodium barbital buffer (pH 8.2) containing 0.01 % NaN_3 as a preservative.

Immunohistological Techniques

Gamma globulin was prepared from the anti-Tachypleus coagulogen serum by ammonium sulfate fractionation at 0 - 40 % saturation, which was repeated twice. Conjugation of fluorescein isothiocyanate (FITC; BBL. Co.) with the gamma globulin was carried out by the methods of Riggs *et al.* (1958). The free dye was removed by dialyzing against a phosphate buffered saline (PBS; 0.01 M phosphate, pH 7.4, 0.145 M NaCl) and by column chromatography on Sephadex G-25

(Lipp, 1961). In order to reduce the non-specific absorption the FITC-conjugated gamma globulin was employed column chromatography on DEAE-Cellulose (DE 52, Whatmann. Co.) by the method of Goldman (1968).

One ml of the Tachypleus blood was withdrawn into a syringe containing absolute ethylalcohol (19 ml) precooled at 4 °C, and then the hemocytes were post-fixed with aceton at -60 °C for 1 hr. The fixed hemocytes blocks (about 4 mm x 4 mm) were dehydrated, cleared, and finally embedded in paraffin and hemocyte sections (4 μm in thickness) were prepared. After deparaffinization they were incubated with the FITC-conjugated antibody at a concentration suitable for specific staining for 20 min at 18 °C in a moist chamber. The excess FITC-conjugated antibody was removed by washing with PBS and the reaction was mounted in buffered glycerol (50 % glycerol in PBS). The stained sections were observed under a Nikon fluorescence microscope (Type FL) with illumination by a high voltage mercury lamp. Control sections were stained with FITC-conjugated antibody absorbed at a slight antigen-excess condition.

Protein Determination

Protein concentrations were determined by the Folin method (Lowry *et al.*, 1951). Absorbance at 280 nm was also used for this purpose.

Apparent Molecular Weight Determination

Molecular weight of the coagulogen was determined by the mobility on the SDS-gel. Bovine serum albumin (Sigma Chemical Co., 68,000 dalton MW), ovalbumin (Sigma Chemical Co., 43,000 dalton MW), and lysozyme (Sigma Chemical Co., 14,300 dalton MW) were used as marker proteins.

RESULTS AND DISCUSSION

The hemocyte of *T. tridentatus* begin to degrade as soon as they are withdrawn from the body as already known with the hemocytes of *L. polyphemus* (Dumont *et al.*, 1966; Solum, 1970) and those of other invertebrates (Ravindranath, 1974a and b). One example before degradation process is shown in

Fig. 1. The hemocytes were round or ellipsoidal and no sign of degradation occurred in all cells. The granules in the cells were rather stationary and the nuclei failed to be seen, being covered with the granules. When blood was prepared without sterilized conditions, the symton of degradation began with swelling of the cell and pseudopodia appeared in any directions, and the hemocytes began to degrade within a minute as reported in the Limulus hemocytes (Dumont *et al.*, 1966; Solum, 1970). In the degradation process, there was one cell type which degraded rather late when compared with others. This cell type may correspond to the so-called stable hemocyte observed by Grégoire (1955). To observe the morphological characteristcs of these hemocytes before they begin to degrade is very difficult. Then, the observation of hemocytes by the cytological and histochemical stainings will be performed to find the different characteristics among hemocytes and to classify them.

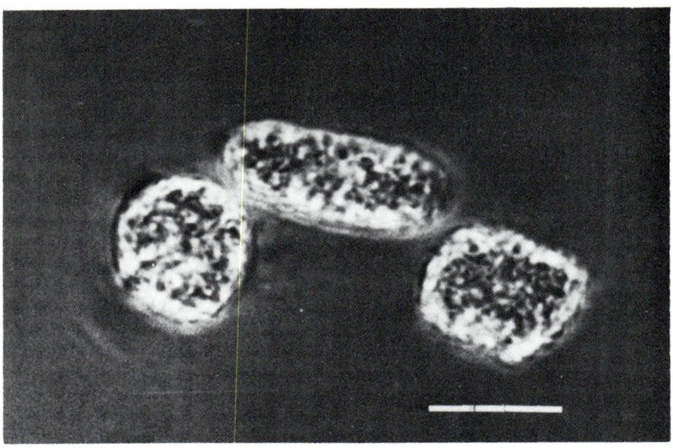

Fig. 1. The Tachypleus hemocytes by phase contrast microscope. After the blood was withdrawn into sterile syringe, a drop of it was placed on a slide glass and observed within 15 sec. Bar indicates 10 microns.

Granules in the invertebrate hemocytes are considered to be suitable criteria for the classification of the cells (Jones, 1960). To describe the properties of the granules the staining with toluidine blue O at pH 8.5 was tried and found to be very useful for this purpose, which is shown in Fig. 2-a. By this staining two kinds of granules were found, one was called bacilliform granules which was stained orthochromatically and the other stained very faintly. Consequently the cells could be classified into two types depending on which kind of granules they contained. The hemocytes which contain bacilliform granules referred to type A cell and those which contained faintly stained granules to type B. Type A cell was a minor population and type B a major one, about 10 % and 90 %, respectively.

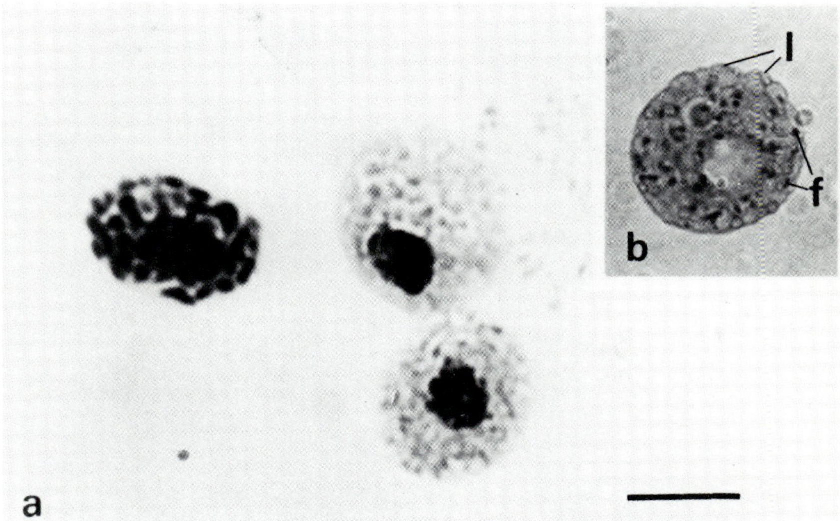

Fig. 2-a. Two types of Tachypleus hemocytes stained with toluidine blue O, at pH 8.5. The cell types observed were one (type A cell) containing bacilliform granules stained orthochromatically, the others (type B cell) containing large granules which were very faintly stained. 2-b. Wright's eosin methylen blue staining of type B cell. One is the fine eosinophilic granules, the other is the large rounded granules not stained. The latter ones are closely packed in the cell. Bar indicates 10 microns.

When Wright' eosin methylenblue solution was used, the other kind of granules shown in Fig. 2-b was found in the type B cell. The granules stained eosinophilic by this solution were named as fine granules. The other granules which closely packed in type B cell were not stained by this solution. The type B cell conatining two kinds of granules seems to be the hemocytes that observed by Ohuye (1938). So far we found three kinds of granules in the Tachypleus hemocytes; bacilliform granules, large rounded granules, and fine granules. Histochemical stainings on these granules were carried out and summerized in Table 1. The bacilliform granules were demonstrated a neutral glycoprotein.

TABLE 1

Histochemical tests of granules in the Tachypleus hemocytes

Histochemical tests	Fine granules	Large granules	Bacilliform granules
Schiff reaction	-	-	-
PAS	-*	-*	+
Acetyl./PAS	-	-	-
Salive/PAS	-*	-*	+
Dimedone/PAS	-	-	-
Alcian blue,pH 1.0/PAS	-	-	-
Alcian blue,pH 2.5/PAS	-	-	-
Hale	-	-	-
M/Hale	-	-	-
M/S/Hale	-	-	-
Sudan black B	-	-	-

*, Stained very faintly; +, positive reaciton; -, negative reaction; M, methylation; S, saponification.

On the contrary, the Limulus hemocytes contain a single kind of granules and recently the protein responsible for the clotting mechanism were demonstrated in them (Mürer *et al.*, 1975). As described above, the Tachypleus hemocytes contained two types of hemocytes containing three kinds of granules in all. This situation tempted us to elucidate the localization of the clotting factors in the hemocytes or in the granules, and to make sure of the difference between *Tachypleus* and *Limulus*.

Coagulogen was prepared by means of twice-repeated preparative polyacrylmide gel electrophoresis shown in Fig. 3. Hemocyte lysate (Pre-gel) of *T. tridentatus* had many protein components, two major and many other minor. The coagulogen was converted into gel by the addition of either clotting enzyme or trypsin.

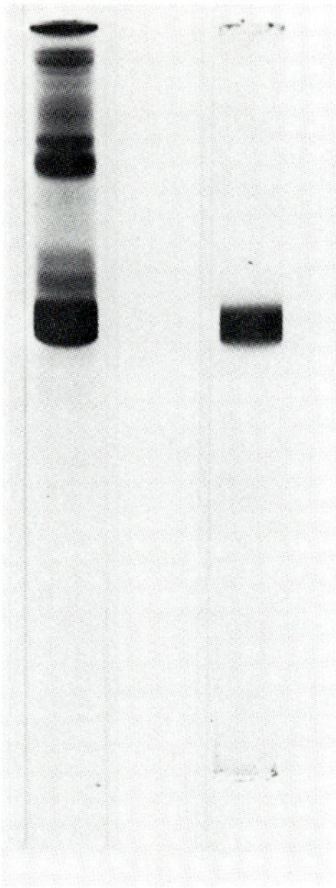

Fig. 3. Polyacrylamide gel electrophoresis of the hemocyte lysate (Pre-gel) and the purified coagulogen. 8 % polyacrylamide gel electrophoresis was carried out in 0.87 M acetic acid. From Shishikura *et al.* (1977), by permission of the publisher, The Wistar Institute Press, Philadelphia.

For the purpose of studying the locality of the coagulogen the high specificity of the serum is one of the most important prerequisitives. The purity of the protein was examined by the rabbit anti-coagulogen serum. The experiment was performed by combining Ouchterlony's double diffusion test with polyacrylamide gel electrophoresis (Fig. 4). After the elecrophoresis of the hemocyte lysate (Pre-gel), the gel was sliced and reacted with the anti-Tachypleus coagulogen serum in agar plates. Pure coagulogen was used as a reference. As shown in Fig. 4-b the reactions gave rise to a precipitine line which was confluent with the line formed by the reference, suggesting the high specificity of the antiserum. The anti-coagulogen serum, therefore, can use to elucidate the localization of the coagulogen in hemocytes and may bring a correct information on the localization of coagulogen.

The Tachypleus hemocytes fixed with cold ethanol were

stained with FITC-conjugated anti-Tachypleus coagulogen antibody. Most of the hemocytes were more or less stained with the antibody.

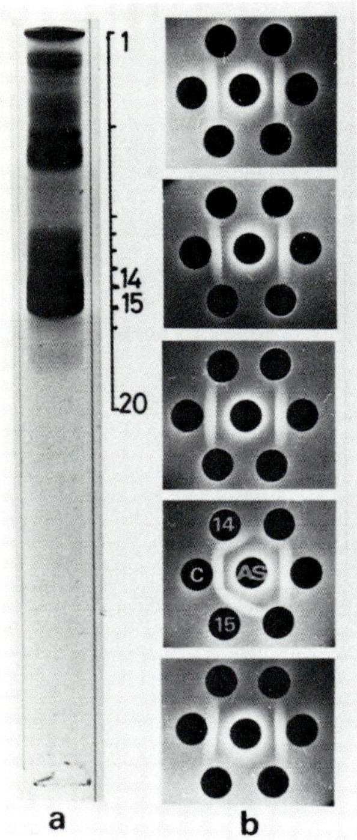

Fig. 4. Specificity of anti-Tachypleus coagulogen serum and detection of the antigen band in polyacrylamide gel. The hemocyte lysate (Pre-gel) was electrophoresed. The gel was cut into slices of 2 mm in thickness. Nuber of gel slices correspond to the number of wells. As a reference, the purified coagulogen was placed in bilateral wells. The antigen in the wells of Nos. 14 and 15 formed a Precipitine line which was confluent with the one formed by the coagulogen. AS, anti-Tachypleus coagulogen serum; C, coagulogen.

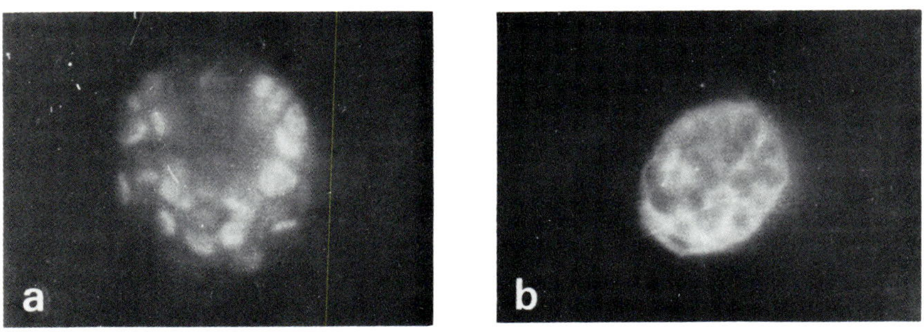

Fig. 5. Fluorescence micrographs of the Tachypleus hemocytes stained with FITC-conjugated antibody. From Shishikura *et al.* (1977), by permission of the publisher, The Wistar Institute Press, Philadelphia.

However, the close inspection revealed that there were two types of staining patterns. In one type the granules were stained strongly but the cytoplasmic matrix was not (Fig. 5-a) and in the other the cytoplasmic matrix of the hemocyte was stained. The granules which looked like the large granules were not stained (Fig. 5-b). The intensity of fluorescence in the latter type varied from cell to cell, probably depending on the concentration of coagulogen. Control experiments using FITC-conjugated anti-Tachypleus coagulogen antibody absorbed with the purified coagulogen failed to stain.

With *L. polyphemus*, the homogenous granules contained all the factors indispensable to the clot formation including coagulogen. Therefore, the granules in the Limulus hemocytes may be homologous ones in type A of the Tachypleus hemocytes localized the coagulogen. The Japanese species, moreover, have another type of cells which was demonstrated the coagulogen in the cytoplasmic matrix. The observation also supported the classification of the Tachypleus hemocytes. To investigate the different characteristics of coagulogens, the biochemical and immunological studies were carried out among the four species of horseshoe crabs.

Hemocyte lysate of the four species of horseshoe crabs were dissolved in 0.87 M acetic acid which inactivated the clotting enzyme, and were electrophoresed in the acid buffer (Fig. 6). Hemocyte lysates of *T. gigas* and *C. rotundicauda* showed, among many bands, one haviest band which had shown almost the same electrophoretic mobility at that of the coagulogen already known of *T. tridentatus* and that of *L. polyphemus*. With all the four species the coagulogen may be the main component in the hemocyte lysate.

Crude coagulogen prepared by means of DEAE Sephadex A-50 column chrmoatography was demonstrated a decernible contamination except for coagulogen on acid gel. These proteins were not retained on the column, which were poorly separated by the following Sephadex G-75 column chromatography. The proteins eluted from the Sephadex G-75 column was concentrated and finally purified by acid preparative polyacrylamide gel electrophoresis which was repeated twice. Each coagulogen obtained finally is shown in Fig. 7 on acid gels (a) and on SDS gels (b). Mobilities of the coagulogens on both electrophoretic systems were found to be the same even when the mixed coagulogen of any combination were electrophoresed. The migration distances of coagulogens were identical to each other, suggesting the same apparent molecular weight (20,000).

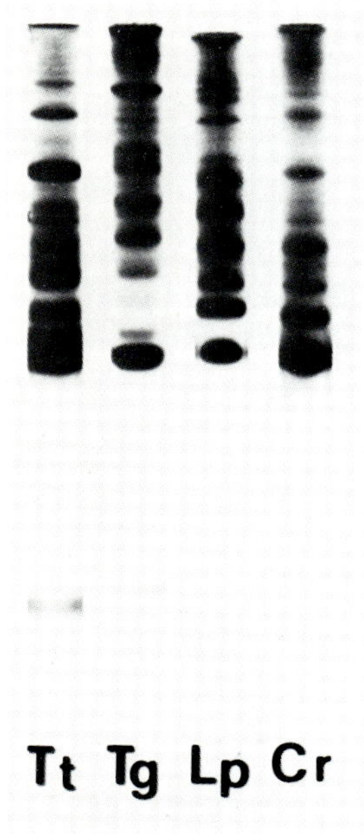

Fig. 6. Electrophoretic patterns of hemocyte lysates from the four species of horseshoe crabs in acid polyacrylamide gel (8 %). Tt, *T, tridentatus;* Tg, *T, gigas;* Lp, *L. polyphemus*; Cr, *C. rotundicauda.*

The crude Tachypleus clotting enzyme obtained form the DEAE Sephadex A-50 column by an application of Tris buffered saline containing 0.3 M NaCl was partially purified by a preparative polyacrylamide gel electrophoresis in alkaline buffer. The fraction was composed of two components with closely similar mobilities on the gel and was proved to convert the 80 °Ctreated Pre-gel into gel. The Tachypleus enzyme fraction was converted not only the coagulogen but aslo those prepared form other three species (Table 2).

The molecular weight of the Limulus coagulogen has been reprted to be 24,500 (Tai *et al.*, 1977) and that of the Tachypleus coagulogen to be about 20,000 (Nakamura *et al.*, 1975; Shishikura *et al.*, 1977). Under the same conditions the four coagulogens in both systems were found to be identical.

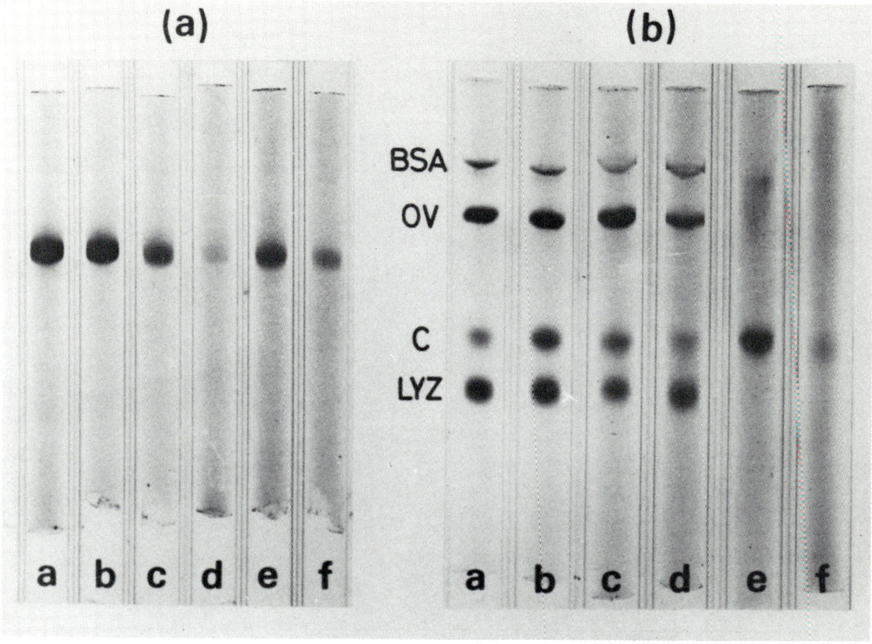

Fig. 7. Comparison of electrophoretic mobility of purified coagulogens in acid polyacrylamide gels (a) and in SDS polyacrylamide gels (b). a, *T. tridentatus*; b, *T. gigas*; c, *L. polyphemus*; d, *C. rotundicauda*; e, a mixed sample a, b, c, and d; f, the same sample of e diluted twice. BSA, bovine serum albumin; OV, ovalbumin; C, coagulogen; LYZ, lysozyme. From Shishikura and Sekiguchi (1978), by permission of the publisher, The Wistar Institute Press, Philadelphia.

On the contrary, the immunological study has presented an interesting results. Fig. 8-a shows that the anti-Tachypleus coagulogen serum formed a heavy precipitine line reacting with the injected antigen. This line was partially identical with the lines formed by the coagulogens of *T. gigas* and *C. rotundicauda*. The antiserum did not form any precipitine line reacting with the coagulogen of *L. polyphemus*. Fig. 8-b shows that the immunological reactions between anti-Limulus coagulogen serum and the four antigens, indicating that the antiserum is to be very specific for the injected antigen and that it does not react with any coagulogen from other three species.

TABLE 2

Clot formation of coagulogens with the partially purified clotting enzyme from *Tachypleus tridentatus*

Substrate (0.1 ml)	Enzyme (ml)	Endotoxin (1 mg/ml)	Clot
80 °C-treated lysate	0.1	0.05	+
80 °C-treated lysate	-	0.05	-
80 °C-treated lysate	0.1	-	-
Coagulogen from *T. tridentatus*	-	0.05	-
Coagulogen from *T. tridentatus*	0.1	0.05	+
Coagulogen from *T. gigas*	-	0.05	-
Coagulogen from *T. gigas*	0.1	0.05	+
Coagulogen from *C. rotundicauda*	-	0.05	-
Coagulogen from *C. rotundicauda*	0.1	0.05	+
Coagulogen from *L. polyphemus*	-	0.05	-
Coagulogen from *L. polyphemus*	0.1	0.05	+

The concentration of coagulogens used were as follows; *T. tridentatus* (140 μg/ml), *T. gigas* (130 μg/ml), *C. rotundicauda* (100 μg/ml), and *L. polyphemus* (140 μg/ml). The lysate (Pre-gel) was dissolved in 1 ml of pyrogen free distilled water. The concentration of enzyme was 410 μg/ml.

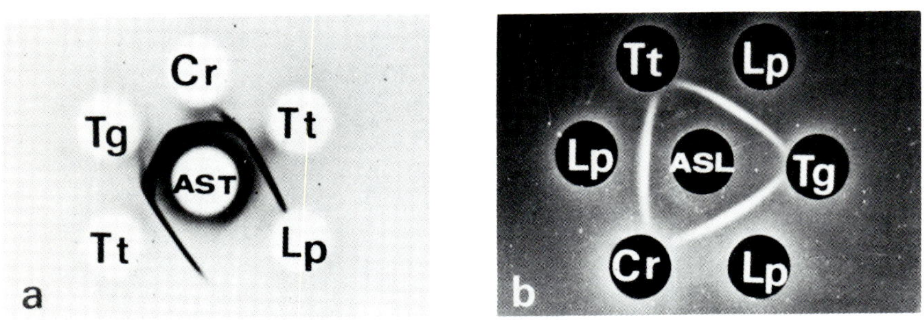

Fig. 8. Immunological comparison of coagulogens using anti-Tachypleus coagulogen serum (a), and anti-Limulus coagulogen serum (b). From Shishikura and Sekiguchi (1978), by permission of the publisher, The Wistar Institute Press, Philadelphia.

The validity of the enzyme for the foreign substrate is considered to have wide ability as even trypsin can convert the coagulogen into a gel in this clotting systems.

The Southeast Asian Species were closely related one another but were very different from the American species, suggesting the phylogenetic relationship. That is, *Limulus polyphemus* belongs to the Subfamily different from *Tachypleus tridentatus*, *Tachypleus gigas*, and *Carcinoscorpius rotundicauda*. Such a relationship considered immunologically as above is also found when the relationship is considerec from the stand point of the hemocyanin molecule (Sugita and Sekiguchi, unpublished data) and the serum proteins (Shuster, 1962). We hope that the analysis of these proteins are carried out to understand in more detail phylogenetic relationship among these animals.

ACKNOWLEDGMENT

We wish to express our sincere thanks to Associate Professor Tamio HIRABAYASHI, the University of Tsukuba, for his earnest guidance and invaluable advice in the course of this work.

REFERENCES

Bang FB (1956). A bacterial desiase of *Limulus polyphemus*. Bull Johns Hopkins Hosp 98:325.

Brummel MC, Montgomery R (1970). Acrylamide gel electrophoresis of the S-Sulfo derivatives of fibrinogen. Anal Biochem 33:28.

Culling CFA (1974). "Handbook of histopathological and histochemical techniques. London: Butterworths.

Davis BJ (1964). Disc electrophoresis II. Method and application to human serum proteins. Ann N Y Acad Sci 121: 404.

Dumont JN, Anderson E, Winner G (1966). Some cytologic characteristics of the hemocytes of Limulus during clotting. J Morph 119:181.

Goldman J (1968). "Fluorescent antibody methods." New York and London: Academic press.

Grégoire C (1955). Blood coagulation in arthropods. VI. A study by phase contrast microscopy of blood reactions in vitro in Onychophora and in various groups of arthropods.

Kobayashi M, Yamamoto M (1974). Preparation of Pre-gel from the amoebocytes of Japanese horseshoe crab (*Tachypleus tridentatus*) and its specificity. Yakugaku Zasshi 94:111.
Levin J, Bang FB (1968). Clottable protein in Limulus. Its localization and kinetics of its coagulation by endotoxin. Thromb Deathes Haemorrh 19:186.
Lipp WJ (1961). Use of gel filtration and polyethylene glycol in the preparation of fluorochrome-labelled proteins. J Histochem 9:485.
Lowry OH, Rosebrough NJ, Farr AL, Randall RJ (1951). Protein measurement with the Folin phenol reagent. J Biol Chem 193:265.
Marchalonis JJ, Edelman CM (1968). Isolation and characterization of a hemagglutinin from *Limulus polyphemus*. J Mol Biol 32:453.
Mürer EH, Levin J, Holme R (1975). Isolation and studies of the granules of the amoebocytes of *Limulus polyphemus*, the horseshoe crab. J Cell Phisiol 86:533.
Nakamura S, Takagi T, Iwanaga S, Niwa M, Takahashi K (1976). A clottable protein (coagulogen) of horseshoe crab hemocytes. Structural change of its polypeptide chain during gel formation. J Biochem 80:649.
Niwa M, Hiramatsu T, Waguri O (1974). The gelation reaction between horseshoe crab amoebocytes and endotoxins studied by the quantitative clot protein method. Jap J Med Sci Biol 27:108.
Ohuye T (1938). On the coelomic corpuscles in the body fluid of some invertebrates. X. Some morphological and histological properties of the granulocytes of various invertebrates. Sci Rep Tohoku Imp Univ Biol 13:593.
Ouchterlony O (1953). Antigen-antibody reactions in gel. IV. Types of reactions in coordinated system of diffusion. Acta Pathol Microbiol Scand 32:231.
Ravindranath MH (1974a). The hemocytes of a scorpion Palamnaeus swammerdami. J Morph 144:1
Ravindranath MH (1974b). The hemocytes of an isopod Ligia exotica Roux. J Morph 144:11.
Riggs JL, Seiwald RJ, Burchhalter JH, Downs CM, Metcalf TC (1958). Isothiocyanate compound as fluorescent labelling agents for immune serum. Am J Pathol 34:1081.
Sano Y (1971). "Histological technics. Theoretical and applied" Tokyo: Nanzando.
Shishikura F, Chiba J, Sekiguchi K (1977). Two types of hemocytes in localization of clottable protein in Japanese horseshoe crab, *Tachypleus tridentatus*. J Exptl Zool 201:303.

Shishikura F, Sekiguchi K (1978). Comparative study on horseshoe crabs coagulogens. J Exptl Zool in publication.
Shuster Jr CN (1962). Serological correspondence among horseshoe "crabs" (Limulidae). Zoologica 47:1.
Solum NO (1970). Coagulation in Limulus-some properties of the clottable protein of *Limulus polyphemus* blood cells. In Macfarlane RG (ed): "The haemostatic mechanism in man and other animals" London and New York: Academic press, p 207.
Sullivan JD, Watson WS (1975). Purification and properties of the clotting enzyme from Limulus lysate. Biochem Biophys Res Commun 66:848.
Tai J, Seid Jr RC, Huhn RD, Liu TY (1977). Studies on Limulus amoebocyte lysate II. Purification of the coagulogen and the mechanism of clotting. J Biol Chem 252:4773.
Weber L, Osborn M (1969). The reliability of molecular weight determination by dodecylsulfate polyacrylamide gel electrophoresis. J Biol Chem 244:4406.
Young NS, Levin J, Prendergast RA (1972). An invertebrate coagulation system activated by endotoxin: Evidence for enzymatic mediation. J clin Invest 51: 1790.

BINDING STUDIES OF RADIOACTIVE LIPOPOLYSACCHARIDE WITH LIMULUS AMEBOCYTES.

F. R. Rickles, P. D. Rick, P. B. Armstrong and J. Levin

VA Med Ctr, Newington,CT;UCONN Health Ctr,Farmington,CT; U Med Serv Sch of Health Sci, Bethesda,MD; U of CA, Davis,CA; The Johns Hopkins U & Hosp, Baltimore,MD; & The Marine Biol Lab, Woods Hole, MA.

The pioneering work of Bang in the 1950's and Levin and Bang in the 1960's at the Marine Biological Laboratory led to the description of a complex defense system in Limulus polyphemus for dealing with an environment filled with gram-negative organisms. (Bang, 1956; Levin and Bang, 1964a; Levin and Bang, 1964b). This system, involves both exocytosis and a sequence of zymogen-enzyme-mediated steps leading ultimately to the formation of a semi-solid gel, whose function it is to immobilize bacteria.

The present study focuses on the first step in this sequence - that is, the interaction of lipopolysaccharide (LPS), with the amebocyte membrane. We have reasoned that this system may serve as a model for the interaction of LPS with a variety of cell membranes and also should allow a better understanding of marine cellular defense mechanisms.

METHODS

In order to prevent cellular aggregation following cardiac puncture, Limulus blood was collected into warm N-ethylmaleimide to a final concentration of 5×10^{-3}M. To avoid exocytosis and gelation in the presence of LPS, the cells were lightly fixed in 30mM formaldehyde. These cells retained the essential appearance of native cells, although the nucleus was somewhat more prominent than in unfixed cells. The cells remain granulated in 30mM formaldehyde indefinitely. The cells were then washed or dialyzed to remove excess formaldehyde and used for binding experiments.

We prepared highly purified, rough lipopolysaccharide

Pages 203—207, Biomedical Applications of the Horseshoe Crab (Limulidae) from the mutant of Salmonella typhimurium G-30, which lacks the enzyme UDP-galactose-4-epimerase and, thus, cannot synthesize galactose (Rickles and Rick, 1977). This preparation, prepared by hot phenol extraction, contained no contaminating lipid A related protein (as determined by polyacrylamide gel electrophoresis) and was labelled with tritium (^3H) by the dry gas method of Wiltzbach and further purified by agarose 6B chromatography (Tomasulo et al., 1975). Endotoxin activity was determined by activation of human mononuclear cell tissue factor and a single major peak of activity corresponding to radioactivity was observed. Analysis of Peak I from the agarose column using paper chromatography and a butanol:ammonium hydroxide solvent system revealed a primary peak of ^3H-G30 in the same position as a ^{14}C-G30 standard, kindly supplied by Dr. Mary Jane Osborn of the University of Connecticut Health Center. This highly purified material was used for binding experiments with the NEM-formaldehyde fixed amebocytes. All materials were handled aseptically and all glassware was rendered pyrogen-free by heating to 180°C for 4 hours. Cells were incubated with the LPS in the presence of 10% native plasma for varying periods of time and then the cells were pelleted by centrifugation, washed 4 times with 3% Nacl and either solubilized for liquid scintillation counting, boiled in 1% SDS for polyacrylamide gel electrophoresis or placed on polylysine-coated glass slides for autoradiography.

RESULTS AND DISCUSSION

A cyclical pattern of so-called "loading and unloading" was observed with the amebocytes and is typical of the pattern seen in mammalian cell systems. It has been postulated by others that this phenomenon may represent shedding of LPS-receptor complexes at least as applied to murine spleen cells (Truffa-Bachi et al., 1977). Alternative explanations for this cyclic fluctuation, seen also in the interaction of LPS with human erythrocytes, rabbit erythrocytes and murine phagocytes, involve: (1) the physical heterogeneity of LPS molecules presented to the cell surface; (2) alteration of charge-charge interaction secondary to LPS induced spatial rearrangements in the cell membrane; or, (3) activation of membrane-bound hydrolytic enzymes initiated by adsorption of LPS. (Bona et al., 1976; Truffa-Bachi et al., 1977; Davies et al., 1978). Regardless, all subsequent experiments were performed after a one hour incubation to minimize this "unloading" phenomenon and at 4°C to maximize binding and minimize internalization.

Using the conditions described, 7 experiments were performed which revealed 13-20% uptake of the radioactive LPS by the amebocytes.

TABLE I. BINDING OF ^{3}H-G30 LPS TO LIMULUS AMEBOCYTES

EXPERIMENT NO.	DPM BOUND	% RADIOACTIVITY BOUND
1	4363	19.8
2	4359	19.9
3	2819	13.9
4	2704	13.2
5	3017	14.8
6	4167	20.0
7	3453	16.6
	$\bar{x}\pm1$ SEM=3554 ±259 DPM	$\bar{x}\pm1$ SEM=16.9±1.1%

This data is quite similar to what has been observed by others with mammalian cells. Heterologous, unlabeled LPS from *E. coli* was capable of inhibiting approximately 30-50% of the uptake of radioactivity by the cells if used in 100 fold excess concentration. Pulse chase experiments with cold LPS revealed a similar reduction in radiolabeled LPS in association with the cell pellet.

Polymyxin B is a cyclic peptide antibiotic which is bactericidal for most gram-negative organisms. PB is also known to bind avidly to the lipid A region of LPS (Morrison and Jacobs, 1976) and is capable of abrogating most, if not all, of the biological activities of LPS. Using polymyxin at 100-fold excess concentration, approximately 30-50% inhibition of LPS binding to amebocytes was demonstrated. It should be noted, however, that this concentration of PB relative to LPS is 10-100x greater than is required to inhibit other biological functions of LPS. In previous studies from our laboratory, PB has been shown to inhibit 100% of human monocyte tissue factor activation by LPS at 10 x less the concentration (Rickles and Rick, 1977). Similar results have been reported for inhibition of the mitogenicity of LPS for mouse spleen cells and activation of both human and rabbit complement. (Morrison and Jacobs, 1976; Wood and Cameron, 1978). This observation suggests that the lipid A end of the molecule may not be critical for the interaction of LPS with the amebocyte.

In an attempt to demonstrate saturation of putative binding sites, increasing concentrations of LPS within a fairly narrow dose range were utilized. Although a progressive decrease in the % of added radioactivity bound was demonstrated, at the highest concentration a large increase in cell-associated radioactivity was observed. This latter observation may have been related to the relative state of aggregation of the LPS molecules presented to the membrane and/or to endocytosis of the particulate LPS. Further studies over a broader concentration range are needed to confirm this observation.

In summary, we have demonstrated interaction of LPS with the amebocyte membrane and with some provisos have established this system as a model for LPS interaction with a variety of cells. Results of autoradiographic experiments should provide further information regarding plasma membrane permeability to LPS and SDS - acrylamide gel experiments of the cell-LPS complex may provide evidence for a specific protein receptor for LPS as has been suggested by Liu and his colleagues and described in detail elsewhere in this symposium. Preliminary results of experiments with the phospholipid phosphatidyl ethanolamine and the organic solvent hexane, however, suggest that "binding" may be less relevant than intercalation of LPS into the lipid bilayer of the cell membrane. Both of these materials appear to interfere with "LPS binding".

Further studies with purified LPS components may also provide a better understanding of the amebocyte degranulation process whereby Limulus insulates itself against the effects of LPS.

REFERENCES

1. Bang FB (1956). A bacterial disease of Limulus polyphemus. Bull Johns Hopkins Hosp 98:325-351.
2. Bona C, Juy D, Truffa-Bachi P, and Kaplan JG (1976). Binding, capping and internalization of lipopolysaccharide in thymic and non-thymic lymphocytes of the mouse. Biological and autoradiographic study. J Microscop Biol Cell 25:47-56.
3. Davies M, Stewart-Tull DES, Jackson DM (1978). The binding of lipopolysaccharide from Escherichia coli to

mammalian cell membranes and its effect on liposomes. Biochim Biophys Acta 508:260-276.
4. Levin J and Bang FB (1964a). The role of endotoxin in the extracellular coagulation of Limulus blood. Bull Johns Hopkins Hosp 115:265-274.
5. Levin J and Bang FB (1964b). A description of cellular coagulation in the Limulus. Bull Johns Hopkins Hosp 115:337-345.
6. Morrison DC and Jacobs DM (1976). Inhibition of lipopolysaccharide - initiated activation of serum complement by polymyxin B. Infect Immun 13:298-301.
7. Rickles FR and Rick PD (1977). Structural features of Salmonella typhimurim lipopolysaccharide required for activation of tissue factor in human mononuclear cells. J Clin Invest 59:1188-1195.
8. Tomasulo PA, March SC and Levin J (1975). Tritiation of endotoxin. Biochim Biophys Acta 400:399-406.
9. Truffa-Bachi P, Kaplan JG and Bona C (1977). The mitogenic effect of lipopolysaccharide. Metabolic processing of lipopolysaccharide by mouse lymphocytes. Cell Immunol 30:1-11.
10. Wood DD and Cameron PM (1978). The relationship between bacterial endotoxin and human B-Cell-activating factor. J Immunol 121:53-60.

A NEW CHROMOGENIC SUBSTRATE METHOD FOR ASSAY OF BACTERIAL ENDOTOXINS USING *Limulus* HEMOCYTE LYSATE

T.HARADA[1],T.MORITA[1],S.IWANAGA[1],S.NAKAMURA[1], and M.NIWA[2]
[1]Institute for Protein Research, Osaka University Suita, Osaka 565, [2]Department of Bacteriology, Osaka City University Medical School, Abeno-ku, Osaka 545, Japan

A hemocyte lysate from horseshoe crab contains a pro-clotting enzyme, which is transformed to the active clotting enzyme in the presence of gram-negative bacterial endotoxin (Levin and Bang, 1964, 1968; Young et al., 1972). The clotting enzyme coagulates a clottable protein, named coagulogen, which is also contained in the hemocyte lysate (Solum, 1973; Mürer et al., 1975), as shown in Fig. 1.

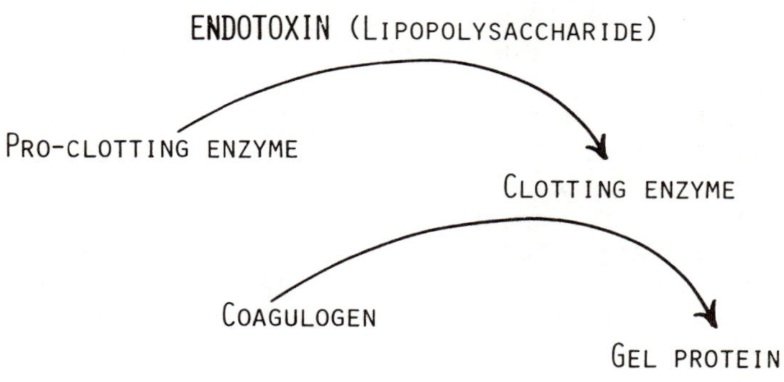

Fig. 1. Principle of the "Limulus test" for bacterial endotoxins.

This gelation reaction of the lysate, so-called *Limulus* test, has been widely employed as a simple and very sensitive assay method for endotoxin (Yin et al., 1972; Hochstein et al., 1973). In a previous paper (Nakamura et al., 1976a,

Pages 209–220, Biomedical Applications of the Horseshoe Crab (Limulidae)

1976d), we found that the gelation of coagulogen by the endotoxin-induced clotting enzyme involves a limited proteolysis. This indicates that the clotting enzyme must have a proteolytic or esterolytic activity like α-thrombin. We wish to report here further investigations on substrate specificity of the clotting enzyme and a new sensitive assay method for bacterial endotoxin.

MATERIALS AND METHODS

Horseshoe crab hemocyte lysate: The hemocyte lysate from Japanese horseshoe crab (Tachypleus tridentatus) was prepared by the method previously described (Nakamura et al., 1976b, 1976c). Limulus polyphemus lysate was a generous gift from Dr. E. B. Seligmann, Jr. Bureau of Biologics, FDA, Bethesda. A commercial preparation of hemocyte lysate from Tachypleus tridentatus, named Pregel[R] (Seikagaku Kogyo Co. Ltd., Tokyo), was also used.

Peptide-p-nitroanilide (PNA) substrates: The synthetic substrates used, see Table I, were products of AB KABl, Mölndal, Sweden, except Z-Gly-Pro-Arg-PNA and Bz-Val-Gly-Arg-PNA, which were obtained from Boehringer Mannheim and Pentapharm AG, Basel, respectively. Boc-Val-Leu-Gly-Arg-PNA, Boc-Leu-Gly-Arg-PNA, Boc-Val-Ser-Gly-Arg-PNA, Boc-Ser(O-Bz)-Gly-Arg-PNA, and Tos-Ile-Glu-Gly-Arg-PNA were newly synthesized by Dr. S. Sakakibara and his group, Institute for Peptide Research, Protein Research Foundation, Minoh, Osaka 562. Purity of all intermediates and products was confirmed by thin layer chromatography, elemental analysis, and amino acid analysis. Bz-Tyr-PNA, BAEE, TAME, acetyl-Gly-Lys-methylester, and Bz-Arg-β-naphthylamide were products of the Protein Research Foundation, Minoh, Osaka 562. Most of the synthetic substrates were dissolved in a pyrogen-free distilled water, but some of the water-insoluble peptides were dissolved in dimethylsulfoxide or dimethylformamide and diluted with Tris-HCl buffer used for the enzyme assay.

Preparation of endotoxin solution: A lipopolysaccharide prepared from E. coli 0111-B4 was a product of Difco Laboratories and used as a standard endotoxin. Heptose-free glycolipid was prepared from Salmonella minnesota R595 (Niwa et al., 1973). A lipopolysaccharide from E. coli 0113 and E. coli UKT-B was kind gifts of Dr. S. Kanoh, National Institute for Public Health, Osaka. Endotoxins from Vibrio cholerae NIH 41 (Ogawa), 569B (Inaba) and NAG 4715 (NAG) were products prepared by Dr. K. Hisatsune, Department of

Pharmacy, Josai University, Saitama Prefecture. These endotoxins were dissolved in a pyrogen-free saline.

Assay procedure of clotting enzyme: As the source of clotting enzyme, the hemocyte lysates prepared from *Tachypleus tridentatus* and *Limulus polyphemus*, and Pregel[R] were used. The reaction mixture containing 0.8 ml of 0.1 mM synthetic substrate dissolved in 0.1 M Tris-HCl buffer, pH 8.0, 50 μl of 0.5 M $MgCl_2$, and 20 μl of a 0.1 % endotoxin solution (*S. minnesota* R595) was preincubated at 37°C for 3 min. Then, 20-50 μl of the lysate was added and the mixture incubated further for 15 min. After incubation, 100 μl of glacial acetic acid was added to terminate the reaction and the PNA released was measured spectrophotometrically at 405 nm.

RESULTS AND DISCUSSION

Figure 2 shows a typical chromogenic substrate for *Limulus* clotting enzyme.

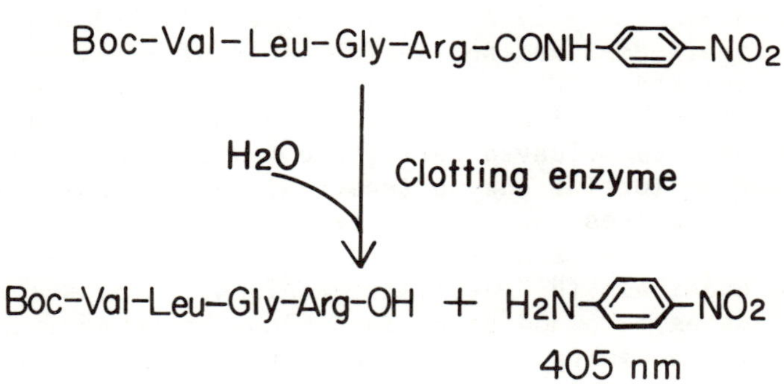

Fig. 2. A typical chromogenic substrate for *Limulus* clotting enzyme. p-Nitroaniline liberated is measured spectrophotometrically at 405 nm (ε =10,500).

This oligopeptide sequence of the substrate was based on the information from the COOH-terminal sequence of the peptide, which newly appears in the conversion of coagulogen into its gel protein (Fig. 3). Because the endotoxin is known to activate a latent clotting enzyme in the *Limulus* hemocyte lysate, the amidase activity seems to be due to the catalytic action of the clotting enzyme. Thus, above principle makes it possible to determine the endotoxin using the lysate.

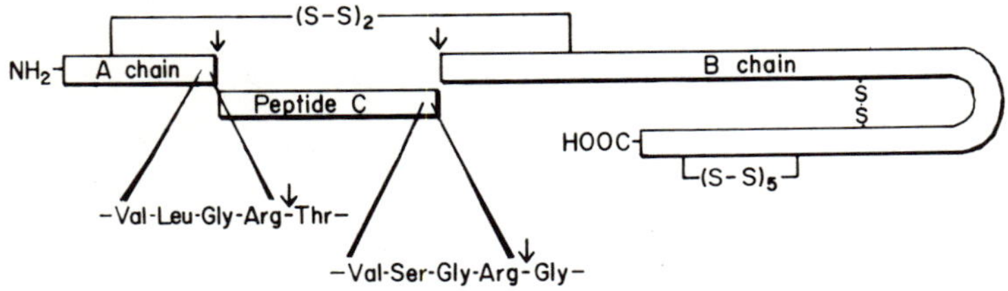

Fig. 3. Peptide linkages of coagulogen cleaved by Tachypleus clotting enzyme (see Takagi et al. in this proceedings).

Table 1 shows the amidase activities of Tachypleus and Limulus hemocyte lysates towards various anilide substrates, which were recently introduced as chromogenic substrates for blood clotting factors and their related enzymes (Witt, 1977). Both lysates showed activity towards the anilide substrates for Factor Xa (Bz-Ile-Glu(γ-OR)-Gly-Arg-PNA) and for urokinase (Bz-Val-Gly-Arg-PNA). Moreover, new anilide substrates, Boc-Val-Leu-Gly-Arg-PNA, Boc-Leu-Gly-Arg-PNA, Boc-Val-Ser-Gly-Arg-PNA, and Boc-Ser(OBz)-Gly-Arg-PNA, were easily hydrolyzed. The tri- and tetrapeptide sequences of these substrates originate also from the sequences located close to the sites cleaved during the gelation of coagulogen by Tachypleus clotting enzyme (see Fig. 3). On the other hand, the substrates, which have been used for α-thrombin (Bz-Phe-Val-Arg-PNA, Z-Gly-Pro-Arg-PNA, and H·D-Val-Leu-Arg-PNA), and urokinase (H·Glu-Gly-Arg-PNA), were unsusceptible to both lysates. In addition, Tachypleus hemocyte lysate showed little activity towards Bz-Tyr-PNA, TAME, and Bz-Arg-β-naphthylamide under the same conditions as used for the oligopeptide substrates (Table 1). These results clearly indicate that Limulus clotting enzyme displays a high specificity towards the peptide PNA having the COOH-terminal Gly-Arg sequence.

The Km and Vmax values estimated by the Lineweaver-Burk plots for Tos-Ile-Glu-Gly-Arg-PNA, Boc-Val-Leu-Gly-Arg-PNA, and Bz-Val-Gly-Arg-PNA, using a partially purified Tachypleus clotting enzyme (Harada et al., 1978), are shown in Table II. The Km values for these substrates were in range of 10^{-4} to 10^{-5} molar.

Table 1. Hydrolysis of various chromogenic substrates by Limulus hemocyte lysates.

Substrates	Nanomoles PNA released per 15 min (Relative activity, %) *Tachypleus*	*Limulus*
Bz-Ile-Glu(γ-OR)-Gly-Arg-PNA	17.7(100)	4.4(100)
Bz-Val-Gly-Arg-PNA	12.4(70)	10.9(227)
Boc-Val-Leu-Gly-Arg-PNA	11.0(62)	19.1(431)
Boc-Leu-Gly-Arg-PNA	17.7(100)	7.7(175)
Boc-Val-Ser-Gly-Arg-PNA	8.3(47)	10.9(227)
Boc-Ser(OBz)-Gly-Arg-PNA	19.6(111)	42.7(971)
Bz-Phe-Val-Arg-PNA	0.6(3)	0.2(5)
Z-Gly-Pro-Arg-PNA	0.4(3)	0.4(9)
H·D-Phe-Pip-Arg-PNA	—	0.3(6)
H-Glu-Gly-Arg-PNA	<0.2	<0.1
H·D-Val-Leu-Arg-PNA	<0.2	<0.1
H·D-Val-Leu-Lys-PNA	<0.2	<0.1
H·D-Pro-Phe-Arg-PNA	<0.2	<0.1
Bz-Tyr-PNA	ND	—
Acetyl-Gly-Lys-ME	4.6 μmole/min/A280=1.0	
TAME	ND	

Bz, benzyl; Boc, tert-butoxycarbonyl; Z, carbobenzoxy; H·D, D-amino acid residue; Pip, Pipecolyl; Acetyl-Gly-Lys-ME, α-N-acetyl-glycyl-lysine methyl ester; TAME, α-N-tosyl-arginine-methyl ester; ND, not detectable.

Table 2. The Km and Vmax values of Tachypleus clotting enzyme towards chromogenic peptide substrates.

Substrate	km	Vmax
P_5 P_4 P_3 P_2 P_1	M	mole/min/A280=1.0
Tos-Ile-Glu-Gly-Arg-pNA	2.1×10^{-4}	2.7
Boc-Val-Leu-Gly-Arg-pNA	5.4×10^{-5}	3.2
BZ-Val-Gly-Arg-pNA	4.6×10^{-5}	3.3

Figure 4 shows the Dixon plot for the inhibition of amidase activity by Tachypleus cogaulogen at the various substrate concentrations. Coagulogen competitively inhibited the amidase-catalyzed hydrolysis of Tos-Ile-Glu-Gly-Arg-PNA with a Ki of 3.8 x 10^{-6}M, indicating that the amidase activity is due to the catalytic action of the clotting enzyme.

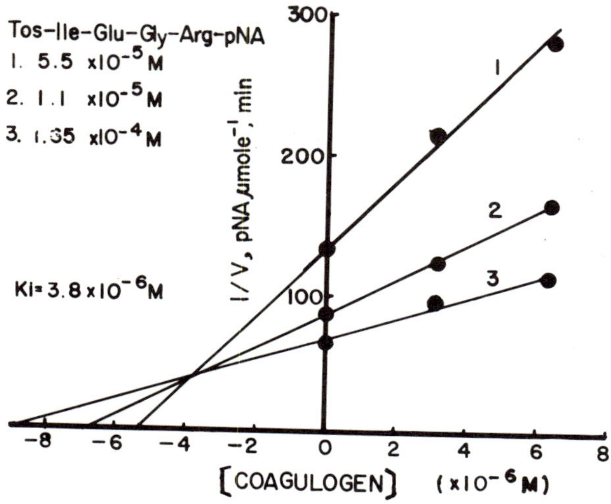

Fig. 4. The Dixon plot of coagulogen for Tachypleus clotting enzyme.

Figure 5 shows the effect of Salmonella minnesota R595 endotoxin on the amidase activity of Tachypleus hemocyte lysate, using the substrate, Tos-Ile-Glu-Gly-Arg-PNA. In the absence of endotoxin, no amidase activity of the lysate was observed. However, the amidase activity appeared after addition of the endotoxin into the reaction mixture. A lag phase observed at the initial stage may indicate that some period is required for the endotoxin-mediated activation of the latent amidase. In fact, the lag phase disappeared when the lysate was pretreated with the endotoxin before addition of substrate.

Using Bz-Ile-Gly-Gly-Arg-PNA as a substrate, the effect of pH on the amidase activity of Tachypleus hemocyte lysate was examined, and the maximum activity was observed at pH 8.0, as shown in Fig. 6.

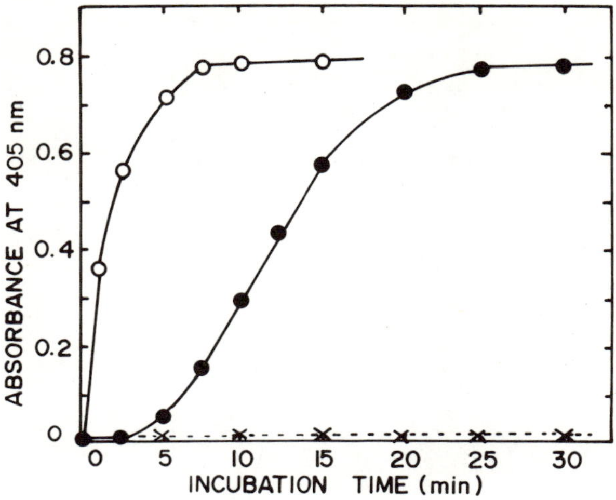

Fig. 5. Amidase activity of _Tachypleus_ hemocyte lysate before (--x--) and after (—●—) treatments with _Salmonella minnesota_ R595 endotoxin. The open circles shows the activity in the lysate pretreated at 37°C with the endotoxin for 15 min. The reaction mixture containing 0.8 ml of 0.1 mM substrate dissolved in 0.1 M Tris-HCl buffer, pH 8.0, 50 µl of 0.5 M $MgCl_2$ and 20 µl of a 0.1 % endotoxin solution, was incubated with 20 µl of the lysate. After incubation, 100 µl of glacial acetic acid was added and p-nitroaniline liberated was measured at 405 nm.

Figure 7 shows the effect of the concentration of _E. coli_ 0111-B4 endotoxin on the amidase activity of commercial _Tachypleus_ lysate, Pregel^R. In this experiment, the reaction mixture containing 50 µl of 2 mM Boc-Val-Leu-Gly-Arg-PNA dissolved in pyrogen-free distilled water, 100 µl of endotoxin solution, 50 µl of Pregel^R solution (one ampule Pregel^R was dissolved in 250 µl of 0.4 M Tris-HCl buffer, pH 8.0, containing 0.04 M $MgCl_2$), in a total volume of 200 µl, was incubated at 37°C. After incubation for indicated times, 0.8 ml of 12.5 % glacial acetic acid was added to terminate the reaction and the PNA released was measured. The result indicates that the amidase activity of the lysate increases with increasing concentration of endotoxin and that a linear relationship between the toxin concentration

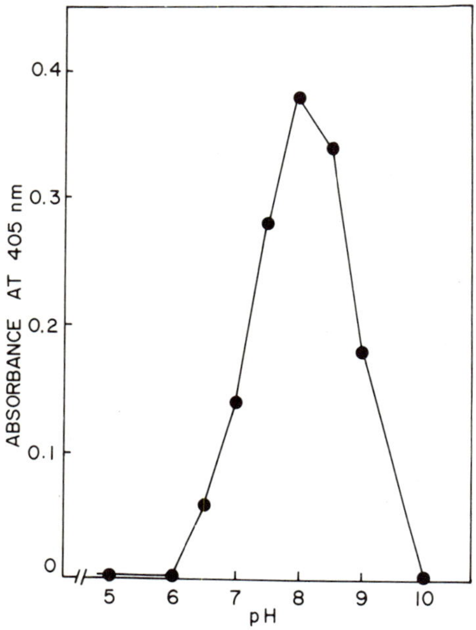

Fig. 6. Effect of pH on amidase activity of Tachypleus hemocyte lysate. The substrate of Bz-Ile-Glu-Gly-Arg-PNA was used. Buffers used were 0.1 M sodium acetate, pH 5.0 to 6.0; 0.1 M potassium phosphate, pH 6.5; 0.1 M Tris-HCl, pH 7.0 to 9.0; and 0.1 M sodium borate, pH 10.0.

and the amidase activity can be observed in the range of 10^{-6} to 10^{-2} μg endotoxin. Figure 8 also shows the effect of various endotoxins extracted from different bacterial strains on the amidase activity of the same lysate as that used in Fig. 7. The amidase activity was induced by all of the endotoxins in the range of 10^{-6} to 10^{-3} μg, except a native polysaccharide from E. coli 0111-B4 and E. coli 0113 strains. Thus, the measurement of amidase activity induced in the hemocyte lysate by endotoxin can be applicable for the detection and the quantitation of various bacterial endotoxins. The method was a ten-times more sensitive than that of the Limulus gelation test and was very reproducible.

Finally, the effects of various synthetic and naturally occurring proteinase inhibitors on the amidase activity of

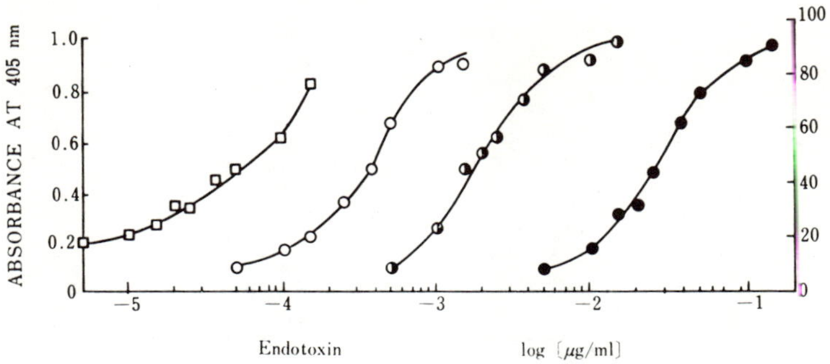

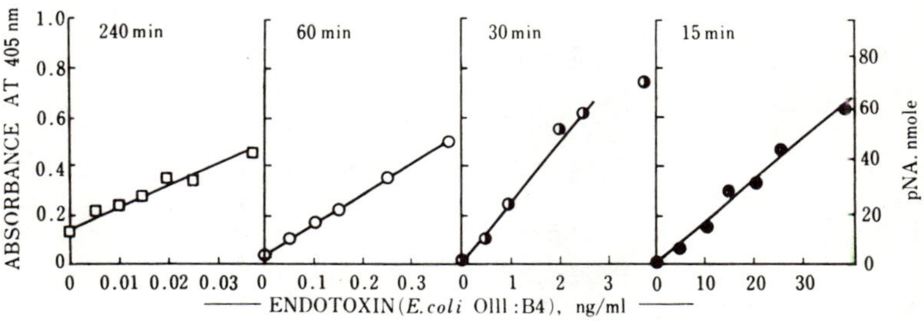

Fig. 7. Effect of different concentrations of E. coli 0111-B4 endotoxin on amidase activity of Pregel[R] (Tachypleus hemocyte lysate). Sterilized and pyrogen-free glass ware and reagents were used. The final concentration of the substrate, Boc-Val-Leu-Gly-Arg-PNA, was 0.5 mM. The reaction mixtures (described in text) were incubated at 37°C for 15 min (●), 30 min (◐), 60 min (○), and 240 min (□).

Tachypleus clotting enzyme were examined after preincubation for 15 min. At 4 to 10 mM concentration, diisopropylphosphofluoridate (DFP) and benzamidine each inhibited the enzyme activity completely, as shown in Table 3. Moreover, PCMB inhibited it partially but EDTA and monoiodoacetic acid had no effect. On the other hand, plasma proteinase inhibitors, such as antithrombin III and α_2-plasmin inhibitor, had a potent inhibitory effect on the amidase activity. These results suggest that Tachypleus clotting enzyme is one of typical "serine"-active site proteinase like blood coagulation factors, such as Factor Xa, Factor XIIa and α-thrombin.

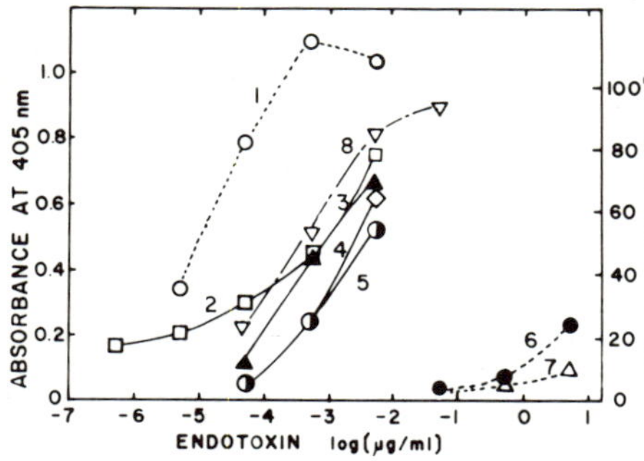

Fig. 8. Effect of various endotoxins on amidase activity of Pregel^R (*Tachypleus* hemocyte lysate). The experimental conditions were almost the same as that used in Fig. 7. 1, *E. coli* UKT-B; 2, *S. minnesota* R595 glycolipid (Re); 3, *E. coli* 0113; 4, *S. minnesota* 114W; 5, *E. coli* 0111-B4; 6, *E. coli* 0111-B4 native polysaccharide, 7, *E. coli* 0113 native polysaccharide; 8, *V. cholerae* NAG (4715).

Moreover, it seems that the assay procedure for bacterial endotoxins described here is not directly applicable for the quantitation of the endotoxin in circulating blood, because the amidase activity of the lysate was strongly inhibited by antithrombin III and α_2-plasmin inhibitors existed in blood plasma. Thus, some pretreatment, such as a heat or a chloroform treatment, on blood plasma samples, could be required for the assay of endotoxins in circulating blood.

ACKNOWLEDGEMENTS

We wish to express our thanks to Drs. S. Kanoh, K. Hisatsune, N. Aoki, and T. Ikenaka, for their kind gifts of some of bacterial endotoxins and proteinase inhibitors. We also thank Dr. M. Yoshida, Marine Laboratory of Okayama University, and Dr. H. Nishii, for assistance in bleeding horseshoe crabs.

Table 3. Effect of various inhibitors on the activity of horseshoe crab clotting enzymes

Inhibitor	Remaining Activity (%)	Inhibitor	Remaining Activity (%)
None	100	DFP (0.4 mM)	45
Antithrombin-III [a] (2 μg/ml)	14	(0.2 mM)	18
(10 μg/ml)	0	(4.0 mM)	2
α_2-Plasmin inhibitor (3 μg/ml)	20	Benzamidine (4 mM)	19
(8 μg/ml)	2	(10 mM)	0
Soybean trypsin inhibitor (20 μg/ml)	44	PCMB (0.3 mM)	75
(50 μg/ml)	26	(1.0 mM)	56
α_1-Antitrypsin (20 μg/ml)	92	(3.0 mM)	20
(50 μg/ml)	83	Antipain (20 μg/ml)	70
Plasminostreptin (72 μg/ml)	100	(60 μg/ml)	53
Limabean inhibitor (54 μg/ml)	100	Leupeptin (60 μg/ml)	100
Bowman-Birk inhibitor (50 μg/ml)	100	Chymostatin (60 μg/ml)	100
Trasylol (66 μg/ml)	100	Polymixin-B (50 μg/ml)	100
Ovomucoid (188 μg/ml)	100	EDTA (25 mM)	100
Hirudine (14.4 ATU/ml)	100	Monoiodo acetic acid (2 mM)	100

[a] In the presence of heparin (5 u/ml).

REFERENCES

Harada T,Morita T,Iwanaga S (1978). A new assay method for bacterial endotoxins using horseshoe crab hemocyte lysate. J Med Enz 3:43-60

Hochstein HD,Ein RJ,Cooper JF,Seligmann EB jr, Wolff SM (1973) Further developments of Limulus amebocyte lysate test. Bull parent Drug Ass. 27:139-148

Levin J,Bang FB (1964). The role of endotoxin in the extracellular coagulation of Limulus blood. Bull Johns Hopkins Hosp 115:265-274

Levin J,Bang FB (1968). Clottable protein in Limulus. Its localization and kinetics of its coagulation by endotoxin. Thromb Diath haemorrh 19:186-197

Mürer EH,Levin J,Holme R (1975). Isolation and studies of the amebocytes of Limulus polyphemus, the horseshoe crab. J cell Physiol 86:533-542

Nakamura S,Takagi T,Iwanaga S,Niwa M,Takahashi K (1976a). A clottable protein (coagulogen) of horseshoe crab hemocytes. Its structural change of its polypeptide chain during gel formation. J Biochem 80:649-652

Nakamura S,Iwanaga S,Harada T,Niwa M (1976b). A clottable protein (coagulogen) from amoebocyte lysate of Japanese horseshoe crab (Tachypleus tridentatus). Its isolation and biochemical properties. J Biochem 80:1011-1021

Nakamura S,Takagi T,Iwanaga S,Niwa M,Takahashi K (1976c). Amino acid sequence studies on the fragments produced from horseshoe crab coagulogen during gel formation. Homologies with primate fibrinopeptide B. Biochem Biophys Res Commun 72:902-908

Nakamura S,Morita T,Iwanaga S,Niwa M,Takahashi K (1977d). A sensitive substrate for the clotting enzyme in horseshoe crab hemocytes. J Biochem 81:1567-1569

Niwa M,Hiramatsu T,Kuwajima Y,Shimouchi H (1973). Micellar structure and endotoxin activities of gram-negative lipopolysaccharides. Jap J Med Sci Biol 26:20-22

Solum NO (1973). The coagulogen of Limulus polyphemus hemocytes. A comparison of the clotted and non-clotted forms of the molecule. Thromb Res 2:55-70

Witt I (1977). New method for the analysis of coagulation using chromogenic substrates (de Gruyter, Berlin)

Yin ET,Galanos C,Kinsky S,Bradshaw RA,Wessler S,Lüderitz O, Sarmiento ME (1972). Picogram-sensitive assay for endotoxin. Gelation of Limulus polyphemus blood cell lysate induced by purified lipopolysaccharides and lipid A from gram-negative bacteria. Biochim Biophys Acta 261:284-289

ENDOTOXIN DETERMINATION IN VISCOUS OPAQUE SOLUTIONS OF IRON DEXTRAN BY LIMULUS AMEBOCYTE LYSATE

Stephen L. Gaffin

Department of Physiology and Biophysics
School of Medicine
Technion-Israel Institute of Technology
POB 9649, Haifa, Israel

SUMMARY

LAL would not form a clot when mixed with a viscous, opaque parenteral preparation of iron dextran spiked with endotoxin. However, recoverable precipitate could be obtained by diluting the LAL iron dextran mixture with PBS and centrifuging. Although the pellet so formed was red colored the protein present could be quantitated by dissolving it in a Coomassie Blue stain solution. The very rapid change in color from reddish black to deep blue was measured quantitatively in a spectrophotometer and was sigmoidally related to the amount of endotoxin used to spike the iron dextran. This method is suggested to be generally useful to measure quantitatively endotoxin concentrations too low to form a clot with LAL but high enough to precipitate recoverable protein from LAL.

INTRODUCTION

The Limulus amebocyte lysate (LAL) technique is increasingly recognized as the most sensitive, rapid and least expensive method for detecting endotoxin in research and clinical laboratories as well as in the parenteral drug industry (Levin and Bang, 1968; Reinhold and Fine, 1971; Weary and Baker, 1977). In this test a clear solution of the lysed white blood cells of the horseshoe crab is mixed with the solution to be tested for one hour at 37°. If the clear

Pages 221–227, Biomedical Applications of the Horseshoe Crab (Limulidae)

solution becomes turbid or forms a gel or "clot" then this is regarded as a positive sign for the presence of endotoxin relative to positive or negative controls run at the same time. However, parenteral preparations of iron dextran can not be tested at present by LAL becasue their very high viscosity and opacity prevent formation of a clot or observation of an increase in viscosity or turbidity of the LAL reaction mixture. The purpose of this study was to modify the usual LAL procedure to detect endotoxin in spiked iron dextran preparations. It was found that while endotoxin did induce the enzymatic reaction in the LAL to form a precipitate of "coagulin" or "beta clot protein" in the iron dextran solution, the precipitated protein could not form a clot (Solum, 1973; Gaffin, 1974,1976). A recently described Coomassie Blue protein stain was used to determine the amount of precipitate formed by the action of endotoxin on the LAL-iron dextran test solution (Bradford, 1976).

MATERIALS AND METHODS

The darkly red opaque viscous test samples of iron dextran were parenteral preparations supplied by TEVA Pharmaceutical Industries, Ltd., Jerusalem. These were 30 % solutions of iron dextran with viscosities of 12 centipoise, according to the manufacturer. Various amounts of endotoxin (DIFCO E coli 0127:B8) in 0.9% NaCl were added to the samples just prior to the LAL assay.

Fresh Coomassie Blue stain solution was prepared weekly according to Bradford (1976) and filtered through filter paper before use. Each preparation was calibrated with standard bovine serum albumin solutions.

LAL was prepared by the method of Reinhold and Fine (1971) from horseshoe crabs obtained in the Woods Hole, Massachusetts area and stored lyophilized at -20°. Some LAL was also prepared from Limuli collected in Florida (Gulf Specimen Co., Panacea) and maintained in a colony in Eilat. Suchlysate was stored as frozen solutions at -20°.

One-tenth ml of LAL plus 0.1 ml sample were incubated at 37° in 12 x 75 mm tubes for one hour. Three ml of PBS solution at 4° were than added to each tube and shaken. The samples were centrifuged for 15 min at room temperature at 2000 x g. The supernatant was discarded and the red pellet was washed three more times with PBS. Three ml of Coomassie Blue stain were added to each tube and then they were vortexed to dissolve completely the pellet. After 20 min incubation at room temperature absorbance was measured at 600 nm.

RESULTS

The fresh Coomassie Blue stain solution appears by eye to be a pale "reddish black" color. However, when more than approximately 10 micrograms protein is added a striking change occurs---the color becomes a deep royal blue. This is reflected by the large change in the visible absorption spectrum in Figure 1. Two small peaks seen at 480 nm and 645 nm are greatly reduced in extinction coefficient when the protein is added. On the other hand a major new peak occurred at 595-600 nm with a very high extinction coefficient.

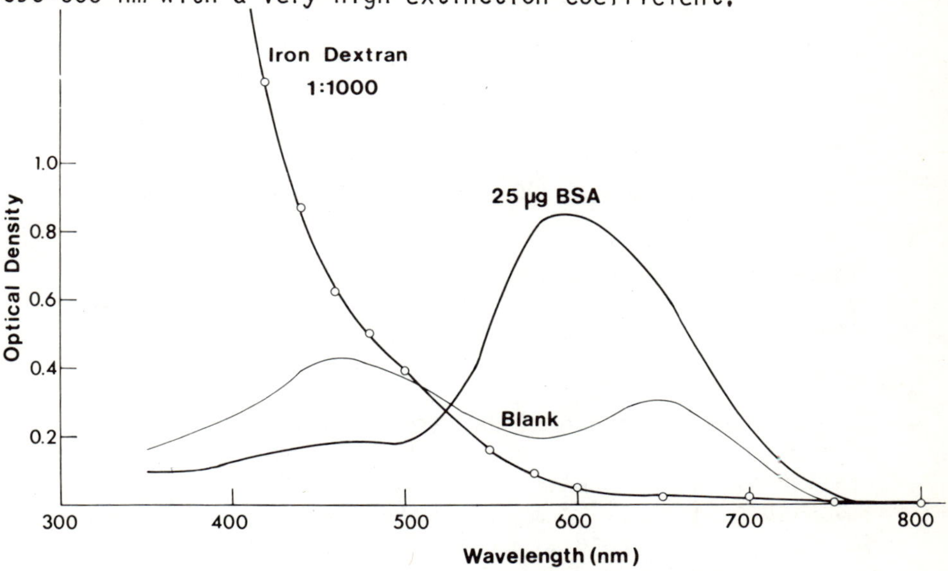

Figure 1. Absorption Spectrum of Coomassie Blue Stain and Iron Dextran. Stock iron dextran was diluted 1:1000 with 0.9% NaCl.

In Figure 2the absorbance of this new peak is plotted as a function of protein added. The absorption depended linearly on the amount of protein until approximately 50 micrograms, but then this slope gradually decreased until saturation at approximately 400 micrograms protein. The maximum slope observed with Coomassie Blue was 6.9 O.D./ mg protein. This was 121 times the slope of the usual Biuret protein reagent (0.057 O.D./mg) used in our laboratory for protein determination.

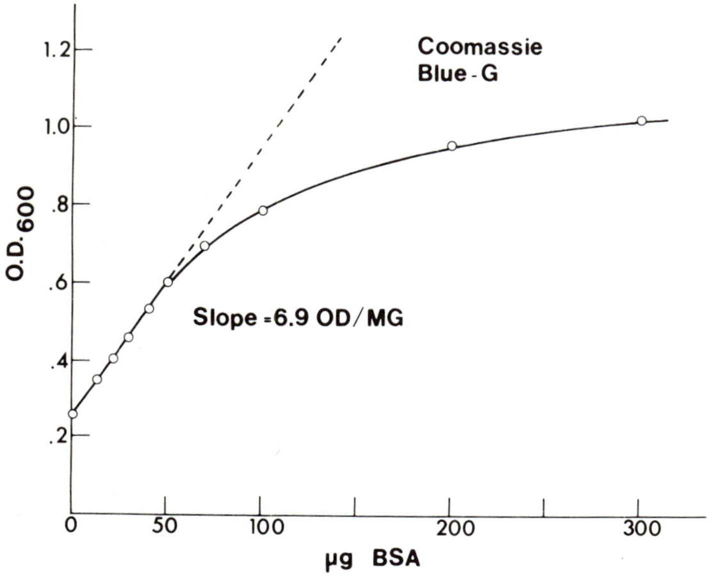

Figure 2. Standard Curve of Bovine Serum Albumin in Coomassie Blue Stain. 0.1 ml samples of BSA were mixed with 3.0 ml stain solution and incubated 20 min at 25°.

Samples of iron dextran were spiked with various amounts of endotoxin and incubated with LAL for one hour. Interpretation of clot formation by inverting the tubes 180° was impossible since all samples flowed downward with gravity. No increase in turbidity of the sample due to clot formation could be seen since the iron dextran LAL mixture was completely opaque. This mixture was diluted with cold PBS solution and centrifuged. The pellet formed was dark red due to the presence of iron dextran entrapped or adsorbed to the proteins. Further rinsing the pellet with PBS did not remove the red color. However, from Figure 1 it is clear that the absorbance of iron dextran is very low at wavelengths greater than about 550 nm. Thus the protein present in the pellet could be determined by the Coomassie

Blue stain with its absorbance maxima at 595-600 nm, even in the presence of some red color. Figure 3 shows the absorbance of the precipitate formed by mixtures of iron dextran and LAL in the presence of various amounts of endotoxin. Note that in three of the preparations (solid lines) a large change in absorbance occurred over a narrow range of endotoxin concentrations. These same LAL preparations could form solid clots when tested against 0.9% NaCl solutions. Significantly, in these cases the clots formed at the same concentrations of endotoxin which caused the large increase in absorbance of Coomassie Blue stain in the iron dextran samples. One LAL preparation (dash line) could not form a solid gel in 0.9% NaCl even in 10^{-6} g/ml endotoxin. In this case the change in absorbance of Coomassie Blue stain was gradual over a wide range of endotoxin concentrations.

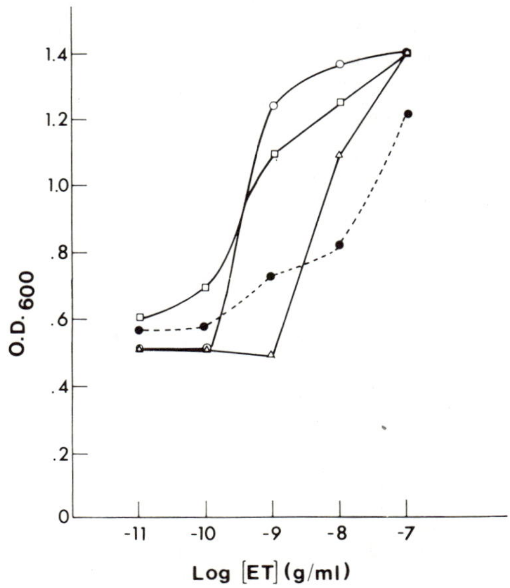

Figure 3. Absorbance of Iron Dextran - LAL Mixture in Coomassie Blue Stain. The precipitate formed by LAL and iron dextran at different endotoxin concentrations was washed four times with PBS solution and dissolved in Coomassie Blue stain. Four different LAL preparations.

DISCUSSION

We have shown that even though LAL does not form a

clot by endotoxin in parenteral preparations of iron dextran, one can modify the standard LAL procedure to still detect endotoxin. A relationship between the rate or extent of LAL clot formation and endotoxin concentration has previously been reported (Young et al, 1972; Sullivan et al., 1976). Since the LAL precipitate formed by endotoxin in the opaque iron dextran solution is strongly red colored, a simple colorimetric assay for protein could not be performed. However, Coomassie Blue stain was found to be useful due to its sensitivity and large change in optical absorbance in a region of the spectrum where the absorbance of iron dextran is low. Occasionally, a particular LAL preparation was "too sensitive" (solid clot in 10^{-11} g/ml endotoxin in saline) and precipitated more protein than could be conveniently measured in the red precipitate. In this case, 0.1 - 1.0 ml aliquots of the Coomassie Blue stain containing the dissolved pellet could be transferred into a second tube containing three ml of fresh stain solution. This effectively lowered the sensitivity of the stain to a more convenient value, while at the same time maintains its property of being unaffected by the red color of the iron dextran. Simply using a more dilute stain would render the method susceptible to interference by the red color.

The Coomassie Blue stain may also be used to advantage in a wide range of solutions with LAL. It can respond qualitatively and quantitatively to concentrations of endotoxin too low to form a solid clot, but high enough to cause an increase in recoverable precipitate from the LAL.

REFERENCES

Bradford M (1976). A rapid method for the quantitation of microgram quantities of protein. Anal Biochem 72: 248.

Gaffin SL (1974). The clotting of the lysed white cells of Limulus induced by endotoxin. Abstr Int Congr Biorheology Rehovot p 26.

Gaffin SL (1976). The clotting of the lysed white cells of Limulus induced by endotoxin. I. Preparation and characterization of clot forming proteins. Biorheology 13: 273.

Levin J, Bang, F (1968). Clottable protein in Limulus: Its localization and kinetics of its coagulation by endotoxin. Thromb Diasth Haemorrh 19: 186.

Reinhold R, Fine J (1971). A technique for quantitative measurement of endotoxin in human plasma. Proc Soc Exptl Biol Med 137: 334.

Solum N (1973). The coagulogen of Limulus polyphemus hemocytes. A comparison of the clotted and non-clotted forms of the molecule. Thromb Res 2: 55.

Sullivan J, Valois F, Watson S (1976). Endotoxins: The Limulus amebocyte lysate system. In Bernheimer A (ed.) "Mechanisms in Bacterial Toxinology" New York: John Wiley and Sons, p 217.

Weary M, Baker B (1977). Utilization of the Limulus amebocyte lysate test for pyrogen testing large volume parenterals, administration sets and medical devices. Bull Parent Drug Assoc 31: 127.

Editor's Summary, Session II, Part A: Biochemistry of the Reaction between Bacterial Endotoxin and Amebocyte Lysate

Robert A. Prendergast

Department of Ophthalmology
Johns Hopkins University School of Medicine
Baltimore, Maryland

The origin of our interest in the response of Limulus polyphemus to gram negative endotoxins was initiated by Dr. Frederik B. Bang in 1956, when he reported in the Bulletin of The Johns Hopkins Hospital a study entitled "A Bacterial Disease of Limulus polyphemus". It is apparent even from the title that this investigator was concerned as a naturalist with the response of this animal to a clearly harmful agent. A great deal of study has been promoted by this initial observation, but it is clear from Dr. Bang's presentation that it would be wise if we went back to study the natural course of various forms of pathology in the living animal. He points out that injection of endotoxin into small animals even as large as 10 cm in length produces a striking change in the circulation of amebocytes which can be clearly visualized in the transparent young Limulus. This was usually followed by recovery of the animal, and therefore the susceptibility, as measured by death, is much lower than would be expected from the extreme sensitivity of the amebocyte lysate test in vitro. Dr. Bang states that therefore many factors must come into play in the intact animal for clearance or at least suppression of the noxious effects of endotoxin, and that these would be amenable to study not only in Limulus but in other marine arthropods susceptible to endotoxin including Emeriti (mole shrimp), where susceptibility to endotoxin is much more marked, than in a small Limulus. He further points

Pages 229–233, Biomedical Applications of the Horseshoe Crab (Limulidae)

to a number of other diseases of Limulus, the investigation of which may give us further clues to the protective and destructive influences associated with this truly amazing and provocative creature.

Previously published reports have demonstrated conclusively that the coagulation system in Limulus polyphemus is dependent on material secreted by the circulating amebocyte. The cytological changes accompanying secretion of clottable material was investigated by Drs. Ornberg and Reese, using apparatus for quick freezing endotoxin-stimulated Limulus amebocytes on a copper block cooled to 4.2°K with liquid helium. It is clear from this report that the earliest evidence of disruption of granule contents is by exocytosis either through the plasmalemma or from one granule into another, the latter fused with the cell membrane. This is an important observation on the cytological changes and, presumably, on the temporal sequence of events in the secretion of granule contents. It should be noted, however, that if the cell membrane is sufficiently disrupted by granule membrane fusion to plasmalemma that indeed residual granules may be released into the surrounding medium, as reported by Bang in 1956. Of particular interest in this report is the striking technique for quick freezing tissue, outlined in the Materials and Methods section. As the authors state, this permits examination of frozen secreting amebocytes "arrested" in approximately 2 milliseconds, giving a superb opportunity of seeing the secretory event in process. Subsequent reports dealing with release of granule contents or of whole granules using this method will be awaited with considerable interest.

While it has been recognized that coagulation of the blood of Limulus polyphemus was detected in the presence of amebocytes since the late 19th century, it is only recently that it has been clear to workers in this field that, in fact, all of the requisite molecules involved in this process were in the amebocyte alone. The presentation of Dr. Levin outlines in striking detail the advances made during the past twenty years which in great measure have helped to clarify the mechanisms involved in this form of coagulation. These include isolation and characterization of the pro-coagulant enzyme activated by endotoxin. This enzyme then acts on the clottable protein (coagulogen) by limited proteolysis, the latter forming a gel which, by electron microscopic examination, is composed of long fibrils which are apparently cross-linked. The molecular weight of the coagulogen is approximately 23,000 to 27,000, as determined by a number of studies of coagulogen isolated from Limulus polyphemus.

Dr. Levin concludes his discussion by commenting on the marked homology between the "simple" coagulation mechanism of Limulus and coagulation of mammalian blood, and points out that understanding the coagulation system of the horseshoe crab may be of considerable help in understanding the biochemical mechanisms involved in coagulation of mammalian blood.

Dr. Liu and his coworkers have succeeded in purifying a number of molecules including the proclotting enzyme, an active form of this enzyme, the coagulogen, and an endotoxin-binding protein from Limulus amebocytes. He has determined the molecular weight of coagulogen to be approximately 24,500, whether in its native state or reduced and carboxymethylated. This is in clear contrast to the results of Mosesson and his group, where simple reduction or reduction with S-carboxymethylation gives rise to a single band with a molecular weight of approximately 12,400. Both of these laboratories used sodium dodecyl sulfate polyacrylamide gel electrophoretic separation for determination of molecular weight, and the differences in their results are not yet clear. Mosesson's group has also examined the effect of another dissociating solvent, 8 M urea, on the apparent molecular weight as examined by column chromatography. Under these circumstances, they have shown that the apparent molecular weight of unreduced coagulogen corresponded to a position of approximately 11,500 daltons. It is clearly of the first importance that structure of coagulogen be established, and future experiments from both of these laboratories should clarify the issue.

Dr. Liu concludes by discussing the physiologic role of

the Limulus clotting system, and notes that all factors which eventually contribute to the clotting are expelled into the blood plasma when the cells degranulate or lyse, and that endotoxin cannot enter the intact amebocyte. His laboratory has succeeded in isolating a protein with a molecular weight of approximately 80, 000 daltons and purified this material to a single band on acrylamide gel electrophoresis. This protein binds with endotoxin and may in fact be the agent on the cell wall which triggers the exocytosis of the coagulation system to trap and isolate endotoxin as it enters the living animal. This defensive maneuver is of considerable importance and, as is stated, endotoxin would be quickly immobilized and engulfed by the clot. Dr. Rickles and his coworkers have also reported on studies of binding of radioactive lipopolysaccharide to Limulus amebocytes. It will be of considerable interest to determine whether the binding site of Rickles and the purified protein isolated from the cell wall by Liu are in fact identical. Future work on this endotoxin-binding membrane protein will be of extreme importance, and should establish the biochemical parameters of the trigger mechanism for release of molecules involved in coagulation.

Comparative studies of Asian and American horseshoe crabs were outlined in the contribution by Shishikura and Sekiguchi. The three Asian species, Tachypleus tridentatus, Tachypleus gigas, and Carcinoscorpius rotundicauda, were compared with the cells and coagulogens of Limulus polyphemus. It was demonstrated that the coagulogen of all of these species has approximately the same molecular weight by polyacrylamide gel electrophoresis, and in fact a mixture of all of these species still demonstrated only a single band. Further, clot formation from each of these species can be induced by the clotting enzyme from T. tridentatus and would imply a high degree of homology between each of the coagulogens studied. On the other hand, as the authors point out, antiserum directed to T. tridentatus coagulogen cross-reacted strongly with the other two Asian species but gave no precipitin line with the coagulogen from Limulus polyphemus. The converse experiment was done, where antibody to Limulus polyphemus coagulogen gave no response to any of the Asian species. It is probable that, while the clotting enzyme from T. tridentatus is effective in forming a Limulus polyphemus coagulogen clot, both the active site of the enzyme and the receptor site must be similar between these species. While Ouchterlony analysis does not show a precipitin line with Limulus polyphemus coagulogen, it is possible that sequence analysis of, let us say, C peptide released from the coagulogen would demonstrate a greater degree of homology than can be ascertained by precipitin reactions. It is clear, however, as the authors point out, that the three Asian species are different from Limulus polyphemus, and that the phylogenetic separation must have occurred eons ago.

The contribution of Takagi and his coworkers on the amino acid sequence studies of coagulogen isolated from T. tridentatus is a striking example of the superb biochemical studies performed by this group. They demonstrate that this coagulogen consists of a single polypeptide chain of 175 residues. The sites cleaved by the active clotting enzyme are shown to be between residues Arg_{18}-Thr_{19} and residues Arg_{46}-Gly_{47}. This cleavage initiates coagulation and the release of C peptide. Of particular interest (and of phylogenetic importance) is that the terminal octapeptides of A chain and C peptide demonstrate remarkable homology and, more importantly, show a striking similarity in sequence to Rhesus and human fibrinopeptide B. Even more stimulating is the experience of these authors with bovine α-thrombin, where this enzyme was able to clot purified coagulogen. These data suggest a relationship in clot formation between species as widely disparate as T. tridentatus and man. Precedents for amino acid sequence homology between widely separated species have long been known. In the case of cytochrome C, for example, there is 60% homology between man and the buttercup love-in-a-mist and 87% between man and the rattlesnake.

THE LIMULUS TEST: A STATUS REPORT

Jack Levin, M.D.
Department of Medicine, The Johns Hopkins
University School of Medicine and Hospital
Baltimore, Maryland

INTRODUCTION

Utilization of the Limulus test has increased greatly during the past few years, as it has been applied in a widening range of applications (Table 1). Many of its uses are the subjects of other reports in this volume and, therefore, will not be discussed here. Clinically, the Limulus test demonstrates an excellent correlation with the presence of gram-negative bacteria in urine and cerebrospinal fluid. The accompanying papers of Jorgensen and Siegel and Nachum review and discuss both these topics in detail. Therefore, it will be the purpose of this report to focus on the detection of endotoxin in blood, the most difficult challenge for any method for the measurement of endotoxin, both because of the presence of inhibitors in blood (Levin, Tomasulo, and Oser, 1970) and because of the apparently great sensitivity required.

Table 1

Applications of the Limulus Test

Pharmaceutical preparations	Biological reagents
Radioisotopes	Body fluids:
Water Supplies	Blood
Intravenous fluids	Cerebrospinal fluid
	Urine

Pages 235–244, Biomedical Applications of the Horseshoe Crab (Limulidae)

THE DETECTION OF ENDOTOXIN IN BLOOD WITH THE LIMULUS TEST

The plasma of normal persons does not produce a positive Limulus test (Levin, et al, 1972). However, many have reported the detection of endotoxin in the blood of patients with gram-negative bacteremia or other gram-negative infections. Three of the studies listed in Table 2 demonstrate a good correlation between gram-negative infection and positive Limulus tests. In some studies, there was a correlation

Table 2

Positive Limulus Tests in Suspected Gram-Negative Sepsis*

Limulus Test No. Positive/Total (%)	Positive Blood Culture (Gram-Neg)	Positive Cultures** (Gram-Neg)	Reference
39/281 (14%)	17	35/39	Levin, et al (1972)
23/218 (11%)	8	19/23	Reinhold, Caridis, and Fine (1972)
75/344 (22%)	28	46/75	Stumacher, Kovnat, and McCabe (1973)
16/54 (29%)	11	16/16	Das, Schwartz, and Folkman (1973)

*These studies were selected because sufficient information was provided to assess the presence of gram-negative infections which did not produce detectable bacteremia.

**Culture from blood, other body fluid, or tissue positive for gram-negative bacteria.

between a positive Limulus test and mortality (Levin, et al, 1972; Das, Schwartz, and Folkman, 1973; Young, 1975). However, the study of Stumacher, Kovnat, and McCabe (1973) reported that some patients with positive Limulus tests had only gram-positive infections, a finding not confirmed by other studies (Levin, et al, 1972; Reinhold, Caridis, and Fine, 1972; Das, Schwartz, and Folkman, 1973).

In some reports, there was a correlation between positive Limulus tests and other potential manifestations of endotoxemia, such as relatively reduced levels of complement (Levin, et al, 1972; Robinson, et al, 1975; Koster, et al, 1978), and decreased systemic vascular resistance and prekallikrein levels (Robinson, et al, 1975). The investigation of Young (1975) showed a striking correlation between low titers of opsonic antibodies against gram-negative organisms and positive Limulus tests, thrombocytopenia, leukopenia, and mortality (Table 3). Fossard, Kakkar, and Elsey (1974) reported positive Limulus tests in 15 of 16 patients with proven gram-negative infections; and in six patients with peritonitis. Although some of their patients without detectable gram-negative infections also had positive Limulus tests, it was noteworthy that all had diseases of the gastrointestinal or biliary tract.

Rowe, Buckner, and Newmark (1975) did not find the Limulus test clinically helpful in their study of children during the post-operative period, although 64% of patients with gram-negative bacteremia had positive tests, while none of five patients with gram-positive bacteremia had

Table 3

Correlation between Limulus Test, Opsonic Titer, and Other Variables*

Group	Opsonic Titer	Cytopenia (%) WBC	Cytopenia (%) Plates.	Mortality (%)	Limulus Test (% Pos.)
I	≤20	74	54	54	78
II	40	55	18	9	55
III	≥80	17	10	13	27

*Modified from Young, 1975

positive Limulus tests. Similarly, Elin, et al (1975) concluded that the test was not useful for the diagnosis of endotoxemia. However, although the authors recognized that the number of circulating bacteria are usually insufficient to produce a positive Limulus test, they did not include an evaluation of the potential role of non-bacteremic, gram-negative infections in their patients. Two reports (Martinez-G, Quintiliani, and Tilton, 1973; Feldman and Pearson, 1974) indicated that few patients with gram-negative bacteremia had positive Limulus tests. Both groups removed inhibitors from their plasma samples using the pH shift method (Reinhold and Fine, 1971), which has not been found satisfactory by some investigators.

There is extensive experience with the use of the Limulus test in diseases of the gastrointestinal and biliary tracts. Since this is the subject of the papers by Liehr and Wilkinson in this volume, only a few selected observations will be mentioned. Koster, et al (1978) demonstrated a relationship between the severity of pseudomembranous colitis in patients with Shigellosis, the development of the hemolytic-uremic syndrome, and the presence of endotoxemia. The Limulus test also correlated with decreased complement levels and mortality. Wilkinson, et al, (1974) detected endotoxemia in 14 of 22 consecutive patients with hepatic failure, although in most cases, there was no evidence of gram-negative infection. However, positive Limulus tests did correlate with renal failure and altered blood coagulation. Wardle (1975) confirmed the association between positive Limulus tests and renal failure. Additional evidence suggesting the liver plays a role in removing endotoxin that enters the portal circulation from the gastrointestinal tract has been provided by studies of children with Reye's syndrome (Cooperstock, Tucker, and Baublis, 1975) and of adults with ascites and cirrhosis (Tarao, et al, 1977; Liehr and Grün, 1977). In both of the latter series, endotoxin also was detected in ascitic fluid by the Limulus test. There are discordant data from patients with typhoid fever (Magliulo, et al, 1976; Butler, et al, 1978).

DISCUSSION

Overall, the Limulus test correlates with evidence of a significant gram-negative infection, in the presence or absence of detectable gram-negative bacteremia. It seems inappropriate to consider all positive Limulus tests, in the absence of bacteremia, as falsely positive results (Stumacher, Kovnat, and McCabe, 1973; Elin, et al, 1975). Bacteremia and endotoxemia are not synonymous terms, but two potentially unrelated pathophysiologic events. There is increasing evidence that other bacterial components (i.e., antigens) may be or remain present in the circulation in the absence of bacteremia (Kenny, et al, 1972; Bartram, et al, 1974). Furthermore, there is considerable evidence that under some circumstances, endotoxin may enter the portal or systemic circulation from the gastrointestinal tract (Caridis, et al, 1972; Wilkinson, et al, 1974; Fossard, Kakkar, and Elsey, 1974; Liehr and Grün, 1977; Koster, et al, 1978). In most instances, a positive Limulus test is probably not produced by endotoxin derived from bacteria concomitantly in the circulation, but rather by endotoxin that has gained access to the circulation from gram-negative organisms in tissue sites or the gastrointestinal tract. Among the patients reported by Fossard, Kakkar, and Elsey (1974) was one patient who had five episodes of intestinal obstruction, two of which were associated with positive Limulus tests (blood cultures were negative). Following resection of a reversed loop, the Limulus test became negative.

Importantly, many studies have demonstrated correlations of positive Limulus tests with clinical abnormalities that can be reasonably attributed to bacterial endotoxins (Levin, et al, 1972; Das, Schwartz, and Folkman, 1973; Young, 1975; Robinson, et al, 1975; Koster, et al, 1978). These clinical correlations are complemented by correlations of the Limulus test with other established assays for endotoxin, some of which are listed in Table 4. Fever, per se, does not produce a positive Limulus test (Glew and Levin, 1975).

The interesting report of Garibaldi, et al (1973) demonstrated an excellent correlation between the presence of urinary tract infection in patients with indwelling catheters and positive Limulus tests; in addition, the greater the concentration of bacteria in the bladder, the higher the frequency of positive Limulus tests. It seems reasonable to conclude that endotoxemia was occurring in

Table 4

Correlation of the Limulus Test with Other Assays for Endotoxin

Pyrogenicity
Mitogenicity
Complement activation
Chick embryo lethality
Dermal Shwartzman reaction
Tissue factor generation

these patients, even in the absence of bacteremia or increased mortality. In this regard, it is important to interpret the significance of a positive Limulus test (and presumptive endotoxemia) within the framework of the clinical setting, just as any diagnostic method must be considered in terms of the individual patient. Accordingly, it is not surprising that in some series, endotoxemia, as detected by the Limulus test, correlated with hypotension and death (Levin, et al, 1972; Das, Schwartz, and Folkman, 1973) while in others, no such serious outcome prevailed (Garibaldi, et al, 1973; Robinson, et al, 1975).

Elin, et al (1978) have suggested that high white blood cell levels in synovial fluid may cause a falsely positive Limulus test. However, the Limulus-positive fluids also produced fever in rabbits, strongly suggesting that an endotoxin-like substance was present. Furthermore, the observations of Hollingsworth and Atkins (1965) demonstrate that the development of synovial inflammation is a very sensitive indicator of bacterial endotoxins. In contrast, Young (1975) has shown that white blood cells decrease the sensitivity of the Limulus test; and Koster, et al, (1978) reported that patients with marked leukemoid reactions (white blood cell count $>70{,}000/mm^3$) had negative Limulus tests.

In summary, the Limulus test reacts with gram-negative, not gram-positive organisms. Using known endotoxins, the Limulus test correlates excellently with other bioassays for endotoxin (Elin, Sandberg, and Rosenstreich, 1976;

Tomasulo, et al, 1977). The Limulus test is not a substitute for a blood culture. However, carefully performed with sensitive amebocyte lysate, it can assist in the diagnosis of gram-negative endotoxemia; and properly interpreted within the framework of the clinical setting, it can aid in the evaluation of the patient. Nevertheless, significant problems remain with its application to the detection of endotoxin in blood (Table 5). Until another more sensitive and specific test for endotoxin becomes available, the ultimate proof that the Limulus test detects only endotoxin in the blood will remain unavailable. A recent report suggested that the measurement of β-hydroxymyristic acid might provide such a test, since this compound is said to be specific for the lipid A moiety of bacterial endotoxins (Maitra, et al, 1978). However, the study demonstrated β-hydroxymyristic acid in the serum of normal persons;

Table 5

Some Problems with Detection of Endotoxin in Blood

Variability of lysate sensitivity
Variability of lysate stability
Endotoxin-free equipment
Positive end point
Inhibitors in blood

therefore, either this compound is not specific for endotoxin or many apparently normal persons have potentially detectable endotoxin in their circulation. Furthermore, the described technology is not as sensitive as the Limulus test.

ACKNOWLEDGMENT

Supported in part by Research Grant HL 01601 from the National Heart, Lung, and Blood Institute, National Institutes of Health, Bethesda, Maryland.

Request for Reprints

Jack Levin, M.D., The Johns Hopkins Hospital, Baltimore, Maryland 21205 USA.

REFERENCES

Bartram CE Jr, Crowder JG, Beeler B, White A (1974). Diagnosis of bacterial diseases by detection of serum antigens by counterimmunoelectrophoresis, sensitivity, and specificity of detecting Pseudomonas and pneumococcal antigens. J Lab Clin Med 83:591.

Butler T, Bell WR, Levin J, Linh NN, Arnold K (1978). Typhoid fever. Studies of blood coagulation, bacteremia, and endotoxemia. Arch Intern Med 138:407.

Caridis DT, Reinhold RB, Woodruff PWH, Fine J (1972). Endotoxaemia in man. Lancet i:1381.

Cooperstock MS, Tucker RP, Baublis JV (1975). Possible pathogenic role of endotoxin in Reye's syndrome. Lancet I:1272.

Das J, Schwartz AA, Folkman J (1973). Clearance of endotoxin by platelets: role in increasing the accuracy of the Limulus gelation test and in combating experimental endotoxemia. Surgery 74:235.

Elin RJ, Robinson RA, Levine AS, Wolff SM (1975). Lack of clinical usefulness of the Limulus test in the diagnosis of endotoxemia. N Engl J Med 293:521.

Elin RJ, Sandberg AL, Rosenstreich DL (1976). Comparison of the pyrogenicity, Limulus activity, mitogenicity and complement reactivity of several bacterial endotoxins and related compounds. J Immunol 117:1238.

Elin RJ, Knowles R, Barth WF, Wolff SM (1978). Lack of specificity of the Limulus lysate test in the diagnosis of pyogenic arthritis. J Infect Dis 137:507.

Feldman S, Pearson TA (1974). The Limulus test and gram-negative bacillary sepsis. Am J Dis Child 128:172.

Fossard DP, Kakkar VV, Elsey PA (1974). Assessment of Limulus test for detecting endotoxaemia. Br Med J 2:465.

Garibaldi RA, Allman GW, Larsen DH, Smith CB, Burke JP (1973). Detection of endotoxemia by the Limulus test in patients with indwelling urinary catheters. J Infect Dis 128:551.

Glew RH, Levin J (1975). Failure to demonstrate circulating endotoxin in malaria. Proc Soc Exp Biol Med 148:508.

Hollingsworth JW, Atkins E (1965). Synovial inflammatory response to bacterial endotoxin. Yale J Biol Med 38:241.

Kenny GE, Wentworth BB, Beasley RP, Foy HM (1972). Correlation of circulating capsular polysaccharide with bacteremia in pneumococcal pneumonia. Infect Immun 6:431.

Koster F, Levin J, Walker L, Tung KSK, Gilman RH, Rahaman MM, Majid MA, Islam S, Williams RC Jr (1978). Hemolytic-uremic syndrome after Shigellosis. Relation to endotoxemia and circulating immune complexes. N Engl J Med 298:927.

Levin J, Tomasulo PA, Oser RS (1970). Detection of endotoxin in human blood and demonstration of an inhibitor. J Lab Clin Med 75:903.

Levin J, Poore TE, Young NS, Margolis S, Zauber NP, Townes AS, Bell WR (1972). Gram-negative sepsis: detection of endotoxemia with the Limulus test. With studies of associated changes in blood coagulation, serum lipids, and complement. Ann Intern Med 76:1.

Liehr H, Grün M (1977). Clinical aspects of Kupffer cell failure in liver diseases. In Wisse E, Knook DL (eds): "Kupffer Cells and Other Liver Sinusoidal Cells", Amsterdam:Elsevier/North-Holland Biomedical Press, p 427.

Magliulo E, Scevola D, Fumarola D, Vaccaro R, Bertotto A, Burberi S (1976). Clinical experience in detecting endotoxemia with the Limulus test in typhoid fever and other Salmonella infections. Infection 4:21.

Maitra SK, Schotz MC, Yoshikawa TT, Guze LB (1978). Determination of lipid A and endotoxin in serum by mass spectroscopy. Proc Natl Acad Sci USA 75:3993.

Martinez-G LA, Quintiliani R, Tilton RC (1973). Clinical experience on the detection of endotoxemia with the Limulus test. J Infect Dis 127:102.

Reinhold RB, Fine J (1971). A technique for quantitative measurement of endotoxin in human plasma. Proc Soc Exp Biol Med 137:334.

Reinhold RB, Caridis DT, Fine J (1972). Diagnosis of clinical endotoxemia. J Reprod Med 8:335.

Robinson JA, Klodnycky ML, Loeb HS, Racic MR, Gunnar RM (1975). Endotoxin, prekallikrein, complement and systemic vascular resistance. Am J Med 59:61.

Rowe MI, Buckner DM, Newmark S (1975). The early diagnosis of gram negative septicemia in the pediatric surgical patient. Ann Surg 182:280.

Stumacher RJ, Kovnat MJ, McCabe WR (1973). Limitations of the usefulness of the Limulus assay for endotoxin. N Engl J Med 288:1261.

Tarao K, So K, Moroi T, Ikeuchi T, Suyama T, Endo O, Fukushima K (1977). Detection of endotoxin in plasma and ascitic fluid of patients with cirrhosis: its clinical significance. Gastroenterology 73:539.

Tomasulo PA, Levin J, Murphy PA, Winkelstein JA (1977). Biological activities of tritiated endotoxins: correlation of the Limulus lysate assay with rabbit pyrogen and complement-activation assays for endotoxin. J Lab Clin Med 89:308.

Wardle EN (1975). Endotoxinaemia and the pathogenesis of acute renal failure. Q J Med 44:389.

Wilkinson SP, Arroyo V, Gazzard BG, Moodie H, Williams R (1974). Relation of renal impairment and haemorrhagic diathesis to endotoxaemia in fulminant hepatic failure. Lancet I:521.

Young LS (1975). Opsonizing antibodies, host factors, and the Limulus assay for endotoxin. Infect Immun 12:88.

INDICATIONS FOR THE USE OF THE LIMULUS LYSATE ASSAY IN BACTERIAL MENINGITIS.

Stuart E. Siegel, M.D. and Ronald Nachum, Ph.D.

Division of Hematology-Oncology, Childrens
Hospital of Los Angeles,
4650 Sunset Boulevard, Los Angeles, CA 90027.

INTRODUCTION.

Since the original description by Levin and Bang of the specific gelation reaction produced by incubating *Limulus* amebocyte or amebocyte lysate, and bacterial endotoxin, (Levin and Bang, 1964), a number of clinical applications of this phenomenon have been explored with varying results. The Limulus Amebocyte Lysate (LAL) assay was first applied to the diagnosis of gram negative bacterial septicemia where its usefulness has been limited by a significant incidence of false negative results (Levin et al, 1970; Levin et al, 1972; Stumacher et al, 1973). Subsequently, the assay has enjoyed a greater specificity when applied to the detection of gram negative urinary tract infections (Jorgensen et al, 1973; Jorgensen et al, 1975), as well as endotoxin contamination of intravenous fluids and drugs for parenteral use (Cooper 1975; Jorgensen and Smith, 1973; Siegel et al, 1976). In this report we will summarize the results of the use of the LAL assay in patients with gram negative bacterial meningitis, perhaps the most useful of the clinical applications for the LAL assay to date.

THE DIAGNOSIS OF GRAM NEGATIVE BACTERIAL MENINGITIS.

Although a number of biochemical and immunologic methods for rapid diagnosis of bacterial meningitis have been studied recently (Fikrig et al, 1973; Coonrod et al, 1972; Edwards et al, 1972), the only generally accepted method remains the demonstration of organisms on gram stain or in cultures of

Pages 245–253, Biomedical Applications of the Horseshoe Crab (Limulidae)

spinal fluid (Swartz and Dodge, 1965; Smith et al, 1973). However, in 1/3 to 1/4 of culture-proven cases, the gram stain is negative and culture results require a minimum of 12-24 hours incubation.

In 1973 we reported the first application of the LAL assay for the diagnosis of gram negative bacterial meningitis (Nachum et al, 1973). Subsequently, the data from several large series of patients have been published confirming our initial conclusions that the LAL assay is a rapid, specific and sensitive test for these infections (Ross et al, 1975; Berman et al, 1976; McCracken and Sarff, 1976; Tuazon et al, 1977; Jorgensen and Lee, 1978; Dyson and Cassady, 1976). A summary of these results is shown in Table 1.

TABLE 1

RESULTS OF LIMULUS LYSATE ASSAY ON INITIAL CSF SAMPLES

DIAGNOSIS (Culture Proven)	No. TESTED (N = 1036)	No. (%) WITH POSITIVE LIMULUS ASSAY	
GRAM-NEGATIVE BACTERIAL MENINGITIS			
Hemophilus influenzae	144	143	(99%)
Neisseria meningitidis	24	24	(100%)
E. coli	12	12	(100%)
Klebsiella-Enterobacter sp.	4	4	(100%)
Serratia sp.	1	1	(100%)
Proteus sp.	1	1	(100%)
Pseudomonas aeruginosa	5	5	(100%)
Citrobacter sp.	2	2	(100%)
E. corrodens	1	1	(100%)
Mima polymorpha	1	1	(100%)
Other Enteric Gram-negative	42	31	(75%)
GRAM-POSITIVE BACTERIAL MENINGITIS			
Streptococcus pneumoniae	33	2	(6%)
Streptococcus (Groups A & B)	25	0	
Staphylococcus aureus	6	0	
Staphylococcus epidermidis	2	0	
Listeria monocytogenes	1	0	
Mycobacterium tuberculosis	3	0	
NON-BACTERIAL			
Cryptococcus neoformans	2	0	
Encephalitis	3	0	
Meningeal leukemia	4	0	
Aseptic or No Meningitis	720	3	(0.4%)

Of the 237 patients with gram negative organisms isolated from initial CSF samples, 225 had positive Limulus assay (5% false negative results). In contrast, only 5 positive assays were noted in the 799 patients without evidence of gram-negative bacterial infection by culture (0.06% false positive results).

Since the gram stain of CSF is the only generally available rapid diagnostic tool presently in use, we compared the sensitivity of this procedure with that of the LAL assay on initial CSF specimens (Table 2).

Clearly, the LAL assay is superior in its sensitivity to the gram stain in detecting the presence of a gram negative organism in the CSF. A clinically useful corollary to this finding is the situation of the patient who presents to the physician having already received partial antibiotic treatment before diagnostic lumbar puncture. Such therapy has been shown to often reduce the number of bacterial organisms present and thereby impair the ability to make a definitive diagnosis (Feldman, 1978). Nachum et al, found that over two-thirds of these patients will still have positive LAL assays after 24 to 48 hours of antibiotic therapy, while less than one quarter will have a positive gram stain (Nachum et al, 1973).

MONITORING ANTIBIOTIC RESPONSE WITH THE LAL ASSAY

A major problem in the clinical management of bacterial meningitis is the decision as to discontinuing antibiotic therapy. At present the criteria for stopping therapy is variable, but usually includes absence of fever for five days, and return of CSF cell count, sugar and protein toward normal levels (Swartz and Dodge, 1965; Smith et al, 1973; Jacob and Kaplan, 1977). Using a tube dilution technique with a known endotoxin standard, Berman et al demonstrated a rapid fall in endotoxin levels following institution of appropriate antibiotic therapy, with no detectable endotoxin after five full days of such treatment (Berman et al, 1976). (Figure 1). A single patient with H. influenzae meningitis has been described to have a persistently positive LAL assay in CSF for 9 days after antibiotic therapy was begun, although the CSF was culture negative after the first two days. (Jorgensen and Smith, 1978). The patient recovered without incident. In contrast, McCracken et al noted a direct correlation between the duration of positive LAL assays

TABLE 2

COMPARISON OF LIMULUS ASSAY AND GRAM STAIN ON INITIAL CSF SAMPLES IN GRAM-NEGATIVE BACTERIAL MENINGITIS

Diagnosis (Culture-proven)	No. Tested*	No. (%) with Positive Limulus Assay		No. (%) with Positive Gram Stain	
Hemophilus influenzae	101	100	(99%)	74	(74%)
Neisseria Meningitidis	15	15	(100%)	10	(66%)
Klebsiella-Enterobacter sp.	3	3	(100%)	1	(33%)
Pseudomonas sp.	5	5	(100%)	3	(60%)
Other Gram-negative	17	17	(100%)	8	(47%)
TOTAL	141	140	(99.5%)	96	(68%)

*Only patients who had both limulus assay and gram stain on initial specimen included.

FIGURE 1.

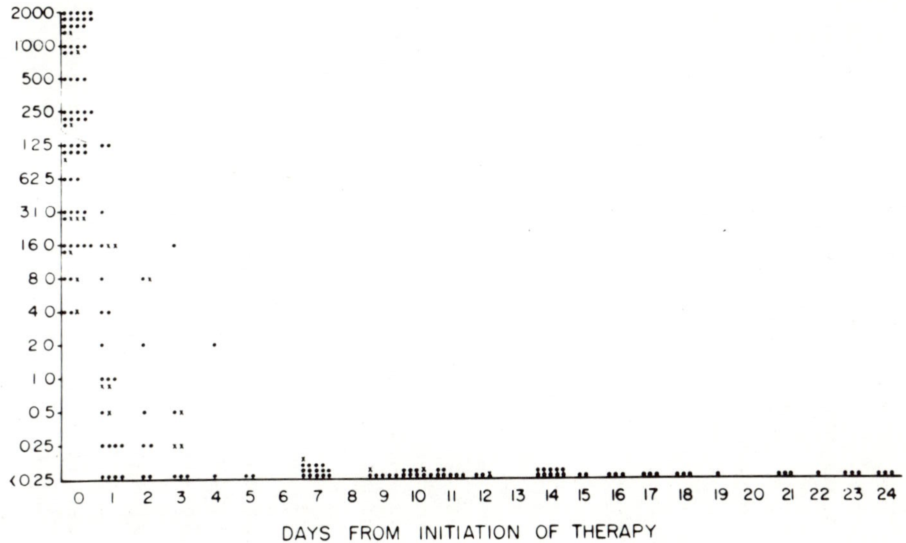

Relative endotoxin concentrations (ng/ml) in initial and serial CSF specimens from 83 patients with documented gram negative bacterial meningitis. Concentrations were obtained by using a standard curve for determination of endotoxin concentration using Limulus assays of serial 2-fold dilutions of E coli (Difco 0127:B8) endotoxin. Assays were read as follows: negative = clear fluid; 3+ = firm gel; 2+ = viscous, adherent granules; 1+ = increased turbidity and viscosity. Negative Limulus assays are indicated by an *X*; 1—2+ assays by o; 3+ assays by ●.

(FROM: Berman NS, Siegel SE, Nachum R, et al (1976). Cerebro-spinal fluid endotoxin concentrations in gram-negative bacterial meningitis. J Pediat 88:553) after the initiation of therapy and outcome of the episode (McCracken and Sarff, 1976).

Several patients have been reported to convert from positive to negative CSF Limulus assays at the onset of therapy, only to note a recurrence of a positive assay 11 to 26 days later, preceded or accompanied by a positive CSF culture for the original gram negative organism (Nachum et al, 1973; Jorgensen and Smith, 1978). Figure 2 illustrates a case in which the LAL assay became positive two days before the organism could be recovered from the CSF. Finally, we recently reported a case in which the LAL assay was positive two days before the initial recovery of a gram-negative organism from the CSF of a child with a pilonidal sinus and clinical signs of meningitis (Siegel and Nachum, 1977).

SUMMARY AND CONCLUSIONS.

Experience with the Limulus assay in over 1000 patients has provided the basis for an accurate assessment of the advantages and limitations of the LAL assay in bacterial meningitis as well as recommendations for the use of this test in various clinical situations.

Advantages of the LAL Assay

1. Rapid, sensitive and specific assay for the presence of gram negative organisms. Less than 1% false negative results and test takes 30 to 90 minutes.

2. Permits quantitation of endotoxin present.

3. Provides most sensitive method for assessing response to antibiotic therapy.

4. May detect recurrent infection prior to the ability to recover organism by routine culture technique.

Disadvantages of the LAL Assay

1. Detects only the presence of gram negative bacteria.

2. Does not differentiate between different gram negative bacteria.

FIGURE 2.

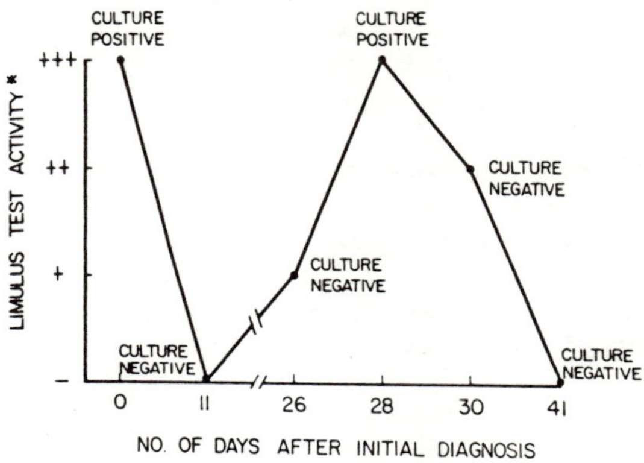

Cerebrospinal-Fluid Culture and Limulus Test Results on a Patient with Recurrent *H. influenzae* Meningitis.

Positive culture indicates recovery of *H. influenzae* from the spinal fluid. At the time of the recurrent episode, the positive limulus test preceded positive culture by two days.

*Scale of limulus activity: negative indicates clear or slight flocculation, + increased turbidity or viscosity, + + adherent granules, and + + + solid gel.

(FROM: Nachum R, Lipsey A, Siegel SE (1973). Rapid detection of gram-negative bacterial meningitis by the Limulus lysate test. N Eng J Med 289:931.

Indications for Clinical Use of LAL Assay in Bacterial Meningitis

1. For rapid detection of gram negative bacterial meningitis on initial diagnositc CSF specimens, particularly in patient populations with high risk of such infections, ie. children, immunosuppressed or debilitated patients, and in the partially treated patient.

2. To monitor response of gram negative bacterial meningitis to antibiotic therapy and to assist in determining when to stop therapy.

3. To detect recurrent gram negative bacterial CSF infection after treatment.

REFERENCES.

Berman NS, Siegel SE, Nachum R, et al (1976). Cerebrospinal fluid endotoxin concentrations in gram-negative bacterial meningitis. J Pediat 88:553.

Coonrod JD, Rytel MW (1972). Determination of aetiology of bacterial meningitis by counter-immunoelectrophoresis. Lancet 1:1154.

Cooper JF (1975). Principles and applications of the Limulus test for pyrogen in parenteral drugs. Bull Parenter Drug Assoc 29:122.

Dyson D, Cassady G (1976). Use of Limulus lysate for detecting gram-negative neonatal meningitis. Pediat 58:105

Edwards EA, Muehle PM, Peckinpaugh RO (1972). Diagnosis of bacterial meningitis by counter-immunoelectrophoresis J Lab Clin Med 80:449

Feldman WE (1978). Effect of prior antibiotic therapy on concentrations of bacteria in CSF. Am J Dis Child 132:672.

Fikrig SM, Berkovich S, Emmett SM, et al (1973). Nitroblue tetrazolium due test and differential diagnosis of meningitis. J Pediat 82:855.

Jorgensen JH, Carvajal HF, Chipps BE, et al (1973). Rapid detection of gram-negative bacteriuria by use of the Limulus endotoxin assay. Appl Microbiol 26:38.

Jorgensen JH, Jones PM (1975). Comparative evaluation of the Limulus assay and the direct gram stain for detection of significant bacteriuria. Am J Clin Path 63:142.

Jorgensen JH, Lee JC (1978). Rapid diagnosis of gram-negative bacterial meningitis by the Limulus endotoxin assay. J Clin Microbiol 7:12.

Jorgensen JH, Smith RF (1973). Rapid detection of contaminated intravenous fluids using the Limulus in vitro endotoxin assay. Appl Microbiol 26:52.

Levin J, Bang FB (1964). The role of endotoxin in the extracellular coagulation of Limulus blood. Bull Johns Hopkins Hosp 115:264.

Levin J, Poore TE, Young NS, et al (1972). Gram-negative sepsis: Detection of endotoxemia with the Limulus test. Ann Int Med 76:1.

Levin J, Poore TE, Zauber NP, et al (1970). Detection of endotoxin in the blood of patients with sepsis due to gram-negative bacteria. N Eng J Med 283:1313.

McCracken GH Jr, Sarff LD (1976). Endotoxin in cerebrospinal fluid. Detection in neonates with bacterial meningitis. JAMA 235:617.

Nachum R, Lipsey A, Siegel SE (1973). Rapid detection of gram-negative bacterial meningitis by the Limulus lysate test. N Eng J Med 289:931.

Ross S, Rodriguez W, Controri G, et al (1975). Limulus lysate test for gram-negative bacterial meningitis. JAMA 233:1366.

Siegel SE, Nachum R (1977). Use of the Limulus lysate assay (LAL) for the detection and quantitation of endotoxin. In Bernheimer W (ed): "Perspectives in Toxinology" New York: John Wiley & Sons, Inc., p.61.

Siegel SE, Nachum R, Leimbrock S, et al (1976). Detection of bacterial endotoxin in antitumor agents. Cancer Treat Rep 60:9.

Smith DH, Ingram DL, Smith AL, et al (1973). Bacterial meningitis. A Symposium. Pediatrics 52:586.

Stumacher RJ, Kovnat MJ, McCabe WR (1973). Limitations of the usefulness of the Limulus assay for endotoxin. N Eng J Med 288:1261.

Swartz MN, Dodge PR (1965). Bacterial meningitis - A Review of selected aspects. 1. General clinical ceatures, special problems and unusual meningeal reactions mimicking bacterial meningitis. N Eng J Med 272:725.

Tuazon CV, Perry AA, Elin RJ, et al (1977). Detection of endotoxin in cerebrospinal and joint fluids by Limulus assay. Arch Int Med 137:55.

DETECTION OF ENDOTOXIN IN URINE: RAPID DIAGNOSIS OF GRAM NEGATIVE BACTERIURIA BY THE LIMULUS ASSAY

James H. Jorgensen, Ph.D.

Associate Professor
Departments of Pathology and Microbiology
The University of Texas Health Science Center
San Antonio, Texas 78284

INTRODUCTION

A quantitative bacteriologic culture revealing greater than 100,000 bacteria per milliliter of urine has been considered the best laboratory procedure available for documenting the presence of urinary tract infection (Kaitz and Williams, 1960). However, a period of 24-48 hours is required for cultivation and characterization of microorganisms from urine.

A simple rapid screening method for the detection of significant bacteriuria would seem to be useful in many situations. There previously have been at least three approaches to development of new screening tests for this purpose. One has been the introduction of miniaturized inexpensive culture devices (Brupacher and Domingue, 1973; Craig and Kunin, 1972). However, these devices still require a conventional culture incubation period of at least 18-24 hours. The second approach has been detection of bacterial products or enzymes in urine by chemical tests such as the Griess nitrate reductase test (Smith, et. al., 1961) or the measurement of tretrazolium reductase (Neter, 1965). While these tests have the advantage of providing more rapid results, they are often much less reliable. A third approach has been the examination of urinary sediment by use of either phase-contrast microscopy (Brody, et. al., 1968) or a Gram stained smear of fresh uncentrifuged urine (Pryles, 1960). Careful examination of a Gram stain has been said to correlate with significant colony counts in 75-95% of cases (Bulger and Kirby, 1963; Kass, 1956). However, considerable judgement

Pages 255–264, Biomedical Applications of the Horseshoe Crab (Limulidae)

must be applied in correlating the number of organisms visualized on a Gram stained smear and the probable colony count.

We first described the use of the Limulus assay for the detection of significant Gram negative bacteriuria in 1973 (Jorgensen, et. al., 1973). We have examined more than 1200 urines from two different patient groups and found the Limulus assay to be a useful screening method for detection of significant bacteriuria (Jorgensen, et. al., 1973; Jorgensen and Jones, 1975). This communication will review the data from our two previous studies and offer certain newer previously unpublished observations.

LIMULUS ASSAY PROCEDURES

The method of preparation of the Limulus lysate utilized in our prior published studies has been described (Jorgensen and Smith, 1973). Lysate used to examine the original 209 urine specimens demonstrated a sensitivity of 1 ng/ml of *Escherichia coli* 055:B5 TCA extracted endotoxin. Lysate used in our subsequent study of more than 1000 urines was of lesser potency, having a minimum detectability of 5 ng/ml of the same endotoxin preparation. A commercial lysate, Difco Pyrotest, has been utilized in more recent studies. Pyrotest demonstrated much greater sensitivity, being able to detect as little as 0.125 ng/ml of endotoxin. Limulus assays of urine samples were performed following appropriate dilution in either pyrogen free physiologic saline or distilled water (Jorgensen, et. al., 1973, Jorgensen and Jones, 1975). Diluted urine samples were reacted with Limulus ameobocyte lysate at 37C for 60-120 minutes. A positive result was considered to be a firm gel that resisted dissolution when the test tube was inverted 180°.

RELATIONSHIP BETWEEN BACTERIAL NUMBERS AND ENDOTOXIN ACTIVITY

In 1974 we described the measurement of bound and free endotoxin occurring during growth of Gram negative bacterial cultures (Jorgensen and Smith, 1974). Figure 1 depicts the growth curves obtained using a culture of *Escherichia coli* grown in a pyrogen free synthetic growth medium.

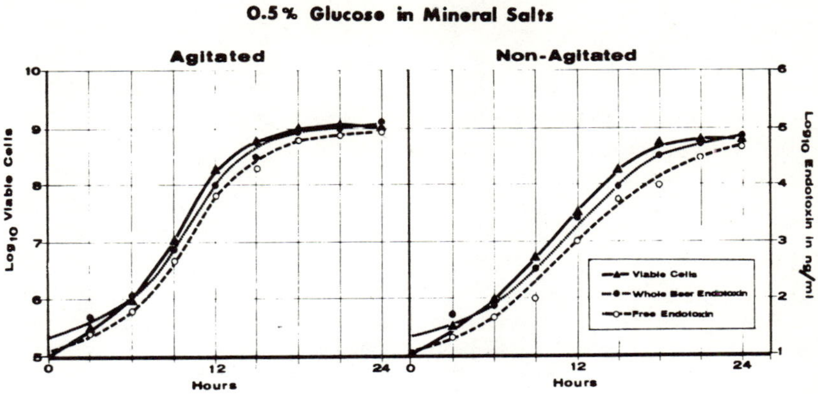

Figure 1. Measurement of bound and free endotoxin during growth

Under these conditions, the amounts of bound or cell associated endotoxin, and free or solubilized endotoxin paralleled the growth of viable bacteria in the medium. Therefore, these data suggested that the classical theory of endotoxin accumulation only during the death phase of Gram negative bacteria was incorrect. Instead, it appeared that the Limulus assay was able to measure endotoxin that was associated with the intact cell walls of viable Gram negative bacteria, and that endotoxin was also solubilized and accumulated readily during growth in a liquid environment. These data further suggested that measurement of the amount of total or bound endotoxin could be predictive of the numbers of viable bacteria present.

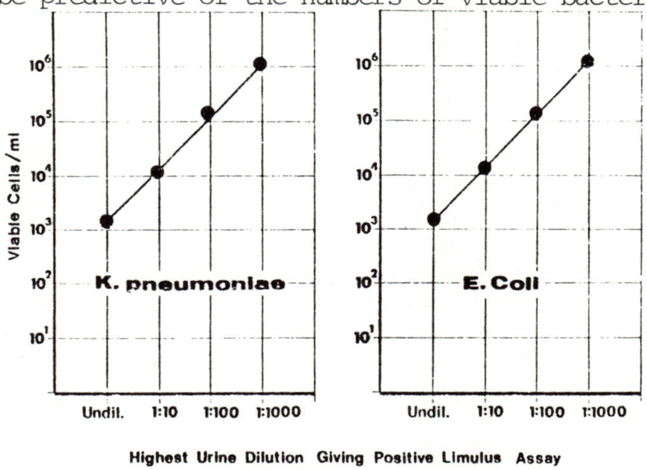

Figure 2. Measurement of endotoxin from bacteria suspended in urine.

These observations were confirmed by carefully suspending varying concentrations of Gram negative bacteria in filter sterilized human urine (Figure 2). Under these experimental conditions, a linear relationship was developed between endotoxin as measurable by the Limulus assay, and viable Gram negative bacteria, over the range of 1000 to 1,000,000 bacterial cells per milliliter of urine. Therefore, by appropriately diluting a urine specimen, it should be possible to establish a threshold of detectability of bacteria in urine. Most authorities would agree that this should be a colony count from a clean voided mid-stream urine specimen of more than 100,000 colony forming units per milliliter (Kass, 1956). Using ameobocyte lysate of 1 ng/ml sensitivity, as in Figure 2, a dilution of urine of 1:100 should correspond to a level of approximately 100,000 bacteria/ml. This hypothesis was tested by examination of 209 urine specimens from children in the Shriner's Burn Institute in Galveston (Jorgensen, et. al., 1973) and by subsequent examination of 1077 urine specimens from patients in the Bexar County Hospital District in San Antonio (Jorgensen and Jones, 1975).

CORRELATION OF URINARY ENDOTOXIN AND BACTERIURIA

The correlation of Limulus assay results and quantitative urine cultures from the Galveston patients are indicated in Table 1. In this initial group of patients there was an overall 99% ability to correctly classify urine specimens as either containing less than 10^5 bacteria/ml or greater than 10^5 bacteria/ml. In the second expanded study performed in San Antonio, there was a total agreement of 94.4% between Limulus assays and urine cultures. If one examines the combined data of both studies, the Limulus assay detected 87% of all urine specimens containing greater than 10^5 bacteria/ml and 99% of those urines containing greater than 10^5 Gram negative bacteria/ml. The specificity of the Limulus assay for Gram negative bacterial endotoxin is exemplified by the almost total detectibility of Gram negative bacteria in significant numbers in urine, while those urines that were incorrectly classified almost always contained Gram positive organisms, which are undetectable by the Limulus assay.

Table 1. Summary of data from two studies to determine ability of Limulus assay to detect significant bacteriuria.

Results	Jorgensen et. al. (1973)	Jorgensen & Jones (1974)	Total of Both Studies
Samples with >10^5 organisms/ml having a positive Limulus assay	23/25 (92%)	175/203 (86%)	198/228 (87%)
Samples with >10^5 Gram-negative bacteria/ml having a positive Limulus assay	23/23 (100%)	159/161 (99%)	182/184 (99%)
Overall ability of Limulus assay to correctly classify urine specimens as <10^5/ml or >10^5/ml	207/209 (99%)	1016/1077 (94%)	1223/1286 (95%)

Table 2 includes the results of Limulus assays and examinations of the direct Gram stains of uncentrifuged urinary sediment for detection of significant bacteriuria in our second study. We were unable to confirm the reported reliability of the direct Gram stain technique. Only 69.5% of urines containing greater than 10^5 organisms/ml could be correctly classified by direct Gram stain examination of urine sediment. A disadvantage of the Limulus assay was the inability to differentiate between a polymicrobic bacteriuria and bacteriuria due to a single species. The Limulus assay detected the total amount of endotoxin present in the urine, irrespective of whether it arose from a single or multiple species.

Table 2. Comparison of Limulus assay and gram stain results on urine specimens containing >10^5 organisms per ml.

Organism	Total No.	Limulus Positive*	Gram Stain Positive*
Escherichia coli	96	94 (97.9%)	78 (81.3%)
Klebsiella pneumoniae	17	17 (100.0%)	14 (82.4%)
Pseudomonas aeruginosa	8	8 (100.0%)	7 (87.5%)
Proteus mirabilis	6	6 (100.0%)	5 (83.3%)
Enterobacter aerogenes	2	2 (100.0%)	0
Enterobacter cloacae	2	2 (100.0%)	1 (50.0%)
Serratia marcescens	1	1 (100.0%)	0
Providencia spp.	1	1 (100.0%)	0
Herellea vaginocola	1	1 (100.0%)	0
Citrobacter freundii	1	1 (100.0%)	0
Staphylococcus aureus	3	0	3 (100.0%)
Group D Streptococcus	2	0	0
Group B Streptococcus	1	0	0
Streptococcus viridans	3	0	1 (33.3%)
Candida albicans	3	0	3 (100.0%)
2 GNR's	21	21 (100.0%)	12 (61.1%)
3 or >GNR's	6	6 (100.0%)	3 (50.0%)
2 GP's	5	0	3 (60.0%)
3 or >GP's	4	2 (50.0%)	1 (25.0%)
1 GNR + 1 GP	12	8 (61.5%)	6 (46.2%)
Mult. GNR's + GP's	7	5 (71.4%)	4 (57.1%)
Total	203	175 (86.2%)	141 (69.5%)

*Percentages indicate % of total number positive.
GNR = Gram-negative rod; GP = Gram-positive organism.

Using a single urine dilution, the Limulus assay does not recognize bacteriuria of 10^4-10^5 organisms/ml (Table 3) the area lying between true bacteriuria and unavoidable uretheral-vaginal contaminants.

Table 3. Comparison of Limulus assay and Gram stain results on urine specimens containing between 10^4 and 10^5 organisms per ml.

	Total No.	Limulus* Positive	Gram Stain Positive*
Escherichia coli	16	7 (43.8%)	2 (12.5%)
Klebsiella pneumoniae	3	2 (66.7%)	1 (33.3%)
Pseudomonas aeruginosa	3	1 (33.3%)	0
Proteus spp.	3	2 (66.7%)	0
Misc. GNR's	5	0	0
Group D Streptococcus	10	0	0
Staphylococcus epidermidis	7	0	0
Group B Streptococcus	2	0	0
Streptococcus viridans	3	0	0
Candida albicans	2	0	0
2 GNR's	7	2 (28.6%)	1 (14.3%)
3 or >GNR's	2	1 (50.0%)	0
2 GP's	18	5 (27.8%)	4 (22.2%)
3 or >GP's	5	1 (20.0%)	1 (20.0%)
1 GNR + 1 GP	13	2 (15.4%)	1 (7.7%)
Mult. GNR's + GP's	12	3 (25.0%)	2 (16.7%)
Total	111	26 (26.4%)	12 (10.8%)

*Percentages indicate % of total number positive.
GNR = Gram-negative rod; GP = Gram-positive organism.

Specimens which demonstrated growth of fewer than 10^4 organisms/ml were rarely positive by the Limulus assay. One explanation for occasional occurrence (<4%) of false-positive results could be the presence of dead Gram negative bacteria from the urine of recently antibiotic treated patients which would still yield a positive Limulus assay. Table 4 depicts the results of an experiment in which Gram negative bacteria were killed by exposure to the antibiotic gentamicin then examined by the Limulus test. The results indicated that viability was not a requirement for the organisms to be detectable by the Limulus assay. Therefore, it is conceivable that the endotoxin level might at time exceed the significant threshold in patients receiving antimicrobial chemotherapy and would reflect antibiotic killed Gram negative bacteria.

Table 4. The ability of the Limulus assay to detect antibiotic-killed cells.

Organism	Colony count before gentamicin*	Colony count after gentamicin	Limulus Reaction @ 1:10	1:100	1:1000
E. coli	1.42×10^6	4	+	+	+
K. pneumoniae	1.20×10^6	0	+	+	+

*Gentamicin, 100 µg/ml. for 2 h at 37C.

Figure 3 describes recent data from examination of 50 selected urine specimens using both a very sensitive Limulus ameobocyte lysate, Difco Pyrotest, and a rather insensitive batch of our own lysate with a sensitivity of only 5 ng/ml.

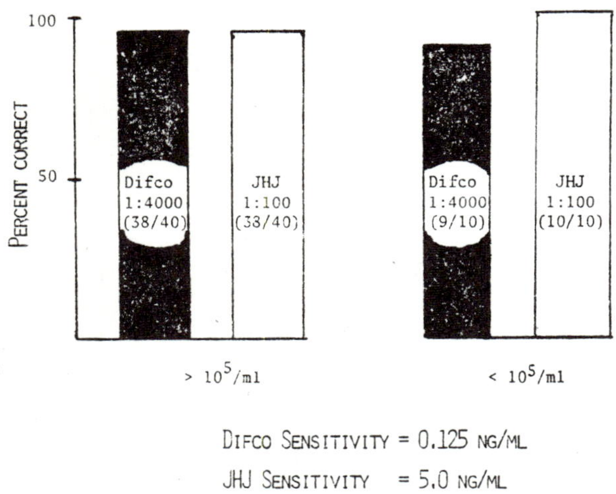

Figure 3. Comparison of lysates of different sensitivities on 50 selected urines.

As can be seen from these more recent data, by appropriate adjustment of the preliminary urine dilution, identical results can be obtained using lysates of vastly differing qualities. Lysates of even lesser sensitivity, i.e., 25-100 ng/ml would seem ideal for the examination of urine. Therefore, this may provide a valuable application for lysate preparations that would not be considered suitable for purposes such as examination of cerebrospinal fluid, or detection of pyrogens in parenteral fluids. It would also seem that use of ameobocyte lysate of 25-100 ng/ml sensitivity would be less costly, and therefore could make the examination of urine economically quite feasible.

VALUE OF ENDOTOXIN DETECTION IN URINE

A simple rapid screening test for the detection of significant bacteriuria would seem to be of importance in a number of situations. The need for periodic screening of certain high risk groups such as pregnant women, school girls, and diabetics for the presence of significant bacteriuria has been well established (Greenberg, et. al., 1965; Smith & Schmidt, 1962). Although methods employing bacterial cultures are the most accurate laboratory procedures for documenting bacteriuria, a significant time lag occurs before results are available. The findings of our two previous studies (Jorgensen, et. al., 1973; Jorgensen and Jones, 1975) demonstrated that the Limulus ameobocyte lysate assay can be used successfully for the detection of significant Gram negative bacteriuria. Results can be available within 1-2 hours of collection of the urine specimens and the performance of the Limulus test does not require the degree of subjective judgement for accurate quantitation of bacteria required by microscopic methods. The predictive value of the Limulus assay for urinary tract infection lies in the normal preponderance of Gram negative uropathogens, which possess endotoxin. The major deficiency of the Limulus test is that Gram positive bacteria and yeasts are undetectable. However, only a small percentage of all urinary tract infections are caused by Gram positive organisms. It must be recognized that the Limulus test cannot provide information regarding the identity and antibiotic susceptibility of organisms in urine or the presence of more than one bacterial species. Therefore, we propose that the Limulus assay is a potentially valuable rapid screening technique for detection of significant bacteriuria in certain patient groups.

REFERENCES

Brody LM, Webster C, Kark RM (1968). Identification of elements of urinary sediment with phase contrast microscopy. JAMA 206:1777-1781.

Brupacher R, Domingue G (1973). Experiences with a screening test for bacteriuria. Am J Clin Pathol 59:203-210.

Bulger RJ, Kirby MM (1963). Simple tests for significant bacteriuria. Arch Intern Med 112:156-160.

Craig WA, Kunin CM (1972). Quantitative urine culture method using a plastic "paddle" containing dual media. Appl Microbiol 23:919-922.

Greenberg ND, Stamler J, Zackler, J, et al (1965). Detection of urinary tract infections in pregnant women. Public Health Rep 80:805-811.

Jorgensen JH, Carvajal HF, Chipps BE, Smith RF (1973). Rapid detection of gram negative bacteriuria by use of the Limulus endotoxin assay. Appl Microbiol 26:38-42.

Jorgensen JH, Smith RF (1973). Preparation, sensitivity and specificity of Limulus lysate for endotoxin assay. Appl Microbiol 26:43-48.

Jorgensen JH, Smith RF (1974). Measurement of bound and free endotoxin by the Limulus assay. Proc Soc Exp Bio Med 146:1024-1031.

Jorgensen, JH and Jones PM (1975). Comparative evaluation of the Limulus assay and the direct Gram stain for detection of significant bacteriuria. Am J Clin Pathol 63:142-148.

Kaitz AL, Williams FJ (1960). Bacteriuria and urinary tract infection in hospitalized patients. New Engl J Med 262:425-430.

Kass EH (1956). Asymptomatic infections of the urinary tract. Trans Assoc Am Phys 69:56-64.

Neter E (1965). Evaluation of the tetrazolium test for the diagnosis of significant bacteriuria. JAMA 192:769.

Pryles CV (1960, The diagnosis of urinary tract infection. Pediatrics 26:441-451.

Smith LG, Schmidt J (1962). Evaluation of three screening tests for patients with significant bacteriuria. JAMA 181:431-433.

Smith LG, Thayer, WR, Malta EM et al (1961). Relationship of the Griess nitrite test to bacterial culture in the diagnosis of urinary tract infection. Ann Intern Med 54:66-72.

THE LAL-BEAD ASSAY FOR ENDOTOXIN

Nick S. Harris, Ph.D. and Robert Feinstein, Ph.D.

Immunology Section, Health Central Inst., Minneapolis, Minnesota 55422 and College of Medicine Texas A & M, College Station, Texas

Studies from our laboratory have demonstrated that certain hydrophobic, aliphatic polymers have a unique specific affinity for binding endotoxin (Harris, 1976). This finding led to the development of an assay procedure (Harris and Feinstein, 1976) that could be used to detect endotoxin in many different biological materials when used in conjunction with the limulus amebocyte lysate (LAL). This new assay procedure (LAL-Bead Assay) can detect endotoxin in materials previously considered inhibitory to the LAL reaction (Elin and Wolff, 1973). Using the LAL-Bead Assay procedure we have found a high correlation among patients with clinically defined sepsis and the presence of free endotoxin in their blood (Harris and Feinstein, 1977a, 1977b).

MATERIALS AND METHODS

The overall design of the LAL-Bead Assay is depicted in Figure 1. More than 75 individual pediatric burn patients (at the Shriners Burns Inst., Galveston, Tx.) with greater than 30% total body burn were selected for study. In addition, more than 25 normals were included as controls. In the performance of the test, 1 cc of either patient or control blood was placed into either a polystyrene tube (Falcon #2003) containing 0.1 cc of pyrogen-free heparin or placed into a 3 cc heparinized-vacutainer. All samples were refrigerated until used and no sample was used later than 24 hours after venipuncture. All studies were performed in duplicate.

Pages 265–274, Biomedical Applications of the Horseshoe Crab (Limulidae)

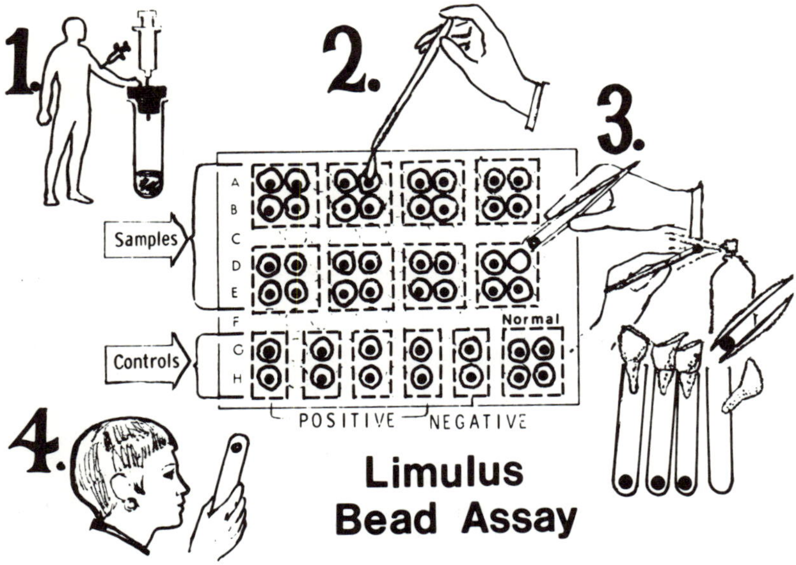

Fig. 1. Design of LAL-Bead Assay.

The whole-heparinized blood (.2 cc) was dispensed into the wells of a PSI bead-tray (Preventive Systems Inc. bead tray distributed by Bio-Concepts Inc., Phoenix, Arizona). Samples were also diluted 1:2 with pyrogen-free water and dispensed into the wells of the PSI tray. In addition to the patient and normal control blood, negative controls were established for water, saline and blank bead. Positive controls consisted of four dilutions (10 ng/ml, 5 ng/ml, 2.5 ng/ml and 1 ng/ml) of endotoxin standard (E.coli endotoxin, Associates of Cape Cod, Woods Hole, Mass.). Each well of the bead-tray contained a PSI-100 bead (pretreated .3 cm diameter hydrophobic poly-α-olefin bead. Since the beads are hydrophobic, it is necessary to push them into the well using a sterile pyrogen-free needle, in order to wet the surface of the bead. The tray was covered and placed in a 37°C incubator for 1 hour. A rocker platform or vibrating table inside the incubator provides a continuous mixing of the bead and sample during the incubation period. Prior to the end of the first incubation period, the limulus amebocyte lysate (LAL) was prepared according to the manufacturers instructions. Each sample tested required 0.1 cc of LAL; this was dispensed into sterile, pyrogen-free 10 X 75 mm flint glass test tubes. While different brands of LAL can and have been used for the bead assay, the most reliable and reproducible results were obtained using Pyrotell (Associates of Cape Cod, Woods Hole, Mass.).

After the primary incubation, the PSI tray was removed from the incubator; using pre-flamed forceps, each bead was washed in a stream of pyrogen-free saline and placed into a tube containing LAL. The tubes containing the LAL plus beads were incubated at 37°C for 90 min. It is imperative that the tubes containing the LAL and beads remain undisturbed during this period. The incubator, itself, must be located in a position in the laboratory with minimum vibration. After the second incubation, the tubes were carefully removed from the incubator and individually observed for either a gellation reaction or an increase in viscosity and turbidity. It is highly recommended that the positive and negative controls be examined first. If the negative controls, saline, water, blank-bead and normal subject do not result in a true negative reaction, i.e., clear and 'run-like water' when inverted, the entire test is voided. The endotoxin standards should exhibit diminishing degrees of viscosity and turbidity in a graded fashion. The test is designed to detect levels of free endotoxin in whole blood of one to 10 ng/ml. By comparing the patients' samples with the various controls and standards, an estimate of the amount of free endotoxin in blood was obtained.

Affinity binding of the bead to different types of endotoxin was compared. *Klebsiella pneumoniae* (Food and Drug Administration), *Pseudomonas aeruginosa* (Parke-Davis and Baxter-Travenol), *Salmonella typhosa* (Tx. Dept. of Public Health and Difco) and *Escherichia coli* (Assoc. of Cape Cod and Difco) all produced similar reactions at comparable dilutions using Pyrotell.

The specificity of the bead for endotoxin was performed as outlined in Figure 2. Samples of either human or sheep blood were separated into two parts; one of which was spiked with endotoxin. The endotoxin-treated sample was then passed down a column containing the PSI-beads. Ten separate trials with human blood and six trials with sheep blood were performed . The assay procedure consisted of a hemogram, SMA-6/60 survey, SMA-12/60 survey, lipogram, electrophoresis, RIA for hormones and clotting factors. The only difference between the paired samples was the removal of endotoxin in the spiked sample. The assay was repeated with non-spiked (normal) blood, instead of endotoxin-treated blood. The results were the same. According to our measurements, the beads were not removing any elements from the blood.

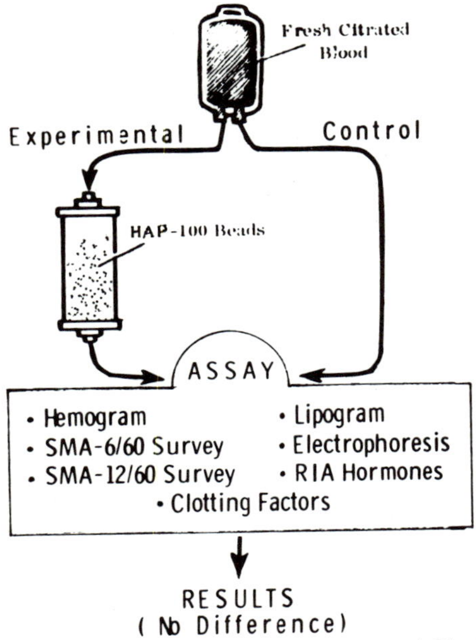

Fig. 2. Design of study to measure specificity of the bead for endotoxin.

RESULTS

This was primarily a retrospective study on various patient samples performed without knowing the clinical status of the patient. At the completion of various stages of the study, the assays were compared with the patients' charts in respect to the clinical judgement of sepsis.

There are various definitions and criteria for sepsis (Wilson,1976 and Rowe et al. 1975). Some investigators believe that bacteremia is necessary for a clinical diagnosis of sepsis (Pruitt and Foley, 1973 and Marvin et al. (1975). However, this criterion was not used in this study. Instead, an operational definition was used. Patients were judged to have clinically-defined septicemia when it was noted on the chart and treatment for the syndrome was initiated. For example, Table 1, gives the results of the blood cultures from twenty clinically-defined septic patients at the Shriners Burns Institute, Galvestion, Tx. Ten of these patients had negative blood cultures, seven had gram-negative organisms and three patients had gram-positive blood cultures.

TABLE 1. Patients with Clinical Septicemia

Blood Culture	Number of Patients
No-growth	10
Gram-negative organism	7
Gram-positive organism	3

Figure 3 presents the relationship of the LAL-bead assay to 10 patients that had gram-negative bacteremia. There was very little correlation between bacteremia and endotoxemia. It did appear, however, that at sometime, either prior to or after the culture was obtained, the blood of the patients did contain endotoxin.

The LAL-Bead Assay had a better corelation when it was compared to the clinical status of the patients. In a group of 61 clinically defined septic patients, 48 had gram-negative infections and 47 of these 48 were LAL positive in the bead assay. The other 23 patients had gram-positive infections. All of these patients were LAL negative with the bead assay.

A typical LAL bead profile is shown in Table 2. This followed the course of a 70% burn patient through the non-septic phase, to a pre-or-suspected stage but not treated, and then a clinically-defined stage of sepsis. Samples were obtained daily when possible. If can be seen from this table that levels of free endotoxin in blood, greater than 1 ng/ml, were detected prior to the time that the patient was clinically defined as septic. It should be noted that

TABLE 2 LAL-Bead Profile of Patient with 70% Burn

Diagnosis[1]	Day[2]	Endotoxin Levels[3]		
		<1 ng	1 ng to 3 ng	>4 ng
	4			
	6			
Not Septic	8	–		
	10			++
	12		+	
	14			
	16		+	
Pre or Suspected	17			++
Septic	18		+	
	19	–		
	20		+	
	22			
	23		+	
	24			
	25		+	
	26			
	27		+	
	28			
Clinically	29	–		
Defined Septic	30		+	
and Treated	31		+	
as Such	32		+	
	33		+	
	34			
	35			
	36			
	37			++
	38			
	39		+	
	40		+	

1 - Diagnosis based on treatment made and chart review
2 - Days post admission
3 - Ng/ml blood related to standards

even during the time period when this patient was a defined septic, he did have an isolated incidence where the endotoxin levels, at the time assayed, were less than 1 ng/ml of blood.

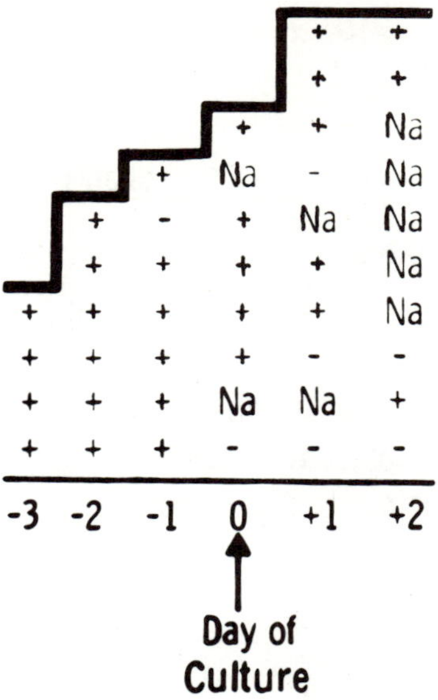

Fig. 3. Relationship of the LAL-Bead Assay to blood cultures.

The results of multiple studies such as this suggest that certain patterns of endotoxemia must be examined in order to relate endotoxin and endotoxemia to sepsis.

A computer model for the predictive value of the LAL-Bead Assay to endotoxemia and sepsis was generated from the results of the LAL-bead profiles. Certain criteria were established: (1) the blood had to contain at least 1 ng of endotoxin per ml; and (2) the endotoxin had to be present in consecutive samples for at least 48 hours.

Using discriminative functional analysis, the data was analyzed for a group of 60 burn patients. Figure 4 presents the results of this analysis. If patient blood was obtained daily, and was positive for endotoxin, the LAL-Bead Assay had greater than a 75% probability of indicating those patients that would become clinically septic. Since this was a retrospective study, it was related to the day that the patient was actually treated for sepsis. Inversely, the LAL-Bead Assay had almost a 100% probability, by virtue of being negative, if the patient's septic status was due to a gram-positive or fungal infection.

- 3 Days : >75% probability

- 2 Days : >80% probability

- 1 Day : >95% probability

100% confirmatory at time patient treated for Gram ⊖ Sepsis.

Fig. 4. Predictive value of LAL-Bead Assay to endotoxemia.

However, studies using blood from normal individuals, spiked with measured amounts of endotoxin, gave inconsistent results. In some cases, if one or two ngs/ml of endotoxin was added to the blood, this amount could be detected. Although, in most cases studied, amounts greatly in excess of this had to be mixed with the blood before 1 or 2 ng of endotoxin could be detected. This was variable depending on the subject and sometimes variable with the same individuals at different times. Likewise, studies of frozen-stored serum from patients with gram-negative bacteremia did not consistently demonstrate endotoxemia. However, endotoxin added to the blood of rabbits and dogs was easily recoverable. The reasons for these observations are not clear, but it may relate to a physiological roll that endotoxin has in homeostasis.

DISCUSSION

The LAL-Bead Assay was developed from the observation

that a certain group of hydrophobic, aliphatic polymers have the unique ability to specifically bind endotoxin. When these polymers are compressed into the form of the bead, they have the ability to bind and remove endotoxin from many different biological materials. The binding of the endotoxin to the bead appears sufficiently strong as to enable the washing off of substances thought to be inhibitory to the LAL reaction. Figure 5 is a schematic of the LAL-Bead Assay. It is assumed that the binding site for endotoxin to the bead is different from the binding site on the endotoxin which is necessary for the LAL gellation reaction. It is therefore possible that different reagents other than LAL could be used to detect the presence of endotoxin on the bead.

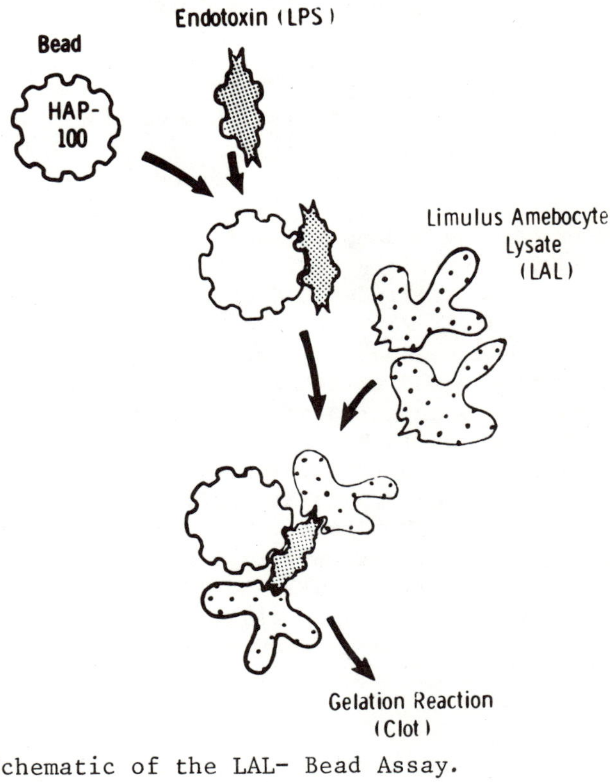

Fig. 5. Schematic of the LAL- Bead Assay.

It was not the purpose of the study to add to the already existing confusion concerning the definition of gram-negative sepsis in relation to endotoxemia. The LAL-Bead Assay did indicate that transient instances of free endotoxin in circulation are not reliable indicators of patient status, but consistent patterns of circulating endotoxin seem to correlate with the clinical course of the patient. (Supported in part by Health Central Institute, Minneapolis, Minnesota and Preventive Systems Inc., Galveston, Texas).

REFERENCES

1. Elin R , Wolff S (1973) Nonspecificity of the limulus amebocyte lysate test: Positive reactions with polynucleotides and proteins. J Infect Dis 128:3.
2. Harris NS (1976) Process for removing endotoxin from biological fluid. US Pat 3,959,128 May 25 Abst in Off Gaz US Pat Office 536:833.
3. Harris NS, Feinstein R (1976) In vitro process for detecting endotoxin in a biological fluid. US Pat 3,944,391 Mar 16 Abst in Off Gaz US Pat Office 536:834.
4. Harris NS, Feinstein R (1977a) A new limulus assay for the detection on endotoxin. J Trauma 17:714-718.
5. Harris NS, Feinstein R (1977b) Relationship of endotoxemia with clinically defined gram-negative septicemia. Surgical Forum 28:23.
6. Marvin J, Heck E, Loebl E, Curreri PW, Baxter C (1975) Usefulness of blood cultures in confirming septic complications in burn patients: Evaluation of a new culture method. J Trauma 15:8.
7. Pruitt BA, Foley FD (1973) The use of biopsies in burn patient care. Surgery 73:887-897.
8. Rowe MI, Buckner DM, Newmark S (1975) The early diagnosis of gram negative septicemia in the pediatric surgical patient. Ann Surg 182:280-286.
9. Wilson RF (1976) The diagnosis and management of severe sepsis and septic shock. Heart Lung 5:422-429.

ASSOCIATION OF ENDOTOXEMIA AND BACTEREMIA IN BURN PATIENTS: A PROSPECTIVE STUDY*

Albert T. McManus, Robert B. Lindberg,
Basil A. Pruitt, Jr., Arthur D. Mason, Jr.
US Army Institute of Surgical Research, Brooke
Army Medical Center, Fort Sam Houston, Texas 78234

We have examined burn patient blood specimens for the occurrence of circulating endotoxin and bacteremia. Free endotoxin was measured with the LAL bead technique (Harris and Feinstein 1977), substituting 12 x 75 mm tubes for the blood absorption step. Incubation times and control tubes were as originally described. Difco E.coli 026:86 endotoxin was used as a standard. A single lot of limulus lysate previously prepared by our Institute was used. The test was capable of measuring 0.625 nanograms of the standard. Endotoxin and blood culture measurements were performed on divided specimens by independent laboratories. Technicians were not aware of individual patient's clinical condition.

A group of 25 consecutively admitted severely burned patients were examined. Specimens were collected daily during the intensive phase of care. A total of 495 paired specimens were obtained. Endotoxin concentrations of 1 ng/ml or greater were detected in 114/495 specimens. Positive blood cultures were observed in 61/495 specimens. Among the isolated organisms, 31 were gram negative; 28 were gram positive; 2 yeast were isolated.

The statistical associations of endotoxin and blood culture were tested by comparing the observed frequencies of association with those predicted by assuming random association. The 4 possible classes of association within a specimen

*Previously presented in part at the 1978 annual meeting of The American Association of Immunologists, Atlanta, Georgia June 1978.

Pages 275–278, Biomedical Applications of the Horseshoe Crab (Limulidae)

were: negative endotoxin - negative blood culture (ET^-C^-), negative endotoxin - positive blood culture (ET^-C^+) positive endotoxin - negative blood culture (ET^+C^-) and positive endotoxin - positive blood culture (ET^+C^+). The predicted value for each class was generated by compounding the observed frequencies of individual responses within a class and multiplying the compound frequency by the total number of paired observations. For example, the predicted value for the number of (ET^+C^-) observed in the total sample (495) given 114/495 endotoxin positives and 61/495 culture positives would be $\frac{114}{495} \times \frac{495-61}{495} \times 495 = 99.95$

The associations of demonstrable circulating endotoxin and blood cultures are presented in Table 1. Chi square analysis was used. As can be seen, the hypothesis that endotoxin and blood culture results were randomly associated was quite true; no significant deviation from the predicted random frequencies could be demonstrated. Table 2 shows that endotoxin and gram negative blood culture were also randomly associated. Tables 3 and 4 show that no predictive relationship existed between endotoxemia and subsequent positive blood culture.

Table 1.

OCCURRENCE OF A POSITIVE ENDOTOXIN TEST
AND A POSITIVE BLOOD CULTURE WITHIN THE SAME SPECIMEN

	ET^-C^-	ET^+C^-	ET^-C^+	ET^+C^+
PREDICTED	334	100	47	14
OBSERVED	337	97	44	17

$X^2_{(1)} = 0.9189$ N.S.

Table 2.

CONCURRENT OCCURRENCE OF ENDOTOXIN AND POSITIVE BLOOD CULTURE FOR GRAM NEGATIVE ORGANISM

	ET^-C^-	ET^+C^-	ET^-C^+	ET^+C^+
PREDICTED	356	107	25	7
OBSERVED	354	109	27	5

$X^2_{(1)} = 1.0584$ N.S.

Table 3.

OCCURRENCE OF A POSITIVE BLOOD CULTURE 24 HOURS AFTER A POSITIVE ENDOTOXIN TEST

	ET^-C^-	ET^+C^-	ET^-C^+	ET^+C^+
PREDICTED	286	98	45	16
OBSERVED	288	96	43	18

$X^2_{(1)} = 0.5615$ N.S.

Table 4.

OCCURRENCE OF A POSITIVE BLOOD CULTURE 48 HOURS AFTER A POSITIVE ENDOTOXIN TEST

	ET^-C^-	ET^+C^-	ET^-C^+	ET^+C^+
PREDICTED	238	96	43	18
OBSERVED	234	100	47	14

$X^2_{(1)} = 1.4656$ N.S.

In summary, while these data indicate moderately frequent occurrence of endotoxemia and bacteremia in severely burned patients, they fail to demonstrate any predictable relationship between these entities; bacteremia is not predicted by endotoxemia nor is endotoxemia necessarily a concomitant of blood culture confirmed gram negative septicemia.

REFERENCE:

Harris NS, Feinstein R (1977). A new limulus assay for the detection of endotoxin. J Trauma 17:714-718.

CLINICAL UTILITY OF THE LIMULUS TEST WITH BLOOD, CSF AND SYNOVIAL FLUID

Ronald J. Elin, M.D., Ph.D.
Clinical Pathology Department, Clinical Center
National Institutes of Health
Bethesda, Maryland 20014

INTRODUCTION

A number of investigators have studied the clinical utility of the limulus amebocyte lysate (LAL) test for the detection of bacterial endotoxins in blood, cerebrospinal fluid (CSF) and synovial fluid. To my knowledge, all of the studies have compared the results of the LAL test with associated culture results. This report attempts to review the results of 25 different clinical studies with LAL. Standard statistical analyses have been performed on these data to assess clinical utility.

STATISTICAL METHODS

The sensitivity, specificity, predictive value of a positive result, predictive value of a negative result and efficiency were determined with LAL and culture results for blood and CSF (Galen and Gambino, 1975). In addition, the chi-square test and the Yates' modification of the chi-square test were used (Cochran, 1954).

LAL RESULTS WITH BLOOD

Reports of 17 different studies in humans with LAL and blood were found in the literature. In each of these studies a blood culture was also obtained and a comparison was made between the results of the LAL test and the blood culture. These 17 studies represent 2,055 LAL assays (Table 1). In these studies, three different methods were used to prepare

Pages 279–292, Biomedical Applications of the Horseshoe Crab (Limulidae)

the plasma for testing with LAL. A chloroform extraction of the plasma was used in twelve studies, pH change in four studies and heating the plasma in one study (Table 1). The majority of reports (13 of 17) primarily selected febrile patients for study. The conclusions of these seventeen studies were mixed. Six studies concluded that the test had potential clinical utility, four studies recommended that additional evaluation of the test was needed, four studies indicated limited or no clinical utility and in three studies it was difficult to draw a conclusion.

Table 1

PLASMA TREATMENT, PATIENT DIAGNOSIS AND CONCLUSIONS OF LAL CLINICAL STUDIES ON BLOOD

Authors	No. of Samples	Plasma* Tx	Patient+ Diagnosis	Conclusions
Caridis	13	pH	Comb	Specific and sensitive
DuBose	22	Chlor	Normal	Need further study
Elin	237	Chlor	Fever	Limited clinical utility
Feldman	154	pH	Fever	Not clinically useful
Fossard	240	Chlor	Fever	Simple, rapid, reliable
Fossard	261	Chlor	Fever	Simple, rapid, reliable
Goldstein	64	Heat	Fever	Needs further study
Jirillo	10	Chlor	Fever	?
Levin	98	Chlor	Fever	Sensitive, reliable, practical
Levin	281	Chlor	Fever	Clinically practical and sensitive
Magliulo	39	pH	Fever	Value and significance questionable
Martinez-G	83	pH	Fever	Needs further study
Oberle	23	Chlor	Fever	Useful adjunct to culture
Stumacher	344	Chlor	Fever	Doubtful utility
Tarao	96	Chlor	Liver	?
Wilkinson	22	Chlor	Liver	?
Young	68	Chlor	Fever	Needs further study
Total	2055			

*Chlor indicates chloroform extraction of plasma and pH indicates a pH change of the plasma to prepare it for the LAL test.

+Comb indicates a combination of diagnoses; normal indicates the study was done on apparently normal individuals; fever indicates the study group was composed primarily of patients with fever, and liver indicates the study group was composed primarily of patients with liver failure.

A summary of the LAL and blood culture results is presented in Table 2. Of the 1,677 negative blood cultures, 320 LAL tests associated with these blood cultures (19%) were positive. Gram-negative bacteria were cultured from 299 blood specimens and 139 associated LAL tests (46%) were positive. Of the 67 blood cultures which grew Gram-positive organisms, 17 had an associated LAL test (25%) which was also positive.

Table 2

LAL AND CULTURE RESULTS WITH BLOOD

	Culture Results					
	No Growth		Gram Neg		Gram Pos	
Authors / LAL Results:	Pos	Neg	Pos	Neg	Pos	Neg
Caridis	7	0	6	0	0	0
DuBose	7	15	0	0	0	0
Elin	36	172	8	15	4	2
Feldman	0	112	2	36	0	3
Fossard	71	161	8	0	0	0
Fossard	81	171	8	0	0	0
Goldstein	0	63	0	0	0	1
Jirillo	4	4	1	1	0	0
Levin	5	66	10	5	0	10
Levin	18	213	17	14	0	16
Magliulo	10	13	12	4	0	0
Martinez-G	1	62	1	15	0	3
Oberle	9	11	3	0	-	-
Stumacher	32	216	28	37	12	15
Tarao	26	70	0	0	0	0
Wilkinson	13	8	0	0	1	0
Young	0	0	35	33	0	0
Total	320	1357	139	160	17	50

The analysis of the data from all seventeen studies is presented in Table 3. The sensitivity (number of patients with a positive culture with an associated positive LAL test divided by the number of patients with a positive culture) of the LAL test is low. For all patients with a positive blood culture (bacteria or fungi), 44% had an associated positive LAL test. If the positive blood culture results are

Table 3

ANALYSIS OF ALL LAL AND CULTURE RESULTS WITH BLOOD*

	Culture Results		
	All Cultures	Gram neg	Gram pos
Sensitivity	44%	46%	25%
Specificity	81%	80%	76%
Predictive value of a positive result	34%	29%	4%
Predictive value of a negative result	86%	90%	97%
Efficiency	74%	75%	75%
χ^2	101	101	0.1
p	<0.0005	<0.0005	>0.7

*The data for this analysis were obtained from the 17 studies listed in Tables 1 and 2.

limited to growth with Gram-negative bacteria, 46% of the LAL tests were positive. If the culture results are limited to growth of Gram-positive bacteria, 25% of the LAL tests were positive. The specificity (number of patients with a negative blood culture and an associated negative LAL test divided by the number of patients with a negative culture) is greater than the sensitivity. For all positive cultures the specificity is 81%. When a positive culture result is limited to Gram-negative bacteria, the specificity is 80% and when positive culture results are limited to Gram-positive organisms, the specificity is 76%. The predictive value of a positive result (true positives divided by total positives) is low. When the LAL test was positive, 34% of all culture results were positive. If positive culture results were limited to growth of Gram-negative bacteria, 29% of the cultures were positive that were associated with the positive LAL test. With positive culture results limited to growth of Gram-positive organisms, 4% of cultures were positive with a positive LAL test. The predictive value of a negative result (true negatives divided by total negatives) showed the highest percentages among the statistical parameters. For all culture results, the predictive value of a negative result was 86%, 90% when the culture results were limited to Gram-negative bacteria and 97% when culture results were limited to Gram-positive organisms. The efficiency (true positives plus true negatives divided by the grand total) is 74% for all cultures. The efficiency is 75% when the results are limited to either Gram-negative bacteria or Gram-positive bacteria. Using the chi-square test, there is a highly significant correlation (<0.0005) between both a positive LAL test and a positive blood culture and between a positive LAL test and a blood culture which grew Gram-negative bacteria (Table 3).

Three different subsets of the data presented in Tables 1 and 2 were selected on the basis of the conclusions of the authors and the method used to prepare the plasma for the LAL test. These additional analyses were performed in an attempt to eliminate a bias that might be present in an individual study or a particular method for the preparation of the plasma. There were six studies that concluded that the LAL test with blood was specific, sensitive, reliable, had clinical utility, etc. (Table 1). An analysis of the data limited to these six studies is shown in Table 4. The sensitivity and predictive value of a negative result have increased for all cultures and positive culture results limited to Gram-negative bacteria compared with the data from all studies (Table 3). However, specificity, predictive value of a positive result and efficiency have remained essentially the same or have decreased slightly. Of interest, the chi-square test shows a significant correlation between culture results limited to growth of Gram-positive bacteria and a positive LAL test ($p < 0.005$).

The second subset limits the data to those studies which used a chloroform treatment of the plasma prior to testing with LAL. Twelve of the seventeen studies used the chloroform method (Table 5). Using this subset the results are essentially the same as those noted for the analysis of all of the data (compare Tables 3 and 5).

The third subset limited the analysis of the data to those studies which used a pH change of the plasma to prepare it for the LAL test. Four of the seventeen studies

Table 4

ANALYSIS OF LAL AND CULTURE RESULTS WITH BLOOD FROM STUDIES WITH A FAVORABLE CONCLUSION*

	Culture Results		
	All Cultures	Gram-neg	Gram-pos
Sensitivity	54%	73%	0%
Specificity	77%	77%	73%
Predictive value of a positive result	21%	21%	0%
Predictive value of a negative result	93%	97%	96%
Efficiency	74%	77%	70%
χ^2	40	85	10
ρ	<0.0005	<0.0005	<0.005

*The data for this analysis were obtained from 6 studies in which the first authors were the following: Caridis, Fossard, Fossard, Levin, Levin and Oberle (see Tables 1 and 2).

used this method (Table 6). For all cultures and those cultures positive with Gram-negative bacteria, the pH method showed the lowest sensitivity and the highest predictive value of a positive result among the four different analyses. This suggests that fewer positive LAL tests occurred with this method but when the LAL test was positive, there was a greater chance that the blood culture would also be positive. Thus, these four different analyses of the data show a maximum sensitivity of 54% with all cultures and 73% if culture results are limited to Gram-negative bacteria, a maximum predictive value of a positive result of 54% and an efficiency that was consistently about 75%.

All of these analyses have compared the LAL test with the blood culture results. The blood culture has been the common denominator that investigators have chosen when assessing the potential clinical utility of the LAL test. This

Table 5

ANALYSIS OF LAL AND CULTURE RESULTS WITH
BLOOD FOR STUDIES USING CHLOROFORM TREATMENT OF PLASMA*

	Culture Results		
	All Cultures	Gram-neg	Gram-pos
Sensitivity	48%	51%	28%
Specificity	79%	78%	74%
Predictive value of a positive result	31%	27%	4%
Predictive value of a negative result	86%	92%	97%
Efficiency	73%	75%	73%
χ^2	85	98	0.2
ρ	<0.0005	<0.0005	>0.6

*The data for this analysis were obtained from 12 studies in which the first authors were the following: DuBose, Elin, Fossard, Fossard, Jirillo, Levin, Levin, Oberle, Stumacher, Tarao, Wilkinson and Young (see Tables 1 and 2).

may be relevant since thirteen of the seventeen studies have selected a patient population for study primarily on the basis of fever (Table 1). If the standard for assessing the clinical utility of the LAL test is the blood culture, the low sensitivity and poor predictive value of a positive result would indicate that a positive LAL test is not synonymous with Gram-negative septicemia.

There is usually not a sufficient amount of endotoxin from the bacteria alone to give a positive test result in Gram-negative septicemia. Quantitative studies of Gram-negative septicemia indicate that most patients have fewer than 100 organisms per milliliter of blood (Finegold, et al, 1969). Since more than 1,000 Gram-negative bacteria per milliliter are required for a positive LAL test (Elin, et al, 1975), and Jorgensen, et al, 1973) and the endotoxin bound

Table 6

ANALYSIS OF LAL AND CULTURE RESULTS WITH BLOOD FOR STUDIES USING pH CHANGE TREATMENT OF PLASMA*

	Culture Results		
	All Cultures	Gram-neg	Gram-pos
Sensitivity	26%	28%	0%
Specificity	91%	92%	86%
Predictive value of a positive result	54%	54%	0%
Predictive value of a negative result	75%	78%	98%
Efficiency	73%	75%	84%
χ^2	14	17	1
ρ	<0.0005	<0.0005	>0.5

* The data for this analysis were obtained from 4 studies in which the first authors were the following: Caridis, Feldman, Magliulo and Martinez-G (see Tables 1 and 2).

to the bacteria and free endotoxin have similar levels of activity in the test (Jorgensen and Smith, 1974), it seems unlikely that the amount of endotoxin present in the bacteria in Gram-negative septicemia could produce a positive LAL test. Thus, a positive LAL test with human plasma, if it is due to endotoxin, may be caused by endogenous liberation of endotoxin into the blood or a combination of endogenous endotoxin and endotoxin from Gram-negative septicemia.

What might be some of the factors related to the low sensitivity and low predictive value of a positive result for the LAL test with blood? For the data analysis in this report, false positive LAL results may have been due to endotoxemia without Gram-negative bacteria in the blood. In addition, two studies have reported positive LAL tests in apparently healthy individuals. In a report by Sweet at al (1978), 4% of healthy individuals who had one or more teeth extracted had positive LAL tests with a concomitant negative blood culture. DuBose et al (1978) found 32.8% of plasma samples from healthy individuals gave a positive LAL test. On the other hand, false negative LAL results would occur with some degree of frequency since the number of Gram-negative bacteria in the blood would usually not be sufficient to effect a positive LAL test. These factors need further study. Thus, the clinical relevance of a positive LAL test with blood cannot be established at this time.

LAL AND CULTURE RESULTS IN CSF

Reports of seven different studies in humans with LAL and cerebrospinal fluid (CSF) were found in the literature. In each of these studies a culture of the CSF was also obtained and a comparison was made between the results of the LAL test and the CSF culture. These seven studies represent 1,346 LAL tests and cultures (Table 7). Six of these seven reports concluded that the LAL test had clinical utility for the detection of Gram-negative meningitis (Table 7).

Table 7

CONCLUSIONS OF LAL CLINICAL STUDIES ON CSF

Authors	No. of Samples	Conclusions
Berman	232	Sensitive, accurate, reliable
Dyson	145	Rapid, reliable, specific
Jorgensen	134	Rapid, sensitive, specific
McCracken	307	Not diagnostic, insensitive
Nachum	112	Sensitive, simple, efficient
Ross	335	Extremely valuable and reliable
Tuazon	81	May be clinically useful
Total	1346	

The data from these seven studies are shown in Table 8. Of the 915 negative cultures, twenty (2%) have an accompanying positive LAL test and all of these occurred in the report by McCracken et al (1976). Of the 364 CSF specimens which grew Gram-negative bacteria, 295 (81%) gave a positive LAL test. On the other hand, only 3% of the CSF specimens which grew Gram-positive bacteria were positive with the LAL test.

Table 8

LAL AND CULTURE RESULTS WITH CSF

Authors	LAL Results:	No Growth Pos	No Growth Neg	Gram-neg Pos	Gram-neg Neg	Gram-pos Pos	Gram-pos Neg
Berman		0	121	86	0	0	21
Dyson		0	135	6	0	0	4
Jorgensen		0	110	23	0	1	0
McCracken		20	110	97	68	0	12
Nachum		0	68	38	0	0	5
Ross		0	284	37	1	0	13
Tuazon		0	67	8	0	1	5
Total		20	895	295	69	2	60

The analysis of all the data is shown in Table 9. With the exception of sensitivity, the other statistical determinants indicate good clinical utility and a very good correlation between a positive LAL test and a culture positive with Gram-negative bacteria. If the study by McCracken et al is eliminated from the data, the results are shown in Table 10. Now all of these statistical determinants for a CSF culture positive with Gram-negative bacteria are 99% or better. This indicates exceptional clinical efficacy for the LAL test in the diagnosis of Gram-negative meningitis.

Table 9

ANALYSIS OF ALL LAL AND CULTURE RESULTS WITH CSF*

	Culture Results: All Cultures	Gram neg	Gram pos
Sensitivity	69%	81%	3%
Specificity	98%	98%	75%
Predictive value of a positive result	94%	93%	1%
Predictive value of a negative result	87%	93%	94%
Efficiency	89%	93%	72%
χ^2	724	915	15
ρ	<0.0005	<0.0005	<0.0005

*The data for this analysis were obtained from the 7 studies listed in Tables 7 and 8.

Table 10

ANALYSIS OF LAL AND CULTURE RESULTS WITH CSF FROM STUDIES WITH A FAVORABLE CONCLUSION*

	Culture Results		
	All Cultures	Gram neg	Gram pos
Sensitivity	80%	99.5%	4%
Specificity	100%	99.8%	80%
Predictive value of a positive results	100%	99.0%	1%
Predictive value of a negative result	94%	99.9%	94%
Efficiency	95.3%	99.7%	76.2%
χ^2	782	1015	8
ρ	<0.0005	<0.0005	<0.005

*The data for this analysis were obtained from 6 studies in which the first authors were the following: Berman, Jorgensen, Nachum, Ross and Tuazon (see Tables 7 and 8).

LAL AND CULTURE RESULTS WITH SYNOVIAL FLUID

Only one study was found that was specifically devoted to the evaluation of the LAL test with synovial fluid (Elin, et al 1978). Seventy-one specimens of synovial fluid from 46 patients were evaluated for reactivity in the LAL test and were concomitantly cultured. All patients with culture-proven septic arthritis or presumptive septic arthritis had joint fluid specimens that yielded a positive result with the limulus assay (12 patients). However, 52.9% of patients (18 or 34) who had a non-septic cause for their joint diffusion also had a joint fluid specimen that was positive in the LAL test. The authors of this study concluded that a positive result in the LAL test is specific for a septic process, but a negative result would be evidence against it.

SUMMARY AND CONCLUSION

The LAL test in its current form has clinical utility for the detection of Gram-negative meningitis. On the other hand, the clinical utility of the LAL test for the diagnosis of Gram-negative septicemia and Gram-negative pyogenic arthritis is marginal. These differences may be related to the composition of these three body fluids and the need to manipulate plasma and synovial fluid by some chemical or physical method prior to performing the LAL test. It may be that future refinements of the LAL test or of the preparation of the test specimen may improve the clinical value of the LAL test for blood and synovial fluid.

RFERENCES

Berman NS, Siegel SE, Nachum R, Lipsey A, Leedom J (1976) Cerebrospinal fluid endotoxin concentrations in gram-negative meningitis. J Pediat 88:553.

Caridis DT, Reinhold RB, Woodruff PWH, Fine J (1972) Endotoxaemia in man. Lancet 24:1381.

Cochran WG (1954) Some methods for strengthening the common χ^2 tests. Biometrics 10:417.

DuBose D, Lemaise M, Brown J, Wolfe D, Hamlet M (1978) Survey for positive Limulus amoebocyte lysate test in plasma from humans and common research animals. J Clin Microbiol 7:139.

Dyson D, Cassady G (1976) Use of Limulus Lysate for detecting gram-negative neonatal meningitis. Pediat 58:105.

Elin RJ, Knowles R, Barth WF, Wolff SM (1978) Lack of specificity of the limulus lysate test in the diagnosis of pyogenic arthritis. J Infect Dis 137:507.

Elin RJ, Lundberg WB, Schmidt PJ (1975) Evaluation of bacterial contamination in blood processing. Transfusion 15:260.

Elin RJ, Robinson RA, Levine AS, Wolff SM (1975) Lack of clinical usefulness of the Limulus test in the diagnosis of endotoxemia. N Engl J Med 293:521.

Feldman S, Pearson TA (1974) The Limulus test and gram-negative bacillary sepsis. Am J Dis Child 128:172.

Finegold SM, White ML, Ziment I, Winn WR (1969) Rapid diagnosis of bacteremia. Appl Microbiol 18:458.

Fossard DP, Kakkar VV (1974) The Limulus test in experimental and clinical endotoxaemia. Br J Surg 61:798.

References (continued)

Fossard DP, Kakkar VV, Elsey PA (1974) Assessment of Limulus test for detecting endotoxaemia. Brit Med J 2:465.

Galen RS, Gambino SR (1975). "Beyond Normality." New York: John Wiley and Sons, p 13.

Goldstein JA, Reller LB, Wang W-LL (1976) Limulus Amebocyte lysate test in neonates. Am J Clin Pathol 66:1012.

Jirillo E, Pasquetto N, Marcuccio L, Monno R, DeRinaldis P, Fumarola D (1975) Endotoxemia detected by Limulus assay in severe malnourished children: plasma effects on leucocytes migration: preliminary investigations. G Batteriol Virol Immunol 68:174.

Jorgensen JH, Carvajal HF, Chipps BE, Smith RJ (1973) Rapid detection of Gram-negative bacteriuria by use of the limulus endotoxin assay. Appl Microbiol 26:38.

Jorgensen JH, Lee JC, Jones PM (1975) Rapid diagnosis of Gram-negative bacterial meningitis by the Limulus endotoxin assay. Texas Rep Biol Med 33:475.

Jorgensen JH, Smith RH (1974) Measurement of bound and free endotoxin by the Limulus assay. Proc Soc Exp Biol Med 146:1024.

Levin J, Poore TE, Young NS, Margolis S, Zauber NP, Townes AS, Bell WR (1972) Gram-negative sepsis: detection of endotoxemia with the Limulus test. Ann Int Med 76:1.

Levin J, Poore TE, Zauber NP, Oser RS (1970) Detection of endotoxin in the blood of patients with sepsis due to gram-negative bacteria. N Engl J Med 283:1313.

Magliulo E, Scevola D, Fumarola D, Vaccaro R, Bertotto A, Burberi S (1976) Clinical experience in detecting endotoxemia with the Limulus test in typhoid fever and other Salmonella infections. Infection 4:3.

Martinez-G LA, Quintiliani R, Tilton RC (1973) Clinical experience on the detection of endotoxemia with the Limulus test. J Infect Dis 127:102.

McCracken GH, Sarff LD (1976) Endotoxin in cerebrospinal fluid. JAMA 235:617.

Nachum R, Lipsev A, Siegel SE (1973) Rapid detection of gram-negative bacterial meningitis by the Limulus lysate test. N Engl J Med 289:931.

Oberle MW, Graham GG, Levin J (1974) Detection of endotoxemia with the Limulus test: preliminary studies in severly malnourished children. J Pediat 85:570.

References (Continued)

Ross S, Rodriquez W, Controni G, Korengold G, Watson S, Khan W (1975) Limulus lysate test for gram-negative bacterial meningitis. JAMA 233:1366.

Stumacher RH, Kovnat MJ, McCabe WF (1973) Limitations of the usefulness of the Limulus assay for endotoxin. N Engl J Med 288:1261.

Sweet JB, Gill VJ, Chusid MJ, Elin RJ (1978) Nitroblue tetrazolium and Limulus assays for bacteremia after dental extraction: effect of topical antiseptics. JADA 96:276.

Tarao K, So K, Moroi T, Ikeuchi T, Suyama T, Endo O, Fukushima K (1977) Detection of endotoxin in plasma and ascitic fluid of patients with cirrhosis: its clinical significance. Gastroenterology 73:539.

Tuazon CU, Perez AA, Elin RJ, Sheagren JN (1977) Detection of endotoxin in cerebrospinal and joint fluids by Limulus assay. Arch Intern Med 137:55.

Wilkinson SP, Arrogo V, Gazzard BG, Moodie H, Williams R (1974) Relation of renal impairment and haemorrhagic diathesis to endotoxaemia in fulminant hepatic failure. Lancet 30:7857.

Young LS (1975) Opsonizing antibodies, host factors, and the Limulus assay for endotoxin. Infect Immun 12:88.

Acknowledgments

My sincere thanks to Mrs. Anna Utter and Mrs. Elsa Elkin for their excellent assistance with the preparation of this manuscript.

CORRELATION OF LIMULUS AMEBOCYTE LYSATE ASSAY WITH ACCEPTED ENDOTOXIN ASSAYS

Peter A. Tomasulo, M.D.

Milwaukee Blood Center, Inc.

P.O. Box 10G, Milwaukee, Wisconsin 53233

Many assays have been utilized to detect endotoxin. They vary greatly in technique, convenience, expense, reliability, specificity, and sensitivity. Most assays of the biological activity of endotoxin depend on the production and assessment of death or morbidity in various living creatures. Death in normal rodents, adrenalectomized rodents, rodents with reticuloendothelial blockade, death in bird embryos, provocation of the localized Schwartzman reaction, production of abortion, production of skin inflammation, damage to rodent sarcomas, tissue factor generation (Rickles FR et al., 1977), and production of fever in rabbits have all been used as endpoints in the assays of the biological effects of gram-negative lipopolysaccharide (Braude AI, 1963). In addition, there are other assays which detect endotoxin, not by its biological effects but rather by its physical characteristics - for example: antigenicity. The assays which depend on antigenicity or other physical characteristics of the molecule are not pertinent to this review.

Assays which depend on living animals or living parts of animals are fraught with difficulty. There is much variability in the response to drugs and toxins. Age, weight, sex, route of administration, time of administration, previous history of the test animal, health of the test animal, and genetic makeup of the test animal all influence the animal's response. If we could control all these factors, there would still be certain biologic variations in the response of identical members of the same species to drugs and toxins. One individual will not respond identically to

Pages 293–302, Biomedical Applications of the Horseshoe Crab (Limulidae)

the same dose of a substance when it is administered on two separate occasions.

Not only must we consider biological variation when we evaluate bioassays, but we must also be aware that not all the observed effects are necessarily caused by the injected drug or the toxin (Braude AI, 1963). Some substances have the capacity to elicit many biological effects directly but also have the capacity to activate various body systems to produce other physiologic and pathologic changes in the organism indirectly. When the phenomena observed in test animals are known to be the result of a combination of direct and indirect effects and when there is the possibility of differing dose/response relationships for the direct and indirect effects of the substance concerned, the results of the assay are very difficult to interpret and to compare with results of other assays.

Many organisms have systems to potentiate or inhibit the action of drugs or toxins. These modifying systems may not be equally powerful in all individuals of the same species. The signs and symptoms manifested by test animals after drug or toxin administration are, then, the sum of the effects of the substance itself plus the effects of the activated potentiating or inhibiting system which existed in the animal prior to drug administration.

All these factors may be active when we administer endotoxin to test animals. Endotoxin can activate the complement system, and it is possible that at some doses of endotoxin we might see a combination of effects due both to the endotoxin directly and to the activated complement system indirectly (Gewurz H et al., 1968). We know that toxic substances and metabolites can be elaborated by tissues damaged directly by endotoxin. There is also in human blood a humoral system which inhibits the effects of endotoxin (Levin J, Tomasulo PA et al., 1970; Rojas-Corona RR et al., 1969; Rosen FS et al., 1958). There is some evidence that these substances are enzymes, but this has not been satisfactorily documented as yet (Braude AI, 1963; Reinhold RB et al., 1971). There is also biologic variation in the response of test animals to endotoxin (Keene WR et al., 1961). Samples must usually be tested in triplicate, and the results of these endotoxin assays are usually expressed as the dose at which 50% of animals manifest an effect. Finally, it should be noted that it is possible to induce tolerance to endotoxin. The mechanism of the production of tolerance is incompletely understood, but investigators must be careful to avoid tolerant animals in their assays.

The rabbit pyrogen assay is an accepted, nearly quantitative assay for endotoxin (Cooper JF, Levin J et al., 1971; Cooper JF, Hochstein HD et al., 1972; Keene WR et al., 1961; Murphy PA et al., 1967) and has been used by the Food and Drug Administration to detect contamination of substances intended for human intravenous use. Three or more rabbits must be injected with each sample, and monitoring must continue for three hours after injection. The variability in the results can be reduced by: 1) using only minimal restraint during the assay; 2) two days of animal training to prevent anxiety during the assay; 3) acclimatization to the stocks and rectal probes for two hours prior to the assay; 4) use of animals of uniform age, weight, and sex; and 5) elimination of animals with abnormal baseline temperatures (Bornstein DL et al., 1963; Keene WR et al., 1961; Murphy PA et al., 1967). Even using these precautions, some results must be rejected because of atypical fever responses. Biologic variation still exists. Not all investigators interpret the results in the same fashion; the effect of recipient weight may not be well documented; (Keene WR et al., 1961) and the assay cannot be used to detect contamination if the substance itself causes fever or has other effects on the physiology of the rabbit making the fever response unreliable. Some substances intended for parenteral use have short biologic or radioactive half lives. The testing of these substances requires having the rather cumbersome resources for the rabbit pyrogen assay close to both the source of production and to the patients who will receive the substances. Finally, it is known that endotoxin is not the only substance that causes fever in rabbits.

Since the observations of Bang (Levin J and Bang FB, 1964) in the late 50s and early 60s, much has been learned about the effects of gram-negative lipopolysaccharide on the blood of the horseshoe crab. In 1968, Levin first reported the use of the limulus lysate system as an assay to detect the presence of endotoxin (Levin J, 1968). The test is now generally accepted as the best in vitro assay for endotoxin. The test is simple and easy to learn, and it requires no sophisticated equipment. Very small samples can be tested easily. The endpoint is the gelation of the clear, colorless, liquid lysate; and as such, it is easy to define and to detect. The lysate is stable and is sensitive to all endotoxins tested (Levin J, Tomasulo PA et al., 1970; Reinhold RB et al., 1971). The limulus system is at least as sensitive as any other assay for the biological effects of endotoxin. It is more sensitive than most assays. As little as 0.0005 μg/ml of endotoxin can be detected in a sample as small as .05 ml. The interpretation and, possibly, the performance of this test might be automated in the near future.

There are disadvantages to the test. One is that the test is new and has attracted a great deal of attention. The major disadvantage of the test, however, is its equisite sensitivity to endotoxin and the seemingly ubiquitous distribution of endotoxin in our environment. Compulsive attention must be paid to absolute sterility and prevention of contamination in the performance of the assay. Discussions about the specificity of the assay must always include a thorough inquiry into the possibility that one sample or another is actually contaminated with endotoxin.

The assay has been found to be quite specific for endotoxin. In a blind study, the lysate identified a series of cultures of gram-negative organisms and the supernatants from gram-negative cultures. In the same blind study, the gram-positive cultures and their supernatants were all negative in the lysate assay (Levin J). The lysate has been shown not to react with cultures of pneumococcus, staphylococcus, and streptococcus, nor does it react with clostridial toxin, streptokinase, streptodornase, streptolysin, thrombin, hemoglobin, calcium, (Reinhold RB et al., 1971) serotonin, histamine, epinephrine, norepinephrine, bradykinin, (Rojas-Corona RR et al., 1969) six species of candida, cryptococcus neoformans, other cryptococcus species, torulopsis glabrata, rhodotorula, and saccharomyces cerevisiae (Jorgensen JH, Smith RF et al., 1973). In addition, human blood from normal blood donors (Levin J, Poore TE et al., 1970) and from patients with DIC, gram-positive infection and other disorders (Levin J, Poore TE et al., 1970), normal cerebrospinal fluid (Nachum R et al., 1973) and normal urine (Jorgensen JH,

Carvajal HF et al., 1973) have been negative in the limulus amebocyte lysate assay.

The limulus lysate assay has not been so specific in the hands of all investigators. In one survey, the limulus amebocyte lysate was found to react with commercial preparations of bovine thrombin and brain thromboplastin, with synthetic nucleotides, and with ribonuclease, but not with human leukocytic pyrogen and commercial dextran fractions (Elin RJ et al., 1973). Yin (Yin ET, 1975) felt this result with thrombin was due not to the thrombin itself but rather to contamination by endotoxin. The results with the other materials might also be explained in the same fashion.

Recently, others have suggested that peptidoglycans (Wildfeuer A, Heymer B et al., 1974) and streptococcal pyrogenic exotoxins (Brunson KW et al., 1976) can gel the limulus amebocyte lysate. Concentrations of peptidoglycans 1,000 to 400,000 times greater than the threshold dose of endotoxin were necessary to produce a positive test, and the concentration of streptococcal pyrogenic exotoxins was 200,000 times the threshold endotoxin dose. In both investigations, an attempt was made to rule out the possibility of endotoxin contamination. As was pointed out by both groups of authors, because whole gram-positive organisms will not lead to a positive result, and because such high concentrations of these substances are necessary to produce a positive result, it is not likely that the effects observed with peptidoglycans and streptococcal pyrogenic exotoxins would ever confuse the correct interpretation of the limulus amebocyte lysate assay. It is known that both peptidoglycans and pyrogenic exotoxin have biologic effects similar to those of endotoxin, and therefore, it is possible that they may possess active sites similar to those on the endotoxin molecule.

Pyrogen Testing

The limulus amebocyte lysate assay correlates very well with the accepted, but probably less specific and less sensitive, assays for endotoxin. Cooper (Cooper JF, Levin J et al., 1971; Cooper JF, Hochstein HD et al., 1972) demonstrated a strong correlation between fever indices and gelation time. Cooper also suggested that the limulus amebocyte lysate was more sensitive than the proscribed Food and Drug Administration protocol to which he compared it. In ranking the effects of 17 different endotoxin preparations, Elin (Elin RJ, Sandberg AL et al., 1976) also demonstrated a strong correlation between the limulus amebocyte lysate and pyrogen tests. Using three preparations of tritiated endotoxin (Tomasulo PA, March SC et al., 1975) and one preparation of native E. coli endotoxin, Tomasulo et al. (Tomasulo PA, Levin J et al., 1977) demonstrated similar ranking of the biological effects of these endotoxins by the rabbit pyrogen assay and by the limulus amebocyte lysate assay. Other investigators have supported this correlation of the limulus amebocyte lysate assay with the rabbit pyrogen assay (Nowotny A et al., 1975).

Complement Activation

The ability of endotoxin to activate complement does not correlate as well with the limulus amebocyte lysate assay as does pyrogen testing (Elin RJ, Sandberg AL et al., 1976). However, Elin (Elin RJ, Sandberg AL et al., 1976) found that complement activation did not correlate well with the pyrogen assay or the mitogenicity test either. Tomasulo et al. (Tomasulo PA, Levin J et al., 1977) showed a moderate correlation in the ranking of the biological effects of the various tritiated endotoxins.

Other Assays

Rojas-Corona (Rojas-Corona RR et al., 1969) et al. have shown that although the limulus amebocyte lysate assay is much more sensitive than an immunoassay and the epinephrine skin sensitivity assay, the limulus amebocyte lysate assay correlates well with those tests when dilutions of native endotoxin and dilutions of detoxified preparations of the lipopolysaccharide are assayed. Reinhold and Fine (Reinhold RB et al., 1971) showed a correlation between the limulus amebocyte lysate assay and the chick embryo lethality test. Nowotny et al. (Nowotny A et al., 1975) demonstrated a strong correlation between the limulus amebocyte lysate assay, mouse lethality, local Schwartzman reaction, chick embryo lethality test, and B cell mitogenicity using standard preparations of endotoxin and some special polysaccharide-rich endotoxin fractions with limited biological activity. Elin et al. (Elin RJ, Sandberg AL et al., 1976) also showed that B cell mitogenicity correlated well with the limulus amebocyte lysate assay and with the rabbit pyrogen assay.

In summary, the limulus amebocyte lysate assay is among the most sensitive and specific assays available to detect the biological activities of gram-negative endotoxin. It is the simplest and one of the least expensive assays to perform. It clearly detects endotoxin in a quantitative fashion. It correlates well with the more accepted endotoxin assays. The specificity of the test is remarkable, and most investigators do not question its usefullness in detecting endotoxin. If the lysate does actually react to substances other than gram-negative lipopolysaccharides, it only reacts to them in such high concentrations that they are not likely to confuse the interpretation of the results of this test. The limulus amebocyte lysate is being studied very carefully in an effort to make it as useful in research and clinical situations as is possible. One wonders whether the possible defects so far discovered in the limulus amebocyte lysate assay would seem significant if each of the other assays for endotoxin was examined as closely. Much of the alleged nonspecificity of the limulus amebocyte lysate test might disappear if there were an absolute mechanism to prevent contamination of samples and test materials with endotoxin.

The question of the specificity of this test, or for that matter of any test, has to remain open until we know what happens on a molecular level. We really can't be sure that any assay is specific for one product or another until we know how the assay works. The issue may be more confused in the future because of the demonstration of at least two sites of biologic activity on the endotoxin molecule (Nowotny A et al., 1975). Until we understand the mechanism of the reaction, those of us with an emotional attachment for the horseshoe crab and a deep and abiding faith in its unique reaction to gram-negative products, will stand by the crab no matter what criticisms are leveled.

REFERENCES

Bornstein DL, Bredenberg C, Wood WB Jr. (1963). Studies on the pathogenesis of fever. XI. Quantitative features of the febrile response to leucocytic pyrogen. J Exp Med 117: 349.

Braude AI (1963). Absorption, distribution, and elimination of endotoxins and their derivatives. In Landy and Braun (eds): "Bacterial Endotoxins," Rahway, New Jersey: Rutgers University Press.

Brunson KW, Watson DW (1976). Limulus amebocyte lysate reaction with streptococcal pyrogenic exotoxin. Infect Immun 14: 1256.

Cook JA, Di Luzio NR, Hoffmann EO (1975). Factors modifying susceptibility to bacterial endotoxin: the effect of lead and cadmium. CRC Crit Rev Toxicol 3: 201.

Cooper JF, Levin J, Wagner HN Jr. (1971). Quantitative comparison of in vitro and in vivo methods for the detection of endotoxin. J Lab Clin Med 78(1): 138.

Cooper JF, Hochstein HD, Seligmann EB Jr (1972). The limulus test for endotoxin (pyrogen) in radiopharmaceuticals and biologicals. Bull Parenter Drug Assoc 26(4): 153.

Cundy KR, Nowotny A (1967). Comparisons of five toxicity parameters of serratia marcescens endotoxins. Proc Soc Exp Biol Med 127: 999.

Elin RJ, Sandberg AL, Rosenstreich DL (1976). Comparison of the pyrogenicity, limulus activity, mitogenicity and complement reactivity of several bacterial endotoxins and related compounds. J Immunol 117(4): 1238.

Elin RJ, Wolff SM (1973). Nonspecificity of the limulus amebocyte lysate test: positive reactions with polynucleotides and proteins. J Infect Dis 128(3): 349.

Gewurz H, Mergenhagen SE, Nowotny A, Phillips JK (1968). Interactions of the complement system with native and chemically modified endotoxins. J Bacteriol 95: 397.

Jorgensen JH, Carvajal HF, Chipps BE, Smith RF (1973). Rapid detection of gram-negative bacteriuria by use of the limulus endotoxin assay. Appl Microbiol 26(1): 38.

Jorgensen JH, Smith RF (1973). Preparation, sensitivity, and specificity of limulus lysate for endotoxin assay. Appl Microbiol 26(1): 43.

Keene WR, Silberman HR, Landy M, with Strelecky KA (1961).

Observations on the pyrogenic response and its application to the bioassay of endotoxin. J Clin Invest 40: 295.
Levin J (1968). A sensitive system to detect endotoxin in vitro. Fed Proc 27: 355.
Levin J. Personal communication.
Levin J, Bang FB (1964). The role of endotoxin in the coagulation of limulus blood. Bull Johns Hopkins Hosp 115: 265.
Levin J, Poore TE, Zauber NP, Oser RS (1970). Detection of endotoxin in blood of patients with sepsis due to gram-negative bacteria. N Engl J Med 283: 1313.
Levin J, Tomasulo PA, Oser RS (1970). Detection of endotoxin in human blood and demonstration of an inhibitor. J Lab Clin Med 75(6): 903.
Murphy PA, Phil D (1967). An assay method for leukocyte pyrogen. J Exp Med 126: 745.
Nachum R, Lipsfy A, Siegel S (1973). Rapid detection of gram-negative bacterial meningitis by the limulus lysate test. N Engl J Med 289: 931.
Nowotny A, Behling UH, Chang HL (1975). Relation of structure to function in bacterial endotoxins. VIII. Biological activities in a polysaccharide-rich fraction. J Immunol 115(1): 199.
Reinhold RB, Fine J (1971). A technique for quantitative measurement of endotoxin in human plasma. Proc Soc Exp Biol Med 137: 334.
Rickles FR, Levin J, Hardin JA, Barr CF, Conrad ME Jr (1977). Tissue factor generation by human mononuclear cells: effects of endotoxin and dissociation of tissue factor generation from mitogenic response. J Lab Clin Med 89: 792.
Rojas-Corona RR, Skarnes R. Tamakuma S, Fine J (1969). The limulus coagulation test for endotoxin. A comparison with other assay methods. Proc Soc Exp Biol Med 132: 599.
Rosen FS, Skarnes RC, Landy M, Shear MJ (1958). Inactivation of endotoxin by a humoral component. III. Role of divalent cation and a dialyzable component. J Exp Med 108: 701.
Skarnes RC, Rosen FS, Shear MJ, Landy M (1958). Inactivation of endotoxin by a humoral component. II. Interaction of endotoxin with serum and plasma. J Exp Med 108(5): 685.
Tomasulo PA, Levin J, Murphy PA, Winkelstein JA (1977). Biological activities of tritiated endotoxins: correlation of the limulus lysate assay with rabbit pyrogen and complement-activation assays for endotoxin. J Lab Clin Med 89(2): 308.

Tomasulo PA, March SC, Levin J (1975). Tritiation of endotoxin. Biochim Biophy Acta 400(2): 399.

Wildfeuer A, Heymer B, Schleifer KH, Haferkamp O (1974). Investigations on the specificity of the limulus test for the detection of endotoxin. Appl Microbiol 28(5): 867.

Yin ET (1975). Endotoxin thrombin and the limulus amebocyte lysate test. J Lab Clin Med 86: 430.

ENDOTOXAEMIA, AS DETECTED BY THE LIMULUS ASSAY, IN FULMINANT HEPATIC FAILURE AND CIRRHOSIS

S.P. WILKINSON

LIVER UNIT
KING'S COLLEGE HOSPITAL,
LONDON, SE 5

Endotoxins, derived from Gram-negative bowel organisms, are probably normal constituents of portal venous blood and filtered by the Kupffer cells of the liver (Ravin et al, 1960; Prytz et al, 1976). There is also evidence that hepatocytes may be involved in endotoxin clearance (Ramadori et al, 1978). Thus with severe liver damage endotoxins might escape into the systemic circulation, because of either Kupffer cell/hepatocyte failure or the development of portasystemic shunts, in the absence of true Gram-negative infection. This could account for some of the renal, haemorrhagic, and haemodynamic changes characteristic of hepatic failure.

FULMINANT HEPATIC FAILURE

Twenty-two patients with fulminant hepatic failure were investigated for systemic endotoxaemia using the Limulus assay of Levin et al (1970) (Wilkinson et al, 1974). The causes for the hepatic failure were presumed viral hepatitis (5 cases), paracetamol (acetaminophen) overdose (10) halothane - associated (3) acute fatty liver of pregnancy (1), acute Budd-Chiari syndrome (1) hypersensitivity to anti-tuberculous drugs (1), and Amanita phalloides posioning (1).

Fourteen of the patients became Limulus-positive and only two of these showed evidence for Gram-negative infection (E.Coli urine infection in both). All of these patients died whilst six of the eight persistently Limulus - negative cases survived ($P<0.001$)

Pages 303–307, Biomedical Applications of the Horseshoe Crab (Limulidae)

RELATION TO RENAL IMPAIRMENT

The glomerular filtration rate (GFR) ranged from 145 to 1 ml/min. The Limulus test was invariably positive when the GFR was less than 25 ml/min, and invariably negative when greater than 40 ml/min. Serial studies in two patients who were initially Limulus-negative showed that the development of renal failure was associated with the Limulus test becoming positive.

In view of the known effects of endotoxins on the kidney, namely renal vasoconstriction (Gillenwater et al, 1963; Cavanagh et al, 1970) and intrarenal fibrin deposition (Beller and Graeff, 1967), it is possible that renal failure was secondary to the endotoxaemia.

RELATION TO INTRAVASCULAR COAGULATION AND HAEMORRHAGIC DIATHESIS

Endotoxins are well known to precipitate intravascular coagulation and a highly significant relationship between laboratory features for this and a positive Limulus test was observed. A highly significant relationship with gastrointertinal mucosal haemorrhages was also found. The latter may be a manifestation of intravascular coagulation (Margaretten and McKay, 1971).

OTHER STUDIES IN FULMINANT HEPATIC FAILURE

In smaller numbers of patients other groups have also reported a high frequency of endotoxaemia as detected by the Limulus assay in fulminant hepatic failure, including Caridis (1972 - two patients), Wardle (1974-two patients), and Liehr and Grun (1976 - three patients).

ATTEMPTS TO REMOVE CIRCULATING ENDOTOXINS

There is increasing interest in the use of entracorporeal haemoperfusion for the removal of circulating toxins in fulminant hepatic failure (Williams and Murray-Lyon, 1975). In vitro studies have been carried out with both activiated charcoal and the resin Amberlite XAD-7.

5 ml of human plasma containing 10 ng/ml endotoxin (Difco E.Coli LPS 026:B6) were shaken with 1g of sorbent for six hours. The plasma was then tested for endotoxin using the Limulus assay. Activated charcoal was found to remove the endotoxin. Unfortunately activated charcoal *per se* is unsuitable for haemoperfusion because of damage to the formed elements of the blood (Weston et al, 1974). This can be considerably reduced by a polyhema coating, but when such charcoal was tested the property for endotoxin removal was lost. The resin XAD-7 also failed to remove the endotoxin. Nolan et al (1975) has also found activated charcoal to effectively bind endotoxin. The strongly basic anion-exchange resin Dowex-1 was equally effective. They also found that Amerblite XAD-2 showed some, though considerably less, binding.

With regard to *in vivo* use in a haemoperfusion circuit, coated charcoal and XAD-7 appear to be of little or no value. Dowex - 1 might be of value, but uncoated resins are bioincompatible (Weston et al, 1974). Possibly coating with albumin might render this resin biocomptabile without undue loss of its adsorbent properties as has been described for XAD - 7 (Hughes et al, 1978).

CIRRHOSIS

Endotoxaemia has also been reported in cirrhosis (Wilkinson et al, 1976; Liehr et al, 1976; Prytz et al, 1976; Clemente et al 1977). Each of these groups have emphasized the absence for other evidence of Gram-negative infection in most patients. These studies have also shown a relationship between a positive Limulus test and renal failure (Wilkinson et al, 1976; Clemente et al, 1977), intravascular coagulation (Liehr et al, 1976), and haemorrhagic gastritis (Clemente et al, 1977)

CONCLUSIONS

Endotaxaemia, as detected by the Limulus assay, in the absence of other evidence for true Gram-negative infection, is common in fulminant hepatic failure and cirrhosis. It is possible that endotoxaemia is responsible for some of the complications of these conditions. Preliminary *in vitro* studies have not yet suggested a satisfactory method for entracorpareal endotoxin removal.

REFERENCES

Beller FK, Graeff H (1967). Deposition of glomerular fibrin in the rabbit after infusion with endotoxin Nature 215 295

Caridis DT, Reinhold RB, Woodruff PWH, Fine J (1972) Endotoxaemia in man. Lancet 1: 1381.

Cavanagh D, Rao PS, Sutton DMC, Bhagat BD, Bacjmann F (1970). The pathophysiology of endotoxin shock in the primate. Amer J Obstet Gynecol 108:705.

Clemente C, Bosch J, Rodes J, Arroyo V, Mas A, Maragall S (1977). Functional renal failure and haemorrhagic gastritis associated with endotoxaemia in cirrhosis. Gut 18:556.

Gillenwater JY,Dooley ES, Rohlich ED (1963). Effects of endotoxin on renal function and memodynamics. Amer J Physiol 205: 293

Hughes RD, Ton HY, Langley PG, Silk DBA, Williams R (1978). The use of an in vitro haemoperfusion circuit to evaluate the blood comptability of albumin - coated Amberlite XAD-7 resin. Int J Artif Org 1: 129.

Levin J, Tomasulo PA, Oser RS (1970). Detection of endotoxin in human blood and demonstration of an inhibitor. J Lab Clin Med 75: 903.

Liehr H, Grun M, Brunswig D, Sautter Th (1976). Endoximamie bei leberzirrhose. Zeits fur Gastroenterol 14:14.

Margareeten W. McKay Dg (1971). Thrombotic ulcerations of the gastrointestinal tract. Arch Intern Med 127:250.

Nolan JP, McDevitt JJ, Goldmann GS (1975). Endotoxin binding by charged and uncharged resins. Proc Soc Exp Biol Med 149: 766.

Prytz H, Holst-Christensen J, Korner B, Liehr H (1976). Portal venous and systemic endotoxaemia in patients with liver disease and systemic endotoxaemia in patients with cirrhosis. Scand J Gastroenterol 11:857

Ramadori G, Hopf U, Meyer Zum Büschenfelde KH (1978). Binding sites for endotoxic lipopolysaccharide (LPS) on isolated rabbit hepatocytes. Ital J Gastroenterol (EASL Supplement) P52.

Ravin H, Rowley D, Jenkins C, Fine J (1960). On the absorption of bacterial endotoxin from the gastrointestinal tract of the normal and shocked animal.

J Exp Med 112: 783.
Wardle, EN, (1974). Fibrinogen in liver disease. Arch Surg 109:741.
Weston MJ, Gazzard Gh, Buxton B, Winch J, Machado A, Flax Williams R (1974), Effects of haemoperfusion through charocal or XAD-2 resin on an animal model of fulminant liver failure. Gut 15: 482
Wilkinson SP, Arrogo V, Gazzard BG, Moodie H, Williams R (1974). Relation of renal impairment and haemorrhagic diathesis to endotoxaemia in fulminant hepatic failure.
Lancet: 1:521
Wilkinson SP, Moodie H, Stamatakis JD, Kakkar VV, Williams R (1976). Endotoxaemia and renal failure in cirrhosis and obstructive jaundice. Brit Med Journal 2: 1415
Williams R. Murray-Lyon IM (1975). "Artificial Liver Support". London: Pitman Medical

THE LIMULUS ASSAY FOR ENDOTOXEMIA, AS APPLIED IN GASTROENTEROLOGY

Heinrich Liehr, M.D., Professor of Medicine,

University of Würzburg, Medical Department,
School of Medicine
Josef-Schneider-Straße 2, 8700 Würzburg, W.Germany

ABSTRACT

Frequency and significance of endotoxemia was investigated in 274 patients succumbing different diseases of the gastrointestinal tract by means of the Limulus Amebocyte Lysate Test (LAL). Circulating endotoxins were found in all patients with fulminant hepatic failure and alcoholic hepatitis. In liver cirrhosis the frequency was 79 %, and ranged between 36 % and 66 % in cases of uncomplicated viral hepatitis, chronic active hepatitis, acute pancreatitis, Crohns' disease and ulcerative colitis. Predominantly, a positive LAL was associated with clinical symptoms known as biological consequences of endotoxin action. Indications are given that an antiendotoxic action is followed by LAL negative blood, and concomitantly endotoxin induced disorders disappear. It is concluded that by the application of LAL a more precise insight into the pathogenesis of gastroenterological diseases is provided.

INTRODUCTION

Since the Limulus assay has become available for detection of endotoxins in biological fluids (18) evidence accrued that endotoxemia may be of pathogenetic significance in different diseases in gastroenterology. Thus, a study was done to evaluate whether 1) endotoxemia is a frequent event in gastrointestinal diseases and 2) whether endotoxemia is merely a symptom or has pathogenetical mean.

Pages 309–320, Biomedical Applications of the Horseshoe Crab (Limulidae)

MATERIAL AND METHODS

The Limulus Amebocyte Lysate Test (LAL) was used for detection of endotoxemia in venous blood by means of the dilution technique (18) which was slightly modified by using an additional plasma dilution of 1:100 with LAL negative saline (19). The procedure consisted in centrifugation of heparinized venous blood (300 g/30 min); 0.1 ml of each undiluted plasma and plasma dilutions of 1:10 and 1:100 were incubated with 0.1 ml Limulus Lysate (Pyrogent[R], Byk-Mallinckrodt, Corp. Dietzenbach, W. Germany) for 24 hours at 37°C. The test was scored positive, if gel granula (+), fluid gel (2+) or a solid gel (3+) had developed. Disposable plastic syringes which were proven pyrogen free by the producer were used for blood sampling and dilution of plasma. The test tubes were conventional pyrogen free glas ware and part of the test kit.

A number of 274 patients was investigated. The patients were hospitalized because of different gastroenterological diseases (Table 1). As indicated later a certain number of patients was investigated morefold. The diagnoses were established by routine physical and laboratory tests including radiological and biopsy studies.

	n	Endotoxin positive
Viral hepatitis	36	13
Fulminant hepatic failure	3	3
Alcoholic hepatitis	6	6
Chronic active hepatitis	12	6
Liver cirrhosis	170	135
Acute pancreatitis	33	22
Crohns' disease	11	7
Ulcerative colitis	4	2
	274	

TABLE 1 Frequency of endotoxemia as detected by the LAL test in gastroenterological diseases.

RESULTS

The overall frequency of endotoxemia is listed in Table 1. The data indicate that endotoxemia in general is a frequent event in various diseases of gastroenterology. Since the total scope is complex the details of the results including those of previous investigations are given by the following paragraphs together with related data of the literature.

ENDOTOXEMIA IN VIRAL HEPATITIS

In uncomplicated viral hepatitis endotoxemia was found in 37 %. All patients with fulminant hepatic failure had endotoxemia. The principle finding is in accordance with observations of other authors (8,37,38), but the figures differ between these studies. This may be explained by differences in the patient groups investigated and also by different methods (i.e. chloroform extraction method (18)) applied. The mechanism of endotoxemia is uniformly thought as consequence of Kupffer cell failure (for review 6,21,28). These cells are the major clearance organ of endotoxins (32) and represent nearly 90 % of the total body reticulo endothelial system (4). Since intestinal absorbed endotoxins are found to be normal constituents of portal blood (30), endotoxemia in viral hepatitis is part of the "spillover phenomenon", the latter recently reviewed by BRADFIELD (6). The clinical significance of endotoxemia in viral hepatitis was investigated by a correlation study (39): Endotoxemia was found in close correlation with occurance of intravascular coagulation and renal disorders. All patients in this study died, if endotoxemia did not disappear spontaneously. By experimental work done in the galactosamine hepatitis model of rats (15,22,23) it was suggested that fulminant hepatic failure may be the consequence of Kupffer cell failure with subsequent systemic endotoxemia. If further prove will be achieved a more precise basis for a rational treatment of fulminant hepatic failure will be established.

ENDOTOXEMIA IN ACUTE ALCOHOLIC HEPATITIS

In acute alcoholic hepatitis endotoxemia was found as a common complication. In most of the cases certain clinical symptoms, such as leucocytosis, fever, hemolysis, and disorders in lipid metabolism improved once endotoxemia had disappeared spontaneously. This observation is of importance since it bears insight into the probable pathogenesis of the disease. It is still unclear why alcoholic hepatitis develops, and why it does so in some alcoholics but not in others (12). It is noteworthy in this context that alcohol is proven to diminish Kupffer cell function (1), thus allowing endotoxins to spillover from the portal blood into the general circulation. The spectrum of the biological activities of endotoxins is broad (33), and comprises most of the extrahepatic symptoms charactarising alcoholic hepatitis (12). Thus, the hypothesis was put forward that alcoholic hepatitis develops if endotoxemia occurs during the course of non-inflammatory alcoholic liver disease (26).

Why the development of cirrhosis is also triggered by endotoxins should be a matter of discussion in the view of experimental work done by BROITMANN et al. (7) and RUTENBERG et al. (31). These authors have demonstrated that nutritional cirrhosis can be prevented by orally administered neomycin (31), and that this beneficial effect can be abolished by additional fed endotoxins (7).

ENDOTOXEMIA IN LIVER CIRRHOSIS

In liver cirrhosis the overall frequency of endotoxemia is about 80 % in our series, and the data of the literature range between 20 % and 100 % (8,9,30,36,37,39). The explanation may be the same as given in connection with viral hepatitis. A further explanation will be the fact that endotoxemia in cirrhotic patients is not a permanent situation but an episodical event. The sequential investigation of 10 patients demonstrats that endotoxemia may be of permanent or of short duration. Since in our study predominantly morefold investigations of the patients were done a higher number of endotoxin positive patients was detected. The stage of the disease has further influence on frequency of endotoxemia. As given elsewhere patients having a marked porta-caval collateral circulation are more often endotoxin positive as opposed to a group of patients without clinical signs of a porta-caval collateral circulation (21). The data reflect Kupffer cell failure due to hemodynamical disorders which are caused by the fact that portal blood is deviated from the Kupffer cells via porta-caval collaterals. Another mechanism leading to endotoxemia may act in addition. Beside the possibility of endotoxin escape into the lymphatic circulation (13) the substances may penetrate transmurally from the intestine into the peritoneal cavity, a mechanism which follows a Michaelis Menton kinetic (29). Clinical prove was given by the detection of endotoxins in ascitic fluid in a frequency of about 80 % of 20 (36), 33 (21), and 21 (9) investigated samples.

The question whether systemically circulating endotoxins merely means endotoxemia or causes endotoxicosis can be answered by correlation studies. By such designed investigations evidence was given that clotting disorders, gastrointestinal bleeding, humoral immune response, and renal disorders both, functionally induced or due to acute tubular necrosis nearly exclusively develop if endotoxins are found circulating in the blood (9,25,38,39). A pathogenetic connection is supported by the fact that an antiendotoxic treatment by using polymyxin B (27) results in disappearance of both endotoxemia and the suggested biological consequences (20,25,34,39).

ENDOTOXEMIA IN CHRONIC ACTIVE HEPATITIS

Patients with chronic active hepatitis were found to be endotoxin positive in 50 %. Unlike fulminant hepatic failure and severe courses of liver cirrhosis signs of endotoxicosis never were observed in these patients. Thus, the significance of these findings can not yet be determined and at least may reflect a state of functional impairment of the clearance system.

ENDOTOXEMIA IN CROHNS' DISEASE

Crohns' disease is accompanied with a considerable high frequency of endotoxemia. It is not thought that endotoxins are involved within the primary mechanisms acting. It seems more reasonable to suggest that the diseased intestine promotes endotoxin escape. AUER et al. investigated in those patients whether there is a correlation between the observed B lymphocytosis (2,3) in previously untreated patients with short duration disease and the observed endotoxemia (3). The results, however, were not conclusive.

ENDOTOXEMIA IN ULCERATIVE COLITIS

The data of endotoxemia in ulcerative colitis are quite preliminary especially because of the small number of patients investigated. JACOB et al. (16) found a 100 % frequency of portal endotoxemia in this disease which points on the fact that increased absorption is present. It seems reasonable to suggest that endotoxemia contributes to the state of toxic colon. In latter cases the hyperdynamic circulation is often causative for the deletariouse outcome of the patients. In one case of toxic colon associated with both, hyperdynamic circulation and endotoxemia the treatment by polymyxin B led to a negative LAL test for endotoxemia and the circulatory disorders normalized (Figure 1).

ENDOTOXEMIA IN ACUTE PANCREATITIS

Circulating endotoxins in acute pancreatitis were described first by FINE (11) and it was suggested that they are the cause of most severe extrahepatic disorders such as pulmonary complications, coagulation disorders, intestinal hemorrhage, and vascular collapse (11). Support is given by the data listed in Table 2. Severe systemical disorders were associated with endotoxemia, and need for intensive care was given predominantly in endotoxin positive patients.

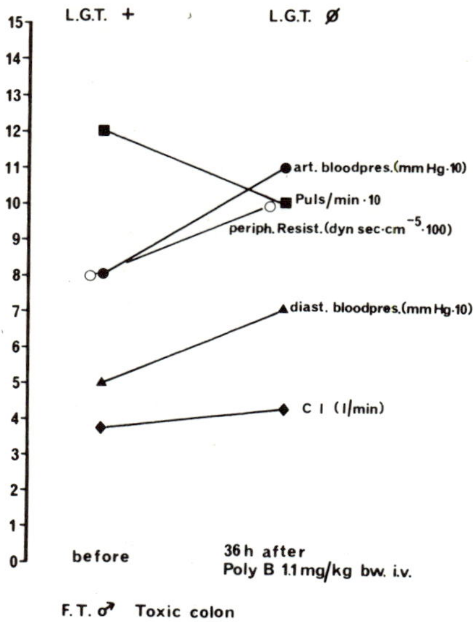

FIGURE 1
Influence of polymyxin B on hemodynamics in a case of toxic colon with positive Limulus Gelation Test (L.G.T.).

Endotoxemia in acute pancreatitis

	Limulus-Gelation-Test	
	positive	negative
Patients (n = 33)	22 (66 %)	11 (33 %)
Renal failure	5	1
DIC	10	3
Pulmonary disorders	2	0
Need for intensive care	15	3
Death	9	0

TABLE 2
Incidence of endotoxemia in acute pancreatitis listed together with severe clinical disorders.

It is important to realize that only those patients died in whom endotoxemia had developed. Endotoxins are known to activate the complement system via the C_3-pathway (14). SEELIG and SEELIG (35) demonstrated experimentally that complement activation is a basic mechanism acting in the pathogenesis of acute pancreatitis leading finally to necrosis of pancreas acinar cells by its cytolytic activity. A clinical study which was done jointly with these authors (24) provided evidence that a decrease of the third component of complement down to 50 % of normal was present only in endotoxin positive patients. The decrease was thought as reflection of complement consumption developing during the first 3 or 4 days after onset of the disease. All patients in whom normalization occured after this period recovered. Half of the patients died in whom this did not happen (Figure 2). Thus, endotoxemia at least determined the clinical picture of acute pancreatitis. Even more, it may be the trigger of a basical acting mechanism.

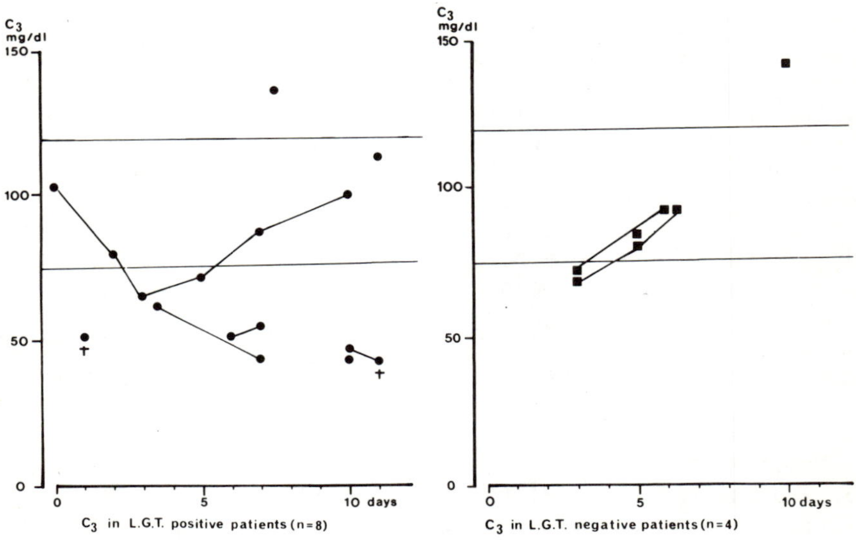

A B

FIGURE 2
Plasma concentrations of complement factor 3 (C_3) in endotoxin positive (A) and endotoxin negative (B) patients in acute pancreatitis. Day 0 = onset of the acute attack.
L.G.T.= Limulus Gelation Test

COMMENT

The results presented provide evidence that the application of the Limulus test in gastroenterology for detection of endotoxemia is of high significance in getting a more precise insight view into the pathogenesis of different diseases. It is realized well that this study is only the scope of the total problem and that further work is needed. In recent time it was a matter of dispute whether the basis of the scope i.e. the Limulus Lysate Test is specific enough to justify the statement that a positive test result is an indication for the presence of endotoxins. By a number of authors (10) the specifity of the test was accepted as giving a qualitative answer. Problems are still present, whether the test is applicable to get a quantitative answer, especially when applied in biological fluids. In the latter case there might be interference with an inhibitor or inhibitors (18), one of which may be cholic acids, especially chenodesoxycholic acid (19).

Although it may be of interest to know the amount of circulating endotoxins under certain conditions, such data are of limited value from a biomedical point of view since the amounts of endotoxins rendered noccious by the organism will differ from subject to subject, and within one subject from day to day dependent on the actual state of the defense system. The question, whether merely endotoxemia is present or even endotoxicosis, can be answered only by correlation studies and therapeutical trials. The latter will be done with purpose to achieve both support of correlation studies, and to find means to help patients to recover succumbing from endotoxicosis. Aspects of both of these goals were given. Beside the number of inherent problems and still open questions the conclusion is justified that endotoxemia is a frequent complication of diseases in gastroenterology and obviously contributes significantly to different basic pathogenetic mechanisms. Thus, an extract of the blue blood of a prehistoric animal the horseshoe crab Limulus which still lives in our time provides us to answer open questions in present clinical medicine. Thus, the theory of "Instestinale Autointoxikation" (5) as formulated in 1928 by L. BOGENDÖRFER, a medical professor of the University of Würzburg, Germany, becomes its support, as does the suggestion of PAWLOV, done in 1893, who concluded from his experiments (17) that "poisonous symptoms appeared, if blood is compelled to deviate from the liver, which is solely due to the fact that the liver no longer plays a protective role against toxins, which are being continually fabricated in the intestinal fluid".

1. Ali, M.V., Nolan, J.P.: Alcohol induced depression of reticuloendothelial function in the rat. J. Lab. Clin. Med. 70, 295-301, 1967
2. Auer, I.O., Wechsler, W., Ziemer, E., Malchow, H., Sommer H.: Immunstatus in Crohns' disease. I. Leucocyte and lymphocyte subpopulations in periferal blood. Scand. J. Gastroent. 13, 561-571, 1978
3. Auer, I.O., Götz, S., Ziemer, E., Malchow, H., Elms, H.: Immunstatus in Crohns' disease. III. Periferal blood B Lymphocytes, enumerated by means of F (ab) 2 antibody fragments, Null and T Lymphocytes. Gut (in press)
4. Biozzi, B., Stiffel, C.: The physiopathology of the reticulo-endothelial cells of the liver and spleen. In: Progress in liver diseases. Popper, H., Schaffner, F. (Edts.) Grune & Stratton, New York, London, pp. 166-191, 1965
5. Bogendörfer, L.: Ueber intestinale Autointoxikation. Berliner Klinik 35, 1-24, 1928
6. Bradfield, J.W.B.: Control of spillover. The importance of Kupffer cell function in clinical medicire. Lancet II, 883-886, 1974
7. Broitmann, S.A., Gottlieb, L.S., Zancheck, N.: Influence of neomycin and ingested endotoxin in the acult rat. J. Exp. Med. 119, 633-640, 1964
8. Caridis, D.T., Woodruff, P.W.H., Reinhold, R.B., Fine, J.: Endotoxaemia in man. Lancet I, 1381-1386, 1972
9. Clemente, C., Bosch, J., Rodes, J., Arroyo, V., Mas, A., Maragall, S.: Functional renal failure and haemorrhagic gastritis associated with endotoxaemia in cirrhosis. Gut 18, 556-560, 1977
10. Cohen, E. (Ed.): Biomedical applications of Limulus polyphemus. Alan, R., Liss, Inc. Scientific Publications, New York, N.Y. (in press)
11. Fine, J.: Acute pancreatitis, Lancet I, 1092, 1975
12. Galambos, J.T.: Alcoholic hepatitis, F. Schaffner, S. Sherlock, C.M. Leevy (Edts.): The liver and its diseases. G. Thieme Verlag, Stuttgart, pp 255-267, 1974
13. Gans, H., Matsumoto, K.: On the escape of endotoxin from the intestine. B. Urbaschek, R. Urbaschek, E. Neter (Edts.) Gram-negative bacterial infections. Springer, Wien - New York, pp 429-435, 1975

14. Götze, O., Müller-Eberhard, H.J.: The C3-activator system: An alteration pathway of complement activation. J. Exp. Med. 134, 90-108, 1971
15. Grün, M., Liehr, H., Rasenack, U.: Significance of endotoxaemia in experimental galactosamine hepatitis in rats. Acta Hepato-Gastroenterol. 24, 64-81, 1977
16. Jacob, A.I., Goldberg, P.K., Bloom, N., Degenshein, G.A., Kozinn, P.J.: Endotoxin and bacteria in portal blood. Gastroenterology 72, 1268-1270, 1977
17. Lancet, Paris correspondent: The antiendotoxic function of the liver. Lancet II, p. 1092, 1893
18. Levin, J., Tomasulo, P.A., Oser, R.S.: Detection of endotoxin in human blood and demonstration of an inhibitor. J. Lab. Clin. Med. 25, 903-911, 1970
19. Liehr, H., Grün, M., Brunswig, D., Sautter, T.: Endotoxinämie bei Leberzirrhose. Ztschr. Gastroenterologie 14, 14-23, 1976
20. Liehr, H., Grün, M., Brunswig, D., Sautter, T.: Endotoxaemia in liver cirrhosis: Treatment with Polymyxin B. Lancet I, 810-811, 1975
21. Liehr, H., Grün, M.: Clinical aspects of Kupffer cell failure in liver diseases. E. Wisse, D.L. Knook (Edts.) Kupffer cells and other liver sinusoidal cells. Elsevier/North-Holland Biomedical Press, Amsterdam-New York - Oxford, pp 427-436, 1977
22. Liehr, H., Grün, M., Seelig, H.P., Seelig, R., Reutter, W., Heine, W.D.: On the pathogenesis of galactosamine hepatitis. Indications of extrahepatocellular mechanisms responsible for liver cell death. Virch. Arch. B. Cell. Path. 26, 331-344, 1978
23. Liehr, H., Grün, M., Seelig, H.P., Seelig, R., Rasenack, U.: Synergistic action of hepatocyte membrane defect and activated complement system in liver cell death. Acta Hepato- Gastroenterol. 25, 105-110, 1978
24. Liehr, H., Seelig, R., Grün, M., Seelig, H.P.: Endotoxemia and complement activation in acute pancreatitis. Proc. XIV. Int. Congr. Int. Med. Rome, 1978 (in press)
25. Liehr, H., Grün, M.: Ruolo delle endotossine nelle alterazioni renali in corso di epatopatia. Min. Med. 69, 3231-3242, 1978
26. Liehr, H., Grün, M.: Endotoxins in liver disease, H. Popper, F. Schaffner (Etdts.): Progress in liver disease. Vol. VI, Grune & Stratton, New York (in press)

27. Neter, E., Gorzynski, A.E., Westphal, O., Lüderitz, O.: The effects of antibiotics on enterobacterial lipopolysaccharides (endotoxins), hemagglutination and hemolysis. J. Immunology 80, 66-72, 1958
28. Nolan, J.P.: The role of endotoxin in liver injury. Gastroenterology 69, 1346-1356, 1975
29. Nolan, J.P., Hare, D.K., Mc Devitt, J.J., Ali, M.V.: In vitro studies of intestinal endotoxin absorption. I. Kinetics of absorption in the isolated everted gut sac. Gastroenterology 72, 434-439, 1977
30. Prytz, H., Holst-Christensen, Körner, B., Liehr, H.: Portal venous and systemic endotoxaemia in patients without liver disease and systemic endotoxaemia in patients with cirrhosis. Scand. J. Gastroenterol. 11, 857-863, 1976
31. Rutenberg, A., Sonnenblick, E., Koven, J., Aprahamian, H.A., Reiner, L., Fine, J.: The role of intestinal bacteria in the development of dietary cirrhosis in rats. J. Exp. Med. 106, 1-13, 1957
32. Rutenberg, S., Skarnes, R., Palmerio, C., Fine, J.: Detoxification of endotoxin by perfusion of liver and spleen. Proc.Soc. Exp. Biol. Med. 125,455-459, 1967
33. Saba, T.M.: Physiology and physiopathology of the reticuloendothelial system. Arch. Intern. Med. 126, 1031-1052, 1970
34. Sautter, T., Liehr, H., Grün, M., Brunswig, D.: Die Behandlung der Endotoxinämie bei Leberzirrhose mit Polymyxin B. Verh. Dtsch. Ges. Inn. Med. 83, 1148-1157, 1977
35. Seelig, R., Seelig, H.P.: The possible role of serum complement system in the formal pathogenesis of acute pancreatitis. Acta Hepato-Gastroenterol. 22, 263 and 335, 1975
36. Tarao, K., Endo, O., Ikeuchi, T., Koi So, Moroi, T., Fukushima, K.: Endotoxin in ascitic fluid of liver cirrhosis in man. Jap. J. Gastroenterology 37, 1366-1372, 1976
37. Wardle, N.E.: Fibrinogen in liver disease. Am. J. Surg. 109, 741-746, 1974
38. Wilkinson, S.P., Gazzard, B.G., Arroyo, V., Moodie, H., Williams, R.: Relation of renal impairment and haemorrhagic diathesis to endotoxemia in fulminant hepatic failure. Lancet 521-524, 1974

39. Wilkinson, S.P., Moodie, H., Stammatakis, J.D., Kakkar, V.V., Williams, R.: Endotoxaemia and renal failure in cirrhosis and obstructive jaundice. Brit. Med. J. II, 1415 - 1418, 1976

CHEMICAL, PHYSICOCHEMICAL AND BIOLOGICAL PROPERTIES OF BACTERIAL LIPOPOLYSACCHARIDES

C. Galanos, M.A. Freudenberg, O. Lüderitz,
E.T. Rietschel and O. Westphal.

Max-Planck-Institut für Immunbiologie,
78 Freiburg, Stübeweg 51, German federal republic.

Introduction.

Lipopolysaccharides (LPS) are pathogenetic agents of gram negative bacteria being responsible for many pathological alterations seen during gram negative infections. In experimental animals isolated LPS produce pathophysiological activities like fever, Shwartzman reaction, depression of blood pressure and in higher concentrations lead to shock and death. On the other hand they increase non-specific resistance to infection, act as adjuvances and induce necrosis and regression of malignant tumours (Weinbaum et al. 1971, Kadis et al. 1971). LPS are not exogenous products of gram-negative bacteria. After their biosynthesis in the cytoplasmic membrane they are transported to the surface where they are retained as vital constituents of the outer membrane of the cell. In this exposed location they also function as the characteristic O antigens of these bacteria. The release of LPS from bacteria takes place after death and lysis of the cell. For this reason, in contrast to the exotoxins which are released from living microorganisms, these toxic substances were named endotoxins. In the present communication the names lipopolysaccharide and endotoxin are used synonymously.

Chemically LPS are made up of a lipid part the lipid A linked

Fig. 1 Schematic Representation of Salmonella Lipopolysaccharides

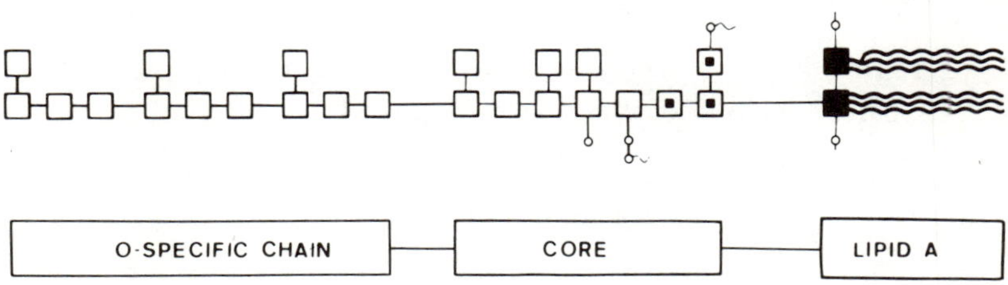

Pages 321–332, Biomedical Applications of the Horseshoe Crab (Limulidae)

to a heteropolysaccharide consisting of the O specific polysaccharide and the core (See Fig.1). The O polysaccharide is made up of repeating oligosaccharide units which carry the serological specificity of the LPS and of the parent bacterial strain. Its sugar composition and structure is highly diverse among gram negative bacteria. The core polysaccharide shows a lower variability, in Enterobacteriaceae its structure and composition is similar among bacterial groups. In Salmonella the core is represented by a branched polysaccharide containing N-acetyl glucosamine, glucose, galactose, heptose and KDO (2-Keto-3-deoxyoctonate).

Lipid A represents the least variable part of the LPS molecule, its structure and composition being very similar among Enterobacteriaceae. Its chemical formula according to up to date knowledge is shown in Fig.2.

Fig. 2

CHEMICAL STRUCTURE OF SALMONELLA LIPID A

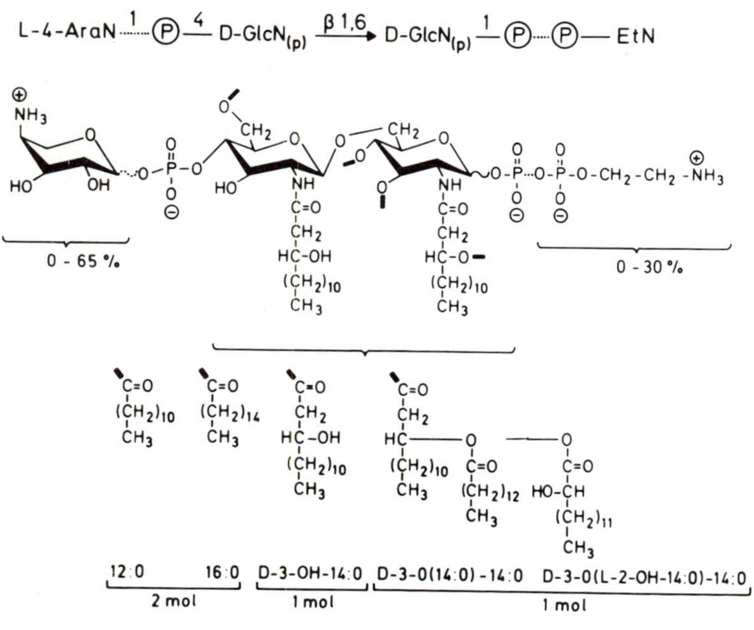

It consists of a central D-glucosaminyl-ß1,6-D-glucosamine disaccharide substituted with phosphate at positions 4' and 1. The amino groups are substituted by D-3-hydroxymyristic acid and available hydroxyl groups are esterified with lauric, palmitic and myristoxymyristic acid. The polysaccharide is linked through KDO to C3 of the non-reducing glucosamine of lipid A .

In Salmonella the two phosphate groups at C4' and C1 are partly substituted by 4-amino-L-arabinose and phosporyl ethanolamine, respectively. These substitutents, however, are not obligatory constituents of lipid A; they are lacking, for instance in E.coli (for review see Galanos et al. 1977).

The above complete LPS structure is synthesized by wild S form (smooth) bacteria, many of which are pathogenic in man and in experimental animals. From many gram negative microorganisms, mutants have been isolated the so called R (rough) mutants. These, as a result of mutational defects synthesize incomplete LPS from which the O polysaccharide is absent and the core may be present in different degrees of completion. Although R form bacteria are usually non-pathogenic, the incomplete LPS they make exhibit all toxic and other biological activities expressed by complete LPS. Their non-pathogenic character is mainly due to their inability to invade and proliferate in the macro-organism.

Originally R mutants were isolated solely from laboratory cultures. As natural source of LPS in different stages of completion, they have been of immense value for chemical, biosynthetic and biological studies on LPS. Recently rough strains of bacteria have also been isolated from patients with urinary tract infections (Hanson, 1973) and have thus become interesting also from the clinical point of view.

In Salmonella five main classes (and a number of sub-classes) of R mutants have been recognized, and designated Ra to Re in

Fig. 3 Schematic Representation of R LPS.

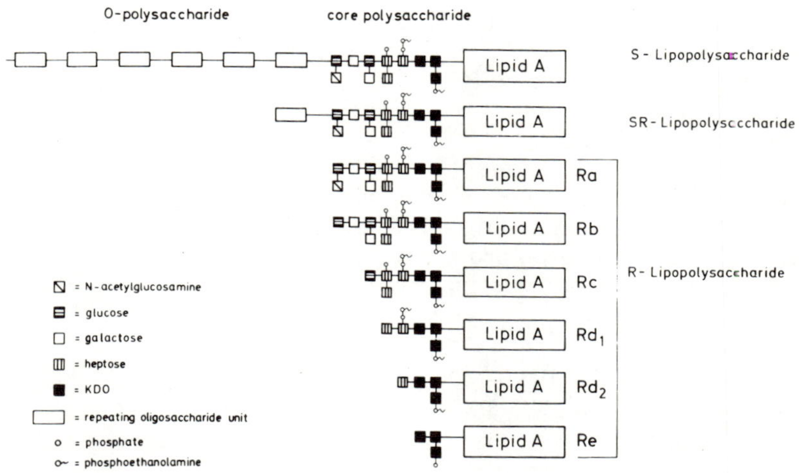

the order of increasing defect. The structure of LPS belonging to the various Salmonella R classes is shown schematically in Fig. 3. The Re class of mutants synthesizing LPS containing lipid A and 3 KDO residues represents the mutant with the highest defect in LPS biosynthesis that is still viable under normal conditions. Lipid A therefore does not occur in the free form on the bacterial cell wall. Mutants defective in the synthesis of KDO have only recently been isolated under special conditions (as temperature

Table 1. Biological Activities of Free Lipid A

Pyrogenicity
Lethal toxicity in mice
Leukopenia
Leukocytosis
Local Shwartzman reaction
Bone marrow necrosis
Embryonic bone resorption
Complement activation
Depression of blood pressure
Platelet aggregation
Hageman factor activation
Induction of plasminogen activator
Limulus lysate gelation
Toxicity enhanced by BCG
Toxicity enhanced by adrenalectomy
Enhanced dermal reactivity to epinephrine
Induction of nonspecific resistance to infection
Induction of tolerance to endotoxin
Induction of early refractory state to temperature change
Adjuvance activity
Mitogenic activity for cells
Tumor necrotic activity
Macrophage activation
Induction of colony stimulating factor
Induction of IgG synthesis in newborn mice
Induction of prostaglandin synthesis
Induction of interferon production
Induction of tumor-necrotizing factor
Induction of mouse liver pyruvate kinase
Type C RNA virus release from mouse spleen cells
Helper activity for friend spleen focus-forming virus in mice
Inhibition of phosphoenolpyruvate carboxykinase
Hypothermia in mice

(for references, Galanos et al. 1977)

- sensitive mutants) (Osborn et al. 1974, Lehman, 1977).

Biological activity of lipid A: While the antigenic specificity of LPS is determined by the polysaccharide, toxicity and other biological activities are expressed by the ubiquitous lipid A. As a rule the polysaccharide as such is biologically inactive. Its role in biological activity is indirect acting as a solibilizing carrier for the insoluble lipid A. The biological activity of lipid A was demonstrated using soluble forms of free lipid A.

Free lipid A is usually prepared by mild acid hydrolysis of LPS whereby the ketosidic linkage between the lipid and KDO is cleaved and lipid A is obtained as an insoluble precipitate. Solubilization of lipid A is achieved either through complexation with solubilizing carriers such as bovine serum albumin or by converting (carrier-free) lipid A into the uniform triethylamine or sodium salt form (Galanos and Lüderitz 1975, see following chapters). Table 1 summarizes the various biological activities demonstrated so far with soluble forms of lipid A.

Immunological properties of lipid A.: Not only the O polysaccharide and the core, but also the lipid A exhibits immunogenic properties inducing the production of specific anti lipid A antibodies in different animals and in man (Galanos et al. 1971). The lipid A's from Enterobacteriaceae due to structural similarities share common determinants and are serologically cross-reacting (see Tab.2). The immunodeterminant group has been identified as a glucosaminyl group in the glucosamine dissaccharide carrying an amide linked 3-D-hydroxymyristic acid (Ługowski and Romanowska, 1974, Jay and Galanos, unpublished). This structure is common to the lipid A of Enterobacteriaceae and other bacterial groups. Other substituents such as phosphate, 4-aminoarabinose or ethanolamine do not seem to play a role in antigenic specificity.

Non-cross-reacting lipid A systems have also been identified recently. Certain species of photosynthetic bacteria Rhodopseudomonas viridis and Rhodopseudomonas palustris synthesize a lipid A, which due to differences in chemical structure compared to lipid A of Salmonella do not show serological cross-reactions with the lipid A of Enterobacteriaceae (Galanos et al. 1977, see Tab. 2).

Anti lipid A antibodies occur naturally in the serum of normal animals and humans, and significantly more frequently in sera of patients with chronic urinary tract infections. Anti lipid A antibodies exhibit biological activities that may be protective or damaging. Under defined experimental conditions anti lipid A antisera suppress the pyrogenic activity of endotoxin (Rietschel and Galanos 1977) and protect mice to the abortive effect of endotoxin (Rioux-Darrieulat et al. 1978) and to infection with S.typhimurium (Galanos et al. 1977). On the other hand lipid A antibodies were shown to

Table 2 Serological Cross-Reactions of Free Lipid A Among Gram-Negative Bacteria

Salmonella	+	R.gelatinosa	+
E.coli	+	R.tenue	+
S.flexneri	+	C.violaceum	+
P.mirabilis	+	N.gonorrhoeae	+
P.vulgaris	+	Rickettsia canada	+
A.tumefaciens	+	R.prowazekii	+
Y.enterocolitica	+	Coxiella burnetii (phase I)	+
Pseudomonas fluorescens	+	R.viridis	-
Y.sinensis	+	R.palustris	-
V.cholerae	+		

+ = Crossreaction with Salmonella anti-lipid A present
- = Crossreaction with Salmonella anti-lipid A absent

mediate the development of kidney damage induced by lipid A (Westenfelder et al. 1977).

Anti lipid antibodies due to their high specificity for lipid A and on account of their wide cross-reactions with lipid A of Enterobacteriaceae and other gram negative microorganisms, represent a potential tool for the detection and measurement of endotoxin in biological fluids.

Physical state and biological activity of LPS: Isolation of LPS from bacteria may be carried out by a variety of extraction procedures. The phenol-water (Westphal et al. 1952) and the PCP (Galanos et al. 1969) procedures are commonly used for the extraction of S and R form LPS, respectively. Regardless of the method of extraction LPS may exhibit large differences in their physicochemical properties.

The factors influencing the physical state (solubility) of LPS are multiple being mainly intrinsic, and related to the chemical properties of the LPS. Thus hydrophobic interactions exerted by the fatty acids in the lipid A moiety are responsible for the formation of aggregates. However, also extrinsic factors such as low and high molecular weight contaminants influence the physical state of LPS and are primarily responsible for the difference in solubility seen with different preparations. Most usual, macromolecular contaminants present in LPS are proteins, lipoproteins, phospholipids and glycans. Their amount may vary according to the isolation procedure employed, and from batch to batch of parent bacterial culture. They are associated physically with the LPS. Their presence in LPS may also influence its biological activity. It is known that gram negative cell wall components other than LPS exhibit their own activities, some of which may resemble those of LPS. Thus the lipoprotein and other cell wall proteins exhibit mitogenic activity for mouse B lymphocytes which could be distinquished from that of LPS (Melchers et al. 1975). Today due to the improvement of LPS isolation and purification methodology it is possible to prepare LPS that are completely free of above macromolecular contaminants (Galanos et al. 1977).

In addition to macromolecular contaminants, LPS regardless of the method of isolation always contain low molecular weight substances, most common being metal cations like Na^{+}, Ca^{2+}, and Mg^{2+} and amines like ethanolamine, putrescine, cadaverine, spermine and spermidine. These basic substances are present in ionic linkage with acidic (phosphate and carboxyl) groups of the LPS molecule and they influence not only the physicochemical but also the biological properties of LPS. Their removal by electrodialysis leads to acidic LPS which by subsequent neutralization with a given base are converted into uniform salt forms of the LPS.

Different salt forms of LPS exhibit characteristic and constant physicochemical properties. Thus in the triethylamine salt form LPS are highly soluble exhibiting a low aggregation as measured by sedimentation coefficient determinations. Solubility decreases and the sedimentation coefficient in creases with the Na^{+} and Ca^{++} form of LPS (see Tab.3).

Electrodialysis of LPS and conversion to the triethylamine form has solved the problem of solubility encountered in the past with many LPS preparations. It has also made possible the preparation of free lipid A in water soluble form, without the use of solubilizing carriers such as BSA which was used in the past.

Biological activities of LPS in uniform salt forms : Studies with uniform salt forms of different LPS showed that individual biological activities may increase or decrease with increasing molecular weight (aggregation) of the LPS (Galanos et al. 1977) (Tab.3).

Lethal toxicity for mice and pyrogenicity for rabbits increase with decreasing S value of the LPS and are highest in the triethylamine salt form (9S). On the other hand, lethal toxicity for rats (Freudenberg and Galanos, 1978) the rate of LPS clearance from the blood, cell surface affinity (Praino et al. 1978) and the property to interact with the complement system increase with increasing S value as long as the LPS remains soluble. They are highest in the Na^{+} or Ca^{2+} form. In the case of mitogenic activity for mouse B lymphocytes no differences in activity of different salt forms were found within the range of molecular weights tested (9S - 200S).

Table 3

PROPERTIES OF THE LIPOPOLYSACCHARIDE OF S. ABORTUS EQUI IN DIFFERENT SALT FORMS

S abortus equi LPS	Sedimentation Coefficient	Solubility (Water)	Lethal Toxicity MICE	Lethal Toxicity RATS	Pyrogenicity (Rabbits)	Rate of clearance from the blood	Interaction with C' in vivo and in vitro	Affinity for cells	Mitogenic activity
Triethylamine (TEN)	9.3	↑	↑		↑		(a)		↕ (b)
Pyridine	↓								
Ethanolamine									
Na									
K									
Putrescine	230								
Ca	partly insol.			↓		↓	↓	↓	

Arrows show direction of increasing activity.
(a) In the TEN form S form LPS is completely non-anticomplementary.
(b) No difference within the above range of (S) values.

Detection and measurement of endotoxin. The question of standard. Since endotoxin was recognized as an important pathogenetic component of gram negative bacteria, much effort has been devoted to the development of methods for its detection and measurement in clinical and laboratory samples.

Of the different methods (chemical or biological) available for the detection of endotoxin the limulus amoebocytelysate (LAL) represents by far the most sensitive assay. As with other biological assays also in the case of LAL the question of specificity is still unsolved. Chemical assays in which a characteristic component of LPS (e.g. D3-hydroxymyristic acid) is measured are usually highly specific, however, so far their sensitivity is relatively low. A further disadvantage of chemical assays is that they make no disclosure of the biological activity of the endotoxin present in a given sample.

Biological assays measure endotoxin activity rather than endotoxin quantity and this relative to the activity of the preparation used as standard.

The choice of the endotoxin used as reference is critical since for reasons elaborated above different preparations may differ significantly in their physical, chemical and biological properties. Such preparations may be perfectly adequate for the research purpose of individual laboratories. The adoption of a LPS preparation as standard for the measurement of endotoxin activities such as the limulus lysate gelation would therefore seem to be of critical importance. In choosing a biological standard it is necessary that quantitative and qualitative aspects of its biological activity are considered and that it is defined chemically and biologically.

The progress made in the last decade in isolation, purification and chemical characterization of LPS has made possible the preparation of a lipopolysaccharide fulfilling the above requirements. This standardized endotoxin preparation is derived from Salmonella abortus equi. It is free of protein and other macromolecular contaminants, and has been prepared in the uniform Na^+ form, exhibiting a sedimentation coeffcient of 80 S (Galanos et al. 1979). The analytical data of its chemical composition is shown in Tab. 4. It is supplied in ampules containing 0.1 ug and 1ug LPS/mg under the name NovoPyrexal and Novo-Pyrexal forte respectively.

The adoption of the above or some other similar preparation as standard for the biological assaying of endotoxin would enable a more direct comparison of the results of different laboratories. In the case of LAL test the use of a common standard would also enable comparison of results obtained with different lysate preparations.

Table 4 Chemical Composition of the S.abortus equi Lipopolysaccharide Na-form

Sugars (unhydro)	%	Fatty acids (unhydro)	%	Other constitutents	%
D-Abequose	13.5	Lauric (C 12)	2.2	Phosphorylresidue	5.0
L-Rhamnose	15.2	Myristic (C 14)	1.8	Sodium	1.8
D-Mannose	15.7	D-3-hydroxymyristic (3-OH-C 14)	5.6	4-amino-L-arabinose	n.d.
D-Galactose	17.8	Palmitic (C 16)	1.1	Ethanolamine	n.d.
D-Glucose	4.0				
KDO	6.8				
D-Glucosamine	4.2				
Total:	82.2		10.7		3.7

Total: 99.7%

References

Freudenberg, M.A. and Galanos, C. (1978). Interaction of lipopolysaccharide and lipid A with complement in rats and its relation to endotoxicity. Infect. Immun. 19: 875-882.

Galanos, C. and Lüderitz, O. (1975). Electrodialysis of lipopolysaccharides and their conversion to uniform salt forms. Eur. J. Biochem. 54: 603-610.

Galanos, C., Lüderitz, O., Rietschel, E.T. and Westphal, O. (1977). Newer aspects of the chemistry and biology of bacterial lipopolysaccharides, with special reference to their lipid A component. Int. Rev. Biochem. 14: 239-335.

Galanos, C., Lüderitz, O. and Westphal, O. (1969). A new method for the extraction of R lipopolysaccharides. Europ. J. Biochem. 9: 245.

Galanos, C., Lüderitz, O. and Westphal, O. (1971). Preparation and properties of antisera against the lipid-A component of bacterial lipopolysaccharides. Europ. J. Biochem. 24: 116-122.

Galanos, C., Roppel. J., Weckesser, J., Rietschel, E.T. and Mayer, H. (1977). Biological activities of lipopolysaccharides and lipid A from Rhodospirillaceae. Infect. Immun. 16: 407-412.

Hanson, L.A. (1973). Host-parasite relationships in urinary tract infections. J. Infect. Dis. 127: 726-730.

Kadis, S., Weinbaum, G. and Ajl, S.J. (1971). Microbiol. Toxins, Vol. V. Academic Press, New York and London.

Lehman, V. (1977). Isolation, purification and properties of an intermediate in 3-deoxy-D-manno-octulosonic acid-lipid A biosynthesis. Eur. J. Biochem. 75: 257-266.

Lüderitz, O., Galanos, C., Risse, H.J., Ruschmann, E., Schlecht, S., Schmidt, G., Schulte-Holthausen, H., Wheat, R., Westphal, O. and Schlosshardt, J. (1966). Structural relationships of Salmonella O and R antigens. Ann. N.Y. Acad. Sci. 133: 349-374.

Lugowski, C. and Romanowska, E. (1974). Biological properties of lipid A from Shigella sonnei. Eur.J. Biochem. 48: 81-87.

Melchers, F. Braun, V. and Galanos, C. (1975). The lipoprotein of the outer membrane of Escherichia coli: AB-lymphocyte mitogen. J. Exp. Med., 142: 473-482.

Osborn, M.J., Rick, P.D., Lehman, V., Rupprecht, E. and Singh, M. (1974). Structure and biosynthesis of the cell envelope of gram-negative bacteria. Ann. N.Y. Acad. Sci. 235: 52-65.

Praino, M., Galanos, C. and Neter, E. (1978). Attachment to erythrocytes of uniform salt forms of lipopolysaccharides from Salmonella abortus equi and its inhibition by various animal sera. Immunol. Commun. in press.

Rietschel, E.T. and Galanos, C. (1977). Lipid A antiserum mediated protection against lipopolysaccharide and lipid A-induced-fever and skin necrosis. Infect. Immun. 15: 34-49.

Rioux-Darrieulat, F., Parant, M. and Chedid, L. (1978). Prevention of endotoxin induced abortion by treatment with antisera. J. Infect. Dis. 137: 7-13.

Weinbaum, G., Kadis, S. and Ajl, S.J. (1971). Microbiol Toxins, Vol. IV. Academic Press, New York and London.

Westenfelder, M., Galanos, C. and Marget, W. (1977). Pathological activities of lipid A. Experimental studies in relation to chronic pyelonephritis. In: Microbiology p. 277-279 (Schlessinger, D. ed.) Washington DC. Am. Soc. Microbiol.

Westphal, O., Lüderitz, O. and Bister, F. (1952). Über die Extraktion von Bakterien mit Phenol/Wasser. Z. Naturforsch. 7b: 148-155.

PYROGEN TESTING OF AN INTRAVENOUS FAT EMULSION

Ruben J. Guzman, M.S. and Harng-Shen Kuo, Ph.D.

Cutter Laboratories, Incorporated

4th & Parker Streets, Berkeley, California 94710

THERMOGENIC RESPONSE IN RABBITS

When fat emulsions are tested in the U.S.P. rabbit test, a false positive may occur as a result of a thermogenic response. The thermogenic response may be due to rapid oxidation of the fat. Lambert (Lambert, Miller, Frost 1951) varied the injection rate from 18 seconds to 30 minutes for a 10 ml/kg dose without affecting the thermogenic response. They repeatedly demonstrated that all components of the emulsion (except oil) were pyrogen-free yet induced a febrile response. When the oil was washed with pyrogen-free water, no pyrogens could be detected in the water wash. The thermogenic response was eliminated when the rate of clearance of fat from the blood was decreased by incorporation of 2% Triton A 20 into the emulsion.

Thompson (Thompson, Jones, Ferrell 1965) studied a number of fat emulsions, including Intralipid® 10% and 20%, in rabbits. The emulsions were administered intravenously 5 days per week for 3 weeks. The daily dose was 15 ml/kg at a rate of 1 ml per 2.5 minutes. All preparations tested produced a daily febrile response for the duration of the experiment. The febrile response usually subsided by the fifth hour after infusion to recur after the following day's infusion. This repeated febrile response is very unlike that expected from a pyrogen, as animals are made fever-tolerant by repeated injections of a pyrogen (Grant 1953).

Pages 333–337, Biomedical Applications of the Horseshoe Crab (Limulidae)

LIMULUS TEST

In order to determine if the Limulus Test could be used as an alternate pyrogen test for Intralipid®, a study comparing the rabbit and Limulus Tests was carried out.

Material and Methods

E. coli endotoxin obtained from Difco (Pyrotrol™ positive control) was used in this study. Limulus lysate was obtained from Haemachem Inc., St. Louis, Mo., Associates of Cape Cod, Woods Hole, Massachusetts, and Mallinckrodt, St. Louis, Missouri. The E. coli endotoxin was standardized against the Klebsiella B.O.B. standard (Lot #1). The reason we used E. coli endotoxin was that this is the most potent commercially available endotoxin.

Intralipid® was spiked with the E. coli endotoxin to result in the following endotoxin levels: 0.03, 0.06, and 0.12 ng/ml. An unspiked sample of Intralipid® was used as control.

All Intralipid® samples were pyrogen tested according to the U.S.P. XIX test procedure. Each rabbit received 10 ml per kg of the solution to be tested. In addition, all the spiked Intralipid® and control samples were tested by the Limulus Test using all three lysates.

Table 1

RESPONSE OF ENDOTOXIN SPIKED INTRALIPID® TO VARIOUS LYSATES

Lot No	Endotoxin ng/ml	SOURCE OF LYSATE					
		1		2		3	
		+++	-	+++	-	+++	-
1	0.12	20	0	20	0	ND	
2	0.06	20	0	20	0	ND	
3	0.06	20	0	0	20	14	6
2	0.03	18	2	0	20	ND	
3	0.03	19	1	ND		0	20

(+++ = Firm Gel; - = No Gel; ND = Not Done)

Results. The test results are shown in Table 1. Of the three lysates evaluated to detect *E. coli* endotoxin in spiked Intralipid® samples, only one was found to be suitable. The other lysates were found to be either marginally acceptable or unsuitable. Twenty aliquots of each spiked sample were tested with each of the three lysates.

Comparison of Rabbit and LAL Pyrogen Testing of Intralipid®. Each test group consisted of three rabbits which were administered 10 ml/kg of Intralipid® spiked with endotoxin. Samples requiring a re-test according to U.S.P. XIX were considered to have failed this test.

The test results of *E. coli* endotoxin dose response in rabbits are shown in Table 2. All *E. coli* endotoxin levels were also tested by the LAL test procedure and gave positive results. A positive Limulus result is a firm gel (+++) which remains intact and does not flow down the sides of the tube when inverted.

Table 2

COMPARISON OF RABBIT AND LAL PYROGEN TESTING OF INTRALIPID®

Lot #	Endotoxin (ng/ml)	Total Temp Rise in 3 Rabbits	Pass/ Fail	20 LAL Tests Positive	20 LAL Tests Negative
1	0.12	3.1°C	Fail	20	0
2	0.06	1.7 2.0	Fail Fail	20 20	0 0
4	0.06	0.6	Pass	20	0
3	0.06	1.3 0.6	Fail Pass	20 20	0 0
2	0.03	1.1 0.6	Pass Pass	18	2
4	0.03	1.3 1.5	Pass Fail	20	0
3	0.03	1.0	Fail	19	1
1	0	0.0	Pass	0	20
2	0	0.0	Pass	0	20
4	0			0	20
3	0	0.0	Pass	0	20

The 0.12 ng/ml level was only tested in one rabbit test since the temperature rise was unequivocal and 20/20 LAL tests were positive. The 0.06 ng/ml levels produced a pyrogenic response in 3/5 rabbit tests and was positive in 100/100 LAL tests. The 0.03 ng/ml produced a pyrogenic response in 2/5 rabbit tests and was positive in 56/60 LAL tests. From these results it was concluded that 0.06 ng/ml was the lowest level that could be detected in a rabbit test and that 0.03 ng/ml could be detected using the LAL test.

The limulus test procedure is, therefore, considered to be an acceptable alternate assay to the rabbit pyrogen test for detection of endotoxin and endotoxin-like material in Intralipid® using a suitable lysate.

PYROGEN TESTING OF THE AQUEOUS PHASE OF INTRALIPID®

Having shown that the limulus test was an acceptable alternate for detecting pyrogen activity in Intralipid®, we then centrifuged Intralipid® spiked with endotoxin and tested the aqueous phase for pyrogenicity. Administration to rabbits of the aqueous phase of Intralipid® whould avoid the thermogenic response caused by the fat and would provide a true test of pyrogenic activity.

Initially, bottles of Intralipid® 10%, Lot 195271 were spiked with Difco E. coli endotoxin Lot No. 509711 at 0, 0.12, 0.06 and 0.03 ng/ml level. Three additional bottles were tested at the 0.03 ng/ml level with a second control. Approximately 350 ml from each bottle were transferred to 500 ml polycarbonate bottles and centrifuged at 13,700 X G at 4°C for two hours. Analysis of the aqueous phase showed that 97% of the fat had been removed and that the concentration of phospholipid had not changed. The aqueous phase was tested for pyrogenicity by U.S.P. rabbit and LAL methods. Samples of the uncentrifuged emulsion were tested at the same time. The results are presented in Table 3.

The results show that pyrogen activity can be detected in the aqueous phase at the same level as in the emulsion. In order to avoid the thermogenic response, an acceptable alternative would be to test the aqueous phase of the emulsion. This procedure could be used for testing lots that have produced a temperature rise in rabbits.

Table 3

PYROGEN TESTING OF INTRALIPID® EMULSION AND ITS AQUEOUS PHASE					
Bottle No.	Sample	Endotoxin (ng/ml)	Temp Rise in 3 Rabbits (°C) (Total)	Pass/ Fail	LAL
1	Emulsion Aqueous	0	0.0 0.0	Pass Pass	Negative Negative
2	Emulsion Aqueous	0.12	2.0 1.9	Fail Fail	Positive Positive
3	Emulsion Aqueous	0.06	1.5 1.6	Fail Fail	Positive Positive
4	Emulsion Aqueous	0.03	0.7 1.2	Pass Fail	Positive Positive
5	Emulsion Aqueous	0	0.0 0.0	Pass Pass	Negative Negative
6	Emulsion Aqueous	0.03	0.5 0.9	Pass Pass	Positive Positive
7	Emulsion Aqueous	0.03	0.4 0.4	Pass Pass	Positive Positive
8	Emulsion Aqueous	0.03	1.7 0.6	Fail Pass	Positive Positive

The results also show that the limulus test is a more sensitive test since it consistently detects 0.03 ng/ml of endotoxin.

REFERENCES

1. Grant R (1953). Am J Physiol 173:246-253.
2. Lambert GF, Miller JD, Frost DV, (1951). Am J Physiol 164:490-496.
3. Thompson SW, Jones LD, Ferrell JR, (1965). Am J Clin Nutrition 16:43-61.

STAGNATION TO ANTICIPATION - AN OVERVIEW OF LAL USE IN THE PHARMACEUTICAL INDUSTRY

E. W. Ferry

ABBOTT LABORATORIES

NORTH CHICAGO, ILLINOIS

It is with mixed feelings that I stand here, and after yesterday's opening sessions, those feelings are even more mixed than I thought they would be when I initially began to gather my thoughts together for this presentation. When Dr. Watson first approached me with the request to appear here and present industry's side of the Limulus story, my immediate impulse was to decline. In fact, I believe I may have even told him something like "Surely you'll be able to find someone more appropriate than I to key-note your program." Whatever reply he made didn't sink in, however, for almost immediately my thoughts had shifted to just who would be here. It appeared that I would be exposed to someone's "complete bibliography" of LAL authors, from the very first one to the present. And, I might add, I wasn't disappointed. Since opportunities like this do not present themselves very often, before I realized the gravity of my decision, I heard the comment "I'll try" issuing forth from my lips.

In the time between that decision and now, I've had time to reflect upon that to which I've committed myself. Represent industry? It's a large industry and I am associated with just one member firm. Wait a minute, - what industry? My association is with the pharmaceutical industry, and that is really only a segment of the larger health industry. What do we have in the health industry? We can start with pharmaceuticals, but then must move on into the fields of devices, equipment, and, yes, even containers. If we choose, we will have no trouble verifying that even the small country hospital is part of the health industry, and from there can stretch out to include academia

Pages 339–344, Biomedical Applications of the Horseshoe Crab (Limulidae)

as an "in-law." Additionally, the water treatment and food industries touch many facets of the health industry, and on-and-on. It becomes very apparent that each of these other areas can be subdivided into groups overseen by specialists whose disciplines I'm not even conscious of. All of these have positive potential needs for LAL. Thus, when the revision of the tentative program arrived and I noted that this section was entitled LAL Assay of Pharmaceutical Products, I immediately breathed a sigh of relief and decided to remain within my own arena. In concentrating on my own personal situation I must remind you that even those of us within the pharmaceutical industry are as varied as we are many, and those who follow me today will have even different concepts. For those that I cannot relate to directly, my apologies and my wish that you will be able to accept this as an example of how it may progress in your own area of interest.

In contrast to the many learned dissertations which we heard yesterday, in my particular association with LAL, I've only briefly touched upon the theoretical, and research and development. I have concentrated primarily upon product release. Thus, I present no eyebrow raising report on the results of an involved study. Those reports, perhaps, will follow. Dr. Watson said I could border on the philosophical and that is more what I will try to do.

To best describe where LAL Assay of Pharmaceutical products stands today, it might be best to utilize the time honored tool of the historical review. In doing this, let me make use of six words; stagnation, serendipity, procrastination, inertia, momentum, and anticipation.

For years, the field of pyrogen testing has been, for all intents and purposes, stagnant. Oh, true, new equipment has been introduced to make the task easier and more precise, but the basic rabbit test has remained unchanged for more than thirty years. In those years, numerous discoveries or attempts to introduce a new method have come and gone. New in-vitro procedures have flared, only to be proven inadequate and not a sufficient replacement for the rabbit. Stagnation reigned.

Then came the serendipity. The dictionary defines serendipity as "that which is found or discovered, which was not sought after." I dare say that on that day when he noticed his first example of gellation in our spike-tailed associate, the furthest thing from Dr. Bang's mind was: "Oh boy, the pharmaceutical industry is saved, for I've found a replacement for that archaic, insensitive rabbit test." No, to the contrary, its value wasn't even realized for several years, until it was known what caused the phenomenon. Even when further study began to elicit the idea of the causative factor, our third word procrastination took over. Weren't there, after all, many in science who had seen "flashes-in-the-pan" of this nature before, and had even, as the term says, "been burned." "Let's not respond too quickly," was their advice.

Luckily, however, there were some who either had faith, or in whom enough curiosity remained to cause them to fan the spark. Inertia, as I learned in high school physics, is that property of matter by which it will remain at rest, or in uniform motion until it is acted upon by an external force. In 1971, those of us in the product release business, who hadn't become aware of LAL though the pure research community, were made aware of a potential new test when we received a market survey from one of the nations larger reagent manufacturers. Inertia had been overcome by that force and LAL had begun to move, even if ever so slowly. I'm sure many companies investigated all of the various aspects of the LAL world, from its isolation and improvement, to its potential use. Often, the comment was heard, that it's "too sensitive, and we'll be discarding otherwise good product," or "It's not as all inclusive as our long eared, furry friends are, and because of that will never become a replacement for them. Maybe if we keep quiet and hide our heads in the sand it'll go away and we'll not be saddled with it as an also test."

The faithful prevailed, however, and the fifth word, momentum, came into play. Data gathering began with most of the larger companies committing substantial capital and effort to this gathering and method comparison. In the pharmaceutical industry, we are pretty much regulated by two "bibles," the USP/NF, and the CFR or Code of Federal Regulations. Both, even today, specify that the rabbit test is the recommended method. Until these two barriers were overcome, therefore, there seemed little hope of an early acceptance of LAL as a widely used procedure for pyrogen testing.

In the pharmaceutical industry, there are a number of cooperative organizations such as the Pharmaceutical Manufacturers Association (PMA) or the Parenteral Drug Association (PDA) which serve as a forum for member companies to exchange ideas and work in a united manner to improve the overall level of health care. From 1972 - 1974, I had the opportunity to serve on a task force created by a now superseded group called the HIA or Health Industries Association. Our charge was a request by the USP revision committee to rework the existing USP section for the pyrogen testing of sets and devices. I bring this up because we were approached by several interests to "perhaps use our influence," to convince the USP that the new procedure should be all new and include LAL as the ultimate testing method. Our decision was - that, that would be exceeding our charge, and thus no recommendation was made. I might add, that when USP XIX appeared, the fruit of our labor had not been included, either.

The first chinks in the armor have appeared, however. As the Good Manufacturing Practices or GMP's have been developed, the trend toward more in-process testing has appeared. FDA approval to employ LAL for this in-process testing has been granted to some companies. The USP has requested a committee to prepare a new general chapter for inclusion in the next revision to incorporate LAL as a test method, and as we learned last evening, only last week a committee met to work on it. The landmark; the red letter day; the date indelibly impressed upon everyone's memory, however, has to be November 4, 1977, for on that day, the Federal Register contained the notice that finally opened the door, and which was also mentioned last evening. That notice carried the word that LAL was finally licensed for use. This was monumental news for the manufacturers of lysate, true, but it was surpassed in impact by the news that manufacturers of biological products and medical devices could now, with prior permission, substitute it as a release test for their product.

Those not associated nor familiar with the pharmaceutical industry and its need for rabbit testing cannot appreciate the simple economic impact this will have. The larger companies have spent in the hundreds of thousands, if not millions of dollars annually in maintaining large rabbit colonies. These will now begin to dwindle. Personnel needed will also diminish. Time required will be shortened and the resultant ability to detect problems earlier will help reduce losses experienced when finished product must be discarded with the added expenses of finishing tied to it. Many previously non-testable items can now be monitored. Radio pharmaceuticals can now be certified in a timely manner, not after the fact, as was often necessary in the past.

To go back to that HIA task force, for a minute, the LAL non-action seems ironic now, for a successor organization of HIA, the Health Industry Manufacturers Association has just published a guide-line for member companies to use in satisfying the BMD requirements for conversion from rabbit to LAL for sets and device testing. The USP, on the other hand, still doesn't have the remodeling of the sets chapter done and as suggested, will most likely include LAL in the next revision. We didn't stop there, however, for later this fall, our HIMA task force is conducting a collaborative study by as many rabbit testing laboratories as will participate to try to establish a bench mark; a ground zero. Both the BOB and BMD have required equivalency data before they'll allow a change-over. It is conceded that rabbit colonies are quite variable from region to region of this huge country and no doubt even more so around the world. If the minimum amount of a reference endotoxin that a rabbit colony can detect can be established, then it will readily follow that a batch of lysate capable of detecting that level or less, will be as, or "more" equivalent than the rabbit test and perhaps help answer, finally, the question of what is passing and or failing in LAL terms.

I have one word left. Anticipation. Or, where do we go from here? Anticipation is high. Of course, further reduction of costs is anticipated, but with an accompanying improvement in accuracy. Additional testing of components and raw materials will be possible. More across the lot sampling and testing will be possible with a resultant addition to the confidence that a manufacturer will have in the product he releases. The big one, however, will be the gradual change in the testing of solutions, both large and small volume, from rabbit to lysate. As Dr. Weiss mentioned, the Bureau of Drugs is working on their follow-up release. As these products are gradually phased in, the previously mentioned rabbit colony reductions, the improvement in accuracy, and the broadened coverage will multiply.

From stagnation to anticipation in a relatively short period of time. It's exciting. It's promising. It will require, still, much more work in developing and polishing. It will still require co-operation between the academic and industrial communities. Regulatory questions must be resolved. We've heard reference to a more sensitive test of the future. Industry will be ready for it. But today, it's still exciting, it's still promising. I'm sure glad that I'm a part of it.

ACCEPTANCE OF THE LIMULUS TEST AS AN ALTERNATIVE PYROGEN TEST FOR RADIOPHARMACEUTICAL AND INTRATHECAL DRUGS

James F. Cooper

Medical University of South Carolina

171 Ashley Avenue, Charleston, S.C. 29403 USA

The use of the Limulus amebocyte lysate (LAL) test as a screening test for endotoxin (pyrogen) represents the most significant biomedical application of *Limulus polyphemus* for the health industries. This new quality assurance measure is being utilized increasingly by the pharmaceutical industry because it is sensitive and specific for endotoxin, is rapid and simple to conduct, and offers great cost savings and flexibility in solving problems of water quality.

EVOLUTION OF THE RABBIT PYROGEN TEST

The inadvertent contamination of intravenous fluids with endotoxin may lead to numerous adverse host responses (12). Such reactions have complicated intravenous therapy from the outset (1). The term pyrogen was applied to this contaminant because the fever-inducing effect was most readily recognized. Decades before the offending agent was characterized, numerous investigators began to reveal the nature and source of this contaminant. The parenteral drug industry is particularly indebted to pyrogen pioneers such as Florence Seibert and her coworkers whose research demonstrated 1) the microbiological origin of pyrogens, 2) the necessity for high-quality water as the vehicle for injectables, and 3) the effectiveness of the rabbit bioassay as a means for detecting pyrogenic contamination (8,16,17). These observations led to the development of the rabbit pyrogen test which was adopted into the *U. S. Pharmacopeia* in 1942. The criteria for this test was based upon the results of a collaborative study by government and industry wherein rectal temperature response was measured following intravenous administration of an

Pages 345–352, Biomedical Applications of the Horseshoe Crab (Limulidae)

endotoxic solution prepared from *Pseudomonas aeroginosa* (18). Therefore, in considering the suitability of the LAL test as a replacement for the rabbit *in vivo* pyrogen test, it is imperative to recognize that the rabbit bioassay was designed to detect the endotoxin of gram-negative bacteria.

Due to its potency, stability, and ubiquity in nature, endotoxin (pyrogen) is the only pyrogenic contaminant that has been identified in sterile products; other potential chemical pyrogens, which are less potent, are detected by various other quality assurance measures (3,6,14,18). Indeed, the drug industry has virtually eliminated pyrogenic reactions from marketed drugs by application of preventative means that destroy pyrogens, *i.e.*, heat sterilization of components and use of production water from properly baffled stills (6,14).

The need for a suitable alternative to the rabbit pyrogen test is evident in consideration of the disadvantages of this *in vivo* approach. The most evident disadvantage is the cost of initiation and maintenance of a pyrogen-test facility and trained personnel. An important limitation is variation in sensitivity to pyrogen from one test facility to another. In a recent instance, a relatively pyrogen-tolerant rabbit colony resulted in the release of a contaminated albumin product that produced pyrogenic reactions in patients. Also, many biologics are tested with difficulty because of the potential for non-specific or hypersensitivity reactions in rabbits (14). Certain drug classes are inappropriately pyrogen tested because of various properties. Examples include short-lived radiopharmaceuticals, cancer chemotherapeutic agents such as *L-asparaginase*, antibiotics such as lincomycin, and hypnotics and sedatives (6). Finally, the rabbit bioassay is unsuitable for screening intrathecal products because of its lack of sensitivity (4).

ADVENT OF THE LIMULUS TEST

The application of LAL testing for drugs arose not because of the recognized need for a better pyrogen test, but because of urgent need for a suitable means for screening radioactive drugs for pyrogenic contamination. Indeed, the advent of LAL testing for sterile products was motivated because radiopharmaceuticals posed an insoluble problem for testing by the official; USP pyrogen test (5). The parenteral drug industry is indebted to Levin and Bang (9,10) who first described the LAL-endotoxin reaction.

The Limulus test represents the only realistic alternative to the rabbit bioassay. It offers the advantages of great sensitivity, simplicity, specificity, rapidity, reproducibility and cost savings (6). Those who hold that the pyrogen problem in drugs is broader than endotoxin view the specificity of this *in vitro* test as its major limitation. In reality, modification or inhibition of the gelation reaction by specific components is the most significant limitation.

APPLICATION FOR RADIOPHARMACEUTICALS

Short-lived radioactive drugs are unsuitably tested by the rabbit pyrogen test because of time, sample size, radiation exposure to personnel, and control of radioactive waste (5,6). The dominant radionuclide used in nuclear medicine is Tc-99m, which has a six-hour half life. This radionuclide may be produced in many radiochemical forms to permit imaging of various organ systems. Although Tc-99m may be administered in its stable, ionic state, it is usually injected in the form of a soluble complex. These Technetium complexes, including DTPA, glucoheptonate, methylene diphosphonate, and pyrophosphate, are more correctly described as chelates. Although these compounds are potentially inhibitory through depletion of divalent cations, inhibition has not been a limiting factor because of the low concentration of these entities in each product. Insoluble forms of Tc-99m, such as colloids and particles, are tested without difficulty. It is notable that colloidal preparations which contain gelatin as a stabilizer frequently will yield a positive LAL test because of the difficulty in eliminating pyrogen from gelatin.

The applicability of the LAL test to radiopharmaceuticals has been surveyed by several authors (3,11,13). Modifiers of the LAL-endotoxin reaction were reported on a few, infrequently used drugs; however, the nature of the inhibition was not verified in rabbits. Usually, inhibition is avoided by neutralization of a product before sampling (6,15). Also, the concentration of benzyl alcohol, the most commonly used preservative, must be 1% or less before optimal testing with LAL reagent (6).

Future radiopharmaceuticals will have a greater need for the LAL test. Within five years, ultra-short lived radionuclides (half lives in minutes) will have widespread usage in the major medical centers. Drugs containing these radioactive tags (C-11, N-13, and F-18) must be processed from a cyclotron target, reacted, and injected within one hour. The only possible pyrogen test will be, at best, a twenty-minute Limulus test; currently, this test approach is applied in existing cyclotron facilities.

APPLICATIONS FOR INTRATHECAL PRODUCTS

The intrathecal route of administration is apparently peculiarly sensitive to the presence of pyrogen (2). The sensitivity of the Limulus test is particularly advantageous for screening drugs designed for intrathecal administration. Drugs which are commonly administered into cerebrospinal fluid (CSF) spaces include contrast media, cancer chemotherapeutic agents, antibiotics, radiopharmaceuticals, and chemical dyes. We recently had the opportunity to study the problem of adverse reactions to intrathecally administered radioactive drugs which we found to contain pyrogenic contamination (4).

Aseptic meningitis is an adverse reaction that was frequently associated with radionuclide cisternography, a nuclear imaging procedure wherein 1-2 ml of a radiotracer is instilled into the subarachnoid spaces. Aseptic meningitis is a patient reaction characterized by life-threatening fever, severe headache, and neck stiffness. Radioactive drugs used for this procedure are radioiodinated human serum albumin (IHSA I-131) and DTPA chelates of In-111 and Yb-169.

During a 15-month period, 39 adverse reactions were associated with intrathecal IHSA I-131 or DTPA In-111. Ten samples of specific lots of the radioactive drugs implicated in 20 of these reactions were obtained, retrospectively; all reacted strongly positive to the Limulus test. The less sensitive rabbit pyrogen test was negative for these preparations when tested on a dose-per-weight basis; however, higher dose levels with one available lot did produce pyrogenic response in the rabbit. The data implicate endotoxin contamination as a cause of these reactions and provide clinical evidence for observations made in animals that endotoxin is at least 1000 times more toxic intrathecally than intravenously (2). We concluded that the rabbit pyrogen test is insufficiently sensitive for intrathecal injectables and should be supplemented by the Limulus test. Further, this study demonstrated how the specificity of the Limulus test could be used to determine the etiology of a previously unexplained phenomena.

For the lack of a better explanation, this reaction had been attributed to the presence of denatured albumin. However, we believe that the following evidence supports endotoxic contamination as the cause of aseptic meningitis following radionuclide cisternography:

1. Limulus test detected endotoxin in all radiotracers that were implicated in meningeal reactions.
2. A single source (supplier) was associated with 24 of 25 reactions to IHSA I-131 although both formulations were similar.
3. Meningeal reactions followed the use of two different radiopharmaceuticals (IHSA I-131 and DTPA In-111).
4. One LAL-positive formulation was pyrogenic in rabbits when retested at a higher dose (vol-per-kg).
5. Endotoxin contamination was traced to components in specific radiopharmaceutical formulations.
6. Incidence of this adverse reaction has decreased sharply following implementation of LAL testing by suppliers of these drugs.

The source of contamination in the radioiodinated albumin was traced to an ion-exchange resin column used during preparation to achieve radiochemical purity. Using LAL, we found that microgram levels of endotoxin were washed off the column initially. We offer the caution that any resin column used in drug processing is a potential source of pyrogen contamination unless washed thoroughly and monitored by Limulus testing. For the In-111 preparation, endotoxin was found in a phosphate buffer. Over twenty batches of drugs for radionuclide cisternography supplied by these three USA suppliers contained less than 1 ng/ml equivalent endotoxin (Difco *E. coli*). No adverse reactions were recognized with this low level of contamination. Quantitative studies showed that aseptic meningitis was associated with endotoxin levels of 1-100 ng per injection.

RELATIONSHIP OF LAL TEST SENSITIVITY TO DRUG SAFETY

It was not until 1969 that Griesman and Hornick (7) demonstrated that rabbit and man were equally responsive to threshold levels of pyrogen (endotoxin) when administered intravenously on a dose-per-kg basis. Therefore, the margin of safety of the LAL test may be related to man by comparison to the rabbit bioassay. The initial study of relative sensitivity demonstrated that the Limulus test was 5 to 10 times more sensitive to purified endotoxin than the rabbit test when a firm-gel endpoint was compared to febrile response (rabbit) to a test dose of 1 ml/kg (5).

Improvements in production and packaging have increased sensitivity and thermal stability of LAL reagent so that licensed material readily detects 0.05 ng/ml concentrations of *E. coli* endotoxins (Difco) whereas threshold pyrogenic doses in rabbits are in the range of 1-5 ng/dose. Comparative studies by drug manufacturers are demonstrating the LAL test (gel endpoint) to be 2 to 5 times more sensitive than a rabbit bioassay at 10 ml/kg (6,15). Therefore, the following assumptions are necessary for relating Limulus test safety for drug products:

1. Rabbit bioassay is primarily *dose* dependent (vol/kg).
2. Limulus test is primarily *concentration* dependent, within a one-hour incubation period.
3. Humans and rabbits are equally responsive to the threshold levels of fever induction by pyrogen.
4. A one hour (incubation) Limulus test is conservatively twice as sensitive as a rabbit test at a 10 ml/kg dose.

CONCLUSION

The LAL test is now established as a suitable alternative for screening parenteral drugs for pyrogen. Although certain materials may modify the LAL-endotoxin reaction, the test is not significantly limited by inhibition or non-specific activation. Although initially applied to pyrogen testing radiopharmaceuticals and biologics, the LAL test is now widely applied in the health industries.

REFERENCES:

1) Bennett IL Jr, Cluff LE (1957). Bacterial pyrogens. Pharmacol Rev 9:427.
2) Bennett IL Jr, Petersdorf RG, Keene WR (1957). Pathogenesis of fever: Evidence for direct cerebral action of bacterial endotoxins. Trans Assn Amer Phy 70:64.
3) Cooper JF, (1975). Principles and applications of the Limulus test for pyrogen in parenteral drugs. Bull Parent Drug Assn 29:122.
4) Cooper JF, Harbert JC (1975). Endotoxin as a cause of aseptic meningitis after radionuclide cisternography. J Nucl Med 16:809.
5) Cooper JF, Levin J, Wagner HN (1971). Quantitative comparison of in vitro and in vivo methods for the detection of endotoxin. J Lab Clin Med 78:138.
6) Cooper JF, Pearson SM (1976). "Detection of endotoxin in biological products by the Limulus test." International Symposium on Pyrogenicity, innocuity and toxicity tests systems for biological products, Budapest. Devl Biol Standard 34, pg 7 (S Karger, Basel 1977).
7) Greisman SE, Hornick RB (1969). Comparative pyrogenic reactivity of rabbit and man to bacterial endotoxin. Proc Soc Exp Biol Med 131: 1154.
8) Hort EC, Penfold WJ (1912). Microorganisms and their relation to fever. J Hyg (London) 12:361.
9) Levin J, Bang FB (1964). The role of endotoxin in the extracellular coagulation of Limulus blood. Bull Johns Hopkins Hosp 115:265.
10) Levin J, Bang FB (1968). Clottable protein in Limulus: Its localization and kinetics of its coagulation by endotoxin. Diath Haemorrh 19:186.
11) McAuley RJ, Ice RD, Curtis EG (1974). The Limulus test for in vitro pyrogen detection. Amer J Hosp Pharm 31:688.
12) Milner KC, Rudbach JA, Ribi E (1971). General characteristics. In Weinbaum G, Kadis S, Ajl SJ, editors. Microbial toxins vol IV, Bacterial endotoxins. New York, Academic Press.
13) Murata H, Kobayashi M, Lio M, Yamada H, Chiba K, Matsui K, Kawaguchi S (1976). Sensitivity of the Limulus test and inhibitory factors in the radiopharmaceuticals. J Nuc Med 17 (12):1088.
14) Personeus GR (1969). Pyrogen testing of small volume parenterals. Bull Parenteral Drug Assn 23:201.
15) Mills DF (1977). Pyrogent, Mallinckrodt Technical Brochure, St. Louis, Mo.

16) Seibert FB (1923). Fever-producing substance found in some distilled waters. Am J Physiol 67:83.
17) Seibert FB (1963). Pyrogens from an historical standpoint. Transfusions 3:245.
18) Varney RF (1962). The Intruders - pyrogens. Parent Drug Assoc 16:6.

APPLICATION OF LAL FOR DETECTION OF ENDOTOXIN IN ANTIBIOTIC PREPARATIONS

Susan J. Harrison, K.Tsuji, and R.M.Enzinger

Control Analytical Research and Development
The Upjohn Company
Kalamazoo, Michigan 49001

ABSTRACT

The *Limulus* amebocyte lysate (LAL) test has been investigated as an alternative method to the USP pyrogen test for the detection of endotoxin contamination in these antibiotic preparations: clindamycin phosphate, lincomycin hydrochloride, neomycin sulfate, and spectinomycin hydrochloride. The antibiotic preparations were tested at the maximum concentrations that would not inhibit gelation of LAL. The USP pyrogen test was also performed for comparison. The LAL and the USP pyrogen tests correlated extremely well when an acceptance criterion based on the minimum pyrogenic dose (MPD) of endotoxin per human dose of drug was used for the LAL test. Use of this MPD provides a practical and feasible acceptance criterion for the LAL test and assures that the LAL sensitivity approximates or surpasses that of the USP pyrogen test.

INTRODUCTION

Several investigators have applied *Limulus* amebocyte lysate (LAL) to test for the presence of endotoxin in parenteral pharmaceutical preparations (1,2,5-8,11). Since many of these products interfered with the gelation of LAL, they were diluted with a sufficient quantity of pyrogen-free water to eliminate the interference (7). Some preparations, e.g. Cloxacillin, required so much dilution to minimize the interference that the sensitivity of the LAL test became less than that of the USP pyrogen

Pages 353–365, Biomedical Applications of the Horseshoe Crab (Limulidae)

test method (7,10). The inhibitory action of the penicillins, which include Cloxacillin, was attributed to their ability to bind the protein required for gelation of LAL.

We have been examining LAL for detection of endotoxin in antibiotic preparations as a possible alternative to the USP pyrogen test. This paper presents historical comparative data between the LAL and the USP pyrogen tests and suggests a practical approach to establishing product acceptance criteria based on a minimum pyrogenic dose (MPD) of endotoxin per human dose of the antibiotic preparations.

MATERIALS AND METHODS

LAL

Pyrotell from Associates of Cape Cod, Inc. (Woods Hole, Mass.) was used. The LAL preparation was reconstituted in the proper quantity of pyrogen-free water (sterile water for injection, The Upjohn Company) as specified by the supplier and kept in an ice container during use.

Endotoxin

Lipopolysaccharide of Escherichia coli 0127:B8, lot no. 629151 from Difco Laboratories was used. The lipopolysaccharide was diluted with pyrogen-free water and tested for LAL end-point determinations within the following range of endotoxin concentrations: 0.01, 0.02, 0.04, 0.06, 0.08, and 0.10 ng/ml.

Test Samples

Clindamycin phosphate, S.S. Cleocin, lincomycin hydrochloride, S.S. Lincocin, neomycin sulfate, and spectinomycin hydrochloride were tested at concentrations of 12, 6, 3, 5, 10, and 50 mg antibiotic base per ml, respectively.

These testing concentrations were determined by adding the E. coli endotoxin to antibiotic solutions at the same endotoxin concentrations used for LAL end-point determinations. The highest antibiotic concentration tested was equivalent to either the human or USP test dose per milliliter solution (not per kilogram body weight). The antibiotic solutions were serially diluted until inhibition was eliminated. All samples except spectinomycin hydrochloride were diluted in pyrogen-free water and pH adjusted to 7.0 ± .3 with 1N pyrogen-free HCl or NaOH. Spectinomycin, however, was diluted in .1M potassium phosphate buffer, pH 8.0 to obtain the desired pH range.

A 0.1 ml quantity of each sample was placed in a pyrogen- free glass test tube (10 x 75 mm) and was mixed with 0.1 ml of the reconstituted lysate. The mixture was then allowed to incubate undisturbed for 1 hr. at 37°C. A sample was judged positive when the gel formed did not collapse upon two 180° inversions of the tube. All weak gel formations were considered negative. All glassware used was depyrogenated by dry-heat treatment at 250°C for more than 1 hr.

USP Pyrogen Test Method

Both New Zealand White and Dutch Belted rabbits were used to determine pyrogenicity of intravenously administered endotoxin by the USP procedure (10). Upon arrival, all rabbits received a 5 ng/kg dose of E. coli endotoxin approximating twice the mean pyrogenic dose. Only those responding with a temperature rise of 0.6°C or more were accepted into the test colony.

In order to establish the endotoxin dose-response curve, 78 rabbits were injected with five graded doses of endotoxin, i.e. 0.5, 1.0, 5.0, 10.0, and 30.0 ng/kg body weight. The temperatures were recorded at hourly intervals for a minimum of 3 hrs. after injection. The average temperature rise was a linear function of the logarithm of the endotoxin dose over the range tested. The pyrogenic concentration of endotoxin per kilogram rabbit body weight that elicited an average febrile response of 0.6°C (mean pyrogenic dose) and the corresponding 95% confidence interval were calculated from the linear regression analysis. The rabbits were always rested for at least two weeks between the pyrogen tests.

RESULTS AND DISCUSSION

Concentration of Antibiotics

Several solutions of the antibiotics were found to interfere with the gelation of LAL. Their ability to bind the protein needed for gelation may be the cause of the problem (7). Therefore, the highest concentrations of antibiotics that did not interfere with the test were determined. Interference was considered eliminated when the gelation endpoints of the endotoxin in the antibiotics were identical to the endpoints in Water for Injection. Table 1 presents an example of these results for lincomycin hydrochloride. The following concentrations were established for testing the antibiotics and their formulations:

clindamycin phosphate	12 mg base/ml
S.S. Cleocin Phosphate	6 mg base/ml
lincomycin hydrochloride	5 mg base/ml
S.S. Lincocin	3 mg base/ml
neomycin sulfate	10 mg base/ml
spectinomycin hydrochloride	50 mg base/ml

Without exception, antibiotics in formulations required higher dilutions than the bulk powders. This is attributed to the presence of excipients in each formula. The highest concentration of antibiotic solution that can be used for the LAL test without interference is the governing factor which limits the ultimate sensitivity of the LAL test.

Another interesting procedural variation occurred in the preparation of spectinomycin hydrochloride. The pH of this antibiotic must be adjusted by dilution in potassium phosphate buffer for dependable results. Erratic interference of lysate gelation resulted when the pH was adjusted by addition of NaOH. A report that NaCl concentrations higher than 0.154 N decrease the LAL sensitivity may explain this discrepancy since the spectinomycin hydrochloride solution requires relatively large amounts of NaOH for neutralization (9). Efforts to minimize the problem through addition of $Ca(OH)_2$ to provide divalent cations proved unsuccessful.

TABLE 1

Effects of Various Concentrations of Lincomycin Hydrochloride on the Sensitivity of the LAL Test

Sample	Concentration	Endotoxin Concentration (ng/ml)						
		0.10	0.08	0.06	0.04	0.02	0.01	0.00
Water		+	+	+	+	+	-	-
Lot								
A	20 mg base/ml	+	+	-	-	-	-	-
B	20 mg base/ml	+	+	+	-	-	-	-
Water		+	+	+	+	-	-	-
Lot								
C	20 mg base/ml	+	+	-	-	-	-	-
D	20 mg base/ml	-	-	-	-	-	-	-
E	20 mg base/ml	-	-	-	-	-	-	-
F	20 mg base/ml	±	±	-	-	-	-	-
Water		+	+	+	+	+	-	-
Lot								
B	10 mg base/ml	+	+	+	+	+	-	-
Water		+	+	+	+	-	-	-
Lot								
C	10 mg base/ml	+	+	+	+	-	-	-
D	10 mg base/ml	+	+	+	-	-	-	-
E	10 mg base/ml	+	+	+	+	-	-	-
F	10 mg base/ml	+	+	+	±	-	-	-
Water		+	+	+	+	+	-	-
Lot								
G	5 mg base/ml	+	+	+	+	+	-	-
H	5 mg base/ml	+	+	+	+	+	-	-
I	5 mg base/ml	+	+	+	+	+	-	-
J	5 mg base/ml	+	+	+	+	+	-	-
A	5 mg base/ml	+	+	+	+	+	-	-

\+ solid gel formed

± weak gel formed but collapsed on inversion

\- no gel formed

Relative Sensitivities of the USP Pyrogen and LAL Tests

As explained above, the LAL sample preparations differ among the antibiotics; thus, the resulting detection limits per mg antibiotic base vary. In order to provide a valid alternate pyrogen test, these limits must be equal to or less than the detection limits of the USP Pyrogen Test.

The sensitivity or minimum pyrogenic dose (MPD) of the USP Pyrogen Test is estimated at the lower 95% confidence limit of the mean pyrogenic dose. Figure 1 depicts the regression analysis of the dose-response data from 78 rabbit injections. A mean pyrogenic dose of 2.21 nanograms endotoxin per kilogram body weight and a 95% confidence interval of 1.06 to 2.42 ng/kg were calculated from this data. Since the sensitivity of USP test approximates 1 ng endotoxin per kg body weight, the detection limits which result when applied to the antibiotics relate to the drug dose per kg.

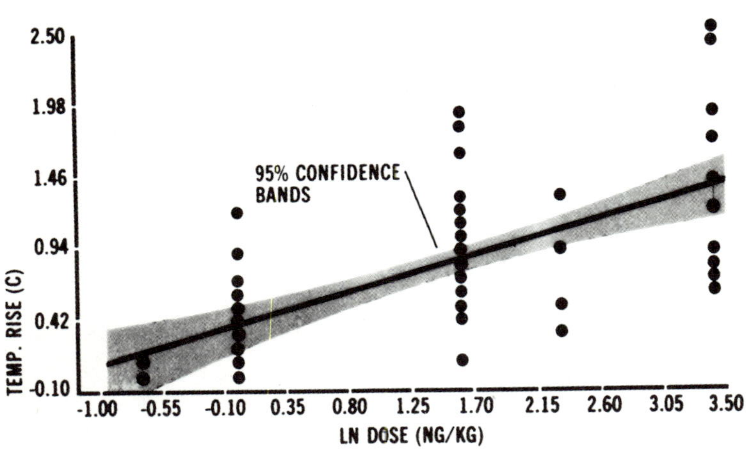

Figure 1
RABBIT FEBRILE RESPONSE VS LN DOSE OF E. Coli ENDOTOXIN
Mean Pyrogenic Dose = 2.21 ng/kg
(95% Confidence Limits = 1.06 - 2.42)

In most cases, the USP rabbit dose equals or approaches that of the human, but as shown in Table 2, the rabbit dose must occasionally be reduced due to the rabbit sensitivity to certain antibiotics.

TABLE 2

Human Dose vs. Rabbit Dose

Product	Human	Rabbit
Clindamycin Phosphate	2700 mg/day (2 doses) = 2700/60 Kg ÷ 2 = 22.5 mg/Kg	24 mg/Kg
Spectinomycin HCl	4 gm = 4000 mg/60 Kg = 66.7 mg/Kg	50 mg/Kg
Neomycin Sulfate	10.5 mg Neo Base/Kg/Day (4 doses) = 3 mg Neo Base/Kg	10 mg Base/Kg
Lincomycin Hydrochloride	600 mg = 600 mg/60 Kg = 10 mg/Kg	.5 mg/Kg

Assume average body weight of 60 Kg.

For example, the USP rabbit dose of 0.5 mg lincomycin base/kg was reduced by a factor of 20 from the general human dose of 10 mg base/kg (600 mg per patient). This reduction eliminates the effects of lincomycin on the rabbit but decreases the sensitivity to only 1 ng endotoxin per 0.5 mg lincomycin base. This detection limit is about twenty times higher than that required for an effective pyrogen test.

The detection limits of the USP and LAL tests for each of the four antibiotics and two products are listed in Table 3. The sensitivities of the LAL test ranged from 2.5 to 100 times higher than those of the USP rabbit method.

TABLE 3

LAL vs. USP Pyrogen Test Detection Limits for E. coli Endotoxin

Products	Detection limits (ng endotoxin/mg antibiotic base)		Ratio of sensitivity
	LAL*	USP pyrogen	(LAL/USP)
clindamycin phosphate	0.10 ng/12 mg	1.0 ng/24 mg	5
S.S. Cleocin, 150 mg/ml	0.10 ng/6 mg	1.0 ng/24 mg	2.5
lincomycin hydrochloride	0.10 ng/5 mg	1.0 ng/0.5 mg	100
S.S. Lincocin, 300 mg/ml	0.10 ng/3 mg	1.0 ng/0.5 mg	60
neomycin sulfate	0.10 ng/10 mg	1.0 ng/10 mg	10
spectinomycin hydrochloride	0.10 ng/50 mg	1.0 ng/50 mg	10

* The least sensitive LAL allowed in the procedure has a gelation endpoint or detection limit of 0.10 ng/ml of the "in-house" endotoxin standard E. coli LPS 0127B8, lot 629151.

Acceptance Criterion

Since the LAL test exhibited varying sensitivities relative to pyrogenicity, a uniform acceptance criterion based on the minimum pyrogenic dose (MPD) of endotoxin per kilogram human body weight is proposed. This endotoxin limit or MPD can be applied to the general human dose of the antibiotic. Since man and rabbit are reported to respond equally on a per kg basis to threshold pyrogenic quantities of endotoxin (4), the MPD was estimated at the USP Pyrogen Test detection limit, i.e. 1 ng endotoxin per kg body weight. These limits relative to the human doses of the antibiotics are given in Table 4 for each product and material. This acceptance criterion assures that the LAL method will provide improved endotoxin detection in samples like lincomycin. For each antibiotic, the LAL sensitivity was sufficient to detect the MPD.

Comparative USP Pyrogen and LAL Test Results

More than 40 production lots of each antibiotic were tested by both the USP Pyrogen and LAL methods. Endotoxin was added to a portion of the LAL test samples in each lot to serve as positive controls. Table 5 compares these parallel test results in terms of the acceptance criteria and demonstrates a correlation of 100% for all samples. These samples, excluding neomycin sulfate, contained only one positive (+) result according to the LAL test and therefore made assessment of the MPD difficult. The neomycin sulfate lots, however, contained many positive samples that provided results both above and below the MPD. The same lot which failed the acceptance criterion according to the LAL test was pyrogenic by the USP Pyrogen method. This correlation indicates that this acceptance criterion is a realistic approach to limiting pyrogenicity. This comparative data supports the theoretical endotoxin limit (MPD) determined statistically from the endotoxin dose-response curve.

Effects of Antibiotics on Endotoxin Pyrogenicity

The majority of the lots of antibiotics tested contained no detectable endotoxin. Since no naturally pyrogenic lots were found in antibiotics other than neomycin

TABLE 4

LAL Detection Limits vs. Minimum Pyrogenic Dose (MPD)

Product	Human Dose (product/body weight)	MPD* (endotoxin/product)	Equivalent LAL Detection Limit per Human Dose (endotoxin/product)
clindamycin phosphate	22.5 mg/kg	1.0 ng/22.5 mg	.19 ng/22.5 mg
S.S. Cleocin, 150 mg	22.5 mg/kg	1.0 ng/22.5 mg	.38 ng/22.5 mg
lincomycin hydrochloride	10 mg/kg	1.0 ng/10 mg	.20 ng/10 mg
S.S. Lincocin, 300 mg/ml	10 mg/kg	1.0 ng/10 mg	.33 ng/10 mg
neomycin sulfate	10.5 mg/kg/dy	1.0 ng/10.5 mg	.105 ng/10.5 mg
spectinomycin hydrochloride	66.7 mg/kg	1.0 ng/66.7 mg	.13 ng/66.7 mg

* MPD = Lower 95% Confidence Limit of the mean pyrogenic dose.

TABLE 5

USP Pyrogen - LAL Test Results

Product	MPD (endotoxin/human dose)	Number of Lots Examined	LAL Results +(≥MPD)	LAL Results +(<MPD)	LAL Results -	USP Results Pyrogenic	USP Results Nonpyrogenic
clindamycin phosphate	1.0 ng/22.5 mg	42	0	1	41	0	42
S.S. Cleocin, 150 mg	1.0 ng/22.5 mg	51	0	0	51	0	51
lincomycin hydrochloride	1.0 ng/10 mg	74	0	0	74	0	74
S.S.Lincocin, 300 mg/ml	1.0 ng/10 mg	82	0	0	82	0	82
neomycin sulfate	1.0 ng/10.5 mg	104	1	80	23	1	103
spectinomycin hydrochloride	1.0 ng/66.7 mg	55	0	0	55	0	55

+ sample preparation formed solid gel

- sample preparation did not form solid gel

MPD minimum pyrogenic dose

sulfate, further testing was needed to assure that a "pyrogenic" USP test would correlate with an LAL value greater than the acceptance criterion. Determining the effects of these antibiotics on the pyrogenicity of endotoxin was also necessary to elucidate possible product synergism or inhibition.

A total of 6 lots of lincomycin hydrochloride and 5 lots of S.S. Lincocin, 300 mg/ml were prepared with two levels of endotoxin, i.e. 0.5 ng endotoxin/.5 mg base or 5 ng endotoxin/.5 mg base, and tested by the USP Pyrogen method. The results confirmed expectations. The rabbits responded pyrogenically to the 5 ng dose, but not to the 0.5 ng dose. Analysis of the data showed no significant difference at the 95% level of confidence between the rabbit febrile responses to endotoxin in water and to endotoxin in the lincomycin solutions. Additional studies on the other antibiotics are necessary for confirmation of their individual acceptance criteria.

SUMMATION

Federal Register Notices have announced that the LAL assay sensitivity must be equivalent to or better than that of the USP Pyrogen method in order to receive approval for testing medical devices (3). No quantitative endotoxin limits have been suggested, however, that would assure comparable sensitivity. The historical and statistical data provided in this report support use of an LAL acceptance criterion based upon the human dose of the drug in question. A limit which approximates the minimum pyrogenic dose of endotoxin per human dose of product assures that the LAL sensitivity equals or surpasses that of the USP Pyrogen Test.

REFERENCES

1. Cooper, J.F. (1976). Principles and applications of the limulus test for pyrogen in parenteral drugs. Indian J. Pharm. 38: 1-6.

2. Ennis, F.A., W.R. Doudle, D.W. Barry, H.D. Hochstein, P.F. Wright, D.T. Karzon, W.M. Marine, and H.M. Meyer (1977). Endotoxin content and clinical reactivity to influenza vaccines. J. Biol. Standard. 5: 165-172.
3. Food and Drug Administration (1977). Licensing of Limulus Amebocyte Lysate, supplementary information. Federal Register 42: 57749.
4. Greisman, S.E., Hornick, R.B. (1969). Comparative pyrogenic reactivity of rabbit and man to bacterial endotoxin. Proc. Soc. Exptl. Biol. Med. 131: 1154.
5. Murata, H., M. Kobayashi, M. Iio, and H. Yamada (1975). Sensitivity of limulus test and inhibitory factors in radiopharmaceuticals. J. Nuc. Med. 16: 552.
6. Nandan, R. and D.R. Brown (1977). An improved <u>in vitro</u> pyrogen test: To detect picogram of endotoxin contamination in intravenous fluids using Limulus amebocyte lysate. J. Lab. Clin. Med. 89: 910-918.
7. Newsome, P.M. (1977). Penicillins and the limulus amebocyte lysate test for endotoxin. J. Pharm. Pharmac. 29: 704-706.
8. Raslogi, S.C., H.D. Hochstein, and E.B. Seligmann, Jr. (1977). Statistical determination of endotoxin content in influenza virus vaccine by the limulus amebocyte lysate test. J. Clin. Microbiol. 6: 144-148.
9. Sullivan, J.D. and S.W. Watson (1974). Factors affecting the sensitivity of <u>Limulus</u> lysate. Appl. Microb. 28: 1023-1026.
10. United States Pharmacopeia XIX (1975). United States Pharmacopeial Convention, Inc., Rockville, Md.
11. Weary, M. and B. Baker (1977). Utilization of the limulus amebocyte lysate test for pyrogen testing large volume parenterals, administration sets, and medical devices. Bull. Parenteral Drug Assoc. 31: 127-135.

LIMULUS AMEBOCYTE LYSATE - A MEANS TO MONITOR INACTIVATION OF LIPOPOLYSACCHARIDE

Kiyoshi Tsuji and Susan J. Harrison

Control Analytical Research and Development
The Upjohn Company
Kalamazoo, Michigan 49001

SUMMARY

A properly selected Limulus Amebocyte Lysate (LAL) has been demonstrated to have a good correlation with the *in vivo* United States Pharmacopeial pyrogen test when testing lipopolysaccharide (LPS) inactivated with ethylene oxide, ^{60}Co irradiation, and dry-heat treatments. Thus, LAL may be used to determine degradation or depyrogenation kinetics of LPS. A criteria for acceptance of an LAL lot for correlation with the USP pyrogen test is described.

INTRODUCTION

It is evident from numerous papers presented at this symposium that Limulus Amebocyte Lysate (LAL) has been established as a sensitive, yet reliable test for the detection of lipopolysaccharide (LPS). Release of pharmaceutical products tested by LAL alone may not be too far in the future (2).

We have been interested in LAL as a convenient means to determine dry-heat destruction kinetics of LPS (8). Dry-heat has traditionally been used to depyrogenate glassware, equipment, heat stable materials, etc, yet basic data on dry-heat destruction kinetics of LPS is lacking. Prior to the use of LAL to determine the kinetics, the specificity of LAL to the region of the LPS molecule responsible for pyrogenic reaction in animals must be demonstrated to assure that any change or degradation in the LPS molecule affecting pyrogenicity is being detected by LAL.

Pages 367–378, Biomedical Applications of the Horseshoe Crab (Limulidae)

MATERIALS AND METHODS

LAL

LAL preparations from Associates of Cape Cod (Woods Hole, Mass.), Difco Laboratories (Detroit, Mich.), Mallinckrodt, Inc. (St. Louis, Mo.), and Sigma Chemical Co., (St. Louis, Mo.) were tested. Those preparations modified for spectrophotometric determinations were not used.

The lyophilized preparations in multitest vials were reconstituted in the proper quantity of pyrogen-free water for injection as specified by the supplier and kept on ice during use.

LPS

LPS preparations of *Escherichia coli* 0127:B8, *Salmonella typhosa* 0901, *Serratia marcescens* lot no. 613721, (all from Difco Lab), *Proteus mirabilis* (The Upjohn Co.), and *Pseudomonas* species ATCC 9229 (Piromen, lot no. P820, Travenol Laboratories, Inc., Morton Grove, Ill.) were used.

LPS Inactivation Treatments

A 0.1 ml aliquot of the *E. coli* LPS solution containing approximately 10,000 or 100,000 ng per ml was pipetted into a glass test tube or into an aluminum cup by use of an Eppendorf or Oxford pipette with a disposable tip. The cup or test tube was then placed under a stream of dry nitrogen for approximately three hours to dry. The LPS samples thus prepared were subjected to the following treatment:

Ethylene Oxide (ETO) treatment. The LPS samples were exposed to ETO gas (12:88, ETO:Freon) for 6.5 hours with 1) no humidity at 50°C, 6.5 psi; 2) 20% relative humidity at 50°C, 6.5 psi; and 3) 50% relative humidity at 60°C, 3.5 psi.

^{60}Co Irradiation. The LPS samples were irradiated by use of a ^{60}Co source at 0.85, 1.8, 3.7, 4.5, and 5.3 Mrads dose.

Dry-heat treatment. The LPS samples were dry-heat treated at various temperatures ranging from 170 to 250°C for various time intervals using a dry-heat destruction oven described by Robertson et. al. (6).

Test Samples

Following the treatments described above, a suitable quantity of pyrogen-free water was added and the LPS was suspended by treating the sample to sonication (38-50 KHz) for 10 minutes.

To determine the quantity of LPS remaining after the treatment, a series of ten-fold dilutions was made for a rough estimation of concentration followed by two-fold dilutions of the suspension in disposable test tubes (20 x 150 mm). Finally, a 0.1 ml quantity of each dilution was transferred to disposable test tubes (10 x 75 mm) for testing by LAL.

A 0.1 ml aliquot of the reconstituted lysate was added to each test tube containing 0.1 ml of LPS and allowed to incubate undisturbed for 1 hour at 37°C. A sample was judged positive when the gel formed did not collapse upon a 180° inversion of the tube. The gelation endpoints of the two-fold dilutions were used to calculate the quantity of LPS remaining after each treatment.

All glassware used was depyrogenated by dry-heat treatment at 250°C for more than 1 hour.

USP Pyrogen Test Method (9)

New Zealand white rabbits were used to determine pyrogenicity of intravenously administered LPS by the USP procedure. Upon arrival, all the rabbits received a dose of LPS of *E. coli* approximately three times that of its PD_{50}, and only those responding with a temperature rise of 0.6°C or more were accepted into the test colony.

Eighteen rabbits were injected with three graded doses of LPS, 1 ml/Kg each, within a predetermined range, and the temperatures were recorded at 12 minute intervals for a minimum of 3 hours after injection. The average temperature rise in each rabbit was a linear function of the log dose over the range tested. The pyrogenic concentration of LPS that elicited an average febrile response of 0.6°C was calculated by a linear regression analysis. Assuming an equal distribution of fever around 0.6°C, this dose would be pyrogenic to 50% of the rabbits tested, hence PD_{50}. The rabbits were always rested for at least 2 weeks between the pyrogen testing.

The percentages of LPS degraded by the various treatments were based on the quantities of LPS in the control. The amount of LPS in all samples were calculated by interpolation from the dose response curve [$y = 0.310\ (\ln x) + 0.454$]. For example, a control LPS sample was diluted with 1,000 ml of pyrogen-free water, and a 1-ml quantity was dosed i.v. per kilogram rabbit body weight. The average febrile response of 8 rabbits was 1.1°C. Thus, the control sample contained 8,000 ng LPS per cup. The sample, dry-heat treated at 190°C for 40 minutes, was diluted with 100 ml of pyrogen-free water, and a 10-ml quantity was dosed i.v. per kilogram of rabbits. The average febrile response of 8 rabbits was 1.2°C. Thus, the dry-heat treated sample contained 111 ng LPS per cup. The percentage of LPS remaining and the percentage of LPS destroyed by the treatment was calculated as 1.4% and 98.6%, respectively.

RESULTS AND DISCUSSION

Correlation Between LAL Preparations and the USP Pyrogen Test

As was stated by Wachtel and Tsuji (10), a correlation between the LAL preparation and the USP pyrogen test method must be established to assure that the LAL is responding to the region of the LPS molecule responsible for pyrogenicity in animals. The correlation was examined from the ratio of the PD_{50} value to the LAL minimum detection limit (10). If an LAL preparation shows good correlation with the USP method, these sensitivity ratios should remain essentially similar for all types of LPS tested.

The results presented in Table 1 indicated that the sensitivity ratios for each LAL preparation to LPS from *E. coli*, *P. mirabilis*, and *S. typhosa* were relatively constant; however, considerable variation was observed with the LPS of Pseudomonas. The LAL preparations from Associates of Cape Cod, Mallinckrodt, and Sigma Chemical showed relatively constant ratios, indicating existance of a good correlation with the USP method. The LAL preparation from Difco, on the other hand, showed a considerable difference, up to 190-fold, between the sensitivity ratios for the LPS of E. coli and the Pseudomonas species. These variations in the ratio indicate that some LAL preparations may lack specificity due to modification of an endotoxin receptor protein to the region of the LPS molecule responsible for pyrogenic reaction in animals. The data of Yin et. al. (11), Elin and Wolff (1), and Sweadner et. al. (7) may suggest that other component(s) or conformational states of LPS, in addition to the lipid A, appear to affect gelation of LAL. The variation may also be attributed to differences in the preparation of LAL (3) and/or the biological variability among the amebocyte (4). Modification of an endotoxin receptor protein during the preparation of LAL could result in LAL losing its specificity to recognize a degraded from an intact LPS molecule.

Since poor correlation between various LAL preparations and the USP pyrogen test were indicated by differences in the sensitivity ratios for LPS of E. coli and the Pseudomonas species, specification criteria for the acceptance of an LAL lot were established which limit ratio variation to not more than a four-fold difference. This wide acceptance limit is needed due to considerable variation in the confidence limits of the USP Pyrogen method (2, 10).

LAL vs. the USP Pyrogen Tests on Degraded LPS

The LAL preparation from Associates of Cape Cod, which met the acceptance criteria, was used to examine the degradation of LPS.

Ethylene oxide (ETO) treatment. ETO treatments with relative humidities of 0 and 20% showed no measurable effect on LPS. The ETO treatment with a relative humidity of 50%, on the other hand, effected 94.3% destruction of LPS (Table 2).

TABLE 1

Correlation between the LAL preparations and the USP pyrogen test

	LPS*			
LAL source	*E. coli*	*P. mirabilis*	*Pseudomonas* spp.	*S. typhosa*
Associates of Cape Cod	17.3	15.7	6.7	23.0
Difco Laboratories	17.3	39.2	332	34.5
Mallinckrodt	34.7	52.2	41.6	34.5
Sigma Chemical	17.3	52.2	66.5	34.5

*expressed in terms of the sensitivity ratio (PD_{50}/LAL min. detection level)

In order to confirm that LAL is monitoring depyrogenation of LPS, the sample treated with ETO in the presence of 50% relative humidity was tested by the USP method. The result presented in Table 2 indicated that the ETO treatment caused 91.2% reduction in pyrogenicity. Thus, the LAL data (94.3% degradation) agreed quite well with that of the USP method (91.2% depyrogenation). However, when LAL of Difco was used, destruction of LPS was not detected. This phenomenon indicates a need for a rigid specification for the selection of LAL.

^{60}Co irradiation treatment. LPS of E. coli was irradiated by ^{60}Co, and samples were taken periodically and tested by LAL to determine the degradation of LPS by ^{60}Co treatment. As shown in Figure 1, the degradation of LPS was linear with a correlation coefficient (r) of -0.956 when the log concentration of LPS remaining was plotted against irradiation dose. The D value, ^{60}Co irradiation dose in Mrad required to destroy 90% of the LPS, is 4.7 Mrads.

The USP pyrogen test was also used to confirm that the LAL is measuring depyrogenation of LPS caused by ^{60}Co irradiation. The LPS sample treated with ^{60}Co irradiation at 4.5 Mrads was used for comparison. The results (Table 2) indicated that 19% of pyrogenic activity remained in the LPS sample after 4.5 Mrads dose. This value (19%) compared favorably to the 12% value determined by LAL (Table 2, Fig. 1).

Dry-heat treatment. Dry-heat sterilization is frequently carried out at 160 to 170°C for a period of 2 to 4 hours (9). Therefore, the dry-heat destruction kinetics of LPS of E. coli were determined by LAL at 170°, 190°, 210°, 230°, and 250°C. In addition, the dry-heat destruction kinetics of LPS of S. marcescens and S. typhosa were determined at 170°C and 210°C. The results are presented in Figures 2, 3, and 4. The dry-heat destruction kinetics of LPS followed the second order (8).

To test the correlation between the LAL and the USP pyrogen tests, samples of LPS of E. coli were subjected to dry-heat treatments at 170°C for 240 minutes and at 190°C for 40 minutes. The results (Table 2, Fig. 2) indicated that the dry-heat treatment at 170°C for 240 minutes effected a 99.9% destruction while the USP test indicated a

TABLE 2

Comparative responses of the LAL and the USP pyrogen tests to LPS degraded by various treatments

Treatment	LPS Destruction (%)	
	LAL	USP
ETO		
0% humidity	0	0
20% humidity	0	0
50% humidity	94.3	91.2
^{60}Co irradiation		
4.5 Mrads	88	81
Dry heat		
170°C for 240 min.	99.9	99.8
190°C for 40 min.	99.6	98.6

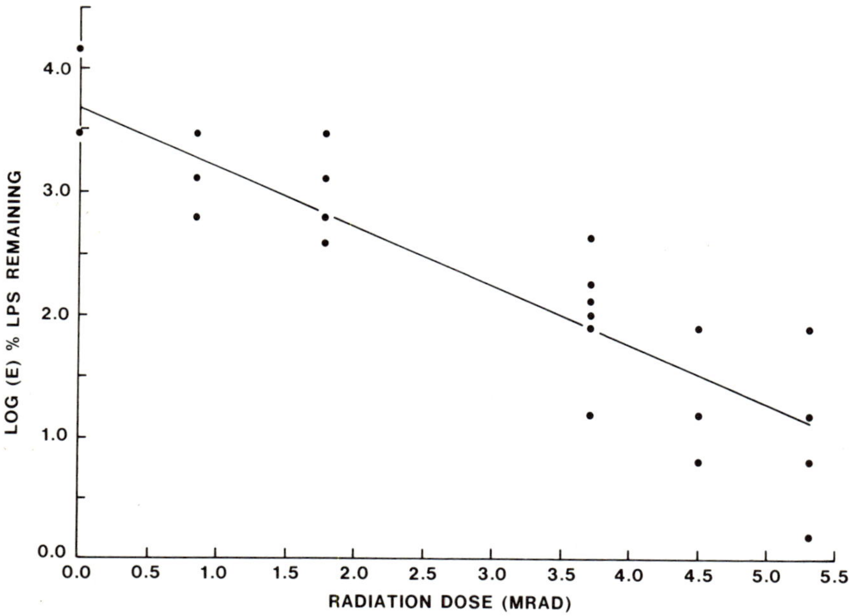

Figure 1. Destruction of lipopolysaccharide of *E. coli* by ^{60}Co irradiation.

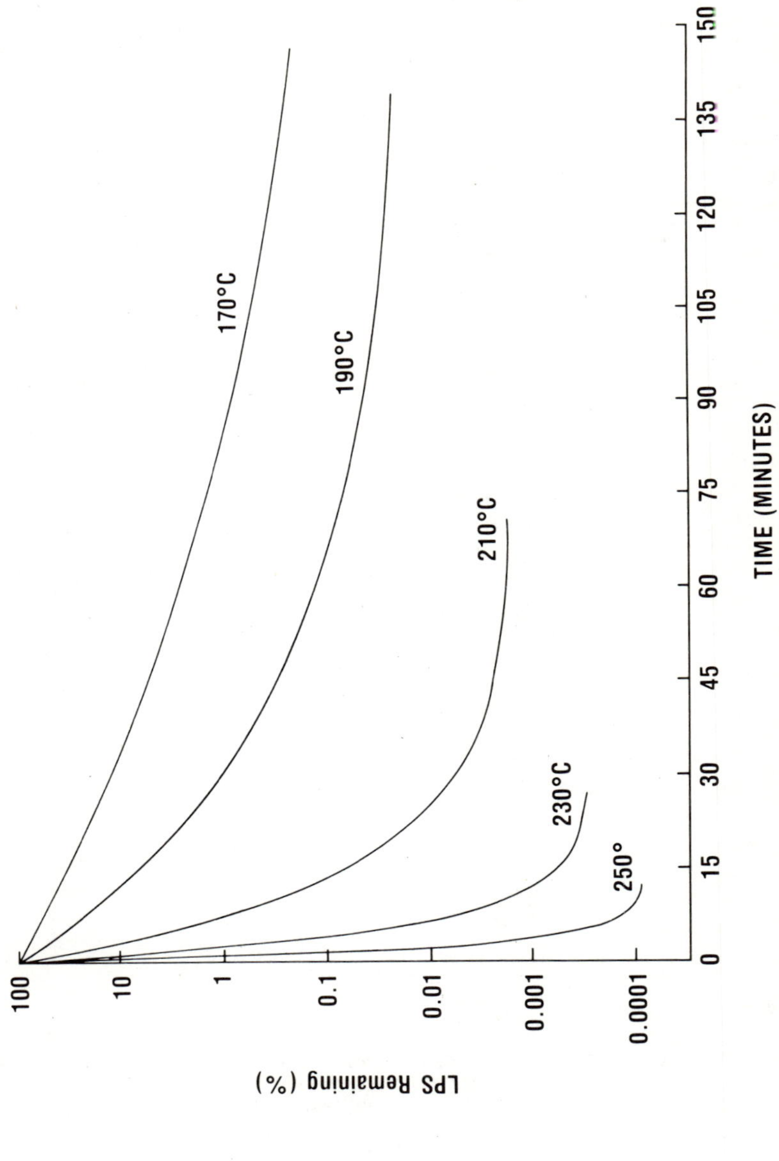

Figure 2. Dry-heat destruction of lipopolysaccharide of E. coli at various temperatures.

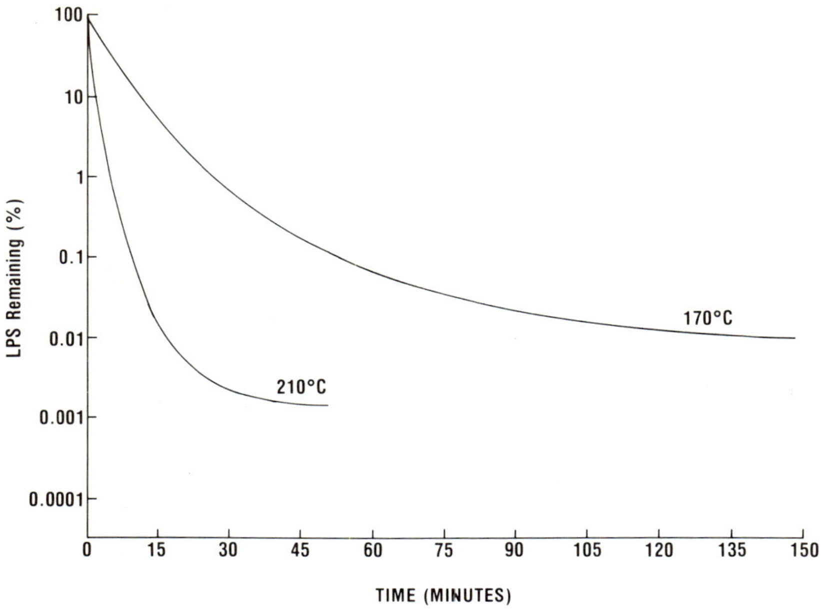

Figure 3. Destruction of lipopolysaccharide of S. typhosa by dry-heat treatments.

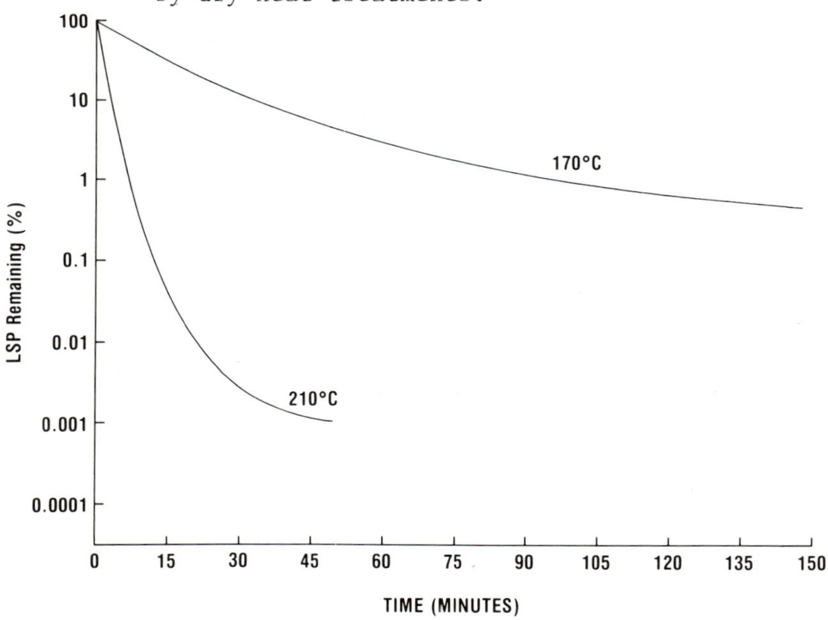

Figure 4. Destruction of lipopolysaccharide of S. marcescens by dry-heat treatments.

99.8% destruction. Results of the two methods were also similar on the LPS sample dry-heat treated at 190°C for 40 minutes (Table 2, Fig. 2). The LAL indicated a 99.6% destruction of the LPS and the USP, a 98.6% depyrogenation of the LPS.

The lipid A portion of the LPS was reported to be responsible for pyrogenicity in animals (5). In our study, samples of LPS degraded by a variety of means, e.g. alkylation (ETO), free radical formation, polymerization, depolymerization, oxidation (irradiation), oxidation, and denaturation (dry-heat), were used to compare specificity of LAL to remaining pyrogenicity in animals (USP pyrogen test). The results obtained indicated that excellent agreement existed between the LAL and the USP pyrogen test data. Thus, it may be concluded that LAL, when properly selected, is specific to the region of the LPS molecule (perhaps Lipid A) responsible for pyrogenicity in animals and that LAL may be used to determine degradation or depyrogenation kinetics of LPS.

LITERATURE CITED

1. Elin, R.J. and S.M.Wolff. 1973. Nonspecificity of the limulus amebocyte lysate test. Positive reactions with polynucleotides and proteins. J. Infec. Diseases. 128: 349-352.
2. Harrison, S.J., K.Tsuji, and R.M.Enzinger. 1978. Application of LAL for detection of endotoxin in antibiotic solution. Proceedings of the Symposium on Biomedical Applications of Limulus polyphemus. Alan R. Liss, Inc., N.Y.
3. Hochstein, H.D., R.J.Elfin, J.F.Cooper, E.B.Seligmann, and S.W.Wolff. 1973. Further development of limulus amebocyte lysate test. Bull. Parenteral Drug Assoc. 27: 139-148.
4. Jorgensen, J.H. and R.F.Smith. 1973. Preparation, sensitivity, and specificity of Limulus lysate for endotoxin assay. Appl. Microbiol. 26: 43-48.
5. Luderitz, D., M. Freudenberg, Ch.Galanos, P.Kiss, V.Lehmann, and E.Th.Rietschel. 1975. Structure of lipopolysaccharides, the pyrogen of gram-negative bacteria. International Symposium on Pyrogens. Univ. College, London. p. 10-19.

6. Robertson, J.H., D.Gleason, and K.Tsuji. 1978. Dry heat destruction of Lipopolysaccharide. I. Design of dry heat destruction apparatus. Appl. Env. Microbiol. (in-press).
7. Sweadner, K.J., M.Forte, and L.L.Nelson. 1977. Filtration removal of endotoxin (pyrogen) in solution in different states of aggregation. Appl. Env. Microbiol. 34: 382-385.
8. Tsuji, K. and S.J.Harrison. 1978. Dry heat destruction of Lipopolysaccharide. II. Dry heat destruction kinetics. Appl. Env. Microbiol. (in-press).
9. United States Pharmacopeia XIX. 1975. United States Pharmacopeia Convention, Inc., Rockville, MD.
10. Wachtel, R.E. and K.Tsuji. 1977. Comparison of Limulus Amebocyte Lysate and correlation with the United States Pharmacopeial pyrogen test. Appl. Env. Microbiol. 33, 1265-1269.
11. Yin, E.Thye., C.Galanos, S.Kinsky, R.A.Bradshaw, S.Wessler, O.Luderitz, and M.E.Sarmiento. 1972. Picogram sensitivity assay of endotoxin: gelation of Limulus polyphemus blood cell lysate induced by purified lipopolysaccharides and lipid A from gram- negative bacteria. Biochem. Biophys. Acta. 261: 284-289.

ENDOTOXIN CONTAMINATION OF SOME COMMERCIAL PREPARATIONS USED IN EXPERIMENTAL RESEARCH

D. Fumarola and E. Jirillo

Institute of Microbiology - Medical School
Policlinico - Bari - Italy

Endotoxin lipopolysaccharide (LPS), a major cell wall constituent of Gram-negative bacteria, is capable of eliciting a wide variety of biological activities both *in vivo* and *in vitro* (Galanos *et al.*, 1977). In addition, because of their ubiquitous and wide distribution in the environment, LPS are able, practically, to influence every system and function both in man and in other species. A confirmation of the above mentioned concept comes from Bennett's words (quoted by Westphal, 1975) "Endotoxins possess an intrinsic fascination that is nothing less than fabulous. They seem to have been endowed by Nature with virtues and vices in the exact and glamorous proportions needed to render them irresistible to any investigator who comes to know them. This spectrum of activities make possible at least one prediction: an investigator in almost any biological field is likely to obtain a "positive" result if he tried entotoxin in the experimental system he is using."

The problem of endotoxin contamination of biological products is not new. Philip *et al.* (1966) have pointed out that bacterial endotoxins with their multiple potent effects on biological systems represent a constant source of variability in experimental biology and pathology. Nowadays, the large employment of biological products in clinical and experimental research has induced some investigators to reexamine the possible presence of endotoxin as contaminant in their system: in fact, unsuspected presence of LPS even in small amounts, could alter the results and complicate interpretation of investigations.

Pages 379–385, Biomedical Applications of the Horseshoe Crab (Limulidae)

In the last years the Limulus assay (L. A.), an easy and sensitive method in detecting bacterial endotoxin or endotoxin-like material even in minute amounts, has been successfully used also to determine LPS contamination of commercial reagents and biological products (Siegel and Nachum, 1977; Fumarola and Jirillo, 1976, 1977; Fumarola, 1977; Bito, 1977; Rickles *et al.*, 1977; Weinberg *et al.*, 1978).

Here, we would like to illustrate our experience regarding endotoxin contamination of some drugs and reagents customarily used in clinical and experimental research, such as products commonly used in immunological studies, erythropoietin preparations, antitumoral agents, coagulation factors and radiographic contrast media.

In the experiments we have used different commercial kits of Limulus amebocyte lysate. Results of Limulus assay on immunological reagents are shown in Table 1.

Table 1. Limulus assay of products employed in immunological investigations and immunotherapy.

Preparation	L.A.
PHA Wellcome	1/3
PHA Difco	2/3
PHA P Wellcome	0/1
PHA P Difco	1/2
PWM Difco	2/3
Con A Pharmacia	1/1
Con A Miles	2/2
Con A KeK	2/2
Hemagel Hoechst	0/3
FC serum Reheis	0/2
Horse Serum Difco	1/2
Antilymphocyte Serum ISM	2/2
Transfer Factor (dr Grob)	3/3
Transfer Factor (dr Kahn)	1/4
Thymosin IV	0/1

We can see that some mitogens (PHA, Con A, PWM), some reagents for tissue culture (Horse Serum) and therapeutic agents (ALS and TF) gave a positive result with L.A.

Immunological field investigators should be aware that low concentrations of endotoxin are capable of stimulating cells to induce a variety of biological transformations. Some recent data of Rickles *et al.* (1977) suggested that contamination of mononuclear cell culture systems may produce various biological effects, which are also seen with some commonly used mitogens and antigens. Weinberg *et al.* (1978) have pointed out in a recent paper, that commercially bought sera and various reagents are frequently contaminated with large amounts of LPS and if this is not taken into consideration when studying their effects on macrophage tumor killing, correct interpretation of the experiments is difficult. In experimental hematology we have examined some erythropoietin preparations. Erythropoietin is well known for its effects on bone marrow compartments (Yoshida *et al.*, 1972; Hirata, 1975). There is evidence that endotoxin is also able to exhibit some analogous effects in this field (Zuckerman *et al.*, 1977). In our laboratory some erythropoietin preparations from sheep plasma and from human urine were examined in L.A. Each assay test was performed in triplicate. The results are illustrated in Table 2.

Table 2. Limulus assay of erythropoietin samples

Source	L.A.	Level (μg/ml)
Step I Epo (Connaught)	+++	10
Step I Epo (Connaught)	+++	10
Step III Epo (Connaught)	+++	1
Step III Epo (Connaught)	+++	1
Human urinary Epo	+++	10
Human urinary Epo	+++	100
Human urinary Epo	+++	50

These observations lead to the conclusion that erythropoietin preparations should be subjected to more precise and accurate methods of purification in order to avoid spurious and erroneous results.

The possible contamination of antitumor agents with bacterial endotoxin must be considered in the evaluation of clinical and experimental results. Several antitumor agents exhibit a high incidence of adverse effects in which pyrogens are likely to play a role; in addition, there is evidence of potentiation between endotoxin and antitumor agents; finally, patients with neoplastic disease undergoing treatment with immunosuppressive agents have the potential for an adverse reaction involving endotoxins; also the malignancy "per se" can lead to the development of endotoxemia. Results of our study are expressed in Table 3.

Table 3. Antitumor agent preparations and L.A.

Drug tested	Samples	L.A.	Endotoxin (ng/ml)
Vincristine	2	+	1 - 1
Daunomycin	2	+	0,1 - 1
Doxorubicin	2	-	- -
Alanosine	2	-	- -
Decarbazine	1	+	10
Bleomycin	2	++	50 - 50
Mitomycin	1	+	10
L-asparaginase	2	++	50 - 50
Glutaminase	1	+	10

The results obtained suggest a careful evaluation of the immunologic and extraimmunologic responses *in vivo* and *in vitro*, induced by these products when they are contaminated (see also Siegel *et al.*, 1976). In addition, this study demonstrates the usefulness of the L.A. to detect endotoxin in drugs for which the rabbit pyrogen test is either insufficiently sensitive or otherwise unsuitable.

In the recent years the description of thromboembolic complications following the use of coagulation factors has led to a discussion of the mechanism of such complication (White *et al.*, 1977). With the aim to contribute information about the problem, we have studied the possible endotoxin contamination of some preparations. Results are in Table 4.

Table 4. Coagulation factor preparations and L.A.

Preparation tested	L.A.
Factorate	0/1
Koate	2/2
Hemofil	1/1
Konine	0/2
Proplex	2/2
Provertin-Um	0/3
Protromplex 200	0/2
Bebulin	1/3
Feiba	0/4

We think that the endotoxin contamination could represent a possible pathogenetic factor of the thrombogenic complications following the administration of some concentrates of coagulation factors. Further extensive studies on several samples could confirm the relevance of these results.

Finally, we have examined the possible role of contaminating endotoxin in radiographic contrast media (RCM) reaction. In a recent paper Slivka *et al.* (1978) remark on the relevance of endotoxemia induced by administration of LPS on the lethality of RCM in rabbits. We think that a contaminating exogenous endotoxin could amplify the types of reaction we see to RCM in clinical practice. Results of our investigation are expressed in Table 5.

Since there is evidence that endogenous endotoxemia can enhance the chemotoxicity of some drugs and exogenous-induced endotoxemia can increase the side effects of RCM, we think that an additional amount of exogenous endotoxin contaminating some samples of RCM can contribure to the pathogenesis of adverse reactions in humans.

In conclusion, endotoxin contamination of some biological products and drugs should be taken into consideration as a factor able to modify results or lead to misinterpretation of artifactual data of experimental and clinical observations. In some samples the LPS content is negligible, but other preparations are heavily charged with endotoxin. The endotoxin content of processed biologic materials should be tested by the investigator before they are used in biological research. The use of endotoxin-contaminated preparations should be avoided in any kind of biomedical research unless it can be shown that the particular system is insensitive to endotoxins.

The L.A. appears to be a sensitive method for detection and quantitation of endotoxin in reagents, biological products and drugs commonly used in clinical and experimental research.

Table 5. Preparation of RCM and L.A.

Samples of RCM	L.A.	Endotoxin (ng/ml)
Methylglucamine salt of ioglycamic acid (6 lots)	++	0,1
	++	1,0
	++	10,0
	++	0,1
	++	1,0
	-	-
Methylglucamine salt of amidotrizoic acid (7 lots)	++	0,1
	++	1,0
	++	0,1
	++	1,0
	++	0,1
	++	50,0
	-	-

REFERENCES

Bito, LZ (1977). Inflammatory effects of endotoxin-like contaminants in commonly used protein preparations. Science 196: 83.

Fumarola D (1977). Possible endotoxin contamination in some antitumor agent preparations: study with the Limulus amebocyte lysate test. Farmaco Ed Pr 32: 444.

Fumarola D, Jirillo E (1976). Contamination of Con A preparations. J. Immunol 116: 1197.

Fumarola D, Jirillo E (1977). Limulus test, parenteral drugs and biological products: an approach. Dev Biol Std 34: 97.

Galanos C, Freudenberg M, Hase S, Jay F, Ruschmann E (1977). Microbiology 4: 269.

Hirata MC (1975). Inhibitory effects of antihistamines and antiserotonins on the bone marrow reactions produced by *E. coli* endotoxin in mice. J Infect Dis 132: 611.

Philip BA, Herbert P, Hollingsworth JW (1966). Separation of contaminating pyrogenic material from commercial bovine serum albumin. Proc Soc Exp Biol Med 123: 576.

Rickles FR, Levin J, Harchin JA, Barr CF, Conrad ME (1977). Tissue factor generation by human mononuclear cells: effects of endotoxin and dissociation of tissue factor generation from mitogenic response. J Lab Clin Med 89: 792.

Siegel SE, Nachum R (1977). Use of Limulus lysate assay (LAL) for the detection and quantitation of endotoxin. In Bernheimer AW (ed): "Perspectives in toxinology" New York: John Wiley & Sons, p. 61.

Siegel SE, Nachum R, Leimbrick S, Karon M (1976). Detection of bacterial endotoxin in antitumor agents. Cancer treatm Rep 9: 60.

Slivka J, Lasser EC, Lang JH, Lyon SG, Hamblin AE (1978). Effect of endotoxemia on contrast media reactions. Invest Radiol 13: 21.

Weinberg JB, Chapman jr HA, Hibbs JB (1978). Characterization of the effects of endotoxin on macrophage tumor cell killing. J Immunol 121: 72.

Westphal O (1975). Bacterial endotoxins. Int Arch Allergy Appl Immunol 49: 1.

White GC, Roberts HR, Kingdon HS, Lundblad R (1977). Prothrombin complex concentrates. Potentially thrombogenic materials and clues to the mechanisms of thrombosis *in vivo*. Blood 49: 159.

Yoshida M, Hirata M, Inada K (1972). Changes in absolute number of red blood cells and nucleated cells in the bone marrow caused by endotoxin. Japan J Med Sci Biol 26: 11.

Zuckerman K, Quesenberry P, Levin J, Sullivan R (1977). Contamination of Erythropoietin (Ep) by endotoxin (Endo): *in vivo* and *in vitro* effects on erythropoiesis and stem cells. Blood 50(suppl 1): 140.

APPLICATIONS AND ADVANTAGES OF THE LIMULUS AMEBOCYTE LYSATE (LAL) PYROGEN TEST FOR PARENTERAL INJECTABLE PRODUCTS

Carmine C. Mascoli and Marlys E. Weary

Travenol Laboratories, Inc.
Morton Grove, Illinois 60053

On September 18, 1973, the United States Federal Register published proposed manufacturing standards for the production of Limulus amebocyte lysate (LAL), which was described as "a sensitive indicator of the presence of bacterial endotoxins." The article went on to state that the LAL test could be used as an "in vitro" test for the detection of bacterial endotoxins (pyrogens) in biological products and other drugs or fluids for parenteral administration in man. At the time of the announcement, Travenol Laboratories had completed enough preliminary experimental research with Limulus lysate to show that the LAL test was as sensitive as and probably more sensitive than the official USP rabbit pyrogen test. We had established a lysate production laboratory at our Kingstree, South Carolina, plant, and we were eager to adapt lysate technology to production situations. The Travenol Optical Density LAL test was introduced into our domestic and international plants early in 1974. It rapidly became our most important quality control tool for in-process pyrogen testing of raw materials, plant waters, and parenteral solutions. For the first time we had a method we could use to monitor nonpyrogenic levels of bacterial endotoxin that may have eluded us when the rabbit test was our only pyrogen test system. During the first year of use, the LAL test enabled us to monitor the gradual elimination of background levels of endotoxin as applications of new technologies were introduced to manufacturing.

Initially, the LAL test was used to evaluate pyrogenicity of plant waters, saline, and simple dextrose solu-

Pages 387–402, Biomedical Applications of the Horseshoe Crab (Limulidae)

tions. However, more chemically complex solutions occasionally inhibited the LAL gel-clotting reaction. We found that it was necessary to qualify new solutions for LAL testing before incorporating them into our LAL test program. This was done by adding graded doses of endotoxin to the solution to check for inhibition. If inhibition of the gel-clotting reaction occurred, experimental procedures were applied to eliminate the inhibitory effect.

In time, procedures were established for qualifying additional solutions for LAL testing, and we now routinely test a wide variety of parenteral solutions. These include most dextrose-saline combinations, mannitol, Ringer's lactate, peritoneal dialysis solutions, sodium lactate, selected physiologic irrigating and electrolyte solutions, ACD, CPD, and sodium citrate. Some of these solutions require pH adjustment or the addition of chemicals prior to testing. Other solutions, which must be diluted for the USP rabbit test, are usually diluted to the same degree for LAL testing. For the most part, however, large volume parenterals can be evaluated by the LAL test without special treatment.

The Travenol Optical Density LAL test as described by Nandan and Brown (1977) requires the following principal steps:

(1) Combine 0.1 ml of test sample with 0.1 ml of lysate in a 10 x 75-mm depyrogenated glass test tube and incubate for one hour at 37°C.

(2) Centrifuge at approximately 1375 x g in a clinical centrifuge for ten minutes.

(3) Carefully remove the supernatant from the precipitated protein in the tube by vacuum aspiration.

(4) Perform a colorimetric determination of the precipitated protein using the Oyama and Eagle (1956) modification of the Lowry protein assay (Lowry, et al, 1951).

(5) Read and record optical density (OD) at 660 nm with a flow-through spectrophotometer.

It is necessary to run positive and negative controls with each test to rule out the possibility of false positive tests due to contamination and false negatives due to inhibition of the clotting reaction by the test sample. Negative controls consist of solutions identical in chemical composition to the test samples that have previously been tested and shown to be nonpyrogenic by the measurement of endotoxin precipitable protein. We have determined that solutions are acceptable for use as negative controls if the optical density of the precipitable protein is less than 0.200. Positive controls consist of three concentrations of an *E. coli* endotoxin standard (055:B5; Difco Laboratories, Michigan) prepared in the solution used for the negative control. The three endotoxin concentrations (200, 50, and 12 pg/ml) form an optical density endotoxin standard curve against which test samples are compared. Test samples with optical density values at or above that of the 50-pg/ml endotoxin control sample are considered to have failed the lysate test.

The most important consideration in adapting the LAL test for pyrogen testing large volume parenterals is the assurance that endotoxin concentrations that fail the rabbit pyrogen test will also fail the LAL test. For this reason, it is essential to use a reference endotoxin standard to establish the equivalency of the two test systems. A valid LAL test system must be equal to or better than the existing rabbit pyrogen test.

Due to the lack of a universally accepted reference endotoxin standard, each laboratory must determine equivalency of the two test systems by using its own reference endotoxin and its own rabbit colony. For this reason, laboratory to laboratory comparisons have been difficult to make. To further complicate matters, rabbit colonies have been shown to respond to the same dose of endotoxin with widely varying febrile responses. An example of this variability was described by Weary and Baker (1977). Table 1 shows the cumulative fever response from eight-rabbit USP pyrogen tests following injections of three graded doses of our *E. coli* reference endotoxin at seven Travenol plant locations. If each of these plants had been asked to define a safe dose of endotoxin for an independent equivalency determination, one (Mountain Home) would have selected a dose below 50 pg/ml. Two others (Alliston and Morton Grove) would have chosen 50 pg/ml. Cleveland and North

Cove would have selected 100 pg/ml and the remaining two plants (Hays and Carolina) 200 pg/ml of endotoxin or more.

Table 1. USP 8-Rabbit Pyrogen Tests on 0.9% Sodium Chloride Spiked with *E. coli* 055:B5 Endotoxin

	Total Temperature Increase (°C)		
Plant Location	200 pg/ml	100 pg/ml	50 pg/ml
Alliston, Canada	6.4 Fail	4.0 Fail	0.1 Pass
Cleveland, MS	4.7 Fail	1.8 Pass	3.1 Pass
Hays, KS	2.6 Pass	1.4 Pass	1.3 Pass
Morton Grove, IL	6.0 Fail	4.3 Fail	2.6 Pass
Mountain Home, AR	4.1 Fail	3.1 Pass	4.1 Fail
North Cove, NC	5.1 Fail	2.6 Pass	2.0 Pass
Carolina, PR	0.9 Pass	1.6 Pass	1.2 Pass

Currently, all official rabbit tests for pyrogen use an arbitrarily selected level of febrile response as a pass/fail reference point. In fact, at this reference point half of the marginally pyrogenic solutions may be expected to pass the rabbit test. Table 2 indicates that there is a positive dose-response relationship between increasing doses of endotoxin and the mean change of rabbit temperature. However, the standard deviations and coefficients of variation suggest highly variable rabbit responses relative to the mean responses at the concentrations and doses under consideration. The importance of this variability is pointed out by the data shown in Tables 3 and 4.

Table 3 shows results of seven marginally failing eight-rabbit tests. By selecting all possible groups of three-rabbit response data from each group of eight in Table 3, we have constructed in Table 4 a list of all possible three-rabbit response data combinations that would have passed an official USP test. If *E. coli* endotoxin concentrations that result in USP eight-rabbit test failures are used for comparison, up to 30% of the selected three-rabbit test groups would have passed the USP three-rabbit test. Given a random selection of rabbits and the

Table 2. Results of USP 8-Rabbit Test After Injection of Saline Containing *E. coli* 055:B5 into Rabbits Weighing from 3 to 5 kg

Endotoxin Concentration (ng/ml)	Volume of Solution Injected (ml/kg)	Total Temperature Increase[a] (°C)	Mean Temperature Increase[b] (°C)	Standard Deviation[b] (°C)	Coefficient of Variation (%)
3.125	1.0	7.80[c]	0.975	0.246	25.2
1.56	1.0	4.75[c]	0.594	0.218	36.7
1.00	1.0	3.70[c]	0.462	0.158	34.2
0.78	1.0	1.40	0.144	0.208	144.4
0.39	1.0	1.00	0.088	0.187	212.5
0.195	1.0	1.20	0.150	0.065	43.3

[a] Negative rabbit temperature values were excluded from total temperature increase determinations according to USP.

[b] Negative rabbit temperature values were included in the determinations of means and standard deviations to properly reflect total variability.

[c] Fail USP test criteria of 3.7°C total increase.

Table 3. Results of Seven USP 8-Rabbit Tests with Failing Total Temperature Increases

Rank of Rabbit Response	Temperature Increase (°C)						
	A	B	C	D	E	F	G
1	0.25	0.35	0.20	0.40	0.30	0.35	0.10
2	0.35	0.35	0.20	0.40	0.30	0.40	0.35
3	0.35	0.40	0.30	0.40	0.50	0.40	0.40
4	0.40	0.40	0.40	0.50	0.55	0.45	0.55
5	0.45	0.45	0.60	0.50	0.60	0.50	0.60
6	0.55	0.50	0.60	0.60	0.65	0.50	0.65
7	0.65	0.65	1.10	0.65	0.70	0.55	0.85
8	0.70	0.70	1.15	0.70	0.85	0.60	0.90
Total	3.70	3.80	4.55	4.15	4.45	3.75	4.40
Average	0.46	0.48	0.57	0.52	0.56	0.47	0.55

Table 4. All Possible Combinations of Passing 3-Rabbit Responses Among Groups Exhibiting 8-Rabbit Test Failure

8-Rabbit Tests			
Test	Total Temperature Increase (°C)	Combinations of Passing 3-Rabbit Responses by Rank	Percent Passing
A	3.70	(1,2,3)(1,2,4)(1,2,5)(1,2,6)(1,3,4)(1,3,5)(1,3,6)(2,3,4)(2,3,5)(2,3,6)(2,4,5)(2,4,6)(3,4,5)(3,4,6)(3,5,6)	26.78
B	3.80	(1,2,3)(1,2,4)(1,2,5)(1,2,6)(1,3,4)(1,3,5)(1,3,6)(2,3,4)(2,3,5)(2,3,6)(2,4,5)(2,4,6)(3,4,5)(3,4,6)(3,5,6)(4,5,6)	28.57
C	4.55	(1,2,3)(1,2,4)(1,3,4)(2,3,4)	7.14
D	4.15	(1,2,3)(1,2,4)(1,2,5)(1,3,4)(1,3,5)(2,3,4)(2,3,5)	12.50
E	4.45	(1,2,3)(1,2,4)(1,3,4)(2,3,4)	7.14
F	3.75	(1,2,3)(1,2,4)(1,2,5)(1,2,6)(1,2,7)(1,3,4)(1,3,5)(1,3,6)(1,3,7)(2,3,4)(2,3,5)(2,3,6)(2,3,7)(2,4,5)(2,4,6)(3,4,5)(3,4,6)	30.36
G	4.40	(1,2,3)(1,2,4)(1,3,4)(2,3,4)	7.14

degree of variability among them, the probability of a false-negative response in a three-rabbit test is uncomfortably high.

Table 5. Pyrogen Test Results for 20 Bottles of 5% Dextrose in Water Containing *E. coli* 055:B5 Endotoxin

Endotoxin Concentration (pg/ml)	Pyrogen Test Failures USP Rabbit Test	LAL Test
35	0/10	0/10
75	1/10	10/10

Compared to the USP rabbit pyrogen test, our LAL pyrogen test results have proven to be far more reproducible and also more sensitive. Table 5 shows the results of a recent study done at our North Cove production facility. A comparative pyrogen test evaluation was made of *E. coli* 055:B5 endotoxin in 5% dextrose in water. During a one week period, 20 blinded samples of 5% dextrose in water were spiked with endotoxin and were submitted to the lysate and rabbit test laboratories for routine pyrogen evaluation. Ten of the samples contained 75 pg/ml of endotoxin and the remaining ten contained 35 pg/ml of endotoxin. Samples containing more than 50 pg/ml of *E. coli* equivalents fail the Travenol LAL OD pyrogen test. Thus, it was expected that the ten solution samples containing 35 pg/ml would pass both pyrogen test procedures. Test results show that this was the case. All ten solution samples containing 75 pg/ml of endotoxin failed the LAL test as expected; however, only one of ten failed the USP rabbit test. The LAL test again showed a superior ability to detect marginally pyrogenic doses of endotoxin.

Having thus demonstrated that the LAL test is more sensitive to commercially available bacterial endotoxin than the rabbit test, we monitored the LAL test in our plants to judge its ability to uncover naturally occurring bacterial endotoxins. We have been particularly interested in comparisons of pyrogen failures detected initially by the rabbit test and confirmed by the LAL test and vice versa.

At the conclusion of each year since 1974, we have compiled an internal report comparing the rabbit and LAL pyrogen test data accumulated by all of our domestic production facilities. A summary of test results from last year is found in Table 6.

Table 6. 1977 Pyrogen Test Results from Travenol Domestic Plants

Description	Number
Rabbit Tests Performed	28,410
LAL Tests Performed	143,196
Products Tested by Both Procedures	37
LAL Test Failures	37
USP Rabbit Test Failures	4

More than 75% of the 37 pyrogenic results in 1977 involved raw materials and plant waters. The four USP rabbit test failures also resulted from raw material pyrogen evaluations. The data again confirmed the greater sensitivity of the LAL test than of the USP rabbit test. It is also significant that in no case did the LAL test fail to confirm a rabbit pyrogen failure.

The LAL test has proved to be a most useful tool for troubleshooting potential and actual pyrogen problems. Travenol's first practical application of this occurred in 1973 when one of our facilities suspected an intermittent pyrogen problem in lots of peritoneal dialysis solution. Within a 48-hour period, over 200 bottles of solution were examined by the LAL test, resulting in statistical evidence that every fourth bottle was contaminated with pyrogens. It was then a simple matter to trace the source of contamination to one of the four filling heads on the solution filling line. It would have been impossible to resolve the problem that quickly, if at all, using the rabbit pyrogen test.

The application of the LAL procedure to biologic products has been slower than for LVP products because many biologics contain substances that inhibit the clotting reaction of the lysate. The test has, however, been successfully applied as a release criterion for embryonated egg-grown, density-gradient purified influenza virus vaccines in the U.S. A product line where the LAL test would be particularly desirable is that of the human blood fractions. These products are inherently antigenic in animals, requiring that pyrogen test rabbits be acclimated, used, and destroyed within a matter of weeks.

Travenol did not begin the development of a LAL test for human blood fractions until the summer of 1976. At that time, a product pyrogenicity problem was encountered with certain lots of Hyland 25% human albumin, all of which had met USP pyrogen test and other criteria upon original production and release. However, upon retest of file samples, some failed the rabbit test. Although no fault could be found with the rabbit pyrogen test records and procedures, it was obvious that the original rabbit tests were controverted by the later tests.

Travenol scientists immediately set out to find a way to quickly evaluate the file samples of all lots of 25% albumin released that year, both before and after the occurrence of the problem. Rabbit testing would have been time-consuming and, practically speaking, too little help, too late. Additionally, it was important to look at new production for an indication of the origin of the pyrogens. Because the Travenol LAL test was shown to be of value for pyrogen testing our LVP line, we tried to adapt it to the 25% albumin product.

Our first attempt involved diluting the product. It was felt that endotoxin, if present, would be detectable in the dilution, and that total content in terms of standard endotoxin equivalents could then be back calculated. It was found that a 1:20 dilution of 25% albumin to which graded doses of endotoxin had been added, could be successfully tested by the Travenol technique. Further, there was a good linear response with increasing graded quantities of purified endotoxin placed into nonpyrogenic product samples. Using the new technique on reserve file samples of product made during the suspect period of time, we attempted to discover whether or not a "pyrogenic" product not detected by official rabbit tests was included in released product in the marketplace, in released undistributed product, or in finished but unreleased product. A LAL pass/fail reference point was chosen (50 pg/ml of E. coli equivalents which is the same as for our LVP line) that was significantly below the level detectable by rabbits. Two new problem lots were uncovered.

Meanwhile, the Bureau of Biologics, FDA, asked for assurance that in the future such lots would not be released. On September 22, 1976, the U.S. Federal Register stated, among other things, that the rabbit pyrogen test, though it had served well for many years, in some instances, namely 25% albumin, failed to detect levels of pyrogen that were later involved in patient reactions. Further, the publication indicated the value of the LAL procedure in detecting bacterial endotoxins as pyrogens in human blood fractions and in drugs for human use.

Hyland proposed to the FDA the introduction of a product pyrogen test scheme that would greatly increase product pyrogen evaluations using both the LAL and rabbit tests. The format is detailed in matrix form in Table 7.

Table 7. Matrix Comparing Current Pyrogen Testing to Plan Existing Prior to July 22, 1976

	Past	Current	
Description of Test Sample	Rabbits Only	Rabbits	LAL
In Process			
A. After reconstitution			X
B. After 2 hours, 50°C heating			X
C. Filter rinse prior to clarification (distilled water flush)			X
D. After clarification			X
Bulk	X	X	X
Beginning of Fill	X	X	X
End of Fill		X	X
Random sample prior to submission to Bureau of Biologics		X	X
Shipping release - beginning		X	X
Shipping release - end		X	X
Past shipping 1, 2, 3, 4 months			XXXX

With the Bureau's approval, Hyland then applied the scheme to its albumin production. A comparison was made of the results of rabbit pyrogen tests and LAL tests on 81 lots of 25% albumin manufactured from November 22, 1976, to October 18, 1976. These data are shown in Table 8. They show that:

(1) The majority of lots produced meet both in-house rabbit requirements and the 50-pg/ml LAL requirement (Quadrant 3). Seventy-four of the 87 points, or 85%, are in Quadrant 3.

(2) None of the lots tested through November 9, 1976, fall into Quadrant 1 (acceptable by LAL but unacceptable by rabbits).

(3) Eleven of the 87 points, 12.6%, were unacceptable in both rabbits and LAL (Quadrant 2).

(4) In only two instances did the LAL test indicate a problem that the rabbits did not detect (Quadrant 4). One of these lots had been discarded for high particulate reject rate at the time of packaging. The other was discarded for high LAL activity.

By this time, the Hyland plant was becoming more and more confident of its ability to detect pyrogens, and an increased effort was made to uncover the source of the pyrogens during albumin processing.

Throughout the fall quarter, LAL test results on the in-process C samples (clarification filter prewash) gave high lysate test results. The filter, composed of celite (diatomaceous earth) material, is used to "clarify" protein fractions.

Early in 1977 we had the first clear-cut evidence that the clarification filters were contributing endotoxin to the product. In January, for the first time, high lysate test results were observed on in-process D samples (final bulk sample after clarification filtration). This was followed by pyrogen failures in bulk by both rabbit and lysate tests. Results are shown in Table 9.

By this time, a computer program had been instituted

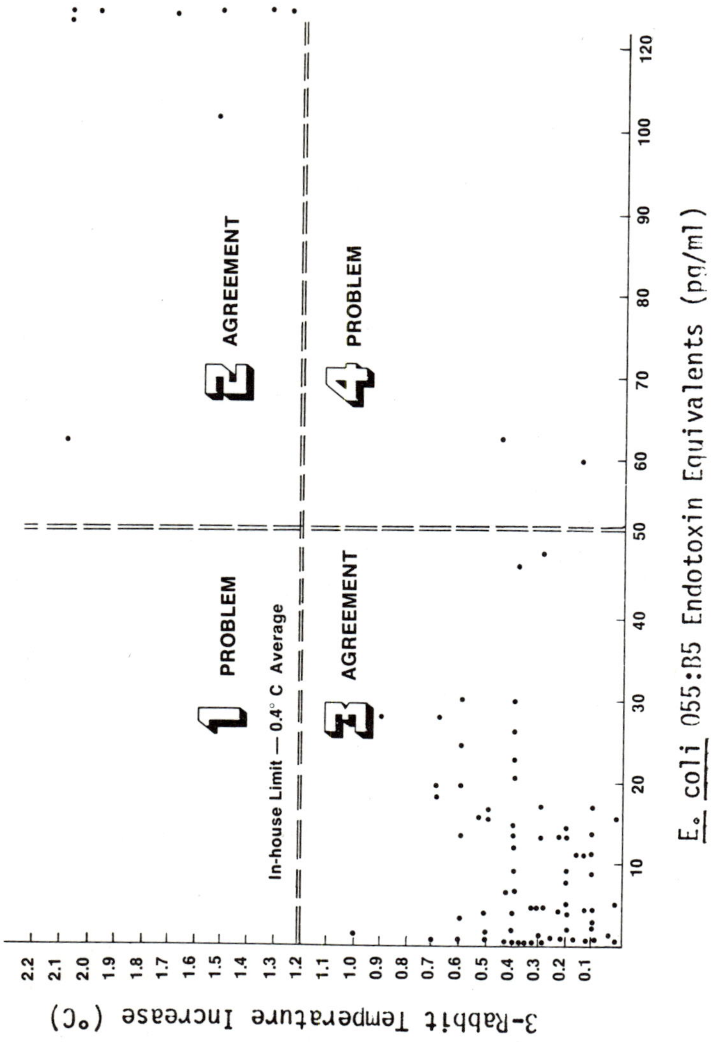

Table 8. Rabbit vs LAL Tests for 81 Lots of 25% Albumin

Table 9. In-Process Testing of Four 25% Albumin Bulks

Lot No.	Sample A After Reconstitution	Sample B After Heating	Sample C Filter Rinse	Sample D After Clarification	Bulk (Rabbits and Lysate)
1	Pass (<50 pg/ml)	Pass (<50 pg/ml)	Fail (∼234 pg/ml)	Fail (∼181 pg/ml)	Fail
2	Pass	Pass	Fail (∼261 pg/ml)	Fail (∼196 pg/ml)	Fail
3	Pass	Pass	Fail (∼352 pg/ml)	Fail (∼129 pg/ml)	Fail[a]
4	Pass	Pass	Fail (∼316 pg/ml)	Fail (∼182 pg/ml)	Fail[a]

[a] These samples failed initial bulk testing but were nonpyrogenic following 14-day heat cycle.

to convert endotoxin levels into pg/ml of E. coli equivalents. In-process and finished product results could now be expressed as pg (+ or -), based on direct optical density readings as compared to standard curve results on 50-pg/ml spiked endotoxin samples. For purposes of comparison, such results obtained for a nonpyrogenic, a mildly pyrogenic, and a strongly pyrogenic lot are shown in Table 10.

Since that time, Hyland has been working closely with the filter vendor to eliminate the endotoxin contamination in their filters. Recently, with our help, the vendor introduced LAL testing as a routine procedure for quality control of filter production and final filter evaluation, with the eventual certification of lots of filters as being suitable for use in clarification filtration of blood fractions.

A recent publication (Steere, et al, 1978) confirmed the failure of rabbits to detect pyrogens in 25% albumin that, in many cases, caused clinical reactions in man. The same paper also confirmed that the Limulus lysate test in all cases was able to detect these pyrogens that ranged in concentration from 2 to 64 ng per ml. In a prospective study, albumin in 433 of 662 vials, infused to patients, gave positive LAL tests, and in 311 of these vials, concentrations of endotoxin ranged from 4 to 64 ng per ml. However, no patient reactions were observed in this study.

In summary, the data presented are derived from literally hundreds of thousands of LAL tests; they represent a vast experience with raw materials and in-process and final product testing of both LVP's and biologic injectables. Far less in number but virtually always correlating with positive LAL tests, were rabbit pyrogen tests. All of our experiences confirm the following:

(1) Pyrogens important in LVP's and devices are endotoxin in nature (positive LAL-positive rabbit).

(2) Never have we seen an unexplained negative LAL-positive rabbit result.

(3) There have been observations of the presence of endotoxin pyrogens detectable by LAL that were not detected by rabbit tests.

Table 10. Pyrogen Test Results Based on Computer Program "CRABBY"

	Nonpyrogenic Lot			Mildly Pyrogenic Lot			Pyrogenic Lot		
Test Sample	Date	Rabbit Test Temperature Increase (°C)	LAL Test (pg/ml)	Date	Rabbit Test Temperature Increase (°C)	LAL Test (pg/ml)	Date	Rabbit Test Temperature Increase (°C)	LAL Test (pg/ml)
In Process									
A			-30			-15			-22
B			-14			-13			-24
C			+140			+170			+234
D			-9			+26			+181
Bulk	12/27/76	+0.6	-6	12/20/76	+0.7	+24	12/27/76	+2.4	+108
Beginning	01/20/77	+0.3	-6	01/14/77	+0.8	+25	01/03/77[a]	+2.7	+112
							01/12/77[b]	+1.3	+95
End	01/20/77	0.0	-10	01/14/77	+0.6	+21	Bulk refractionated		
R Protocol	01/24/77	0.0	-5	01/17/77	+0.6	+13			
Ship - beginning	02/11/77	+0.1	-13	02/03/77	+0.9	+25			
Ship - end	02/11/77	+0.3	-11	02/03/77	+0.8	+21			
Released date	02/28/77			02/17/77					
Ship + 1 month	03/28/77		-15	03/22/77		+24			
Ship + 2 months	04/28/77		-13	04/15/77		+31			
Ship + 3 months	05/27/77		No Record	05/17/77		+19			
Ship + 4 months	06/28/77		-17	06/17/77		+14			

[a] Samples heated seven days in an effort to reduce pyrogenicity.
[b] Samples heated 14 days in an effort to reduce pyrogenicity.

(4) In some cases, reliance only upon official USP rabbit test failed initially to detect pyrogens which were sometimes confirmed later by rabbit tests, but always detected and confirmed by LAL tests.

In conclusion, it is our belief that those concerned with detection of pyrogens in parenteral injectable products or medical devices, manufacturers and regulatory officials alike, should be far more concerned with the relative insensitivity of rabbits to detection of endotoxin pyrogens than with the possible but rarely, if ever, seen cases of nonendotoxin pyrogens.

In the current era of improving GMP and with the present concern for potential pyrogenic reactions, manufacturers, regulatory officials, and consumers of parenteral injectable products and medical devices can best be served by the LAL and not the rabbit test.

Limulus amebocyte lysate: Additional standards (1973). Fed Regist 38: 26130-26132.

Lowry OH, Rosebrough NJ, Farr AL, Randall RJ (1951). Protein measurement with the Folin phenol reagent. J Biol Chem 193: 265-275.

Nandan R, Brown DR (1977). An improved in vitro pyrogen test: To detect picograms of endotoxin contamination in intravenous fluids using Limulus amebocyte lysate. J Lab Clin Med 89: 910-918.

Oyama VI, Eagle H (1956). Measurement of cell growth in tissue culture with phenol reagent (Folin-Ciocalteau). Proc Soc Exptl Biol Med 91: 305-307.

Steere AC, Rifaat MK, Seligmann EB Jr, Hochstein HD, Friedland G, Dasse P, Wustrack KO, Axnick KJ, Barker LF (1978). Pyrogen reactions associated with the infusion of normal serum albumin (human). Transfusion 18: 102-107.

Weary M, Baker B (1977). Utilization of the Limulus amebocyte lysate test for pyrogen testing large volume parenterals, administration sets, and medical devices. Bull Parenter Drug Assoc 31: 127-133.

QUANTITATIVE DETECTION OF ENDOTOXIN BY NEPHELOMETRY

John A. Dubczak, Richard Cotter, and
Frank R. Dastoli
Travenol Laboratories, Inc.
Morton Grove, Illinois

INTRODUCTION

The amebocyte lysate of Limulus polyphemus is used in the most sensitive assay for the presence of gram negative endotoxin (Levin and Bang, 1968). This lysate has been used to assay a variety of biological, radiopharmaceutical, and parenteral solutions (Cooper et al, 1972; Nandan and Brown, 1977). The Limulus amebocyte lysate (LAL) test is enzymatic and involves the activation of a specific proclotting enzyme by endotoxin in the presence of divalent cations. This activated enzyme cleaves clottable protein called coagulagen into polypeptide subunits that aggregate to form a gel protein matrix. The endpoint of this assay is usually a solid gel clot that maintains its integrity following inversion. Most commercially available LAL assays rely on the solid clot endpoint. Although sensitive, these clot procedures are highly subjective and are referred to as semiquantitative. As the applications of the LAL test expand into clinical areas, a more quantitative method becomes mandatory.

A quantitative approach using the clot method was reported by Reinhold and Fine (1971). In their system, serial two-fold dilutions of known concentrations of Serratia marcescens endotoxin were made, and the resulting pattern of positive and negative tests was used to demonstrate the reliability of the LAL test above and below the end point. The end point was best discerned at endotoxin concentrations of 100 to 500 pg. Niwa et al (1974) reported on a quantitative LAL method for assaying endotoxin

Pages 403–414, Biomedical Applications of the Horseshoe Crab (Limulidae)

concentration by using the Lowry assay (Lowry et al, 1951) to measure the protein in the precipitated clots. They used purified endotoxins from Serratia minnesota R595 and Escherichia coli 0113 in their LAL test to produce sigmoid curves in which the optical densities of the samples were roughly proportional to the endotoxin concentrations. However, native polysaccharide of E. coli 0113, K antigen of Vibrio parahaemolyticus Iwate 14, and Wax D of Mycobacterium tuberculosis Aoyama B curiously were all negative at 1 μg/ml. An improved method was reported by Nandan and Brown (1977) that showed increased sensitivity by measuring quantities of endotoxin as low as 6 pg.

A variation of the above spectrophotometric procedures was used by Hollander and Harding (1976) to measure endotoxin concentrations in mouse plasma. Combined endotoxin and lysate mixtures were placed in a spectrophotometer, and turbidity readings at 380 nm were recorded. The course of the reaction was shown by a series of complex curves that were dependent on the concentration of E. coli endotoxin. More linear curves were obtained by plotting 1/t, where t = time, against endotoxin concentration. While sensitivity was reported to be as low as 0.12 pg, both t and the slope of the line were found to vary on different days.

In this report, a new quantitative LAL test method for analyzing low concentrations of endotoxin is described. This method employs the Hyland Laser Nephelometer PDQ that is designed to quantitate specific protein concentrations. This instrument differs from conventional nephelometers and spectrophotometers in that its light source is a highly intense, angled, collimated laser beam. In this method, sodium dodecyl sulfate (SDS) is used to terminate the enzymatic reaction at its optimum point, and to ensure uniformity in the particle size of the precipitated coagulagen complex. Sodium carboxymethylcellulose (CMC) is subsequently added as a suspension stabilizer. The nephelometric procedure is rapid, having an assay time of approximately 60 minutes; sensitive, detecting picogram quantities of endotoxin; lysate saving, using only 50 μl; and simple, consisting of only 3 steps. Finally, this procedure has a high degree of reproducibility demonstrated by an 8.6% coefficient of variation at the 50-pg/ml endotoxin concentration.

MATERIALS AND METHODS

A laser nephelometer was obtained from Hyland Diagnostics, Costa Mesa, CA. Pyrogen-free water and Limulus lysate were products of Travenol Laboratories. Endotoxin was an E. coli 055:B5 standard purified by Bovin's TCA extraction procedure and purchased from Difco Laboratories, Detroit, MI. Sodium dodecyl sulfate was obtained from MC and B Manufacturing Chemists, Norwood, OH. Sodium carboxymethylcellulose was manufactured by Hercules, Wilmington, DE. Borosilicate test tubes (10 x 75 mm) were products of Corning Glass Works, Corning, NY. Swinex filter housings and filters (0.45 and 1.2 µm) were products of Millipore, Bedford, MA.

Glassware was depyrogenated with dry heat at 250°C for one hour. Filter assemblies were made pyrogen-free by flushing with copious amounts of sterile, nonpyrogenic water. Lyophilized Limulus lysate was reconstituted at a pH of 7.4 and an ionic strength of 0.07 moles/liter with 5.0 ml of 0.05M Tris/HCl buffer containing 0.23% $MgCl_2$. The SDS was prepared at a concentration of 1.0% in water, and the CMC at 0.08% in saline. The lysate was reconstituted and filtered through a sterile 0.45-µm filter just before use. The SDS was filtered in advance with a 0.45-µm filter. The viscosity of CMC necessitated filtration with a 1.2-µm filter. CMC was filtered as it was being delivered to the test tubes by attaching the Swinex filter to a Corning automatic repipetter. Reference standard curves were prepared with known E. coli endotoxin concentrations in pyrogen-free water.

The reaction mixture consisted of a 2:1 ratio of test solution to lysate (0.1 ml of test sample to 0.05 ml of lysate). Sample mixtures were incubated in a 37°C water bath for 30 minutes. Following the 30-minute incubation, the reaction was stopped by adding 0.2 ml of 1.0% SDS. Samples were then incubated at room temperature for 2 to 3 minutes to assure the complete binding of SDS to the protein precipitate. One milliliter of the 0.08% CMC preparation was then added. Samples were mixed with a vortex mixer, and the turbidity immediately determined.

RESULTS AND DISCUSSION

The development of a nephelometric LAL pyrogen assay system was undertaken in our laboratories as a means of obtaining a rapid, accurate, and sensitive method of measuring endotoxin. In the original system, lysate was reconstituted in 3 ml of 0.5% $MgCl_2$. One-tenth milliliter was then incubated with an equal volume of the test sample for 60 minutes at 37°C. Following incubation, 1 ml of normal saline was added to the reaction mixture to obtain the minimum volume necessary for nephelometric quantitation. The coagulagen clot was then mechanically disrupted and assayed by light scattering.

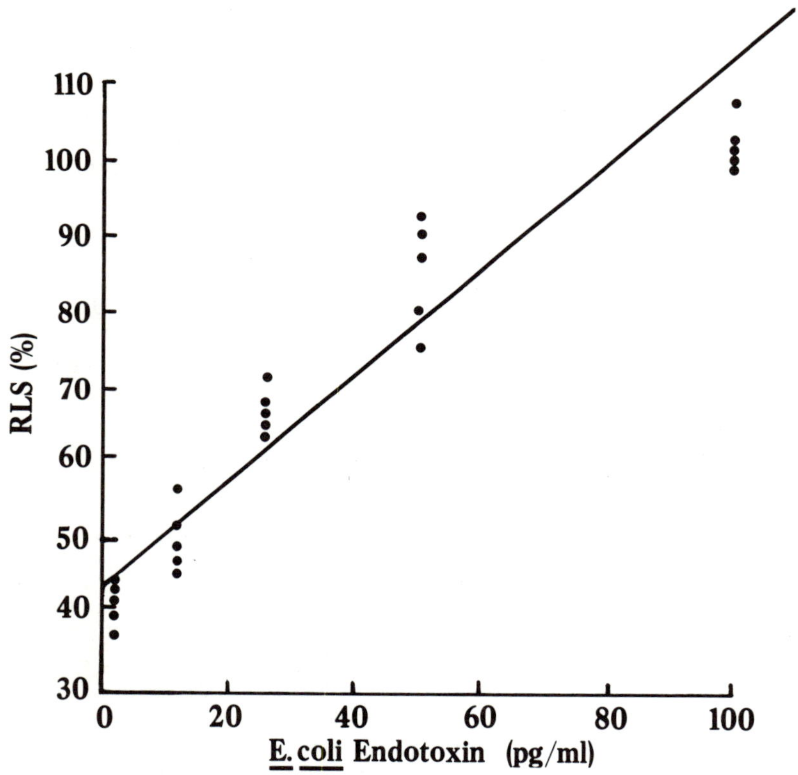

Figure 1. Standard Endotoxin Curve. Lysate was reconstituted with 3 ml of 0.5% $MgCl_2$, and 0.1 ml of lysate was incubated with 0.1 ml of endotoxin for 60 minutes at 37°C. The lysate clot was mechanically disrupted and then quantitated by laser nephelometry.

The data in Figure 1 were obtained with the original LAL nephelometric assay system and indicate large variations, lack of linearity, and a high background turbidity.

In an attempt to overcome these assay problems, the volumes of test sample and lysate were reduced to 0.05 ml, and 1 ml of 0.08% CMC, a suspension agent, was added after incubation. Figure 2 gives the results of this modification. The high background was corrected; however, the variation at a given endotoxin concentration and the lack of linearity between concentrations still persisted because of the wide distribution of particle size.

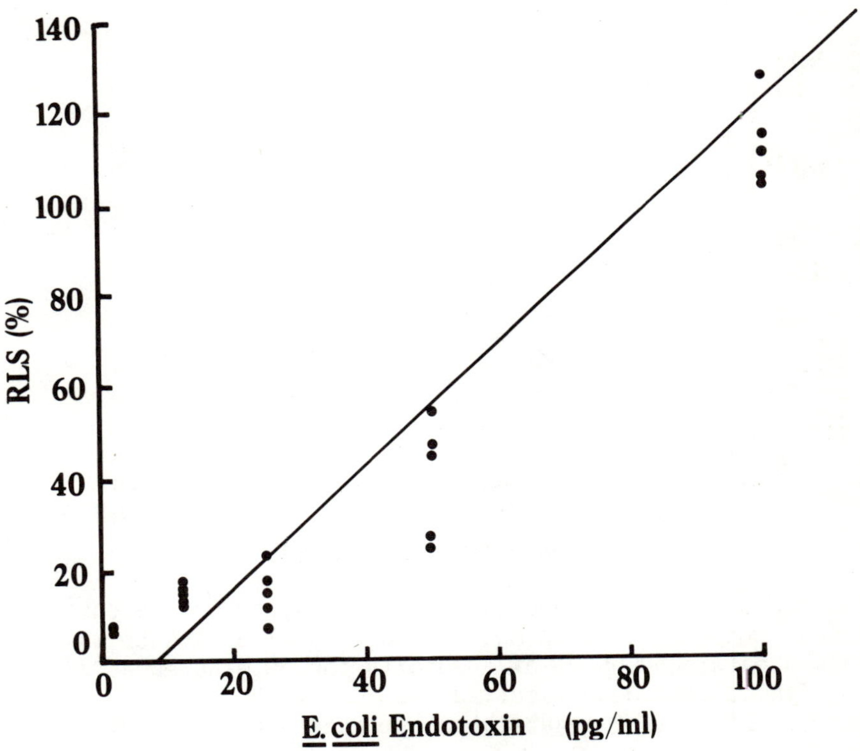

Figure 2. Standard Endotoxin Curve. Lysate was reconstituted with 3 ml of 0.5% $MgCl_2$, and 0.05 ml of lysate was incubated with 0.05 ml of endotoxin for 60 minutes at 37°C. The lysate clot was mechanically disrupted, combined with 1 ml of 0.08% carboxymethylcellulose, and quantitated by laser nephelometry.

Replacement of mechanical clot disruption by the addition of 1 ml of 1.0% SDS, a known ionic detergent with the ability to dissolve the LAL endotoxin coagulagen clot (Solum, 1973), was then tested. The data in Figure 3 indicate much less variation in the presence of SDS. However, there is still a lack of linearity in the assay.

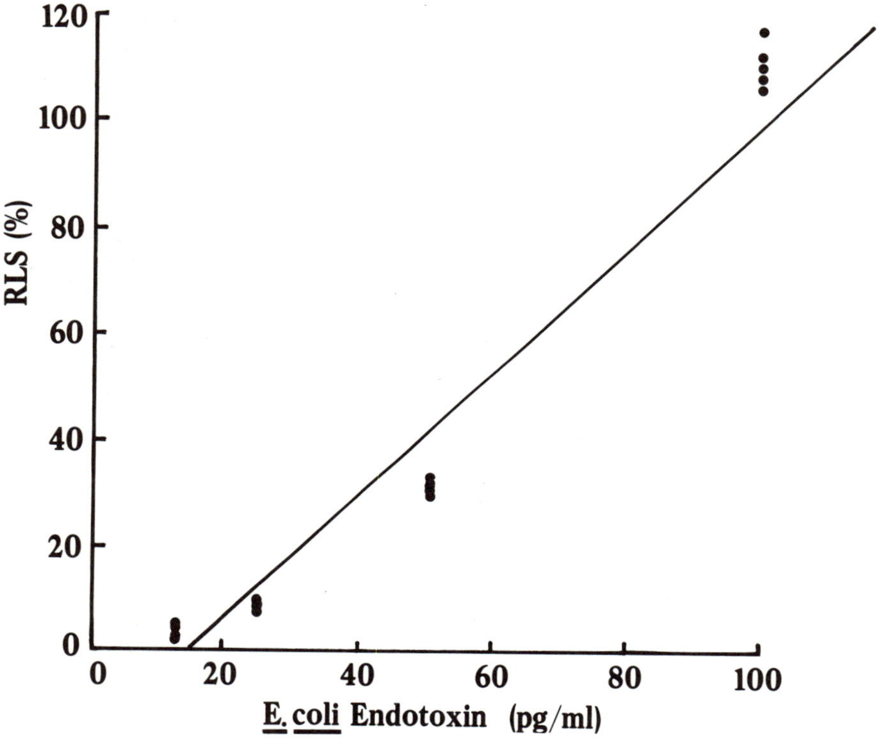

Figure 3. Standard Endotoxin Curve. Lysate was reconstituted with 3 ml of 0.5% $MgCl_2$, and 0.05 ml of lysate was incubated with 0.05 ml of endotoxin for 60 minutes at 37°C. The lysate clot was dissolved with 0.2 ml of 1.0% sodium dodecylsulfate and quantitated by laser nephelometry.

The assay system next tested involved combination of SDS and CMC. This system optimized SDS's dissolution of the clot into light scattering complexes that were dispersed and suspended by CMC. This combination (Figure 4) results in small variation and good linearity over the

0- to 100-pg endotoxin range.

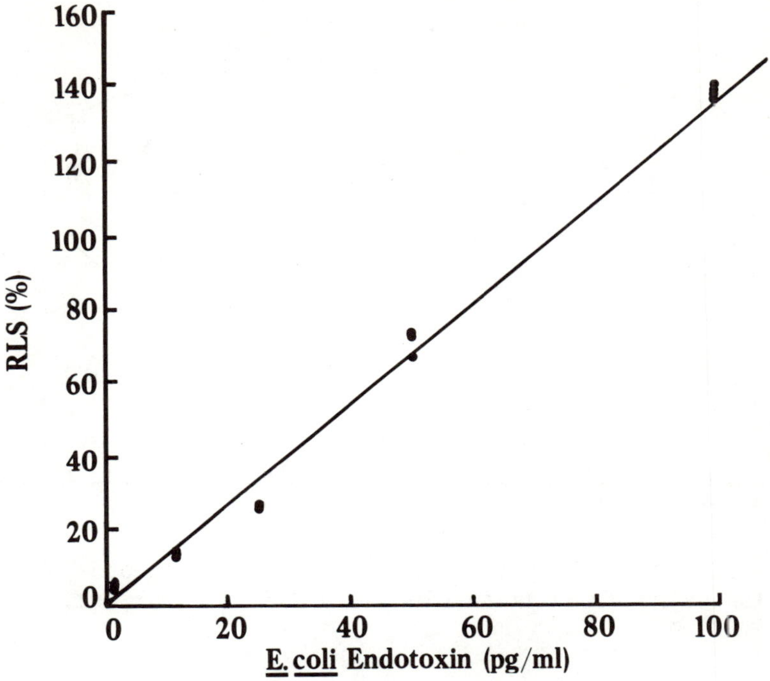

Figure 4. Standard Endotoxin Curve. Lysate was reconstituted with 3 ml of 0.5% $MgCl_2$, and 0.05 ml of lysate was incubated with 0.05 ml of endotoxin for 60 minutes at 37°C. The lysate clot was dissolved with 0.2 ml of 1.0% sodium dodecylsulfate, suspended in 1 ml of 0.08% carboxymethylcellulose, and quantitated by laser nephelometry.

The nephelometric assay was further improved by optimizing the chemical parameters that affect both the interaction of endotoxin with LAL and the subsequent clotting mechanism. The introduction into the LAL system of a buffer, 0.05M imidazole or 0.05M Tris, at an ionic strength of 0.05 moles/liter indicated an optimal pH range of 7.2 to 7.5 for LAL activity (Figure 5). The effect of ionic strength on this buffered LAL system is shown in Figure 6. These curves represent LAL-endotoxin interaction at a constant buffer molarity and pH, while the ionic strength

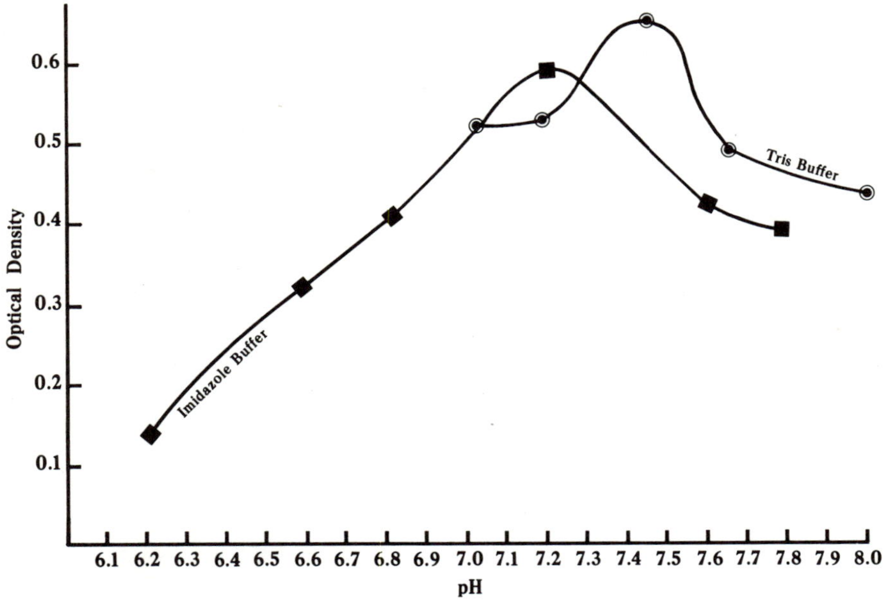

Figure 5. The pH activity profile of the LAL-endotoxin reaction using 0.05M imidazole and 0.05M Tris buffers.

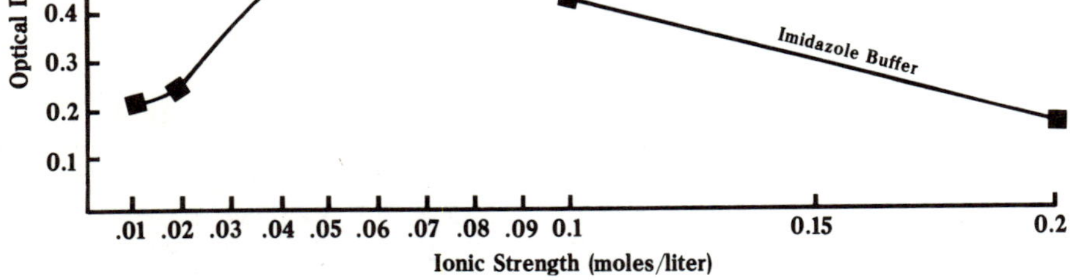

Figure 6. The ionic strength activity profile of the LAL-endotoxin reaction using 0.05M imidazole and 0.05M Tris buffers. The ionic strength was controlled by varying the amount of $MgCl_2$ present in the reaction mixture.

of the reaction mixture is varied with $MgCl_2$. The optimum ionic strength of this system was found to be in the range of 0.065 to 0.085 moles/liter. The observed ionic strength optimum may vary to some extent because of specific ion effects.

Using this optimized system, the lysate reconstitution volume of 3.0 ml was reexamined. Lysate reconstitution volumes from 1 to 5 ml were evaluated, and no significant difference was shown in the 3- to 5-ml range. Five milliliters (see Figures 7 and 8) was chosen as the optimum reconstitution volume.

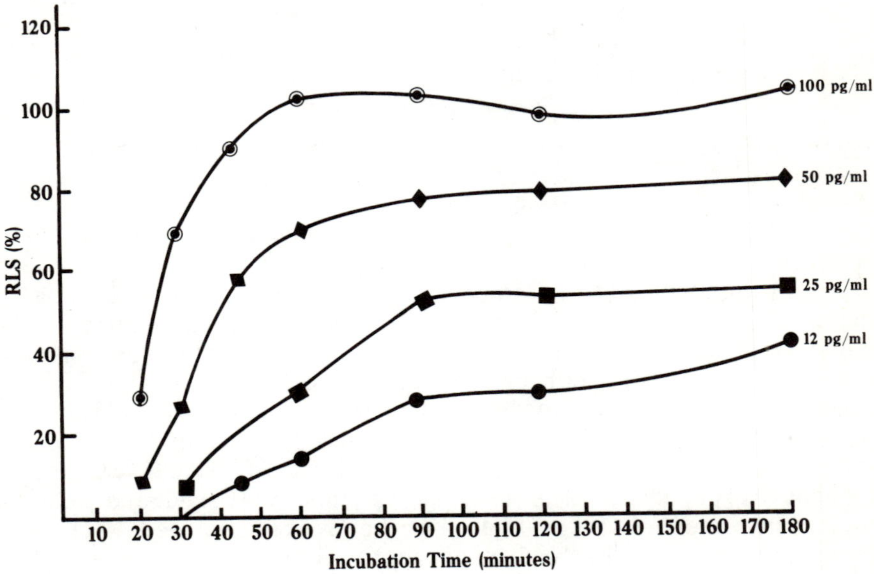

Figure 7. The effect of incubating four concentrations of endotoxin at 37°C from 0 to 180 minutes with lysate reconstituted in 5 ml of imidazole-buffered 0.5% $MgCl_2$.

The final assay modification was a change in the sample to lysate volume ratio from 1:1 to 2:1. This modification enabled good quantitation of endotoxin levels after 30 minutes of incubation (Table 1).

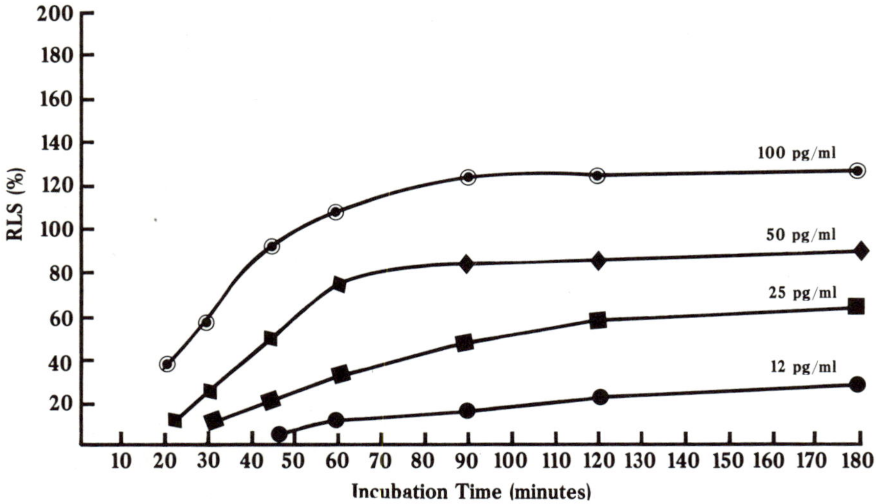

Figure 8. The effect of incubating four concentrations of endotoxin at 37°C from 0 to 180 minutes with lysate reconstituted in 5 ml of Tris-buffered 0.5% $MgCl_2$.

The nephelometric procedure that has evolved from these studies is as follows: 0.1 ml of test sample is added to 0.05 ml of lysate that has been reconstituted with 5 ml of 0.05M Tris buffer containing 0.23% $MgCl_2$ to give a pH of 7.4 and an ionic strength of 0.07 moles/liter. This reaction mixture is incubated for 30 minutes at 37°C, after which 0.2 ml of 1.0% SDS and 1 ml of 0.08% CMC are sequentially added. The mixture is then vortexed and read in a laser nephelometer that has been standardized with a known endotoxin. Table 1 represents typical data obtained when known endotoxin concentrations are assayed using this final nephelometric LAL system. As can be observed from the data obtained on three consecutive days, the assays are highly reproducible with an average slope of 1.55 and a coefficient of variation of 8.6% at the 50-pg/ml endotoxin concentration. Data of this nature produced consistently in our laboratory indicate that the nephelometric LAL assay system is a substantial improvement over existing test methods for the quantitation of endotoxin.

Table 1. Comparison of Results from Three Consecutive Days Using Optimized Nephelometric LAL Pyrogen Assay System

		Endotoxin Conc (pg/ml)					
	Sample	0	25	50	100	Slope	R^2 [a]
Day 1	1	1.1	17.9	68.3	153.2	1.583	0.9799
	2	4.7	25.6	72.7	150.5	1.505	0.9882
	3	0.4	18.2	69.5	157.3	1.631	0.9805
Day 2	1	0.7	25.7	75.0	149.4	1.528	0.9921
	2	0.0	26.1	79.4	154.5	1.590	0.9912
	3	0.1	26.8	75.2	149.3	1.529	0.9938
Day 3	1	1.7	31.1	80.2	151.8	1.533	0.9948
	2	1.3	33.2	86.4	150.2	1.519	0.9898
	3	0.0	29.7	87.3	151.14	1.555	0.9860
Mean			26.0	77.1	151.9	1.55	0.9879
Standard Deviation			5.2	6.7	2.6	0.04	0.0055
Coeff of Variation			20%	8.6%	1.7%	2.5%	0.55%

[a] R^2 = correlation coefficient indicating the linearity of the assay (1.0 suggests an ideal line).

Cooper JF, Hochstein HD, Seligmann EB Jr (1972). The Limulus test for endotoxin (pyrogen) in radiopharmaceuticals and biologicals. Bull Parenter Drug Assoc 26: 153-162.

Hollander VP, Harding WC (1976). A sensitive spectrophotometric method for measurement of plasma endotoxin. Biochem Med 15: 28-33.

Levin J, Bang FB (1968). Clottable protein in Limulus: Its localization and kinetics of its coagulation by encotoxin. Thromb Diath Haemorrh 19: 186.

Lowry OH, Rosebrough NJ, Farr AL, Randall, RJ (1951). Protein measurement with the Folin phenol reagent. J Biol Chem 193: 265.

Nandan R, Brown DR (1977). An improved in-vitro pyrogen test: To detect picograms of endotoxin contamination in intravenous fluids using Limulus amoebocyte lysate. J Lab Clin Med 89: 910.

Niwa M, Hiramatsu T, Waguri, O (1974). The gelation reactions between horseshoe crab amoebocytes and endotoxin studied by the quantitative clot protein method. Jpn J Med Sci Bio 27: 108-111.

Reinhold RB, Fine J (1971). A technique for quantitative measurement of endotoxin in human plasma. Proc Soc Expt Biol Med 137: 334-340.

Solum NO (1973). The coagulogen of limulus polyphemus hemocytes. A comparison of the clotted and non-clotted forms of the molecule. Thromb Res 2: 55.

The authors would like to thank Mr. R. Clark Brown, Ms. Kathy Hughes, and Mr. Noah Sackrison for their excellent technical assistance in providing the pH and ionic strength data presented in this report.

QUANTITATIVE METHOD FOR DETERMINING LESS THAN A PG/ML OF LPS

Frederica W. Valois

Woods Hole Oceanographic Institution
Woods Hole, Massachusetts 02543

The objective of our investigation was to modify the Limulus amebocyte lysate (LAL) method for the quantitation of bacterial biomass in oceanic environments. Although we could detect as little as 15 pg/ml of lipopolysaccharide (LPS) with the clot method, exact quantitation was not possible. To meet our needs we designed a spectrophoto-metric assay for use with seawater which allowed quantitation of LPS and in so doing found we could increase the sensitivity of the test to detect less than 0.1 pg LPS/ml. This same method can also be used for quantitation of LPS in other substances including body fluids and pharmaceuticals.

With the clot method equal quantities of LAL and sample are normally used (Levin and Bang, 1964). With the spectrophotometric method one part of LAL was added to five parts of sample (e.g., 0.2 ml LAL to 1 ml sample or 0.1 ml LAL to 0.5 ml sample). A turbid solution rather than a clot was formed after incubation of this mixture for 1 hr at 37°C. The turbidity which was directly proportional to the LPS concentration was measured in a spectrophotometer at 360 μm. Sample values were calculated from a standard curve which was run under the same experimental conditions as the sample.

All LAL used was frozen dried and obtained from the Associates of Cape Cod. The clotting sensitivity, based on the FDA endotoxin standard, for the lots employed ranged from 0.03-0.25 ng/ml. The endotoxin standard used was an

Pages 415–422, Biomedical Applications of the Horseshoe Crab (Limulidae)

Escherichia coli standard also obtained from the Associates of Cape Cod. The frozen dried LAL, which was packaged 5 ml/vial, was reconstituted with either 5 ml of pyrogen-free water (DPW) or depyrogenated seawater (DPSW) depending upon the experiment. The LAL used in our experiments had been extracted with chloroform to remove an inhibitor. It also contained 0.01 M $CaCl_2$, 3% NaCl and 0.35% human albumin. No buffering system was added since the LAL itself exhibits considerable buffering action. The most convenient spectrophotometer for reading turbidity with this test was found to be the Zeiss PM 2DL equipped with a rapid sampling device.

The LPS standard curve was generated by using concentrations of endotoxin ranging from 0.05-10 pg/ml, however the range of concentrations varied depending upon the sensitivity of the lot of LAL used. The standard curve was designed to include concentrations which bracketed the linear portion of the curve. Optimal linearity should cover a 10-fold range and this could be achieved with most lots of LAL tested. Figure 1 shows standard curves from a variety of LAL lots. Lots 207, 211, 114 and 359 which were

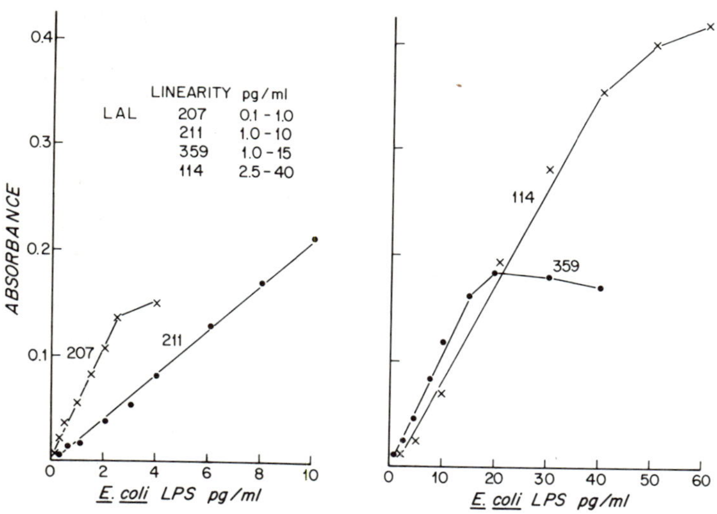

FIGURE 1. Comparison of sensitivity and linearity of 4 lots of LAL.

reconstituted with DPSW, had a range of linearity from 0.1 to 2.5 pg/ml to 2.5 to 40 pg/ml. Once past the linear portion of the reaction a plateau occurs which will eventually drop off with increasing concentration of LPS.

The sensitivity and linearity of each lot of LAL can be controlled to some degree by varying experimental conditions. Table 1 summarizes the conditions which affected the sensitivity of lot 207. The highest sensitivity was achieved when the LAL was reconstituted with DPSW. Addition of 0.1 M Mg^{++} to DPSW did not enhance the sensitivity but when 0.1 M Mg^{++} was added to LAL reconstituted in DPW the sensitivity of the LAL was increased 10-fold. Normally all dilutions of the endotoxin standard and samples were made with 3% NaCl. When the samples were diluted with saline rather than 3% NaCl, the sensitivity dropped slightly.

TABLE 1. FACTORS AFFECTING SENSITIVITY OF LAL

LAL Reconstituted With:	LPS Standard Diluent	Range of Linearity pg LPS/ml
DPSW	3% NaCl	0.1-1.0
DPSW	DPSW	0.1-1.0
DPSW with 0.1 M Mg^{++}	3% NaCl	0.1-1.0
DPSW	Saline	0.5-5.0
DPW	3% NaCl	3.0-20.0
DPW with 0.1 M Mg^{++}	3% NaCl	0.2-2.0

Figure 2 shows the changes that can be effected by increasing the ratio of LAL to sample. The greater the ratio the more sensitive the assay becomes. Also, the range of linearity increases as the ratio becomes larger.

The sensitivity and reproducibility of this assay is dependent on several factors including the ions present, time of incubation, dilution of LAL, and the lot of LAL used. Although the standard curve varies from lot to lot, reproducible standard curves for each lot can be obtained provided standard experimental conditions are employed.

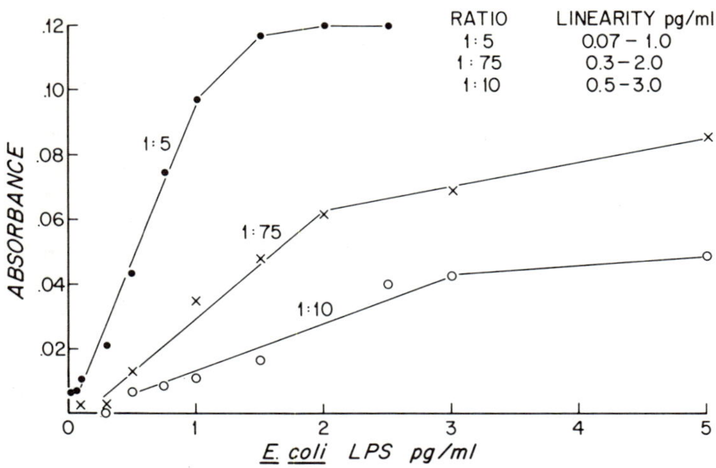

FIGURE 2. Sensitivity and linearity of the LAL test with varying LAL/sample ratios.

The spectrophotometric method is usually 13-20 times more sensitive than the clot method with the same lot of LAL, but on occasion has been over 100 times more sensitive than the clot method. Normally, the LAL that is used clots between 0.03-.25 ng/ml. However, some lots have clotted with as little as 7 pg/ml *E. coli* 2 reference endotoxin standard.

The smallest concentration of endotoxin measured with the spectrophotometric method is 20 fg/ml or 2 x 10^{-14} gm/ml. An *E. coli* cell in log phase has about 40 fg LPS and in theory we should be able to detect the presence of one cell/ml. However, in practice we have not been able to detect less than 5 bacteria/ml.

In running this spectrophotometric method a few precautions must be observed. Because of absorption problems LPS dilutions for the standard curve must be made just prior to use and kept on ice. These dilutions must not be kept longer than 2 hours. Reconstituted LAL will retain its activity if kept on ice during the day. It can be frozen for up to a week but should not be thawed more than once.

The fact that incubation time is critical is shown in Figure 3. Turbidity continues to increase even after a 60-minute incubation at 37°C. If the samples are placed in ice after the incubation period, the turbidity will still slowly increase. It is important that the incubation time be the same for all samples in a given set. Optimally the samples must be read within a 2 minute time period. If a large number of samples are being assayed, LAL addition must be staggered so all are incubated for the same period of time. For this reason we recommend that a spectrophotometer with a rapid sampling device be employed so up to 20 samples can be read in a 2 minute period.

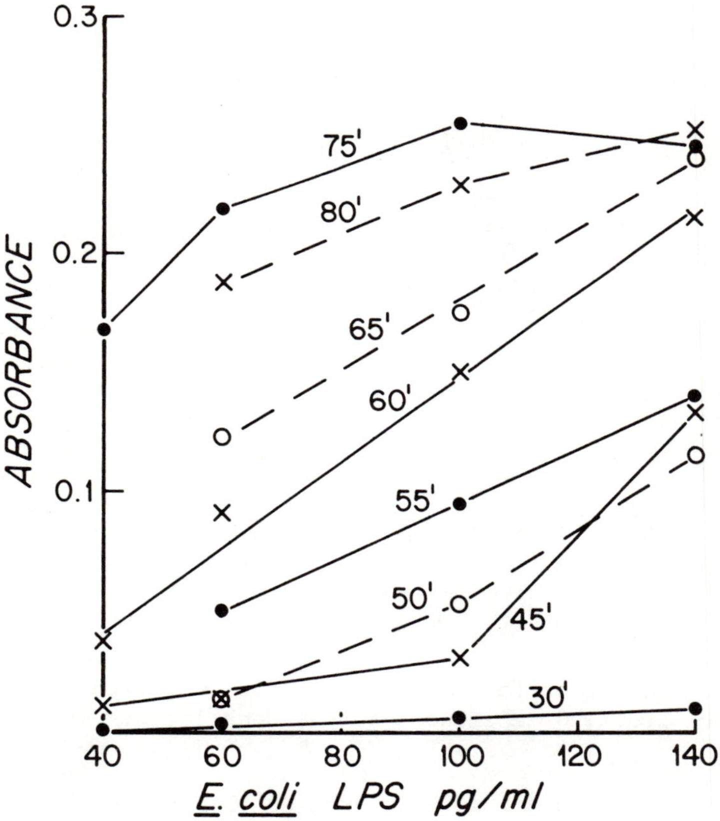

FIGURE 3. Effect of incubation period on sensitivity and linearity of LAL test.

To ascertain if the turbidimetric method could be used to measure the concentration of bacteria or bacterial biomass, the growth of *E. coli* was followed by direct counts, using a fluorescent technique, and by the LAL method as shown in Figure 4. Only the cellular bound LPS was measured in these experiments. Cellular carbon content was analyzed through various stages of the growth phase and a C/LPS ratio of 6.35 was derived (Watson, Novitsky, Quinby and Valois, 1977). A significant correlation between LPS concentration and bacterial number during the log phase of growth was found, however in the stationary phase of growth the cells became too fragile to measure accurately bound LPS.

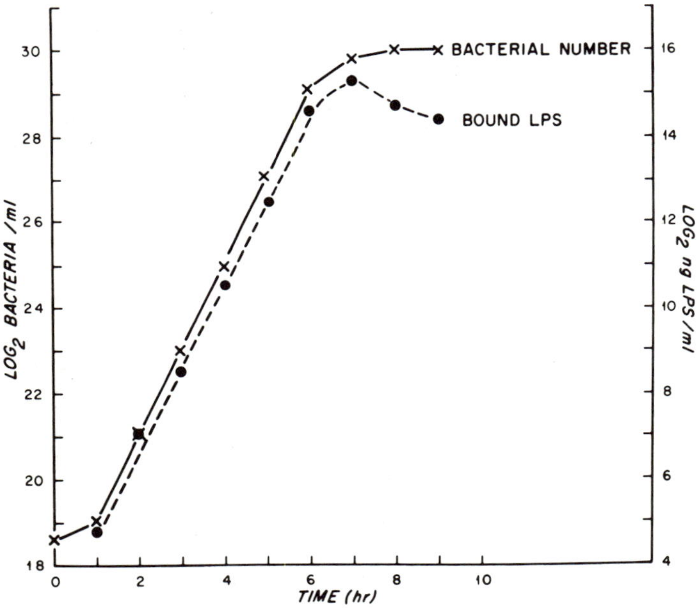

FIGURE 4. Growth curve of *E. coli* as followed with the LAL and direct counting (epifluorescence) techniques.

Figure 5 shows how we have applied this technique to determine the vertical distribution of bacteria throughout a 4200 m water column. In this study we compared direct counts of bacteria, again using a fluorescent technique, to the concentration of bound LPS in seawater samples and found a good agreement between the two methods. We feel that the LPS technique can be employed in ecological work to study the concentration and distribution of bacteria in aquatic environments.

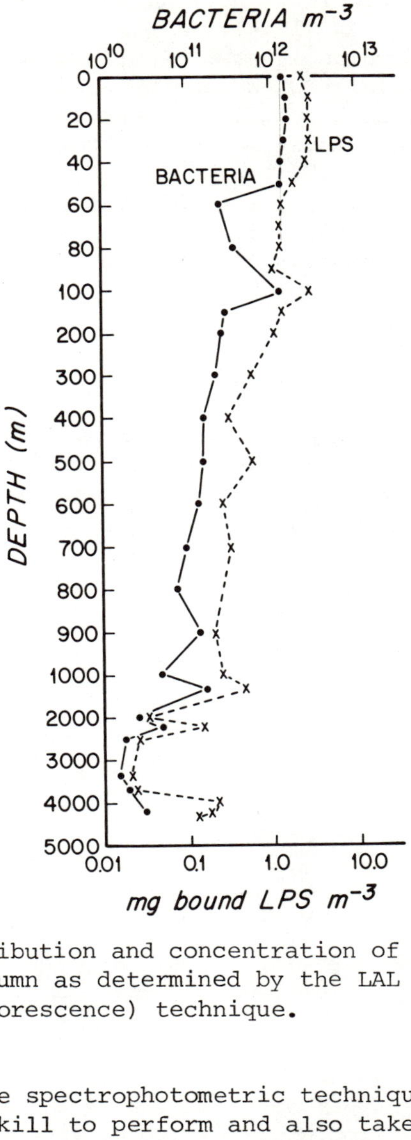

FIGURE 5. Distribution and concentration of bacteria in a 4200 m water column as determined by the LAL and direct counting (epifluorescence) technique.

Although the spectrophotometric technique requires more technical skill to perform and also takes more time to assemble than the clot technique, it has features which are useful. It can be used to measure quantitatively LPS and since it can achieve a high degree of sensitivity, femtogram concentrations of LPS can be determined.

REFERENCES

Levin J, Bang FB (1964). A description of cellular coagulation in the Limulus. Bull Johns Hopkins Hosp 115:337.

Sullivan J, Watson SW (1974). Factors affecting the sensitivity of Limulus lysate. Appl Microbiol 28:1023.

Sullivan JDJr, Valois FW, Watson SW (1976). Endotoxins: the Limulus amebocyte lysate system. In Bernheimer AW, (ed): "Mechanisms in bacterial toxinology," John Wiley & Sons Inc, p. 217.

Watson SW, Novitsky TJ, Quinby HL, Valois FW (1977). Determination of bacterial number and biomass in the marine environment. Appl Environ Microbiol 33:940.

ACKNOWLEDGEMENTS

This work was supported by a grant from the Hartford Foundation. Figure 4 was reproduced from Watson, Novitsky, Quinby, and Valois 1977. Appl. Environ. Microbiol. 33: 942.

A TURBIDIMETRIC *LIMULUS* AMEBOCYTE ASSAY FOR THE QUANTITATIVE DETERMINATION OF GRAM NEGATIVE BACTERIAL ENDOTOXIN

Joseph D. Teller and Kristine M. Kelly

Worthington Biochemical Corporation

Freehold, New Jersey 07728

INTRODUCTION

Our laboratory became involved with the Limulus lysate test five years ago when we were developing a process for the isolation of a bovine liver protein to be used as a parenteral anti-inflammatory drug. Since the early stages of the process were conducive to pyrogen formation, rigorous efforts had to be made to remove them and monitor for their presence at every stage of the process thereafter. This necessitated ordering many rabbit tests, incurring inordinate expense and delay. Notwithstanding these disadvantages, rabbit pyrogen testing has been one of the most valuable and reliable bioassays in all of pharmacology and we are in great debt to the early workers, Hart and Penfold in England and Florence Seibert in Philadelphia and to those far-seeing regulatory biologists who instituted the official rabbit testing of parenteral drugs in the early nineteen-forties. By setting the temperature limits which today are to be found practically unchanged in every pharmacopeia in the world, millions of patients have been protected against the unpleasantness of chills and fever and the danger of endotoxic shock. The beauty of the test resides in the remarkable similarity in response of rabbit and man as illustrated in the figure taken from Greisman and Hornick (1969) (Figure 1), showing that the threshold dose, while varying in absolute amount depending on the bacterial source of endotoxin, is essentially the same for both species. Yet, because it is a bioassay, great care must be taken to standardize animals and conditions and even at best, extraneous effects can cause problems. One unexpected variable

Pages 423–433, Biomedical Applications of the Horseshoe Crab (Limulidae)

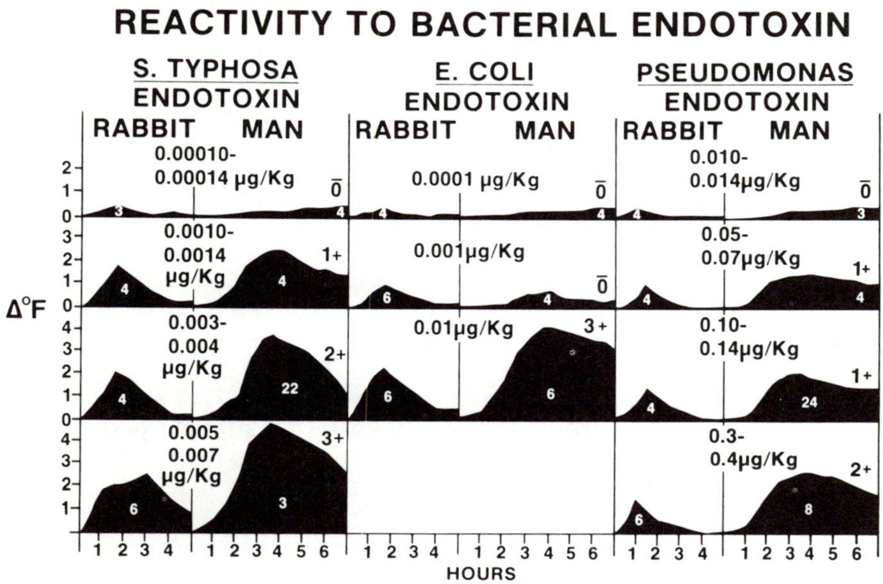

Figure 1

Taken from Greisman and Hornick, 1969. The numbers in white refer to the total number of subjects injected.

was the season of the year, reported in a study from Budapest by Simon et al (1976). Table 1 summarizes their results showing that as the year progressed through the season, singificantly different temperature elevations were obtained with the International Pyrogen Standard.

Who would have ever thought that an improvement over this excellent, albeit, imperfect test would come from the blood of the horseshoe crab? And isn't it a beautiful example of the application of fundamental biological research to an important problem in medicine and pharmacology?

Although the original form of the Limulus lysate test based on the formation of a gel is simple to perform, we were

SEASONAL VARIATION PYROGEN TESTING
DOSE 3.5ng/kg

NO. RABBITS	DATE	MEAN TEMP.
36	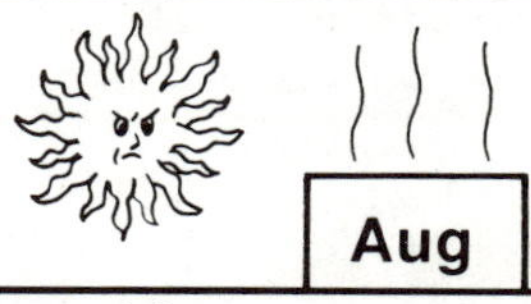 Aug	.286
36	 Jan	.478
27	 March	.574

Table 1

Adapted from Simon et al., 1976. "Mean Temp" refers to increase in deg. C.

interested in a more quantitative procedure, one that did not depend on the somewhat subjective decision of gel formation. We reasoned that if the lysate were diluted and reacted with endotoxin, it should become turbid rather than form a gel and that conditions could be established that would result in a graded turbidity directly related to the concentration of endotoxin.

MATERIALS AND METHODS

Limulus lysate was prepared essentially according to the original method of Levin and Bang (1968). All of the usual precautions were taken to maintain sterility of solutions and glassware. *E. coli* endotoxin preparations were obtained from Difco. FDA endotoxin standard was provided by the Bureau of Biologics and disposable plastic ware was used for endotoxin dilutions and lysate reaction tubes.

Lysate was diluted approximately five times with water and 0.1 ml endotoxin was added to 1.0 ml diluted lysate. After an hour at 37°C. a turbidity developed which was read in a spectrophotometer. The readings indicated that there was some relationship between turbidity and endotoxin concentration. However, examination of the suspension revealed a flocculent precipitate which did not lend itself to uniform dispersion.

Attempts were made to improve the milieu of the precipitate by adding various salts. 0.05M phosphate at pH 7.0 greatly improved the readings, lowering the blank substantially and improving the graded response. (Table 2) Halving the phosphate concentration improved the turbidity curve and lyophilized vials of lysate were prepared with phosphate such

LIMULUS LYSATE TURBIDITY
H_2O vs. PO_4 - pH 7.0

Endotoxin	Water		0.05M PO_4, pH 7.0	
	O.D.	Less Blk.	O.D.	Less Blk.
.000 μg/ml	.641	--	.283	--
.001 '	.645	.004	.329	.040
.005 '	.664	.023	.448	.165
.01 '	.660	.019	.496	.213
.05 '	1.137	.496	.606	.323

Table 2

that on reconstitution with 10.0 ml of water the final concentration was .025M phosphate. 1.0 ml aliquots when reacted with 0.1 ml of endotoxin at 37°C for one hour produced suitable standard curves after reading at 360 nm which was determined to be the optimum wavelength at which to read the absorbancy. While this formulation gave an acceptable endotoxin curve, when the lyophilized vials were stored at 40°C for three days (a heat stress test to allow for adverse shipping conditions) the reagent lost all of its endotoxin sensitivity. It was restored by substituting pipes buffer for phosphate and a combination of .025M phosphate and pipes was seen to improve the standard curve and lower the blank. (Table 3).

EFFECT OF PIPES BUFFER ON STABILITY OF LIMULUS LYSATE

Endotoxin ng/ml	ABSORBANCY AT 360nm					
	0.025M Phosphate		0.025M Pipes		0.0125M Phosphate Pipes	
	5°C	40°C	5°C	40°C	5°C	40°C
0.0000	.055	.147	.158	.167	0.123	.114
0.0625	.131	.146	.184	.182	0.167	.150
0.125	.190	.141	.212	.210	0.245	.202
0.25	.309	.145	.267	.300	0.406	.433
0.50	.431	.145	.498	.441	.506	.499

Table 3

Each vial of lysate is run with its own standard curve and unknowns and each lot must be standardized to pre-determined specifications for maximum blank reading and minimum absorbancies for corresponding endotoxin concentrations.

A typical acceptable standard curve and its application to the determination of endotoxin in an unknown sample is shown in Figure 2.

If a more sensitive reagent is required, the reaction may be amplified by using a more concentrated lysate. A lyophilized vial is reconstituted to only 1.0 ml and to 0.1 ml aliquots of concentrated lysate 1.0 ml endotoxin dilutions are added. An example of analyses performed on various samples of a laboratory water supply by this procedure is shown in Figure 3.

Several forms of injectable drugs obtained through retail commercial sources were tested. It was obvious that any materials which have an inherent turbidty or color should be diluted with water or preferably .025M phosphate - pipes buffer and the absorbancy value read and subtracted from

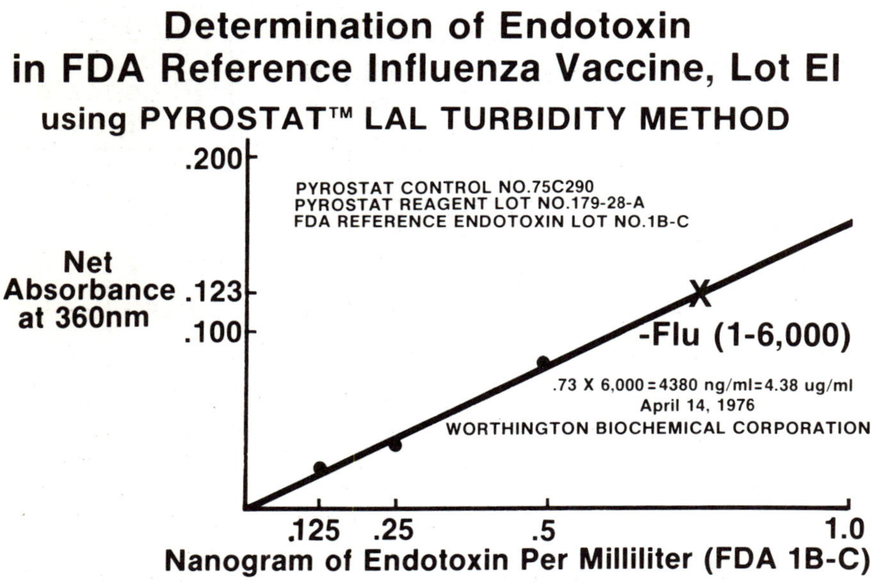

Figure 2

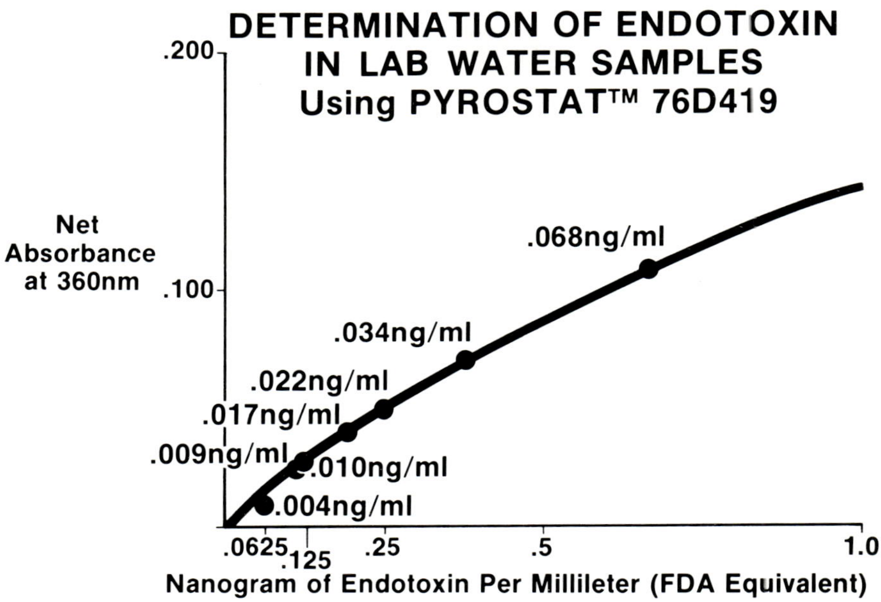

Figure 3

The values for endotoxin on the abscissa for the expanded method should be divided by 10.

the net lysate reaction turbidity reading. Four different lots of insulin were tested in this manner. Results as shown in Table 4 indicate a wide range of apparent pyrogen levels.

Returning to our original motive for this investigation, the preparation of a parenteral drug, we learned of some interesting findings during a human clinical trial. Two manufacturing lots, despite having met all of the required specifications including sterility and freedom from pyrogens by rabbit testing were reported as giving some patients local reactions of erythma and itching at the injection site. When these lots were tested by the Limulus turbidity method, they were found to be higher in endotoxin than any of the

PYROGENS IN INSULIN

Sample	Type	ng/ml (FDA)
A	U-40 GZ	18
B	U-80 Reg.	< 0.1
C	U-40 Reg.	< 0.1
D	U-40 Reg.	48

Table 4

lots that gave no reactions. The results are seen in Figure 4.

Thus, for this product we were able to use a reference point of 1.0 ng per ml to ensure freedom from a reaction that was unattainable with conventional pyrogen testing. Since it is unlikely that the endotoxin, per se, at this low value would give a local reaction, it is interesting to speculate that in light of the proposed mechanism of the Limulus lysate reaction, i.e. activation of a proteolytic zymogen, the endotoxin was activating either a trace of zymogen in the drug itself or an endogenous protease in the local tissue.

Since it was established rather early that the gel test was of significant value in the determination of cerebrospinal fluid (CSF) endotoxin, we wished to determine the usefulness of the the turbidity test with this type of

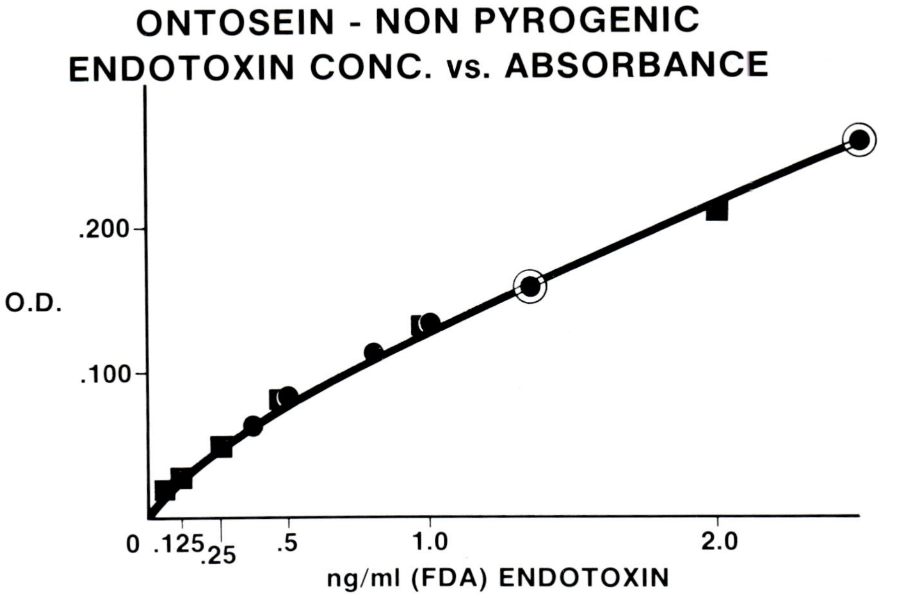

Figure 4

The squares are the standard endotoxin values, the circles are the values for six different drug lots. The double circles are the two lots that gave local reactions.

material. Since an inhibitor had been shown to be present in blood plasma, we looked for one in CSF by adding endotoxin to a series of patients' spinal fluids. We were surprised to find that in all cases our recovery figures revealed, on the contrary, an apparent activator of the reaction. Table 5 illustrates that the theoretically expected absorbancy value of 0.166 was increased approximately 2 fold in 9 out of 10 samples. To insure that this phenomenon was not peculiar to one type of purified endotoxin we tested the effect of CSF on five different endotoxin samples with essentially the same results (Table 6). What this activator may be and whether it is significant in the determination of the true CSF level by this method has not yet been determined.

Net O.D. 360nm of Different Spinal Fluids with and without 0.66ng/ml Endotoxin

Endotoxin Control (0.66ng/ml in water) O.D. - 0.144

Spinal Fluid	Without	With	Difference
1.	0.090	0.368	0.378
2.	0.018	0.344	0.326
3.	0.024	0.332	0.308
4.	0.197	0.388	0.191
5.	0.028	0.333	0.305
6.	0.035	0.353	0.318
7.	0.040	0.318	0.278
8.	0.063	0.386	0.323
9.	0.095	0.367	0.272
10.	0.051	0.361	0.310

TABLE 5

Reaction mixture contained 0.05 ml CSF which when reacted with lysate alone gave an absorbancy of only 0.015.

Effect of Spinal Fluid on Various Endotoxins

Interaction with LAL

		Net O.D. 360nm	
	ng/ml	without spinal fluid	with spinal fluid
E. coli 0127:B8	0.125	0.246	0.493
E. coli 0111:B4	0.125	0.175	0.307
S. typhosa 0901	0.25	0.244	0.461
FDA Klebsiella	1.0	0.272	0.435
Mallinckrodt's E. coli	0.25	0.294	0.680

Table 6

In the case of urine, recovery experiments also revealed appreciable activation, but again, its relationship to urine samples containing endotoxin derived from an infectious process remains to be investigated.

REFERENCES

1. Greisman, S. E. and Hornick, R.B. (1969) Comparative Pyrogenic Reactivity of Rabbit and Man to Bacterial Endotoxin. Proc. Soc. Exp. Biol. Med., 131:1154

2. Simon, S., Toth, M., Wallerstein, GY., Remeny, ZS (1976) Studies with the International Pyrogen Standard on the Sensitivity and Reproducibility of Pharmacopoeial Pyrogen Testing. J. Pharm. Pharmac. 28:111

3. Levin, J. and Bang, F.B. (1968) Clottable Protein in Limulus: its localization and Kinetics of its Coagulation by Endotoxin. Thrombos, Diathes, Haemorrh. 19:186

PROPERTIES OF LIMULUS AMEBOCYTE LYSATE AND THE TURBIDIMETRIC ASSAY FOR THE QUANTITATIVE DETERMINATION OF GRAM NEGATIVE BACTERIAL ENDOTOXIN

R.J.L. Bondar, J.D. Teller, A. Bowanko, K.M.Kelly

Worthington Biochemical Corporation

Freehold, New Jersey 07728

INTRODUCTION

Levin and Bang (1964) first demonstrated the use of horseshoe crab blood components to measure endotoxin. Subsequently Levin and Bang (1968) demonstrated that the endotoxin activated Limulus reaction can be followed by measuring a change in optical density (OD), however, the semi-quantitative gel method has won popular acceptance and extensive use. Although other quantitative methods intended for specific purposes have been published (Niwa *et al.*, 1974, Nandan and Brown, 1976, Hollander and Harding, 1976) they have not been used routinely. Pyrostat™ LAL (Limulus amebocyte lysate) reagent is a simple turbidimetric method for the quantitative detection of endotoxin wherein the turbidity formed is measured spectrophotometrically at 360nm. A standard curve, constructed by plotting OD vs. standard endotoxin concentration, and a blank, are used to provide precise and accurate results independent of vial to vial variation or alterations in activity due to aging or minor contamination. The versatility, reliability and some properties of this assay system are presented.

METHODS AND MATERIALS

All glassware was purchased pyrogen-free or was depyrogenated by heat treating at 175°C in a hot air oven for 2-8 hrs. Sterile disposable pipettes and culture tubes were purchased from Scientific Products and found to be pyrogen-free. Pyrogen-free water (McGaw) was used in all work.

Pages 435–451, Biomedical Applications of the Horseshoe Crab (Limulidae)

Pyrostat™ LAL reagent is a lyophilized preparation of *Limulus* amebocyte lysate containing 0.011 M PIPES (piperazine-N,N'-bis(2-ethanesulfonic acid)(Calbiochem) and 0.011 M sodium phosphate (Baker) buffers (pH 6.8) when fully reconstituted.

The detection of endotoxin with Pyrostat™ LAL reagent involves the addition of 0.1 ml sample or standard concentrations of endotoxin (prepared from Difco, W *E. coli* 127:B8 and included in the Pyrostat™ LAL reagent package) to 1.0 ml of the reconstituted LAL reagent and incubating the reaction mixture for one hour at 37°C. The turbidity formed at the conclusion of the incubation period is dependent upon the endotoxin concentration. The concentration of endotoxin in unknown samples is determined from a plot of OD_{360} vs. the standard endotoxin concentrations.

To determine the effect of various materials on endotoxin determination, each substance was added to the reaction mixture as indicated in the figures below. In all cases the LAL reagent composition is the same in the presence or absence of effectors in the final reaction mixture. The effectors used were commercially available disodium carbenicillin normal human serum albumin (U.S.P., salt-poor) and sodium chloride. None of these materials were further purified but preliminary studies indicated that they were relatively endotoxin-free.

To determine the effect of unwashed bacterial cells of *E. coli* K37, *S. aureus* Foggi and *S. facalis* (ATCC 12984) on the standard curve, 0.1 ml of the cell suspension and 0.1 ml of the endotoxin standards were added to 0.9 ml of the Pyrostat™ LAL reagent reconstituted to 9.9 ml instead of 11 ml which is the recommended procedure for routine use.

Precision studies were conducted with reference endotoxin (Worthington Biochemical Corp.) reference influenza vaccine lot E1 (FDA, Bureau of Biologics), globin zinc insulin USP (U-40) (Eli Lilly) and Iletin® insulin regular injection USP (U-40) (Eli Lilly).

RESULTS AND DISCUSSION

Time vs. OD as a Function of Endotoxin Concentration

The response to endotoxin of different LAL preparations can vary in the rate of turbidity formation, lag phase time, and maximum turbidity formed. The lot to lot differences in lag phase time and rate of turbidity formation can be seen in Figure 1. It shows the turbidity increase as measured by the OD at 360 nm as a function of incubation time at various levels of endotoxin for 3 different lots of LAL.

Lot C (Figure 1C) achieves maximum OD approximately twice as fast as lot A at 0.4 ng/ml endotoxin. Lot B (Figure 1B) is intermediate in its activity. The difference could be due to an inhibitor (Sullivan and Watson, 1974), greater concentration of reactive components such as the proenzyme or to activators in the LAL.

Figure 1 demonstrates that standard curves generated at 60 min for lot A and lot C would be unacceptable. In the first case the OD value for 0.05 ng/ml endotoxin would not be significantly different from the blank value and in the second the difference between 0.2 and 0.4 ng/ml would not be easily discriminated. The standard curve for lot B at 60 min would be acceptable by both criteria. On the other hand lot A would yield quantitative data at longer incubation times and lot C at shorter incubation times.

All three lots show that the lag phase and linear portion of the curves but not the maximum OD is dependent upon the endotoxin concentration. Given sufficient time and barring secondary reactions, the maximum OD for all endotoxin concentrations would be the same for a given lot. Some of the possible secondary reactions are inactivation over long periods of time at low endotoxin concentrations or resolublization of the turbid material.

Effect of Various Concentrations of PIPES and Phosphate Buffers on LAL

Buffer concentration has a marked effect on both the rate of turbidity formation and the maximum turbidity formed. Figure 2 shows that the amount of turbidity formed increases with higher buffer concentration. The effect, however, is endotoxin concentration dependent. At 0.4 ng/ml endotoxin the OD at 0.036 M is higher after 60 min. than at 0.045 M buffer. (See also Figures 3 and 4 discussed below).

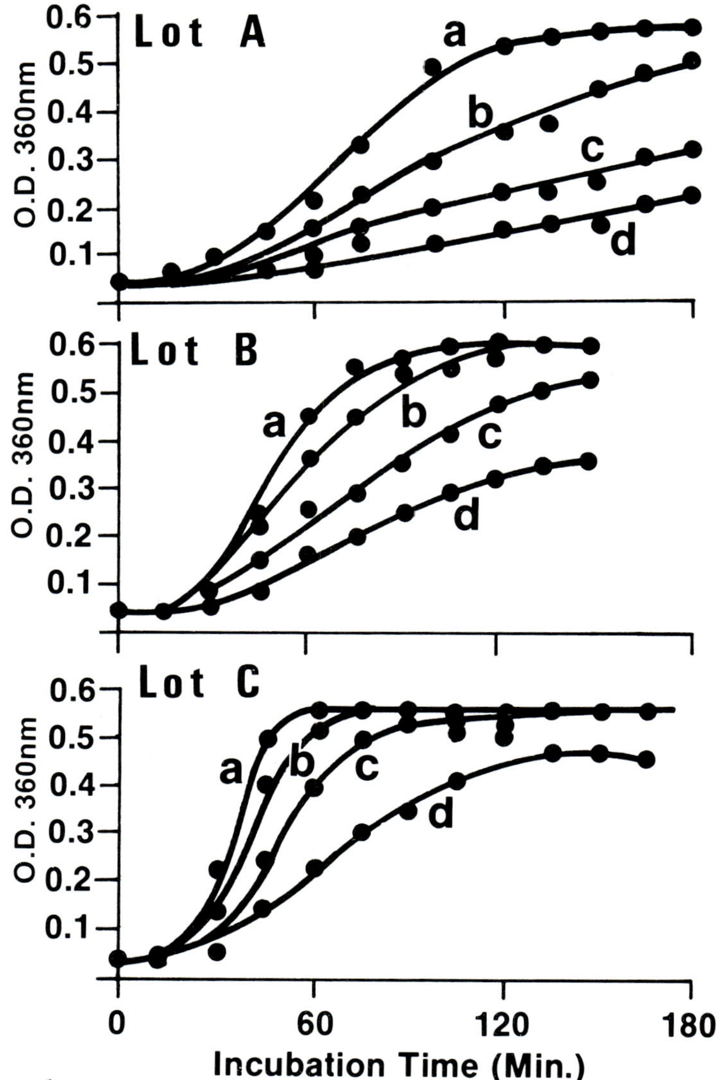

Figure 1

Time vs. OD as a Function of Endotoxin Concentration. Three lots of Pyrostat™ LAL reagents A, B, and C were incubated for various time periods at 37°C with standard endotoxin concentration (a = 0.4 ng/ml, b = 0.2 ng/ml, c = 0.1 ng/ml, d = 0.05 ng/ml). The assay conditions are described in Materials and Methods. The lot to lot variation is seen in the lag phase and rate of reaction as measured by the increase in OD at 360 nm.

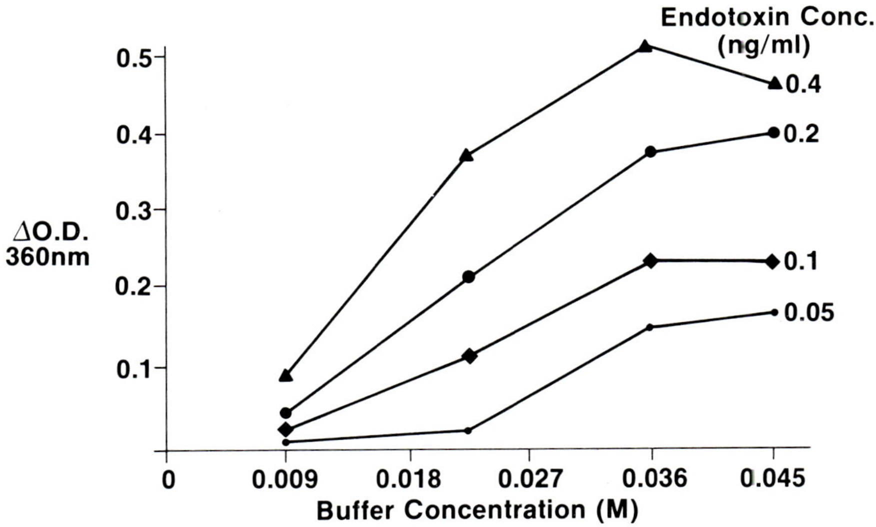

Figure 2

Effect of Buffer Concentration on Endotoxin Analysis. A 1:1 ratio of PIPES and phosphate buffers (pH 6.8) were incubated with endotoxin at the levels indicated. The assay conditions are described in Materials and Methods.

Effect of NaCl on Pyrostat™ LAL Reagent

Similar effects observed with increasing buffer concentration may also be seen with increasing NaCl concentration (Figures 3 and 4). Figure 3 represents a lot of Pyrostat™ LAL reagent with activity between that represented for lots B and C, (Figure 1) and Figure 4 would be between lots A and B, (Figure 1). In both cases the OD first increases with increasing concentration of NaCl and then decreases to the point where complete inhibition is observed at 0.3 M final test concentration. It is interesting to observe that the maximum OD obtained shifts to lower concentrations of NaCl as the concentration of endotoxin increases. Because results in Figure 2 imply a similar effect it is possibly a general ionic strength effect as opposed to a specific ion effect. This question is currently under investigation. Precisely what the mechanism is cannot be gleaned from this data. The important observation is that depending upon the salt concentration, both an activation and an inhibition can be observed with increasing or decreasing concentrations.

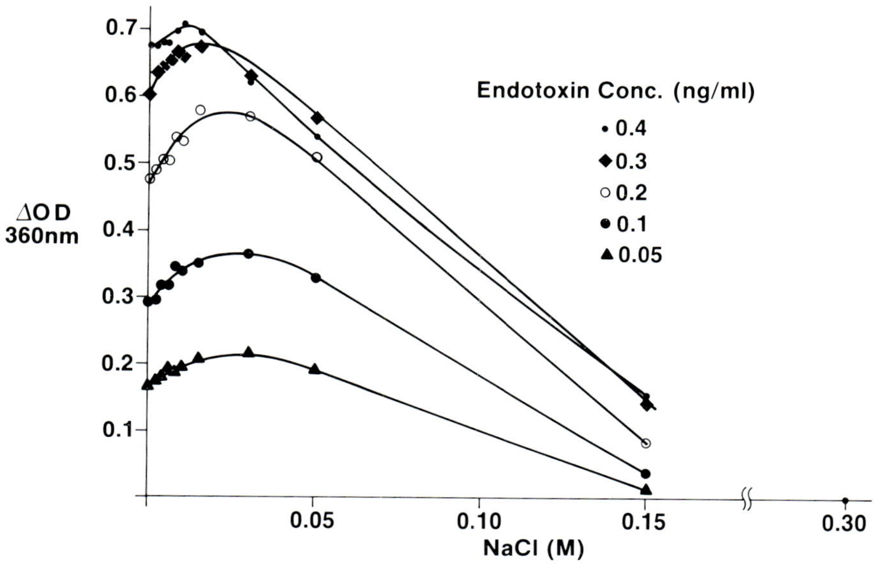

Figure 3

Effect of NaCl on Endotoxin Analysis. Endotoxin at the concentrations indicated in 0.1 ml and 0.1 ml of the appropriate NaCl concentration were added consecutively to 0.9 ml of Pyrostat™ LAL reagent reconstituted to 9.9 ml instead of 11 ml. The reaction mixture was incubated for 1 hr. at 37°C.

Effect of Disodium Carbenicillin on Pyrostat™ LAL Reagent

It has been reported (McCullough and Scolnick, 1976) that disodium carbenicillin inhibits the LAL gel test at a concentration (weight/total reaction mixture) of 0.5 to 1.0% even when 5 ng of endotoxin is present in the final reaction mixture. With the Pyrostat™ LAL reagent, sample concentrations as high as 4% can be analyzed within the range of 0.05 to 0.4 ng/ml of added *E. coli* endotoxin (Figure 5). Although ionic strength again may contribute to the decreased activity, the greater dilution of the

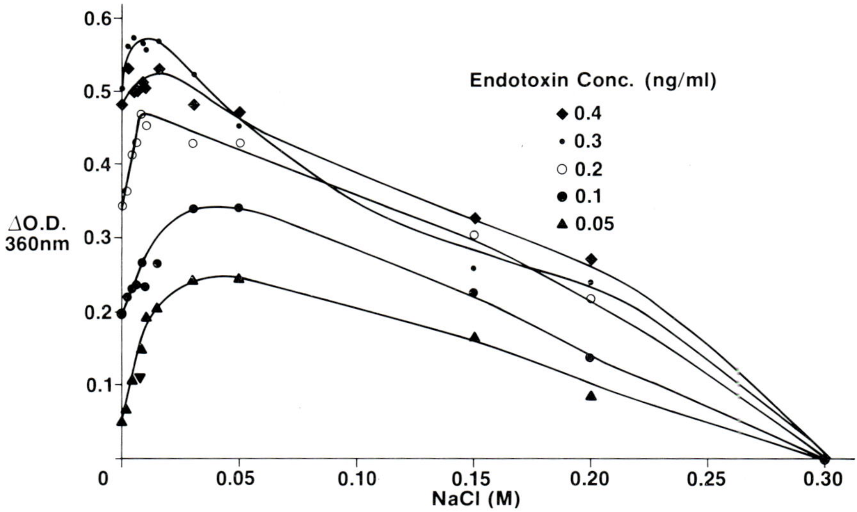

Figure 4

Effect of NaCl on Endotoxin Analysis. The conditions are indicated in Figure 3. Although the degree of activation and inhibition is different between Figure 3 and Figure 4 the qualitative effects are the same and complete inhibition occurs at 0.3 M NaCl in the reaction mixture in both cases.

sample for assay with the Pyrostat™ LAL reagent compared to the LAL gel test probably accounts for the higher concentration of carbenicillin assayable.

Effect of Human Serum Albumin on Pyrostat™ LAL Reagent

There are several reports (personal communication) that human serum albumin inhibits the LAL gel test. Again because of the sample to reagent ratio even 25% albumin can be assayed with the Pyrostat™ LAL reagent (Figure 6) albeit inhibition is clearly observed. This inhibition is partially reversible as noted in Figure 7. Endotoxin dissolved in pyrogen-free water and preincubated in an equal volume of 25% injectable human serum albumin for up to two days shows an apparent partial reversal of the inhibition which

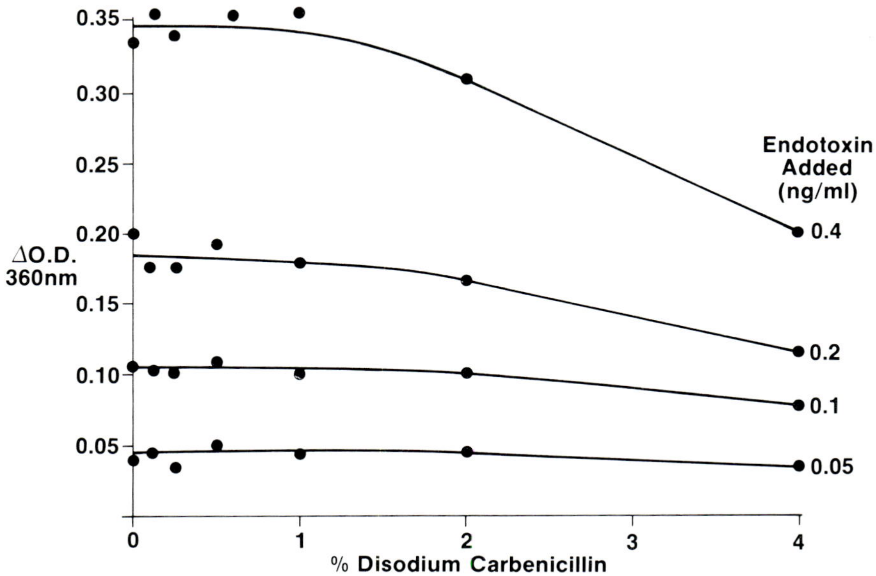

Figure 5

Effect of Disodium Carbenicillin on Endotoxin Analysis. The concentrations of disodium carbenicillin are expressed in % (W:V) in the sample volume. The assay condtions are described in Materials and Methods.

appears to plateau. This is graphically demonstrated for the 0.4 ng/ml level in Figure 7B. Precisely what is taking place and what effect the other components in the albumin solution have is unknown but it clearly indicates that the apparent level of endotoxin measured is subject both to matrix and temporal effects.

Effect of Gram Negative and Gram Positive Bacteria

It is well known that LAL reacts to Gram negative endotoxin. We wanted to investigate the potential effect of unwashed viable Gram negative and Gram positive bacterial cells. The data is shown in Tables 1, 2 and 3. In each case standard *E. coli* endotoxin was assayed at the levels

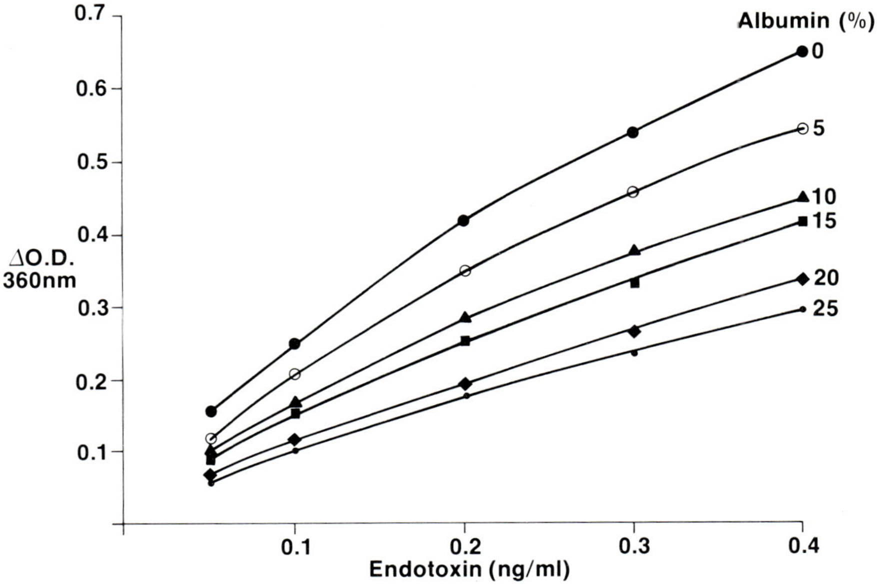

Figure 6

Effect of Human Serum Albumin on the Standard Curves. Endotoxin at the concentrations indicated in 0.1 ml and 0.1 ml of the appropriate albumin concentration in water were added consecutively to 0.9 ml of Pyrostat™ LAL reagent reconstituted to 9.9 ml instead of 11 ml and incubated at 37°C for 1 hour.

indicated in the presence of increasing concentrations of cells. With *E. coli* cells (Table 1) endogenous endotoxin at 10^3 cells/ml raises the blank above the limits set for the lot indicating that the assay should not be relied upon. Nevertheless an acceptable standard curve is obtained. At 10^4 cells/ml sufficient endogenous endotoxin is present to carry the reaction to completion preventing the analysis of additionally added endotoxin. This is shown as ΔOD values for all added endotoxin levels to be the same and close to zero. With *S. faecalis* (Table 2) and *S. aureus* (Table 3) contamination with up to 10^6 cells/ml shows no effect on the blank or generation of a standard curve. *S. aureus* at

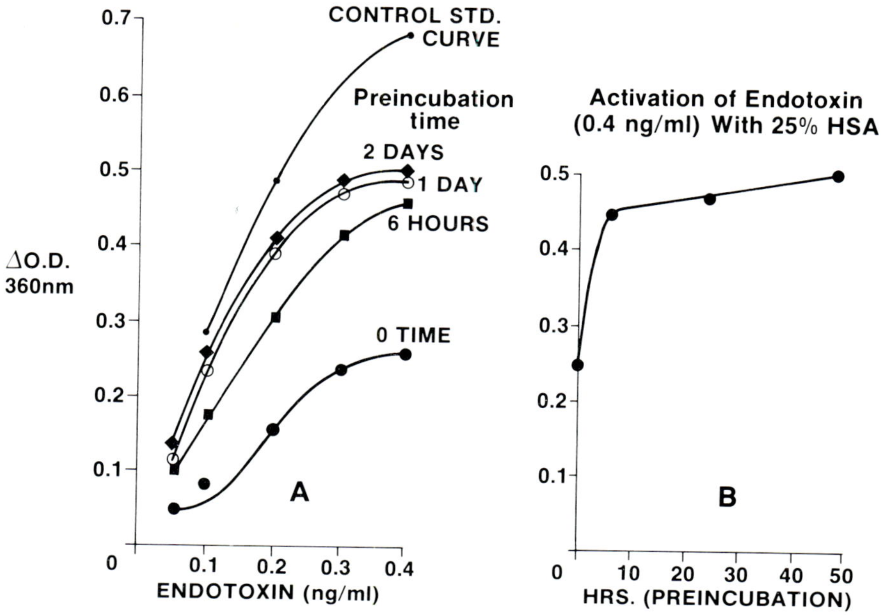

Figure 7

Effect of Preincubation of Endotoxin with 25% Albumin. The activating effect of preincubation of endotoxin with 25% albumin on the standard curves is seen in plot A. Plot B shows the progress of the activation as a function of preincubation time with 0.4 ng/ml endotoxin. Albumin at a concentration of 25% was preincubated in an equal volume of each of the endotoxin standard concentrations for various periods of time at 2-8°C. A volume of 0.2 ml of the mixture was added to 0.9 ml of Pyrostat™ LAL reagent reconstituted to 9.9 ml instead of 11 ml and incubated at 37°C for 1 hr.

higher levels, particularly 10^9 cells/ml, shows a positive effect on the blanks and a deleterious effect on the generation of the standard curve. This effect may be due to endogenous turbidity of the sample rather than actual reaction of Gram positive endotoxin with Pyrostat™ LAL reagent.

Precision of Pyrostat™ LAL Assay

TABLE 1

THE EFFECT OF *E. COLI* ON PYROSTAT™ LAL
Δ O.D. 360 nm

ENDOTOXIN (ng/ml)	Cells/ml									
	0		10		10^2		10^3		10^4	
	1*	2*	1*	2*	1*	2*	1*	2*	1*	2*
Blank	0.089	0.077	0.074	0.082	0.087	0.083	0.181	0.181	0.589	0.564
0.05	0.083	0.088	0.100	0.085	0.109	0.100	0.076	0.077	0.003	0.028
0.10	0.169	0.186	0.187	0.189	0.205	0.190	0.146	0.132	0.004	0.031
0.20	0.312	0.300	0.316	0.295	0.311	0.290	0.270	0.272	0.020	0.044
0.40	0.505	0.510	0.507	0.502	0.497	0.505	0.414	0.405	0.031	0.045

*These numbers indicate duplicate determinations.

TABLE 2

THE EFFECT OF *S. FAECALIS* ON PYROSTAT™ LAL

Δ O.D. 360 nm

ENDOTOXIN	Cells/ml						
(ng/ml)	0	10	10^2	10^3	10^4	10^5	10^6
Blank	0.072	0.072	0.072	0.072	0.071	0.072	0.071
0.05	0.092	0.103	0.103	0.116	0.117	0.122	0.129
0.10	0.198	0.198	0.197	0.201	0.196	0.202	0.196
0.20	0.291	0.302	0.301	0.320	0.308	0.323	0.318
0.40	0.483	0.468	0.482	0.478	0.465	0.473	0.475

TABLE 3

THE EFFECT OF *S. AUREUS* ON PYROSTAT™ LAL

Δ O.D. 360 nm

ENDOTOXIN (ng/ml)	Cells/ml 0		10^6		10^7		10^8		10^9	
	1*	2*	1*	2*	1*	2*	1*	2*	1*	2*
Blank	0.085	0.085	0.094	0.090	0.108	0.115	0.303	0.295	0.846	0.835
0.05	0.163	0.165	0.163	0.156	0.164	0.165	0.087	0.093	0.036	0.051
0.10	0.231	0.238	0.232	0.239	0.238	0.232	0.172	0.190	0.034	0.071
0.20	0.441	0.439	0.424	0.430	0.428	0.425	0.333	0.339	0.084	0.084
0.40	0.563	0.564	0.545	0.550	0.554	0.547	0.395	0.402	0.089	0.102

*These numbers indicate duplicate determinations.

The turbidimetric method has a distinct advantage over the gel test method in precision and accuracy. The gel test accuracy is limited by the extent of the dilution factor of standard endotoxin concentration whereas with the turbidimetric method the actual value is determined from the standard curve. The precision in determining the OD at different levels of endotoxin standards is shown in Figure 8. Next to each point is recorded the coefficient of variation of the OD measurement for five replicates. A single pool of LAL reagent and endotoxin was used to minimize vial to vial variation.

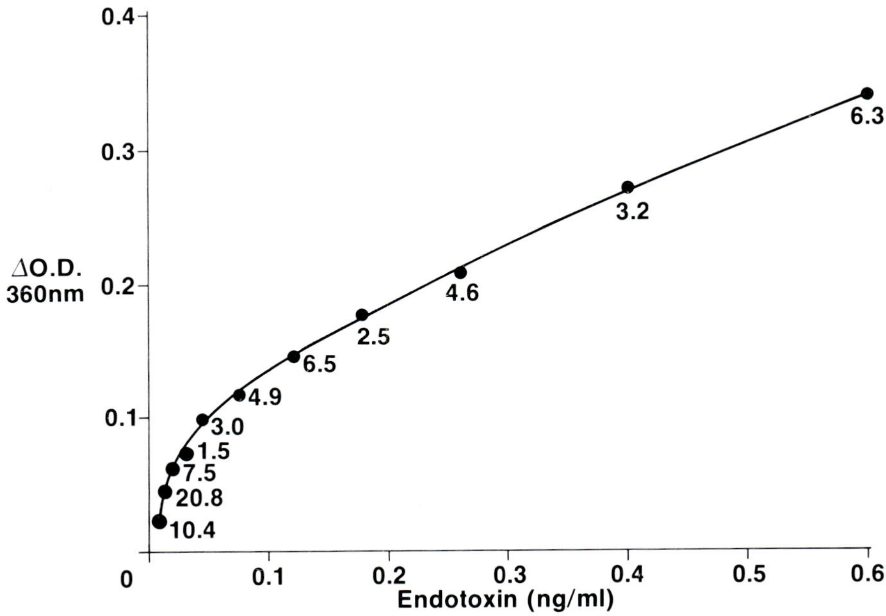

Figure 8

Coefficient of Variation of Standard Endotoxin. Endotoxin concentration is plotted as a function of ΔOD. Concentrations were assayed randomly five times. The number next to each point is the coefficient of variation for the measurement of optical density. All data was collected from a single pool of LAL reagent and endotoxin to minimize vial to vial variation. Each determination contained 0.1 ml endotoxin sample and 1.0 ml Pyrostat™ LAL reagent and was incubated at 37°C for 1 hr.

TABLE 4

PRECISION STUDY ON ENDOTOXIN CONTAMINATION
IN INSULIN SAMPLES

SAMPLE[1]	ng/ml ENDOTOXIN[2]					$\overline{X}$	S.D.	%C.V.
GZI_1[3]	0.90	0.90	0.84	0.81	0.80	0.850	0.0479	5.64
GZI_2	0.46	0.47	0.46	0.44	0.46	0.458	0.0109	2.39
I_1[4]	1.95	1.98	2.03	1.91	2.24	2.022	0.129	6.40
I_2	1.03	0.95	0.95	0.98	1.03	0.988	0.0402	4.07
I_3	0.59	0.54	0.52	0.52	0.52	0.538	0.0303	5.64

1. Subscripts designate different dilutions of the same test insulin
2. _E. coli_ results expressed in terms of FDA equivalent concentration
3. Globin zinc insulin injection, USP (U-40)
4. Iletin[R] insulin regular injection, USP (U-40)

The precision of the assay of endotoxin in unknown samples is presented for the case of insulin in Table 4. Five dilutions of two different kinds of insulin were each assayed five times. The precision varied from 2.4% coefficient of variation (CV) at 0.5 ng/ml to 6.4% at 2.0 ng/ml FDA equivalent endotoxin. A similar experiment was performed on three successive days. The overall average CV was 6.5% and varied from 1.2% to 14.6%. Although the standard curve generated from run to run, vial to vial, and lot to lot will vary, the endotoxin content of an unknown sample will be assayed consistently. Table 5 shows that the endotoxin level in FDA influenza vaccine lot E1 assayed with six different lots of Pyrostat™ LAL reagent yielded a mean value of 3 ng/ml with a 16% coefficient of variation.

TABLE 5

ENDOTOXIN CONTAMINATION IN FDA REFERENCE
INFLUENZA VACCINE LOT E1

PYROSTAT™ LAL REAGENT

Lot	μg ENDOTOXIN*/ml VACCINE
1	3.0
2	3.4
3	3.0
4	2.6
5	3.7
6	2.4

Mean = 3.0 μg/ml
Standard Deviation = ± 0.48
CV = 16%

*These endotoxin values are expressed in terms of FDA equivalent values.

Acknowledgements

We would like to thank Elizabeth Baluyot, Richard Rahfield and Lorraine Barolia for their invaluable technical assistance.

REFERENCES

Hollander VP, Harding WC, (1976). A sensitive spectrophotometric method for measurement of plasma endotoxin. Biochem Med 15:28.

Levin J, Bang FB, (1964). The role of endotoxin in the extracellular coagulation of Limulus blood. Bull John Hopkins Hosp 1:265.

Levin J, Bang FB, (1968). Clottable Protein in Limulus: Its localization and kinetics of coagulation by endotoxin. Thromos Diathes Haemrh 19:186.

McCullough KZ, Scolnick SA (1976). Effect of Semisynthetic Penicillins on the Limulus Lysate Test. Antimicrob Agents Chemother 9:856.

Niwa M, Hiramatsu T, Waguir O, (1974). The gelation reaction between Horseshoe crab amoebocytes and endotoxins studied by the quantitative clot protein method. Jap J Med Sci Biol 27:108.

Nandan R, Brown DR, (1976). An improved in vitro program test: To detect picograms of endotoxin contamination in intravenous fluids using Limulus amebocyte lysate. J Lab Clin Med 89:910.

Sullivan Jr JD, Watson SW (1974). Factors Affecting the Sensitivty of Limulus Lysate. Applied Microbiol 28:1623.

USE OF THE *LIMULUS* AMOEBOCYTE LYSATE ASSAY SYSTEM FOR DETECTION OF BACTERIAL ENDOTOXIN IN FLUIDS ASSOCIATED WITH HEMODIALYSIS PROCEDURES

Loretta A. Carson, Norman J. Petersen and Martin S. Favero
Hepatitis Laboratories Division, Center for Disease Control, 4402 North Seventh Street, Phoenix, Arizona 85014

Pyrogenic reactions in patients undergoing hemodialysis procedures are typically characterized by shaking chills, fever and hypotension. Other symptoms include headache, myalgia, nausea and vomiting, and leukopenia followed by leukocytosis. Onset of symptoms generally occurs within 1-3 hrs after the initiation of dialysis and, unless very severe, symptoms usually abate shortly after termination of dialysis. Laboratory and field studies conducted over the past decade have extensively documented the association between the occurrence of pyrogenic reactions in patients on maintenance dialysis and the levels of Gram-negative bacterial contaminants in fluids used in dialysis procedures (1,3,4,5,8,12,13,15,17). The use of the *Limulus* amoebocyte lysate (LAL) test has enabled us to more clearly define some of the factors compounding problems associated with microbial contamination of dialysis systems.

Hemodialysis systems currently used in hospital centers or for home patient care are basically of two types (Figure 1): single-pass systems which utilize a parallel-plate or hollow fiber membrane dialyzer, and recirculating single-pass systems which use a coil dialyzer. In dialysis procedures, incoming source water is heated and proportioned with a concentrated salt solution to obtain dialysate with electrolyte concentrations similar to plasma. In the single-pass system (Figure 2) dialysate is circulated through the dialyzer at a flow rate of approximately 300-500 ml/min. The blood compartment of the dialyzer is connected by arterial and venous blood lines to the patient via a fistula or shunt mechanism. Waste solutes from the

Pages 453–464, Biomedical Applications of the Horseshoe Crab (Limulidae)

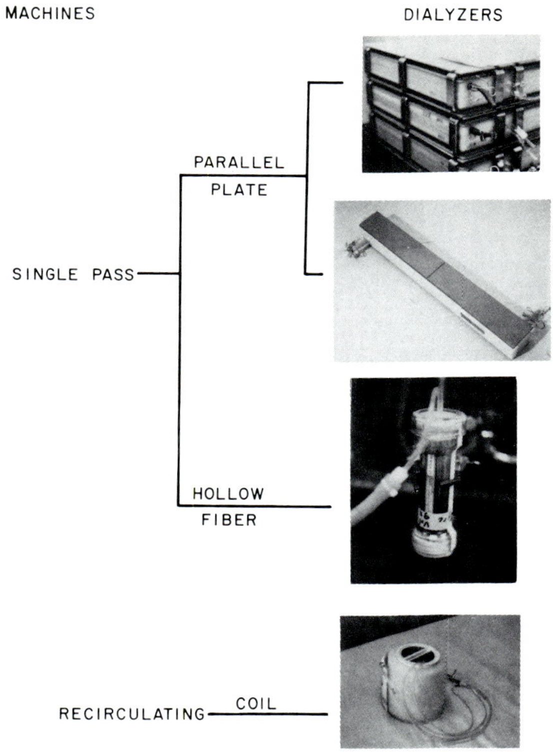

FIGURE 1. Types of machines and dialyzers used in hemodialysis procedures.

patient's blood pass by diffusion across the membrane, and dialysate is then exhausted to drain. In the recirculating single-pass system (Figure 3), dialysate is pumped first to a recirculating cannister. A portion of the dialysate is exhausted after passage through the coil dialyzer, and a portion spills back into the cannister to be recirculated through the dialyzer.

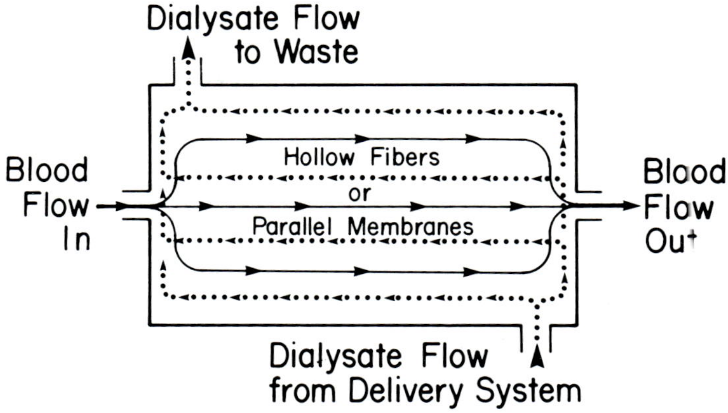

FIGURE 2. Diagram of dialysate flow in a single-pass hemodialysis system.

A. HEMODIALYSIS FLUIDS

Gram-negative "water bacteria" multiply readily in dialysate enriched by patient waste materials (3,4,5). Table 1 shows results obtained with *Pseudomonas, Flavobacterium, Bordetella* and *Yersenia* organisms grown in filter-sterilized dialysate collected during a patient hemodialysis treatment, in which population levels of 10^7-10^8 cells/ml were obtained.

TABLE 1. Comparative Growth of Gram-Negative Bacterial Organisms in Dialysis Fluids from a Patient Hemodialysis System

	Log Number of Viable Cells/ml				
Time (hr)	*P. cepacia*	*P. aeruginosa*	*Flavobacterium*	*Bordetella*	*Yersenia*
0	2.0	1.5	2.4	3.3	1.6
5	2.0	1.6	2.4	3.4	2.5
10	2.4	3.6	3.8	4.5	3.9
25	4.3	8.1	4.9	8.2	8.1
50	8.1	8.7	7.9	8.7	8.6

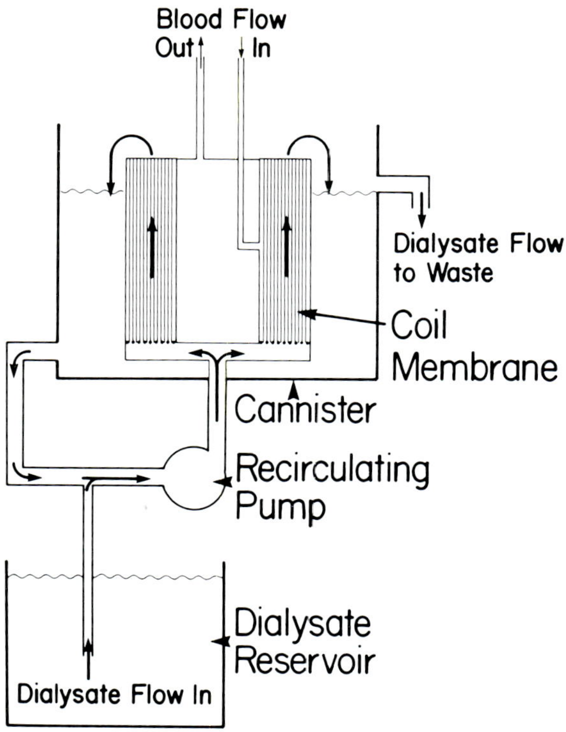

FIGURE 3. Diagram of dialysate flow in a recirculating single-pass hemodialysis system.

Endotoxin concentrations also were determined for these organisms, using the LAL clot formation assay procedure (18). Lysate utilized in these studies (Pyrotell; Associates of Cape Cod, Inc.) was capable of detecting 0.06-0.125 ng/ml of endotoxin equivalents of reference *E. coli* lipopolysaccharide (LPS; Difco). The data points shown in Figure 4 represent overall growth patterns and endotoxin production without reference to individual strains, and endotoxin concentrations are expressed as increases over baseline levels present in the filter-sterilized dialysate. Control spiked samples showed that the dialysate fluid

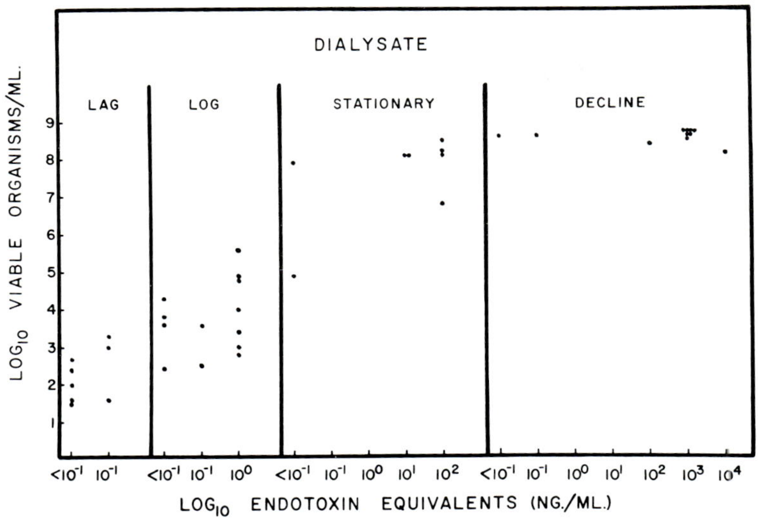

FIGURE 4. Production of endotoxin by Gram-negative bacterial organisms during growth in hemodialysis fluids.

itself was not inhibitory to lysate gelation. In general, endotoxin concentrations correlated with viable cell counts, although some organisms with maximum stationary population levels of 10^8 cells/ml showed <0.1 ng/ml increase in endotoxin equivalents. Levels of endotoxin as high as 10^4 ng/ml of endotoxin equivalents were ultimately attained. Griesman and Hornick (7) have reported that endotoxin doses as low as 1-10 ng/kg of body weight were capable of eliciting pyrogenic reactions in man, and the presence of endotoxin in dialysis fluids at levels of 10^3-10^4 ng/ml constitutes a greatly increased risk of pyrogenic reactions for patients undergoing hemodialysis. Free LPS, measured as the difference between endotoxin concentrations of culture filtrates and whole culture fluids, accounted for 40-50% of the total detectable endotoxin. Other investigators (9,10,16,18,19, 20) have reported that free LPS concentrations in cultures of bacteria grown in conventional laboratory media ranged from 20-60% of the total measurable endotoxin, and that free LPS may be present in the form of low molecular weight (10,000-50,000) subunits. Studies were conducted to further explore problem areas associated with sources of bacterial and endotoxin contamination in dialysis fluids.

B. WATER SOURCES AND WATER TREATMENT SYSTEMS

Most dialysis centers use water from public supplies, although occasional home units may operate on private well water. Deluzio and Friedman (2) found little or no endotoxin contamination in water systems which utilized deep wells as their source water, whereas endotoxin levels ranging from 1-800 μg/ml were found in surface waters. Jorgensen and coworkers (11) reported endotoxin concentrations as high as 500 ng/ml in samples from 10 public drinking water systems throughout the U.S. Assays of tap water samples collected over a 3-month period at our laboratories and other locations in the Phoenix metropolitan area revealed that endotoxin was invariably present at levels up to 100 ng/ml. However, endotoxin concentrations did not appear to be related to total viable cell counts. We found levels of endotoxin ranging from 5-40 ng/ml in samples containing $<10^3$ viable cells/ml and, conversely, <1 ng/ml in samples with $>10^3$ cells/ml. It was further found that although treatment of raw water by coagulation, sedimentation, filtration and chlorination effected a 3-log reduction in the mean total viable count, the median endotoxin concentration was reduced by only 1 log.

Typically, source waters in dialysis centers are further treated to remove chemical and other impurities. The treatment varies from center to center and consists of one or more of the following: filtration, ion-exchange and reverse osmosis (R.O.). Based on the results of studies conducted in a variety of hemodialysis centers in Phoenix as well as in six other states, some general observations can be made: 1) Gram-negative bacteria are capable of growth in the treated waters used to prepare dialysis fluids, although when compared to levels observed for dialysate, lower maximum population levels of 10^5-10^6 cells/ml are typically obtained in softened, deionized or R.O. waters. As we observed previously for patient dialysate fluids, concentrations of bacterial endotoxin in these treated waters will range from 1-1,000 ng/ml, depending on the levels, and types, of contaminants present. 2) Filtration is essentially ineffective in reducing levels of endotoxin. Particulate filters (prefilters) notably do not remove bacteria or bacterial endotoxin, but in fact become colonized and actually amplify levels. The majority of filters examined which contain activated carbon similarly increase levels of Gram-negative bacteria and do not remove endotoxins. Absolute filters will reduce bacterial levels but not endotoxin, and when not routinely disinfected these filters also become colonized, and can thus become foci for dissemination of organisms. 3) Water softeners and deionizers are readily colonized, allowing for rapid growth and endotoxin production by Gram-negative water bacteria. 4) R.O. appears to be the most effective means presently in use for treating water for dialysis procedures.

R.O. units used in dialysis centers employ either spiral wound membranes or hollow fibers as semipermeable barriers to reject unwanted chemicals and other contaminants in supply water. Studies of the two types of units were conducted in our laboratories, and Figure 5 shows the mean values of total viable cell counts from microbiological assays performed on supply water just prior to entering the R.O. modules and product water leaving the modules. The mean bacterial rejection rate for the two units tested was calculated based on the level of bacteria in the supply and product water after disinfection, when colonization of the units would be minimal. The respective rates of rejection when challenged with high levels of bacteria (10^5-10^6 cells/ml) in the source water were 99.998% for the membrane unit and 99.965% for the fiber unit. Translated into more practical terms this means that the treatment of a water supply containing from 10^2-10^3 bacteria/ml will theoretically produce a product water containing 1.7-17 bacteria per liter. However, the hollow fiber unit contained an activated carbon prefilter to prevent chlorine deterioration of the fibers, which appeared to amplify the bacterial level in the incoming source water. While the bacterial rejection rates appear comparable, product water from the hollow fiber unit invariably contained significantly higher bacterial levels than product water from the membrane unit. Incidentally, rejection of bacteria was not related to rejection of total dissolved solids (TDS). Mean rejection rates of TDS, which averaged 650 ppm during the period of the experiment were 97.5% for the hollow fiber unit and 91.5% for the membrane unit. It can also be seen that when these units are not regularly or adequately disinfected, "grow-through" or

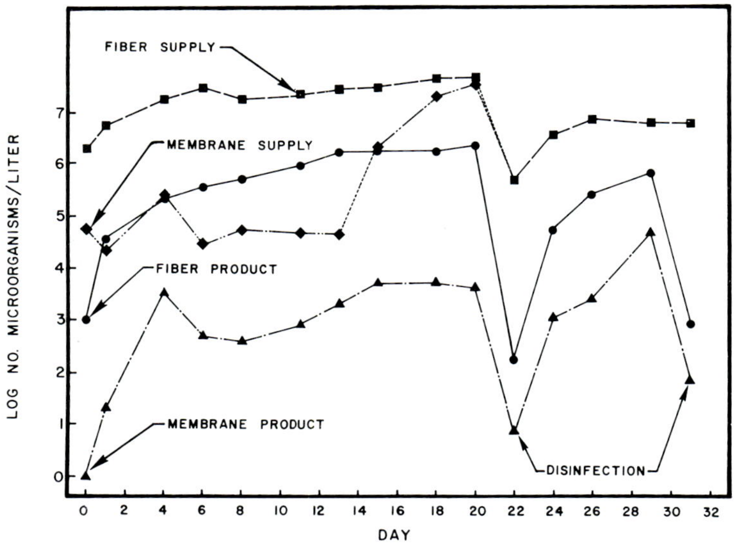

FIGURE 5. Comparison of the microbiological quality of product waters from hollow-fiber and spiral-wound membranes of reverse osmosis units.

penetration of the membranes can readily occur. The presence of even small numbers of bacteria on the downstream side of the membrane can constitute a potentially serious problem because of the ability of these microorganisms to multiply rapidly in water environments.

Assays for endotoxin in supply and product water were performed in conjunction with total viable count assays. In 30 samples of water supplying either R.O. unit, levels of endotoxin ranged from 1-100 ng/ml. No endotoxin was detected in any of the product water samples from either the hollow fiber or spiral-wound membrane units when units were adequately disinfected. When the downstream side of the membrane or fiber units became colonized in the absence of disinfectant, bacterial endotoxin ranging from 1-100 ng/ml was detected in the product water.

C. FLUID DISTRIBUTION SYSTEMS

To determine the influence of distribution piping on the microbiological quality of water, 65-foot lengths of ¼" and 1" PVC pipes were colonized with Gram-negative bacteria. Water inoculated with natural contaminants from dialysis fluids was allowed to remain stagnant in the pipes for several days. Levels of 10^6-10^7 cells/ml were attained. The pipes were then attached to the product water line of a disinfected R.O. unit, and the system was operated 6 hrs/ day, 5 days/week for 3 weeks, with each pipe discharging 500 ml/min. Results of microbiological assays showed that the mean level for 8 samples from the small pipe was 2.7 x 10^6 bacteria/liter compared to a value of 8.3 x 10^6 for the large pipe. Figure 6 shows the frequency distributions of endotoxin in the pipe influent and effluents. The influent graph shows only an occasional occurrence of endotoxin, which originated in the absolute filter. The small pipe shows the greatest tendency to discharge water containing endotoxin and the fewest samples free of endotoxin. These data show that colonized pipes, even when flushed with product water from R.O. units containing no endotoxin or low levels of endotoxin, can constitute a major source of endotoxin in dialysis systems.

Questions concerning the etiology of pyrogenic reactions in patients on maintenance dialysis still have not been fully resolved. Although bacteremias and septicemias associated with pyrogenic reactions in hemodialysis patients have been reported, these have been assumed to be the result of membrane leaks during dialysis in which the integrity of the dialysis membrane or its housing was impaired. It has also been largely assumed that intact dialysis membranes provide adequate physical barriers to the passage of bacteria or bacterial endotoxins. Results of studies reported here have shown that these assumptions are no longer tenable. Quarles and coworkers (14) also showed that products from growing cultures of *Serratia marcescens* were able to pass intact membranes of a commercial capillary cartridge dialyzer, causing acute signs of toxemia, pyrexia and leukopenia in goats, and Gazenfeldt and Eliahou (6) found that patients on maintenance hemodialysis developed high titers of specific antibodies to the bacterial flora in dialysis fluids.

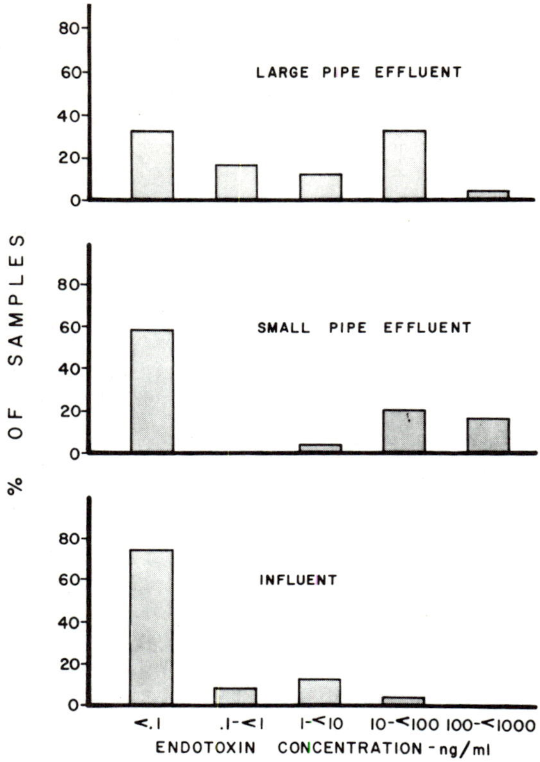

FIGURE 6. Effect of inside diameter on the level of bacterial endotoxin in water discharged from colonized 65-foot plastic pipes.

Microbiological assays of hemodialysis fluids aimed only at enumerating viable organisms may thus fall far short of reflecting the quality of dialysis fluids in terms of their potential for eliciting pyrogenic reactions. The LAL test has provided a significant tool both in a qualitative and a quantitative sense among procedures used to define and reduce potential health risks for hemodialysis patients.

REFERENCES

1. Dawids SG, Vejlsgaard R (1976). Bacteriological and clinical evaluation of different dialysate delivery systems. Acta Med Scand 199:151.
2. DeLuzio NR, Friedmann TJ (1973). Bacterial endotoxins in the environment. Nature 244:49.
3. Favero MS, Petersen NJ, Boyer KM, Carson LA, Bond WW (1974). Microbial contamination of renal dialysis systems and associated health risks. Trans Amer Soc Artif Int Organs XX:175.
4. Favero MS, Carson LA, Bond WW, Petersen NJ (1974). Factors that influence microbial contamination of fluids associated with hemodialysis machines. Appl Microbiol 28:822.
5. Favero MS, Petersen NJ, Carson, Bond WW, Hindman SH (1975). Gram-negative water bacteria in hemodialysis systems. Health Lab Sci 12:321.
6. Gazenfeldt-Gazit E, Eliahou HE (1969). Endotoxin antibodies in patients on maintenance hemodialysis. Israel J Med Sci 5:1032.
7. Greisman SE, Hornick RB (1969). Comparative pyrogenic reactivity of rabbit and man to bacterial endotoxin. Proc Soc Exp Biol 131:1154.
8. Jans H, Bretlau P, Nielsen B (1978). Bacteriological contamination of dialyzers. Nephron 20:10.
9. Kass EH, Wolff SM (1973). Bacterial lipopolysaccharides: Chemistry, biology and clinical significance of endotoxins. J Infect Dis 128:Supplement.
10. Jorgensen JH, Smith RF (1974). Measurement of bound and free endotoxin by the *Limulus* assay. Proc Soc Exp Biol Med 146:1024.
11. Jorgensen JH, Lee JC, Pahren HR (1976). Rapid detection of bacterial endotoxins in drinking water and renovated wastewater. Appl Environ Microbiol 32:347.
12. Petersen NJ, Carson LA, Favero MS (1977). Microbiological quality of water in an automatic peritoneal dialysis system. Dial Transplant 6:38.
13. Petersen NJ, Boyer KM, Carson LA, Favero MS (1978). Pyrogenic reactions from inadequate disinfection of a dialysis fluid distribution system. Dial Transplant 7:52.
14. Quarles JM, Belding RC, Beaman TC, Gerhardt P (1974). Hemodialysis culture of *Serratia marcescens* in a goat-artificial kidney-fermentor system. Infect Immun 9:550.

15. Raij L, Shapiro FL, Michael AF (1973). Endotoxemia in febrile reactions during hemodialysis. Kidney Internat 4:57.
16. Ribi E, Anacker RL, Brown R, Haskins WT, Malmgrin B, Milner KC, Rudbach JA (1966). Reaction of endotoxin and surfactants. J Bact 92:1493.
17. Robinson R, Rosen S (1971). Pyrexial reactions during hemodialysis. Br Med J 1:528.
18. Sullivan JD, Valois FW, Watson SW (1976). Endotoxins: The *Limulus* Amoebocyte Lysate system. In Bernheimer AW (ed): "Mechanisms in Bacterial Toxinology," New York: John Wiley & Sons, Inc., p 218.
19. Sweadner KJ, Forte M, Nelsen LJ (1977). Filtration removal of endotoxin (pyrogens) in solution in different states of aggregation. Appl Environ Microbiol 34:382.
20. Work F (1971). Production, chemistry and properties of bacterial pyrogens and endotoxins. In Wolstenholme GEW, Birch J (eds): "Pyrogens and fever," Edinburgh: Churchill Publishing Co, p 23.

CHARACTERIZATION OF AN ENDOTOXIN-LIKE SUBSTANCE RECOVERED FROM AN INTRAVENOUS ANTIBIOTIC SOLUTION INVOLVING PYROGENIC REACTIONS

Anita K. Highsmith, Roger L. Anderson,
C. Michael West and Richard E. Dixon
Bacterial Diseases Division, Bureau of Epidemiology, Center for Disease Control, Public Health Service, U. S. Department of Health, Education, and Welfare, Atlanta, Georgia 30333

On the evening of July 22, 1977, an outbreak of pyrogenic reactions occurred at a large referral hospital. The reactions were associated with intravenous (IV) administration of a commercial preparation of cephalothin. Because of the possibility that a commercial product which is distributed nationwide might be contaminated, the Center for Disease Control (CDC) was invited to participate with the Food and Drug Administration (FDA) in an investigation. This paper will present the epidemiologic aspects and the laboratory findings of the outbreak.

EPIDEMIOLOGIC INVESTIGATION

The outbreak was initially recognized by a nurse on a surgical unit. She was notified that a patient had suddenly developed fever and chills. While attending to this individual, who had become hypotensive and appeared to be in shock, similar changes were noted in another patient in a nearby room who had also received cephalothin. Within the next hour, 5 of 25 postoperative patients on this ward had abrupt onset of fever, rigors, and hypotension. The only common factor was that each had just received an IV infusion of cephalothin.

A case was defined by the sudden onset of rigors, fever >102°F, tachycardia >120 beats per minute, and hypotension, all of which were unexplained by underlying illness. There were 16 cases, including 15 in adults and 1 in a child, located throughout the hospital. Many of the

Pages 465–471, Biomedical Applications of the Horseshoe Crab (Limulidae)

affected patients were without symptoms by the day after the outbreak, and all patients were fully recovered at the time of the hospital discharge.

An endotoxin-like substance was detected in the IV antibiotic infusion fluids implicated, in the febrile response of these 16 patients. A series of laboratory analyses were designed to characterize this substance and evaluate its possible sources.

Products to which case patients had been exposed are listed in Table 1. Commercial Sterile Water for Injection, USP* (SWI) (Travenol Laboratories, Inc., Deerfield, Illinois) was used to reconstitute 20-g vials of powdered cephalothin (Eli Lilly & Co., Indianapolis, Indiana). Aliquots of the resulting concentrated cephalothin solution were then added to 50-ml bags of 0.9% Sodium Chloride, Injection, USP (NS) (Travenol Laboratories, Inc., Deerfield, Illinois).

Table 1. Products Examined as Possible Sources of Contamination

Sterile water for injection
Cephalothin (unreconstituted)
0.9% Sodium chloride solution in 50 ml bags
Cephalothin in normal saline prepared for administration

LIMULUS AMEBOCYTE LYSATE ASSAY AND RABBIT PYROGENICITY TEST

Samples of implicated and non-implicated IV fluids were examined for activity in the *Limulus* amebocyte lysate assay (LAL) and the rabbit pyrogen test (RPT). There was complete correlation between the results of both the LAL and the RPT which also agreed with clinical data from case patients.

LAL tests were performed using Pyrotell (Associates of Cape Cod, Inc.),[1] and *Escherichia coli* 0111:B4

---

*Use of trade names is for identification only and does not constitute endorsement by the Public Health Service, U. S. Department of Health, Education & Welfare.

lipopolysaccharide as a positive control. Fluid to be tested was incubated for 1 h at 37°C. A test result was read as positive only when the resulting clot remained firm after 180° inversion.

Rabbit temperature elevations ranged from 0.6°C to 1.2°C fulfilling criteria for positive RPT.(2) After these initial tests demonstrated that the LAL reaction was equal to the RPT for detecting contamination, an additional 182 bags of prepared, but unused, antibiotic IV fluids were examined by the LAL. Thirty bags were found to be positive for an endotoxin-like substance. LAL-positive samples contained from 4 to 100 ng/ml of this endotoxin-like substance. Three trials with fluid samples examined blindly over a 2-month period demonstrated complete reproducibility of the LAL.

Cephalothin in the test samples inhibited clot formation in the LAL test, and the degree of inhibition was related to cephalothin concentration (Table 2). Sixty-six samples were *Limulus* positive and were marked as containing 250 mg or 2 g of cephalothin. However, if these samples had been screened using the undiluted cephalothin IV fluid, as available from the pharmacy, only 13 (20%) would have been positive. A rapid screening of undiluted IV fluid would not have been an adequate procedure due to the apparent inhibitory activity of cephalothin at these concentrations. Therefore, all samples examined in this study were diluted to at least 1:8 using 2-fold dilutions.

Table 2. Effect of Diluting Cephalothin for IV Administration on the Results of *Limulus* Lysate Assay

No. Samples Tested	Conc.	Dilution 1:2	1:4	1:8
13 (20%)	+	+	+	+
42 (63%)	-	+	+	+
9 (14%)	-	-	+	+
2 (3%)	-	-	-	+

Total 66 (100%)

Endotoxin activity was evaluated in control IV fluid with known cephalothin concentrations, with and without commercially prepared endotoxin. IV fluid was assayed at the time of preparation and at intervals of up to 9 months. Cephalothin alone did not respond to the LAL in these control tests. Cephalothin inhibited LAL in 53 (80%) of known endotoxin-positive specimens.

LAL-positive samples were heat stable when treated at 56°C for 5 or 15 minutes, and at 121°C (autoclaved) for 15 minutes.

MICROBIOLOGIC CULTURING OF INFUSION FLUIDS

All available Limulus positive infusion fluids available from the outbreak, and the sterile distilled water, cephalothin and normal saline were examined for bacterial growth. Aliquots of equal volumes were inoculated into sterile distilled water and brain heart infusion broth, and incubated at 4, 25, and 37°C. Ten ml aliquots were inoculated as TSA pour plates, incubated at 37°C, and 2 milliliters of fluid were streaked directly onto chocolate agar with 2% IsoVitaleX and incubated 37°C, 2-5% CO_2. All cultures were incubated a minimum of 14 days and found to be negative for bacterial growth or visible turbidity.

Gas-liquid chromatography determinations for the presence of volatile fatty acids and amines of low molecular weight did not reveal any differences between the LAL-positive and -negative samples.

Gram stains of LAL positive IV fluid were prepared. When examined directly and after centrifugation, all Gram stains appeared negative for bacterial forms. However, when aliquots were concentrated by membrane filtration, using a .22 μm Nuclepore filter (Nuclepore Corp., Pleasanton, CA), Gram-negative rods were readily visible. Flagella staining of these materials was negative.

Electron microscopy studies were conducted on LAL-positive and -negative IV fluids, commercial endotoxin, and saline in various combinations using a double blind code system. Forty-ml aliquots of each sample were concentrated using a .22 μm Nuclepore filter. LAL-negative

IV fluid containing 2 g cephalothin did not reveal bacterial forms. Occasionally, remnants of lysed cells were observed in LAL-positive IV fluid containing powdered commercial endotoxin standard and 2 g cephalothin, but no intact bacteria were seen. Bacterial forms were present in LAL-positive IV fluid containing 2 g cephalothin associated with a patient reaction. These forms appeared to be degenerated, approximately 2 μm in length, and the loss of cellular detail suggested nonviability. Masses of fibrous material were seen occasionally.

DISCUSSION

Characteristics of the endotoxin-like substance found in the contaminated IV fluids to date are reviewed in Table 3. From the clinical manifestations of the reactions in patients, combined with these positive laboratory findings, we concluded that the pyrogenic substance is probably endotoxin produced by gram-negative bacteria. Indirect fluorescent-antibody studies are in progress to determine if it is possible to identify the organisms in the contaminated fluid.

Table 3. Characteristics of Endotoxin-like Substance

Test	Reaction
Human pyrogenic response to IV infusion	+
Limulus lysate assay	+
Rabbit pyrogen test	+
Heat stable	+
Ability to withstand pH change	-
Growth in broth, agar	-
Growth in water	-
Growth at various temperatures	-
Gas-liquid chromatography	-
Indirect fluorescent-antibody test	Not complete

The LAL assay has proven to be a reliable and reproducible laboratory marker, making it possible to examine rapidly all IV products prepared for the 60 patients. A total of 29 bags of IV fluids were found to be *Limulus* positive. Sixteen were associated with patients who had reactions and another 13 would have almost certainly produced reactions had they been administered.

From the epidemiologic investigation we determined that contamination of the products was unlikely to have been present upon receipt from the manufacturers but was probably introduced in the hospital during formulation or other processing. Had the sterile water, normal saline mini-bags, or cephalothin been contaminated, all prepared vials of cephalothin on July 22 should have been contaminated. As additional evidence against product contamination, laboratory and animal tests conducted by CDC and FDA have shown that none of the unopened commercial products remaining in pharmacy supplies are contaminated.

Therefore, we considered the possibility of extrinsic contamination in the pharmacy. However, from observations of the procedures and techniques used to reconstitute the vials and prepare infusion bags, no sources of contamination were apparent. So, we subjected the IV fluid to preliminary screening for a variety of standard clinical laboratory determinations. One of the substances which was detected in significant concentrations in the contaminated IV fluid compared to the uncontaminated IV fluids was urea-nitrogen (Table 4).

Table 4. Results of Other Laboratory Tests

Samples[a]		Test	
		Urea	Creatinine
Cephalothin	LAL	mg%	mg%
2 g	+	18	29
2 g	-	0	31
Controls			
Saline + 2 g cephalothin endotoxin	+	0	27
Saline + endotoxin		0	0

[a]Mean determination on IV fluid from 5 infusion bags

The meaning of this finding is at present unclear. Determinations for other substances, for example, creatinine, were also found in high concentrations; however, the same levels were detected in both contaminated and uncontaminated IV fluids and in control samples. This suggested that the cephalothin was interferring with the creatinine determinations.

In conclusion, although the ultimate source of the contaminating material has yet to be determined, we believe that the problem was limited to this hospital and that continuing laboratory testing is necessary to identify the source and the mechanism of contamination.

ACKNOWLEDGMENTS

We thank Joseph O. Randolph, Chief, Antibiotics Biological Branch, NCAA, FDA, Washington, D.C. for performing the Rabbit Pyrogenicity Tests; Carry Calloway, Pathology Division, Bureau of Laboratories, CDC, for electron microscopy examination; and C. Wayne Moss, Ph.D., Bacteriology Division, Bureau of Laboratories, CDC, for GLC determinations.

REFERENCES

1. Sullivan, J.D., Jr., F.W. Valois, and S.W. Watson. 1976. p. 217-236. In A. W. Bernheimer (ed), Mechanisms in bacterial toxinology. John Wiley and Sons, New York.

2. United States Pharmacopeial Convention, Inc. 1974. Pyrogen Test. p. 613. In United States Pharmacopeia XIX. Mack Publishing Co., Eastern, PA.

LIMULUS AMEBOCYTE LYSATE ASSAY USED IN EVALUATING IMMUNOREGULATORY EFFECTS OF ENDOTOXIN IN TYPE-C RNA VIRUS REAGENTS

ROBERT M. DONAHOE and ROBERT L. PETERS

MICROBIOLOGICAL ASSOCIATES
5221 RIVER ROAD, BETHESDA, MD. 20016

INTRODUCTION

The Limulus amebocyte lysate (LAL) assay has become a widely accepted method for detection and quantitation of endotoxin. The LAL assay is extremely sensitive and has distinct advantages over the classical rabbit pyrogen assay for endotoxin. Because of convenience and economy of the LAL assay, many recent applications have been reported. Applications in the area of product quality control have proven to be highly desireable. The LAL assay has been used to detect endotoxin in various laboratory and clinical reagents (Moody, EEM et al. 1975, Bito, LZ, 1977). Because endotoxin can affect the usefulness of certain reagents, a quality control decision must be reached concerning what level of endotoxin in a given reagent is acceptable. In making such a decision, consideration must be given to the effects of endotoxin on product utility and to the ability to avoid or eliminate the contaminant. Under these circumstances, one can only reach a sound decision by being aware of potential problems endotoxin might cause.

While studying cell-mediated immune (CMI) responses to murine type-C RNA viruses, we encountered a situation which exemplifies problems that can occur when endotoxin-contaminated reagents are used. During application of a previously developed lymphocyte transformation assay (LTA, Donahoe, RM et al., 1976) and footpad swelling assay (FSA, Peters, RL et al. 1975), we noted interassay variation that seemed attributable to the use of different virus reagent lots. Being aware of the facts that viruses raised in cell culture

Pages 473–483, Biomedical Applications of the Horseshoe Crab (Limulidae)

often contain endotoxin (Moody, EEM et al. 1975) and that endotoxin can have dramatic as well as subtle immunologic and physiologic effects (Landy, M and Braun, W, 1964, Kass, EH and Wolff, SM, 1973), we used the LAL assay to examine our virus reagents for endotoxin. Endotoxin in varying concentrations was found. The consequence of this virus reagent contaminant on performance of the LTA and FSA is the subject of this report. A preliminary report on some of these observations has already been published (Donahoe, RM and Peters RL, 1978).

MATERIALS AND METHODS

Viruses

Rauscher murine leukemia virus (R-MuLV) and Gross murine leukemia virus (G-MuLV) were obtained through the Resources and Logistics Branch of the Virus Cancer Program, National Institutes of Health, Bethesda, MD. Each virus had been banded twice in sucrose and then pelleted, and was received suspended in 0.0167M sodium citrate. The final virus concentration was 1000-to 5000-fold that of the original. Protein concentrations were determined for each virus preparation by the method of Lowry et al. (1951).

Virus reagent production

Vaccines were prepared from the virus preparations by inactivation with formalin as described elsewhere (Donahoe et al., 1976) and UV-irradiated R-MuLV was used as an eliciting antigen in the in vitro phase of the LTA as described in that report.

Mice

BALB/cCr mice were obtained from Microbiological Associates, Walkersville, MD.

Cell-mediated immunological assays

The LTA for measurement of sensitivity to MuLV has been

described previously (Donahoe et al., 1976). Fourteen days after vaccination of mice with formalin-inactivated virus, spleens are removed from 5 mice from each experimental group, pooled, and mashed with the blunt end of a sterile syringe in RPMI-1640 medium supplemented with 10% human serum and antibiotics (cell culture reagents from Microbiological Associates, Bethesda, MD). The cells are filtered through gauze, washed and suspended to 5 x 10^6 leukocytes/ml culture medium. Replicates (0.1 ml) of each suspension are cultured in #3040 Microtest II tissue culture plates (Falcon Plastics, Oxnard, CA) along with 0.05 ml of antigen, mitogen or, as a baseline control, culture medium. One day prior to harvest, each well is pulsed with 0.5-µCi ^{3}H-Thymidine (New England Nuclear Corp., Boston, MA; sp. act. 2 Ci/mM). Replicate sets of cultures are harvested after 3 or 5 days. Incorporation of radioactive label into DNA is determined. Stimulation of cell division is calculated as a mean stimulation index by the following formula:

$$\text{Mean Stimulation Index (SI)} = \frac{\text{Mean cpm of radioactive label incorporated by leukocytes from test or control mice exposed to antigen or mitogen}}{\text{Mean cpm of radioactive label incorporated by leukocytes from test or control mice exposed to culture medium}}$$

To compare responses from mice in test groups to those from mice in control groups stimulation ratios were calculated from stimulation indices as follows:

$$\text{Stimulation Ratio (SR)} = \frac{SI_T \text{ (from test groups)}}{SI_C \text{ (from control groups)}}$$

In this calculation an SR of 1 indicates no difference in stimulation of spleen cells from test and control animals.

The footpad swelling assay (FSA) for detection of delayed-hypersensitivity responses to MuLV has been described previously (Peters et al, 1975). Seven days after vaccination with formalin-inactivated virus, mice are challenged in the left hind footpad with 15 µl of antigen or control preparation. Right hind footpads are inoculated with diluent as controls. The hind footpads are then measured with calipers 4 and 24 h after challenge. A percent mean increase in footpad diameter per test or control group (5-6 mice each) is calculated by dividing the mean difference in left and right footpad diameters of each group by the mean diameter of the control footpad.

The standard endotoxin used in these assays was from Difco, Detroit, MI, Escherichia coli lipopolysaccharide 0127:B8, Westphal extract.

Endotoxin determination

The Limulus Amebocyte Lysate test kit (Microbiological Associates, Walkersville, MD), with a sensitivity of 0.25-ng endotoxin/ml, was used for measurement of endotoxin.

RESULTS

Table 1 shows levels of endotoxin detected with the LAL assay in various lots of R-MuLV and G-MuLV. Concentrations ranged from 5- to 666-ng endotoxin per mg of virus protein.

Table 1. Endotoxin in Type-C RNA Virus

Virus	Endotoxin levels in representative virus lots
R-MuLV	500[a], 400, 400, 40, 40, 40, 5, 5, 5
G-MuLV	666, 500, 500, 400, 400, 400, 400, 40, 5

[a] ng endotoxin per mg virus protein.

We examined the possibility that virus might interfere with LAL-assay measurements of endotoxin. Endotoxin and R-MuLV were mixed and then assayed for endotoxin content. The results in Table 2 show that R-MuLV did not interfere with the ability to quantitate endotoxin.

We next examined the effect of endotoxin in R-MuLV on the LTA. A R-MuLV lot, containing a low level of endotoxin (2.5 ng per mg virus protein), was divided and one portion inactivated with formalin while the other was UV-irradiated as previously described (Donahoe et al., 1976). Some portions were then supplemented with endotoxin so that final endotoxin concentrations were either 50-or 500-ng per mg virus protein. These preparations were used in a standard LTA as defined by

Donahoe (1976). Formalin-treated virus was used for vaccination of mice and UV-irradiated virus as antigen for in vitro stimulation of spleen cells from the vaccinated mice. Stimulation ratios were calculated from the results of this test and are shown in Figure 1 to illustrate the effects of endotoxin present in R-MuLV. Vaccination involved administration of three doses of viral vaccine subcutaneously on days 0, 7 and 14. The first dose of vaccine was mixed 1:1 with complete Freund's adjuvant (CFA) before injection.

Table 2. Noninterference of R-MuLV on endotoxin determination with the Limulus amebocyte lysate assay

Sample tested: Virus present[a]	Sample tested: endotoxin concentration/ml[b]		Sample log dilutions tested for endotoxin in LAL assay[c]				
	Present originally	Added	10^{-1}	10^{-2}	10^{-3}	10^{-4}	10^{-5}
-	0	50 ng	+	+	-	n t	n t
+	33 ng	0	+	-	-	n t	n t
+	33 ng	50 ng	+	+	-	n t	n t
-	0	500 ng	+	+	+	-	n t
+	33 ng	500 ng	+	+	+	-	n t
-	0	5000 ng	n t	+	+	+	-
+	33 ng	5000 ng	n t	+	+	+	-
-	0	0	-	n t	n t	n t	n t

[a]R-MuLV, original concentration = 2 mg protein/ml; + = included in sample; - = not included.

[b]The R-MuLV had 33 ng of endotoxin/ml when originally measured in the LAL assay. In some instances, as indicated, endotoxin was added to the virus sample. This was done by incubating virus and endotoxin at room temperature for 45 minutes before testing the sample in the LAL assay.

[c]A 0.2-ml sample was tested for each dilution; + = positive for gelation of Limulus amebocyte lysate; - = no gelation; n t = not tested.

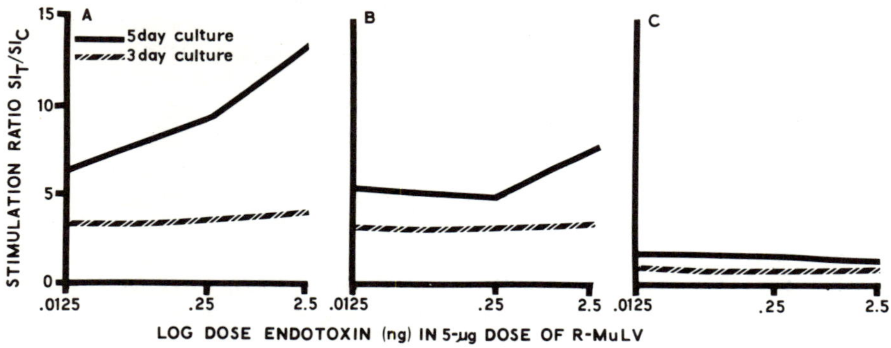

Figure 1. Stimulation ratios from a lymphocyte transformation assay. Spleen cell cultures were established from mice vaccinated with R-MuLV that contained endotoxin or endotoxin alone. Replicates of these cultures were exposed to 5-µg R-MuLV that contained endotoxin at the indicated doses. Stimulation ratios were determined by dividing the stimulation indices obtained with groups of mice that were vaccinated with virus plus endotoxin by those obtained when mice were vaccinated with endotoxin alone in doses equivalent to those found in the corresponding virus vaccines: 2.5-ng endotoxin per mg virus protein (Figure 1A), 50 ng (Figure 1B) or 500 ng (Figure 1C). Separate sets of cultures were held for 3 or 5 days as indicated.

Spleen cells from mice vaccinated with R-MuLV, that contained 2.5- or 50-ng endotoxin (Figure 1A and 1B respectively), obviously were stimulated by exposure to R-MuLV. Stimulation was appreciably greater after 5 days versus 3 days of culture. Five day, but not 3 day, cultures were increasingly stimulated in relation to increases in endotoxin concentration. This latter trend indicates that immunoenhancement of virus-specific responses by endotoxin occurred. In contrast, spleen cells from mice vaccinated with R-MuLV that contained 500-ng endotoxin were not appreciably stimulated by R-MuLV whether cultured for 3 or 5 days (see Figure 1C). This observation indicates that sufficiently high doses of endotoxin in the R-MuLV vaccine administered in vivo suppressed subsequent development of secondary antiviral immune responses in vitro.

We also examined the effects of endotoxin in virus on a FSA. A 30-μg dose of either R-MuLV vaccine that contained 5-ng endotoxin or G-MuLV that contained 500-ng endotoxin per mg virus protein was injected subcutaneously into each young adult BALB/c mouse tested. These mice and appropriate control mice were held for 6 days and then tested for sensitivity to virus or endotoxin in a FSA. Results of this study appear in Table 3.

The data in Table 3 indicate that endotoxin in association with virus can affect the outcome of a FSA. Footpad swelling was highly variable in this test. Multiple factors seemed to influence host responses. Both the type and dose of the sensitizing material used for vaccination influenced the way animals responded to footpad injections. Mice sensitized with virus vaccine that contained relatively low levels of endotoxin exhibited a typical delayed response to homologous viral antigen injected into their footpads. Mice vaccinated with virus that contained a relatively high level of endotoxin showed no appreciable delayed response to virus antigen injected into their footpads. Higher footpad swelling at 4 h versus 24 h was indicative of endotoxin stimulation. In some instances CFA alone seemed to sensitize mice to respond to footpad injections with endotoxin. Endotoxin alone appeared to cause footpad swelling, expecially at 4 h.

Further studies were done to document the ability of endotoxin alone to induce footpad swelling. Representative data of several such studies appear in Figure 2. In this test, some mice were injected subcutaneously with 0.1-ml doses that contained 5-μg endotoxin while others remained untreated. Groups of these mice were footpad challenged 7 days later with titrated doses of endotoxin and footpad swelling determined 4 h (Figure 2A) and 24 h (Figure 2B) after challenge.

Table 3. Footpad swelling response of mice vaccinated and challenged with MuLV vaccines containing relative high or low amounts of endotoxin

Vaccine[a]	Mean percentage increase in footpad diameter in response to challenge antigens[b]							
	15-µg R-MuLV and 0.08-ng endotoxin		15-µg G-MuLV and 7.5-ng endotoxin		endotoxin 0.08 ng		endotoxin 7.5 ng	
	4 h	24 h	4 h	24 h	4 h	24 h	4 h	24 h
30-µg R-MuLV, containing 0.15 ng endotoxin + CFA	2±2	17±7	8±3	10±5	8±5	8±5	5±2	3±2
30-µg G-MuLV, containing 15.0 ng endotoxin + CFA	7±4	0	8±3	3±2	3±0	1±2	5±2	1±2
0.15-ng endotoxin + CFA	4±2	0	6±4	7±4	4±2	8±4	2±0	3±2
15.0-ng endotoxin + CFA	5±2	2±3	5±4	3±0	3±1	1±0	10±6	4±3
CFA (control)	5±3	6±5	8±4	2±1	5±3	9±4	5±2	1±2
Diluent (control)	0	2±1	8±1	5±4	2±1	2±2	5±3	3±3

[a]Vaccines were suspended in sodium-citrate diluent and mixed 1:1 with CFA (complete Freund's adjuvant). Groups of 20, 8-10 week old BALB/cCr mice were inoculated subcutaneously with 0.1 ml of each vaccine.
[b]Five or six mice from each vaccine group were tested per footpad challenge antigen 1 week after vaccination.

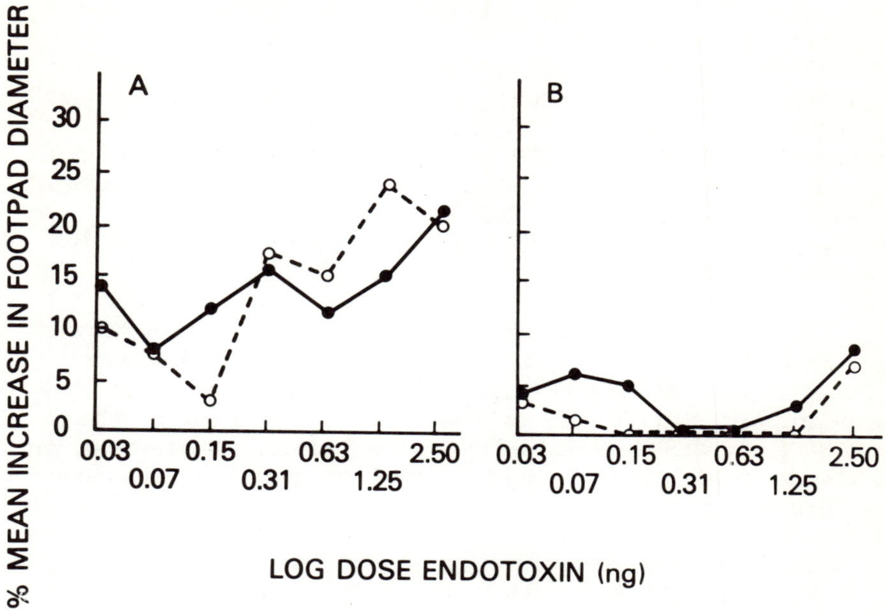

Figure 2. Footpad swelling assay with mice challenged with endotoxin alone. Ten week old mice were injected with 0.5-μg endotoxin (•——•) or were untreated (o---o). One week later groups of 5-6 each of these mice were challenged with the indicated doses of endotoxin. Footpad swelling was determined 4 h (Figure 2A) and 24 h (Figure 2B) after challenge.

Endotoxin induced greater 4-h than 24-h footpad swelling. The extent of swelling was dependent on endotoxin dose and biphasic kinetics were indicated. It is especially notable that very low (subnanogram) doses of endotoxin-induced appreciable footpad swelling.

DISCUSSION

The results indicate that contaminating endotoxin can alter responses in our CMI assays. There are several reasons to believe that the immunoregulatory effects we observed were due to endotoxin: 1) experimental controls indicated that endotoxin was involved, 2) the immunoregulatory effects were kinetically related to endotoxin dose, 3) other possible contaminating components of our virus reagents like cell wall fragments or serum proteins are unlikely candidates as immunoregulators, 4) endotoxin is well known for its physiologic and immunologic regulatory properties (Landy, M and Braun, W, 1964, Kass, EH and Wolff, SM, 1973), 5) others have detected the presence of endotoxin in virus reagents similar to ours (Moody, EEM et al., 1975). Further discussion on these points has been reported previously (Donahoe, RM and Peter, RL, 1978).

The type of immunoregulation observed depended upon experimental conditions. Immunoenhancement and immunosuppression was noted, the latter in the presence of relatively high levels of contaminating endotoxin, the former in the presence of very low levels. These observations should have impact on quality control decisions whenever tolerable levels of reagent contamination with endotoxin are being considered. One could presume that the undesireability of relatively high levels of endotoxin contamination is obvious and subsequently the best quality control decision would be to avoid or eliminate such contamination. Our immunosuppression data supports this reasoning. In contrast, one must give greater consideration to the potential problems posed by low-dose endotoxin reagent contamination before deciding on reagent utility. The results from our LTA studies show that low doses of endotoxin enhanced lymphocyte transformation. This activity does not seen attributable to mitogenic effects alone. Rather, endotoxin appears to have contributed to enhanced virus specific responses through a secondary helper effect. Weinberg, JB et al (1978) have reported similar observations from studies of macrophage tumoricidal activity. They too recorded immunoenhancing effects attributable to the use of reagents contaminated with endotoxin. Secondary helper effects were suggested because prior activation of macrophages was required before endotoxin could enhance responses. In our other assay, the FSA, we found footpad swelling to be easily induced with low endotoxin doses whether given alone or with virus. This effect was probably due to the well known inflammatory properties of endotoxin. Collectively these observations help define circumstances where product utility should be questioned when endotoxin is present. More objective quality control decisions about endotoxin contamination should be possible with this knowledge. Manufacturer, clinician and researcher alike need to be aware of the potential problems reagent contamination with endotoxin poses. With the availability of a conventient economical assay for endotoxin detection like the LAL assay, the quality control task of screening reagents for endotoxin contamination so that potential consequential problems can be addressed should be simplified.

REFERENCES

Bito LZ, (1977). Inflammatory effects of endotoxin-like contaminants in commonly used protein preparations. Science 196:83.

Donahoe RM, Peters RL, Van Vleck R, Huebner RJ, Kelloff GJ (1976). Development of a lymphocyte transformation microassay for murine leukemia virus. J Nat Cancer Inst 56:51.

Donahoe RM, Peters RL (1978). Inconsistency of cell-mediated assays when viral reagents contain endotoxin. Proc Soc Exp Biol Med (in press).

Kass EH, Wolff SM (1973). "Bacterial Lipopolysaccharides." Chicago: University of Chicago Press.

Landy M, Braun W (1964). "Bacterial endotoxins." New Brunswick: Rutgers University.

Lowry OH, Rosebrough NJ, Farr AL, Randall RJ (1951). Protein measurement with the Folin phenol reagent. J Biol Chem 193:265.

Moody EEM, Trousdale MD, Jorgensen JH, Shelokov A (1975). Bacteriophages and endotoxin in licensed live-virus vaccines. J Infec Dis 131:588.

Peters RL, Donahoe RM, Kelloff GJ (1975). Assay in the mouse for delayed-type hypersensitivity to murine leukemia virus. J Nat Cancer Inst 55:1089.

Weinberg JB, Chapman Jr HA, Hibbs Jr JB (1978). Characterization of the effects of endotoxin on macrophage tumor cell killing. J Immunol 121:72

ENDOTOXIN CONTAMINATION OF BIOLOGICAL REAGENTS: DETECTION AND REMOVAL OF ENDOTOXIN WITH LIMULUS AMEBOCYTE LYSATE

F.R. Rickles, J. Levin, E. Atkins and P. Quesenberry

VA Med Ctr,Newington,CT; UConn Health Ctr,Farmington,CT; The Johns Hopkins U Sch of Med, Baltimore,MD; Yale U Sch of Med, New Haven, CT, Harvard Med Sch & Peter Bent Brigham Hosp,Boston,MA & The Marine Biol Lab, Woods Hole, MA.

Endotoxin, the lipopolysaccharide portion of the cell wall of gram-negative bacteria, is a ubiquitous substance which is highly reactive in a wide variety of biological systems. As a minor contaminant, endotoxin is capable of producing either false positive or false negative results. In vitro assays which measure such functions as cytotoxicity, mitogenicity, cell surface reactions or cell release reactions can be influenced positively by the presence of endotoxin (Gimber and Rafter, 1969; Vaheri et al., 1973; Luderitz et al., 1973; Ozato et al., 1975; Schmidke and Najarian, 1975; Bradley et al., 1975; Smith, 1976; Elin et al., 1976; Mergenhaven et al., 1976; Hawiger et al., 1977; Rickles et al., 1977) while some in vivo assays, dependent upon physiologic responses in experimental animals, can be suppressed by endotoxin (Spivak and Levin, 1977; Quesenberry et al., In Press; Zuckerman et al., Submitted for publication. Two excellent examples of the difficulties involved in the evaluation of contaminated biological reagents were provided by the availability of two preparations of the jack-bean lectin concanavalin A (Con A), one contaminated by endotoxin and one free of LPS, and a preparation of the erythrocyte stimulating hormone, erythropoietin, which could be rendered free of LPS. Detailed studies of the preparations of Con A have been performed which allowed resolution of conflicts over several of its biological properties (Rickles et al., 1979). This report represents a brief survey of those studies. In addition, removal of contaminating endotoxin from both Con A and erythropoietin has been accomplished using limulus lysate with the recovery of functionally intact proteins.

Pages 485–499, Biomedical Applications of the Horseshoe Crab (Limulidae)

METHODS

The preparations of Con A were evaluated in four separate biological systems: (1) induction of endogenous pyrogen release from rabbit blood cells; (2) gelation of the limulus lysate; (3) activation of human monocyte tissue factor, a cell surface coagulation protein; and (4) stimulation of DNA synthesis in human mononuclear cells. The complete methodology and sources of all materials are detailed elsewhere (Rickles et al., 1973; Rickles et al., 1977; Rickles and Rick, 1977; Rickles et al., 1979). Initially, Con a preparations were screened for pyrogenicity by direct injection into rabbits. More sensitive assays for pyrogenicity utilized blood cells or mesenteric lymph node cells from normal rabbits. Supernatants from these stimulated cells were injected into intact female rabbits and febrile responses were reported as maximum differences in temperature from the baseline. The limulus assay was performed in the standard manner and gelation of dilutions of Con A preparations was compared to a reference lot of *E. coli* endotoxin. Tissue factor activation was performed by measuring the procoagulant activity of sonicated human monocytes after incubation with Con A, and comparing the clotting times of the preparations with a tissue factor standard (a rabbit brain extract). Synthesis of DNA was measured in human lymphocytes 72 hours after Con A stimulation by quantifying the uptake of tritiated thymidine after pulse labelling. Each of these four assays was performed in the presence and absence of either alpha-methyl-D-glucoside (MAG) or alpha-methyl-D-mannoside (MAM), specific inhibitors of Con A. Human erythropoietin, extracted from the urine of a child with congenital red cell hypoplasia, was obtained from the Erythropoietin Committee of the National Institutes of Health and was tested for activity in the hypertransfused polycythemic mouse assay before and after adsorption with limulus lysate. The details of this assay for erythropoietin, which is based on the stimulation of radioactive ^{59}Fe incorporation by the bone marrow of the recipient animal, are published elsewhere (Quesenberry et al., 1978).

RESULTS AND DISCUSSION

Earlier experiments with Con A had suggested that the mitogen was pyrogenic and considerable speculation arose regarding the specificity of this observation (Atkins and Francis, 1973; Rosenthal, 1975). Figure 1 shows the average febrile responses to three, ten-fold increases in dosage of Con A from the pyrogenic lot #210073. Of note are the short latent period (15 min) before onset of fever and the biphasic characteristic of the response (peaking at 1 1/2 and 4 hours after injection) to the highest dose. The similarity of these responses to those produced by a standard endotoxin can be seen by comparison to the effects of three, ten-fold increments in dosage of filtered typhoid vaccine, shown in the same figure. Similar broadly monophasic fevers, peaking at 1 1/2 hours, were observed with the intermediate dosage of both pyrogens. In contrast, another lot of Con A (#8744) was not pyrogenic even when tested with the highest dose (Fig. 1)

When increasing amounts of pyrogenic Con A, Lot #210073, were incubated with normal blood cells (NBC) for 18 hrs. an increasing febrile response was observed when the cell-free supernatant was injected into rabbits (Fig 2). This fever was characterized by rapid onset (15 minute latency) and was monophasic, as are the fevers produced by endogenous pyrogens.

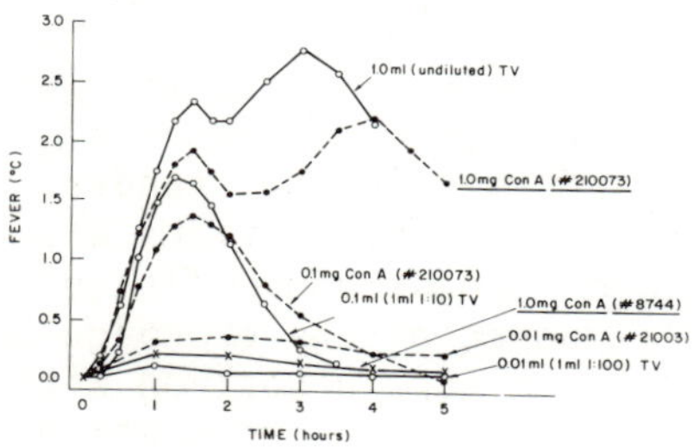

Figure 1. Similarity of mean febrile responses (2-8 recipients each) to three 10-fold increments in dosage of a pyrogenic lot of Con A (#210073 and filtered Typhoid vaccine (TV), given intravenously. The mean febrile response of 3 rabbits to the highest dose of nonpyrogenic lot of Con A (8744) is shown for comparison (reproduced with permission from reference 14).

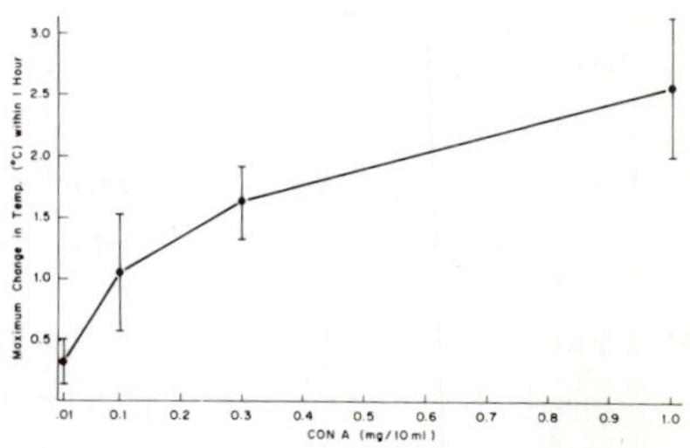

Figure 2. Mean maximum febrile responses to EP released in vitro by NBC incubated with increasing doses of pyrogenic Con A (210073) (reproduced with permission from reference 14).

As noted in Table 1, the febrile response could not be inhibited by preincubation of Con A with MAG, thus suggesting that endogenous pyrogen release was not a function of Con A binding to cellular oligosaccharides.

TABLE I. MAXIMAL FEBRILE RESPONSES TO ENDOGENOUS PYROGEN RELEASED BY NORMAL BLOOD CELLS INCUBATED WITH CONCANAVALIN A* AND α METHYLGLUCOSIDE (MAG)

Con A (mg/ml)	0.003	0.003	0.003	0.01	0.03
MAG (mg/ml)	0.2	0.5	1.0	10.0	10.0
Recipient‡	ΔT(°C)	ΔT(°C)	ΔT(°C)	ΔT(°C)	ΔT(°C)
R_1	1.10	1.15	1.25	0.60	0.95
R_2	1.50	1.20	2.00	1.45	1.65
R_3	1.80		1.90		1.75
Average	1.47§	1.18	1.72	1.03	1.45

* Con A #210073 ("pyrogenic" Con A - see Text)

‡ Each animal received 10 ml/dose.

§ Difference in values between different groups (2-3 recipients/group) not statistically significant (p=0.79)

Control injections (NBC or MAG incubated in media alone) were negative (<0.3°C). (reproduced with permission from reference 14).

Similar results were obtained when Con A was used in an attempt to stimulate the release of a putative pyrogen-producing lymphokine from rabbit mesenteric lymph node cells. (Table 2). Only the pyrogenic preparation of Con A was able to induce a lymphocyte supernatant capable of activating blood cells. As noted, the non-pyrogenic prep failed to induce any response even at the highest dose.

Table 2

ENDOGENOUS PYROGEN GENERATED BY NORMAL BLOOD CELLS INCUBATED 18 HOURS WITH CELL SUSPENSIONS OR SUPERNATANTS OF MESENTERIC LYMPH NODE (MLN) CELLS EXPOSED TO VARIOUS DOSES OF NON-PYROGENIC CON A (#8744)

Mesenteric Lymph Node Cells*	Con A (dose)				
	None	5μg	25μg	100μg	500μg
Cell Suspension	0.20† ± 0.05	0.19 ± 0.08	0.20 ± 0.06	0.23 ± 0.07	0.23 ± 0.04
Supernatant (18 h)	0.16 ± 0.02	0.24 ± 0.03	0.35 ± 0.11	0.24 ± 0.08	-
Supernatant (42 h)	0.23 ± 0.05	0.25 ± 0.05	0.24 ± 0.07	0.33 ± 0.01	-

* MLN = 2×10^8 cells/dose

† Each value represents the mean maximum change of temperature in degrees centigrade ± 1 SEM, achieved by 4-5 rabbits in 60 minutes.

reproduced with permission from reference 14).

In view of these disparate results the two preparations of Con A then were evaluated for the presence of exogenous pyrogen using the quantitative limulus assay.

In 4 separate experiments the pyrogenic Con A consistently produced a positive limulus amebocyte lysate test while the non-pyrogenic material remained negative. The specificity of this observation was confirmed by the failure of MAM to inhibit the gelation reaction induced by the contaminated Con A. If gelation had been due to properties of Con A itself, complete inhibition should have been accomplished at 50-100mM MAM. This effect is illustrated more dramatically in Table 4 where a dilution experiment is detailed. Even at very low levels of contamination, the gelation of the lysate by the Con A was not inhibited by MAM.

Table 3.

COMPARISON OF ACTIVITY IN LIMULUS ASSAY
OF PYROGENIC AND NON-PYROGENIC CONCANAVALIN A

Concanavalin A (1 mg/ml) (Code Number)	Endotoxin Concentration by Limulus Assay (μg/ml)		
	α METHYLMANNOSIDE (MAM) CONCENTRATION		
	0	50mM	100mM
210073 (pyrogenic)	0.5*	0.5	0.5
8744 (non-pyrogenic)	0.0	0.0	0.0

* Each number represents the mean of at least 4 blinded experiments. No significant difference in endotoxin concentration was found in any of the replicate samples ($p>0.1$). E. coli endotoxin was used as a standard (reproduced with permission from reference 14).

Table 4.

LACK OF EFFECT OF α METHYLMANNOSIDE (MAM) UPON DETECTION OF ENDOTOXIN CONTAMINATING A PREPARATION OF CONCANAVALIN A

Concanavalin A Concentration Tested* (μg/ml)	Endotoxin Concentration Detected by the Limulus Assay (μg/ml)	
	with MAM (100mM)	without MAM
100	0.01	0.01
10	0.001	0.001
1	0.0001	0.00005-0.0001
0.1	0.00001	0.00001
Endotoxin Concentration Tested‡ (μg/ml)		
0.01	0.01	0.01
0.001	0.001	0.001
0.0001	0.0001	0.0001-0.0005
0.00001	0.00001	0.00001-0.00005

* Concanavalin A - Code #1579

‡ *E. coli* LPS standard diluted with pyrogen free saline to equal presumed concentration of endotoxin contaminating the Concanavalin A.

The results of a representative experiment are shown in this table.

All assays were performed on coded samples (reproduced with permission from reference 14).

The magnitude of the contamination problem became clear when we surveyed 7 other commercial preparations of Con A for endotoxin as noted below.

Table 5.

ENDOTOXIN CONTAMINATION OF DIFFERENT PREPARATIONS OF CONCANAVALIN A

Preparation	Code Number*	Source	Endotoxin Concentration by Limulus Assay (μg/ml)
1.	210073	Calbiochem.	0.5‡
2.	1579	ICN	0.1
3.	1981	ICN	0.1
4.	115C-7060	Sigma	0.05
5.	16C-7210	Sigma	0.025
6.	112517	Aldrich	0.01
7.	141	Miles-Yeda	0.01
8.	90	Miles-Yeda	0.0005
9.	8744	Nutritional Biochem	0.0

* Code Number corresponds to the lot number of each product. All preparations were tested at a concentration of 1 mg of Con A/ml of tissue culture medium (RPMI 1640, Microbiological Associates, Bethesda, MD.).
‡ E. coli endotoxin was used as a standard (reproduced with permission from reference 14).

Since the presence of exogenous endotoxin appeared to have caused initial misinterpretation of the results of EP release experiments we felt it important to examine the interaction of endotoxin with Con A in other experimental systems in which endotoxin might alter or mimic the apparently inherent properties of the lectin.

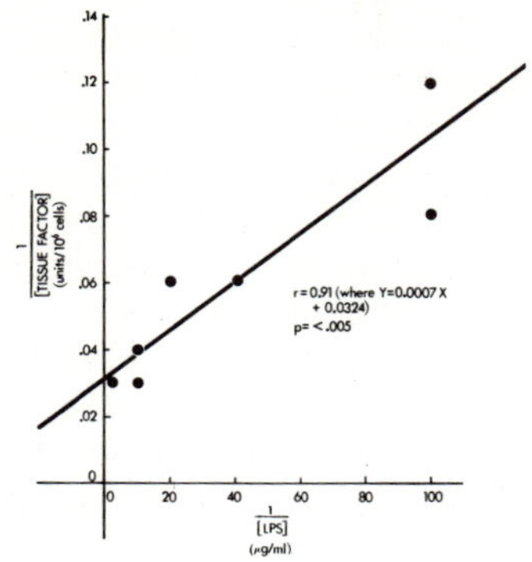

Figure 3. Correlation of tissue factor activity generated by mononuclear cells in response to different preparations of Con A and the concentration of endotoxin (LPS) detected in each preparation by the limulus assay. Con A preparations 1-7(cf. Table 5) were each tested at 2 μg/ml with 1×10^6 ml human mononuclear cells in the tissue factor assay and at 1 μg/ml in the limulus assay. A double reciprocal plot of the data is shown. Each point represents the mean of at least 3 experiments (reproduced with permission from reference 14).

The generation of tissue factor activity (TFA) by human mononuclear cells is a very sensitive technique for the detection of endotoxin contamination. In figure 3, results of parallel assays of 7 of the Con A preparations are graphed. A good correlation was found between the results of the limulus assay plotted on the horizontal axis, and the results of the TF assay, plotted on the vertical axis. Tissue factor activity can be measured directly in the presence of most mitogens and antigens tested (Rickles et al., 1973; Rickles et al., 1977) with the exception of Con A. The apparent failure of Con A to activate tissue factor is now known to be due to the ability of Con A to bind to carbohydrate residues in tissue factor and inhibit the expression of coagulant activity (Zacharski et al., 1974; Pitlick, 1975). However, this inhibitory effect is reversible. Therefore, we next compared the generation of tissue factor by pyrogenic and non-pyrogenic preparations of Concanavalin A.

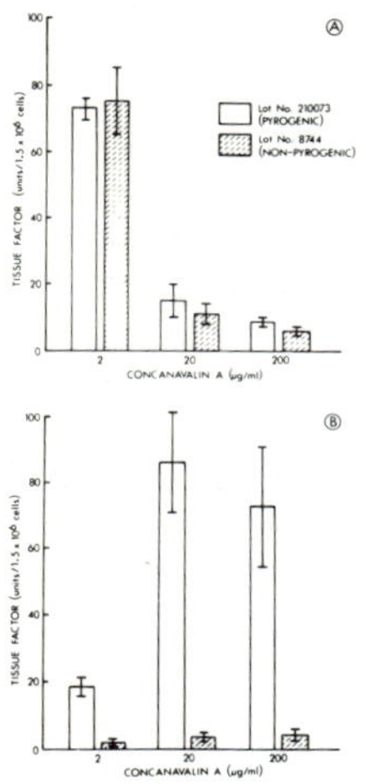

Figure 4. Tissue factor generation by human mononuclear cells _in vitro_, in response to pyrogenic and non-pyrogenic Con A. Each set of bars represents the mean tissue factor units (± 1 SEM) of at least 4 cultures (4A). The effects of the addition of MAM (50-100mM) to otherwise identical cultures are shown in Figure 4B. All of the cultures (with and without MAM) were harvested by sonication, after 16-24 hours of incubation with Con A. The apparent paradox of finding greater tissue factor activity in cultures stimulated with the lower concentrations of Con A is due to the binding of tissue factor by the higher concentrations of Con A (reproduced with permission from reference 14).

In the absence of MAM no distinction could be made between the preparations (4A) but addition of MAM to the system led to inhibition of TF generation by the non-pyrogenic Con A only. Therefore, the persistence of TFA generation in response to the pyrogenic preparation of Con A confirmed the presence of endotoxin in the material by an independent assay.

On the other hand, no modulating effect of contaminating endotoxin could be found when Con A was used to stimulate lymphocyte DNA synthesis. Moreover, functional Con A has been recovered from a contaminated preparation after adsorption to the limulus lysate. In preliminary experiments with small batches of radiolabelled Con A, 70% of the material became incorporated into the gel after treatment with limulus lysate. Pretreatment of the lectin with MAM, however, permitted the recovery of 70-91% of Con A in the supernatant of the gel. Four separate lots of Con A have been treated in this manner and all 4 have retained complete mitogenic activity without any evidence of persistent endotoxin contamination (Table 6).

TABLE 6. Recovery of Mitogenic Activity of Concanavalin A After Adsorption to Limulus Lysate

	PEAK MITOGENIC ACTIVITY*			
	BEFORE ADSORPTION		AFTER ADSORPTION	
Concanavalin A Preparation	cpm‡ ± 1 SEM	Stimulation Index (cpm-stimulated/ cpm-control)	cpm ± 1 SEM	Stimulation Index (cpm-stimulated/ cpm-control)
Miles-Yeda #141	8692 ± 3262	71.4	15222 ± 2529	125.7
Sigma #115-C	9985 ± 762	82.5	7248 ± 1067	60.0
Sigma #16-C	1199 ± 175	9.9	877 ± 212	7.2
Aldrich #112547	9160 ± 906	75.7	6562 ± 1084	54.2

* The peak mitogenic activity for each preparation was determined from a 4 log dose response curve.

‡ Each value represents the mean of at least 4 cultures.

Injection of the hormone eyrthropoietin into a hypertransfused, polycythemic mouse should result in proliferation and differentiation of precursor cells into mature erythroid cells. Recent studies utilizing various preparations of erythropoietin have demonstrated suboptimal stimulation of erythroid maturation in the experimental animal and other effects on hematopoiesis which suggested contamination of the preparations with endotoxin. Indeed, endotoxin was found in all erythropoietin preparations tested and has been successfully removed without loss of erythropoietic activity, by the use of limulus amebocyte lysate (Spivak and Levin, 1977; Quesenberry et al., 1978; Zuckerman et al. submitted for publication). Adsorbed material was found to be significantly more active in the stimulation of erythropoietic activity and no longer produced endotoxin related changes in bone marrow granulopoiesis (Figure 5).

In summary, the effects of contaminating endotoxin on the properties of Con A were variable in these studies. We examined 4 systems and found the following:

1) only endotoxin-contaminated preparations of Con A were capable of inducing fever, activating leukocytes to produce endogenous pyrogen <u>in vitro</u>, or clotting limulus lysate. Since these activities were uninfluenced by the specific inhibitors of Con A, MAG or MAM, they are presumably due to endotoxin alone.

2) in low dosage, both contaminated and uncontaminated Con A released similar amounts of TFA. However, at higher doses, endotoxin exerted an independent effect that could be separated, by the use of MAM, from that of Con A.

3) both types of preparations of Con A were equally mitogenic at the two doses tested and this activity was completely suppressed by MAM. The concentration of endotoxin in the contaminated samples was evidently insufficient to initiate DNA synthesis independently or act synergistically with Con A.

4) Con A treated with limulus lysate retains its mitogenic properties in the absence of demonstrated endotoxin contamination.

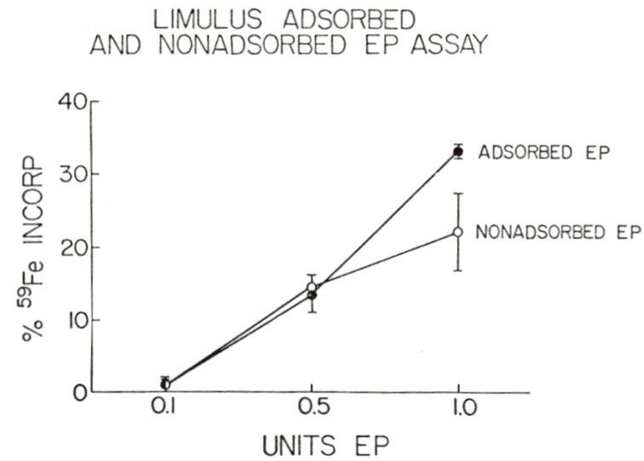

Figure 5. Erythropoietin (EP*) assay of urinary EP preparation before and after adsorption with limulus lysate. The non-adsorbed preparation was diluted during the limulus lysate adsorption procedure. Units of EP per ml of adsorbed EP were calculated from the dilution involved in the adsorption procedure with the assumption that EP activity was not lost during adsorption. Each experimental point represents the mean % ^{59}Fe incorporation ± 1 SEM of 5 mice. The adsorbed EP preparation demonstrated more activity at the highest dose, suggesting that endotoxin in the EP preparation may have inhibited the erythropoietic response in the assay mice. (reproduced with permission from reference 10). *Unfortunately, the standard abbreviations for both endogenous pyrogen and erythropoietin are EP.

Erythropoietin, treated in a similar fashion, becomes more effective in the stimulation of erythropoiesis than endotoxin-contaminated material. Inadvertent contamination of proteins must be considered as a potential explanation for disparate experimental results in many biological systems. The availability of limulus lysate should permit specific testing for and removal of contaminating endotoxin.

References

1. Atkins E and Francis L (1973). Role of lymphocytes in the pyrogenic response: comparison of endotoxin with specific antigen and the nonspecific mitogen Concanavlin A. J Infect Dis Suppl 128:S277-S283.
2. Bradley SG, Marecki NM, Bond JS, Munson AE and John DT (1975). Enhanced cytotoxicity in mice of combinations of concanavalin A and selected antitumor drugs. In: Concanavalin A. Ed. Chowdhury TK and Weiss AK. Adv Exp Med Biol Plenum Press. New York, NY 55: 291-307.
3. Elin RJ, Sandberg AL and Rosenstreich DL (1976). Comparison of the pyrogenicity, limulus activity, mitogenicity and complement reactivity of several bacterial endotoxins and related compounds. J Immunol 117:1238-1242.
4. Gimber PE and Rafter GW (1969). The interaction of Escherichia coli endotoxin with leukocytes. Arch Biochem Biophys 135:14-20.
5. Hawiger J, Hawiger A, Steckley S, Timmons S and Cheng C (1977). Membrane changes in human platelets induced by lipopolysaccharide endotoxin. Brit J Haematol 35: 285-299.
6. Lüderitz O, Galanos C, Lehmann V, Nurminen M, Rietschel ET, Rosenfelder G, Simon M and Westphal O (1973). Lipid A: Chemical structure and biological activity. J Inf Dis Suppl 128:S17-S29.
7. Mergenhagen SE, Rosenstreich DL, Wilton M, Wahl SM and Wahl LM (1976). Interaction of endotoxin with lymphocytes and macrophages. In: The Role of Immunological Factors in Infections, Allergic and Autoimmune Processes. Ed Beers, RF and Bassett EG. Raven Press, New York, NY p 17.
8. Ozato K, Adler WH and Ebert JD (1975). Synergism of bacterial lipopolysaccharides and Concanvalin A in the activation of thymic lymphocytes. Cell Immunol 17: 532-541.
9. Pitlick FA (1975). Concanavalin A inhibits tissue factor

coagulant activity. J Clin Invest 55:175-179.
10. Quesenberry P, Levin J, Zuckerman K, Renricea N, Sullivan R and Tyler W (1978). Stem cell migration induced by erythropoietin or hemolytic anemia: the effects of actinomycin and endotoxin contamination of erythropoietin preparations. Brit J Haematol, In Press.
11. Rickles FR, Hardin JA, Pitlick FA, Hoyer LW and Conrad ME (1973). Tissue factor activity in lymphocyte cultures from normal individuals and patients with hemophilia A. J Clin Invest 52:1427-1434.
12. Rickles FR and Rick PD (1977). Structural features of Salmonella typhimurium lipopolysaccharide required for activation of tissue factor in human mononuclear cells. J Clin Invest 59:1188-1195.
13. Rickles FR, Levin J, Hardin JA, Barr CF and Conrad ME Jr. (1977). Tissue factor generation by human mononuclear cells: Effects of endotoxin and dissociation of tissue factor generation from mitogenic response. J Lab Clin Med 89:792-803.
14. Rickles FR, Levin J, Rosenthal DI and Atkins E (1979). Functional interaction of concanavalin A and bacterial endotoxin (lipopolysaccharide): effects on the measurement of endogenous pyrogen release, human mononuclear cell tissue factor activation, lymphocyte DNA synthesis and gelation of limulus amebocyte lysate. J Lab Clin Med, In Press.
15. Rosenthal DI (1975). The role of the lymphocyte in experimental fever: Use of the mitogen concanavalin A. MD Thesis. Yale University School of Medicine
16. Schmidtke JR and Najarian JS (1975). Synergistic effects on DNA synthesis of phytohemagglutin or concanavalin A and lipopolysaccharide in human peripheral blood lymphocytes. J Immunol 114:742-746.
17. Smith GL (1976). Synergistic action of bacterial lipopolysaccharides on serum-stimulated DNA synthesis in mouse embryo fibroblasts. Proc Soc Exp Biol Med 153: 187-192.
18. Spivak JL and Levin J (1977). Endotoxin contamination of some erythropoietin preparations: Reply No. 2 Blood 50:549-550 (Letter).
19. Vaheri A, Ruoslanti E, Sarvas M and Nurminen M (1973) Mitogenic effect by lipopolysaccharide and pokeweed lectin on density-inhibited chick embryo fibroblasts. J Exp Med 138:1356-1364.
20. Zacharski LR, Rosenstein R and Phillips PG (1974).

Concanavalin A inhibition of tissue factor (thromboplastin) activity. Blood J Hematol 44:783-787.

21. Zuckerman KS, Quesenberry PJ, Levin J and Sullivan R(1978) Contamination of erythropoietin by endotoxin: in vivo and in vitro effects on murine erythropoiesis, Submitted for publication.

Limulus Lysate Testing of Biologics Requirements for The Rabbit Pyrogen Test

H. D. Hochstein, Dr.P.H.
and Edward B. Seligmann Jr., Ph.D.
Bureau of Biologics/Food and Drug Administration
8800 Rockville Pike
Bethesda, Maryland 20014

During the last eight years, experience has accumulated in using Limulus Amebocyte Lysate or LAL test for testing a variety of kinds of drugs and devices so that now most manufacturers of drugs and devices realize that it is a useful tool for quantitating the purity of products in terms of endotoxin content. Also, during this time there have been several Federal Register notices (January, 1973; September, 1973; May, 1977; November 4, 1977; August 11, 1978) published relating to the control of the lysate. For the most part, questions of specificity have been laid at rest or at least can be evaluated by use of proper controls. It is recognized that there are some substances that will interfere with the true endpoint when performing the test. Generally these are substances that prevent gel formation. However, false positives with respect to the rabbit pyrogen test may occur. Fortunately, these occurrences are rare. As yet they are not fully understood. By means of properly devised negative and positive controls, false positives and false negatives should not be a problem in routine testing.

Before a new preparation can be tested by the LAL test, it is necessary to determine if it will interfere with the LAL test. For the manufacturer of a drug or a device, this is a relatively simple procedure which needs to be done once for a given product but must be repeated each time the manufacturing procedure is changed. However, when the LAL test is used for testing unknown materials or samples that may vary unpredictably in their composition, it is necessary with each test to employ suitable controls that will detect the presence of any interfering substances.

Pages 501–506, Biomedical Applications of the Horseshoe Crab (Limulidae)

Four manufacturers of Limulus Amebocyte Lysate have been licensed to date. Several others have pending license applications still under consideration for licensing by the Bureau of Biologics. The four licensed manufacturers are: Associates of Cape Cod, Inc.; Mallinckrodt, Inc.; Microbiological Associates, Division Dynasciences Corporation and Travenol Laboratories, Inc. One of these is manufacturing lysate only for in house use.

The LAL test is a very sensitive detector of endotoxin. The active ingredients in the lysate that combine with endotoxin to form a solid gel can be altered irreversibly by a number of factors resulting in a destruction of the gel formation reaction. Because it is a sensitive test, any error introduced by the technician will affect the endpoint significantly. Therefore, it is important to adhere to a very exact technique every time the test is performed. Furthermore, the test should be designed so that an estimate of variability of each test can be made and compared to an established upper limit of this variability.

If it is found that the standard deviation of a test exceeds this limit, the test should be repeated. The results of all tests performed must be used to determine a new standard deviation, and the geometric mean endpoint computed only after the standard deviation is found to be below the established limit for the number of replicate tests performed. If the variability remains high, the cause should be determined before doing additional tests (Rastogi, May-June 1977). Because the sensitivity of a batch of lysate can be altered by unfavorable storage conditions, irregularities in rehydration, contamination with extraneous endotoxins, vibration during incubation, and other factors, each test should be accompanied by a titration of a known endotoxin in order to reconfirm the sensitivity of the lysate used in the particular test. Use of a single concentration of endotoxin does not provide an adequate positive control. In our experience the test is easily reproducible to plus or minus one two-fold dilution. To achieve this sensitivity, it is important to establish the actual sensitivity of the lysate used in a particular test by assaying a series of two-fold dilutions of endotoxin in replicate. We recommend four replicate titrations of the endotoxin, accompanied by four replicate titrations of each test sample, as a minimum block for a single test. Such a block provides the means for determining a reliable standard deviation as a measure of the test variability and a reliable geometric mean endpoint. As a laboratory worker becomes more experienced in performing the test, variability is reduced and it should be possible to reduce the number of replicates to three.

The LAL test has become very useful to manufacturers of drugs and devices as an inprocess control of components and of various stages of manufacture. For this purpose, it serves as a purity control with respect to bacterial endotoxin content. As long as the sensitivity of the lysate used is controlled as already discussed, it is possible to use the absolute dilution endpoint of the sample as an indicator of purity. For instance it may be established that to produce a final product that is safe clinically, a 1:2 dilution of the product must yield a negative LAL test. It may be expected that some lots of such a product when tested undiluted will be negative while others will be positive. As long as the sensitivity of the lysate used in each test is unchanged a positive test on an undiluted sample of that product would not be considered to be a failure. In our experience from testing a variety of biological products by both the LAL and rabbit pyrogen tests, the latter will usually not be positive unless the LAL test on the same sample is postive at a dilution of 1:64 or greater when using lysate of a sensitivity of approximately 0.25 ng/ml titrated against the reference endotoxin distributed by the Bureau of Biologics. This value may appear high and indeed we have found some samples that failed the rabbit pyrogen test with a positive LAL endpoint of 1:8. Therefore, when using a lysate with a sensitivity of 0.25 ng/ml we consider a positive endpoint of 1:16 plus or minus one two-fold dilution to be indicative of a borderline sample that should be confirmed by a rabbit pyrogen test. Whenever dilution is used to express an LAL test endpoint the sensitivity of the lysate in terms of a reference endotoxin must be determined and recorded.

Each manufacturer should establish workable limits for the endpoint of the LAL test for each product when it is used as an inprocess test. These limits should be established so there is a reasonable degree of assurance that when the product reaches the final stage and is subjected to the official rabbit pyrogen test every lot will pass. Manufacturing processes should be modified until all components of the final product pass at this level. Several manufacturers of blood fractionation products have established suitable limits so that every final lot passes the rabbit pyrogen test where previously a percentage of final lots failed the rabbit pyrogen test. The speed in performing the LAL test and its relative low cost as compared to maintaining a large colony of rabbits and the labor involved in a rabbit pyrogen test, make it feasible to do the amount of testing necessary to evaluate variations in manufacturing procedures that will improve the purity of the product.

A modification of this type of in-process testing with the LAL test may be made by performing a comparative test of a test sample in parallel with an established reference preparation of the same type of material (Rastogi, August 1977). The manufacturer can select a lot having low LAL activity. By this method the dilution endpoints of the test material and the reference are compared and a ratio that has been established as being acceptable is used as the criterion for acceptance. Again replicate testing should be used to monitor test variability and increase confidence in the results. While in this kind of testing the absolute sensitivity of the lysate being used is less important than in the method described previously, it would be prudent to test the sensitivity of the lysate periodically to establish that no changes have taken place.

In the Federal Register of November 4, 1977 it was announced that the LAL test could be used as a substitute for the official rabbit pyrogen test when testing licensed biological products and devices providing approval is first obtained from the appropriate Bureau of FDA. For biological products this requires submission of a license amendment by each manufacturer for each product to be tested by the LAL test. To date the Bureau of Biologics has not received any license amendments for this substitution. However, several manufacturers are collecting data and have indicated they plan to amend their product licenses for specific products. We have been asked for the type of data that we would expect in support of such an amendment. We have refrained from presenting a protocol for such testing because the type of data will be dependent upon the particular product being tested and because it is the responsibility of manufacturers to make such proposals. The proposal should include at least the following:

1) Data to demonstrate that the product does not interfere in any way with the endpoint of the LAL test. Such data should be collected on multiple lots of the particular product.

2) Data demonstrating the degree of precision obtainable in the manufacturer's laboratory when performing the LAL

test.

3) A test protocol for the LAL test as a pyrogen test. This should include positive and negative controls and sufficient replication so that the degree of variability can be estimated and the endpoint established with certainty.

4) A proposed level of LAL activity in the product that would be considered to be unacceptable; any activity detected below that value would be considered acceptable. Ideally this value should be based upon clinical data, but it may be difficult to acquire such data. Preferably the level should be set below that which is detectable by the rabbit pyrogen test. The LAL test is considerably more sensitive than the rabbit pyrogen test and patients will benefit by making use of this greater sensitivity in testing products subject to the pyrogen test.

The LAL test is also useful as a purity test for a large variety of products that are not now subject to the rabbit pyrogen test. This use is similar to that of the in-process quality control testing described previously. It may be desirable to apply the LAL test to products that have an inherent endotoxin concentration. In these cases a maximum level of endotoxin activity should be set. This will standardize the endotoxin content from lot to lot. When testing a product that has a relatively high endotoxin activity it may be difficult to determine if the product interferes with the LAL test. Generally with such products, a ten-fold series of dilutions is used to establish a preliminary endpoint. The sample is then retested starting at the lowest ten-fold dilution that gave a positive and make a two-fold dilution series to the next higher ten-fold dilution. Sometimes in products having a high endotoxin level a prozone will be noted in which the undiluted product or even the product diluted 1:2 will yield a negative LAL test and higher dilutions will be positive. An unknown sample found to be negative when diluted through 1:100 should be tested for interference before concluding that no endotoxin is present. It is probable that products having endpoints exceeding 1:100 need not be tested for interference with the LAL test. However, our experience is still limited in using the LAL test to control the purity of such products. We believe it would be worthwhile to investigate more extensively the use of the LAL test as a purity test on vaccines produced from organisms yielding endotoxin. The LAL test is certainly more sensitive and less variable than establishing antigen concentration by cell count. It is possible the LAL test could be used to estimate the potency of certain vaccines if a correlation can be established between potency and LAL activity. This is an area that should be investigated.

References

Federal Register (January 12, 1973) Vol. 38, No. 8 p 1404

Federal Register (September 18, 1973) Vol. 38, No. 180 Pp 26130-26132.

Federal Register (May 6, 1977) Vol. 42, No. 88 p 10107.

Federal Register (November 4, 1977) Vol. 42, No. 213 p 57749.

Federal Register (August 11, 1978) Vol. 43, No. 156 Pp 35731-35734.

Rastogi, SC, Hochstein, HD, and Seligmann, EB Jr. II Quality Control Procedures for testing Limulus Amebocyte Lysate Sensitivity Using Reference Endotoxin lb. Bull. Parenteral Drug Assoc., 31, 120-127 (1977).

Rastogi, SC, Hochstein, HD, and Seligmann, EB Jr. Statistical Determination of Endotoxin Content in Influenza Virus Vaccine by the Limulus Amebocyte Lysate Test. Journal of Clin. Micro. (1977) 6, Pp 144-148.

Key Words

Limulus Amebocyte Lysate
Endotoxin
Pyrogen
LAL

VIEWS ON THE RELIABILITY OF THE LIMULUS TEST

Peter J. Weiss

NCAA, Bureau of Drugs, Food and Drug Administration, 200 C St., S.W., Washington, D.C. 20204

INTRODUCTION

The *Limulus* Amebocyte Lysate (LAL) test has been used for the diagnosis of endotoxemias and meningitis (Clumeck et al, 1976) and for the detection of endotoxins in blood samples (Goldstein et al, 1976). It has been shown to give results in good agreement with those by culture method for detection of Gram-negative bacteria (Musiani et al, 1976) and for detection of Gram-negative bacterial meningitis (Ross et al, 1975).

This discussion, however, will be restricted to the use of the LAL test as a substitute or alternative for the official pyrogen test utilizing rabbits. A general question, whether posed by compliance personnel in connection with GMP inspectional evaluations or by drug quality compliance personnel responsible for assuring that the finished product is safe, effective and of proper purity, is whether any test method is acceptable as an alternative to an existing official method, be it biological or chemical.

The basic question is whether a method is scientifically valid and equivalent or superior to another method. An equivalency test for the *Limulus* method versus the rabbit pyrogen method must be based on experiments in all the solvents and on all the products for which one intends to use the substitute method.

To be presented at the Limulus Biomedical Symposium, Marine Biological Laboratory, Woods Hole, Mass., Oct. 7-9, 1978.

Pages 507–512, Biomedical Applications of the Horseshoe Crab (Limulidae)

We recognize, and it has been stated in the literature, that inaccurate results are obtained with the rabbit pyrogen test in certain situations (Bangham, 1971). It is the official test for the determination of pyrogens in drugs, however, and an evaluation of the reliability of the LAL test will have to be based on a comparison with the rabbit pyrogen test.

Standardization of the LAL test

After the licensing of *Limulus* Amebocyte Lysate (FR Nov. 4, 1977) for detecting pyrogens in biological products and medical devices on a product-by-product basis, additional standards for LAL were published August 11, 1978, in the Federal Register. These standards are essential for testing the potency of the LAL material.

Contradictory claims have been made about the power of the LAL test. Some claim that it is 5 to 50 times more sensitive than the rabbit test. Other statements that the LAL test can detect 0.03 to 0.06 nanograms of *E. coli* endotoxin per ml can be misleading, since such estimates have to be based on the specific lot of LAL and endotoxin used. Other aspects also need to be considered. The LAL test has been deemed equivalent to the rabbit test for bacterial endotoxins derived from *Klebsiella pneumoniae* but approximately 10 times as sensitive for endotoxins derived from *Escherichia coli*, *Serratia marcescens* and *Proteus vulgaris* (van Noordwijk and de Jong, 1976). The interpretation of the use of the LAL test also needs to be standardized. The earliest detection was based on measurement of clotting in a given time and the degree of gelation. Thus the reading of the *Limulus* test by this method is still subjective in several aspects. An improved method, using turbidimetric measurements, eliminates the individual observer's errors.

Interferences in the LAL test

The standardization of the LAL test needs to be fully clarified. For example, the question of whether several endotoxins can be utilized for standardization, as well as the question of interference by various substances with the LAL test need to be considered. Limited experience (Cooper et al, 1972) with an early preparation of LAL showed agreement in 130 of 155 radiopharmaceutical and biological products tested. Two preparations were positive in the rabbit only (those containing 5% normal serum albumin), and 23 were positive in the *Limulus* test only. We believe it is essential that before the LAL test is accepted for drugs, qualification studies must be done for each product by spiking with at least several endotoxins and testing serially diluted samples along with unspiked samples.

Performance of these challenge tests revealed a large number of products for which the LAL appears applicable (M. J. Kalm, private communication, 1976). Also mentioned in this same study were many products for which the LAL test was not acceptable, mostly because not all spiked samples gave positive responses. These included 20% dextrose injection, 10% invert sugar in various electrolyte solutions, 50% dextrose USP, 24.5% potassium acetate injection and bacteriostatic water for injection with benzyl alcohol as a preservative.

LAL was also examined for use in the detection of pyrogens of radiopharmaceuticals (McAuley et al, 1974). Test results identical to those by the rabbit test were found on most products. End-point interference was demonstrated by technetium-99m microspheres, iodine-131 19-iodocholesterol and others.

Other radiopharmaceuticals were found to interfere in the work using amebocyte lysate of the Japanese horseshoe crab, *Tachypleus tridentatus* (Murata et al, 1976). These authors found that benzyl alcohol at concentrations of more than 1% had an inhibitory effect on the gelation reaction, and several radioactive compounds showed intense inhibition of the gelation reaction. It has not been established whether such lysate can be used interchangeably with the LAL from *Limulus polyphemus*.

The LAL test was also used to detect the presence of bacterial endotoxin in a variety of antitumor agents (Siegel et al, 1976). Endotoxin prepared from *E. coli* was used for comparison. Concentrations from 0.1 to 63 ng in individual lots of six different antitumor agents were detected.

Thrombin, thromboplastin, polyriboinosinic-polyribocytidylic acid and ribonuclease were found to give positive LAL tests (Elin and Wolff, 1973), but this was claimed to be inaccurate (Thye Yin, 1975) when highly purified fractions isolated from the *Limulus* lysate were used. Other studies (Suzuki et al, 1977) indicated that a variety of polysaccharide derivatives such as dextrans give a positive *Limulus* test.

The cause and extent of inhibition of the LAL test by various penicillins was described in some detail (Newsome, 1977). The author hypothesized that the greater inhibitory effect of cloxacillin may be due to its higher serum binding capacity. A correlation of serum binding potential of various penicillins with their ability to inhibit the LAL reaction was assumed. It must be borne in mind that only samples that were highly pyrogenic in the rabbit caused gelation in the LAL test, and some that were negative in the limulus test gave pyrogenic responses in the rabbits.

Comparison of the *Limulus* test for endotoxins with the rabbit test for pyrogens of the European Pharmacopoeia was reported by van Noordwijk and de Jong (1976). They showed that bivalent calcium ions at concentrations of 0.34 M or higher caused negative LAL reactions and that a positive LAL reaction was not obtained even after the addition of *E. coli* endotoxin to pure samples of tetracycline HCl.

In a comparison of the *Limulus* Amebocyte Lysate test with the USP pyrogen test, Wachtel and Tsuji (1977) evaluated six LAL preparations obtained from five different suppliers for sensitivity, dependability, cost and convenience of use. Not only were major differences found among the LAL preparations with respect to sensitivities to different endotoxins, but also differences of up to 100-fold existed in the sensitivity of LAL preparations to the same endotoxin. Although the LAL and USP pyrogen test correlated reasonably well with pure and intact endotoxin, large discrepancies in correlation were found when partially degraded endotoxins were compared. Some LAL preparations responded equally to both the intact and degraded whereas others responded only to the intact endotoxins.

Finally, Sullivan and Watson (1975) question whether heparin inhibits the *Limulus* test for endotoxin. These authors feel that addition of 0.05 M calcium chloride and 0.154 M sodium chloride counteracts such inhibition.

In another study, Elin et al (1976) found good overall correlation between LAL reactivity and pyrogenicity in rabbits when these two parameters and mitogenicity and complement reactivity were compared in 17 endotoxins, modified endotoxins and related compounds. Although this correlation was highly significant the rank order of some compounds varied with respect to the parameters under discussion here. The rank order is an *E. coli* lipopolysaccharide>phthalylated lipopolysaccharide>phthalylated saccharide>polysaccharide in the LAL test, but this order was reversed in terms of pyrogenicity in rabbits.

In conclusion, let me express my views on the reliability of the LAL test. The test is unquestionably a valuable tool for the detection of endotoxin biocontamination. In order for it to be applicable to a large number of drug products, however, three conditions need to be fulfilled.

1) The test needs to be rigidly standardized as to use and to interpretation using the U.S. Reference *E. coli* endotoxin or other reference endotoxin, as is presently being done by the Bureau of Biologics, Food and Drug Administration.

2) For each product for which the *Limulus* test is to be used as a measure of endotoxin contamination, rigidly controlled experiments must be conducted in which samples of several lots are spiked with endotoxins from at least two species of microorganisms and are tested by the LAL test in comparison with unspiked control samples. If any of the spiked samples do not show positive responses, LAL should not be used in the determination of endotoxin contamination in that type of product.

3) Finally a decision must be reached as to what LAL positive indicator is equivalent to a tolerable degree of pyrogenicity.

REFERENCES

Bangham DR (1971). The dilemma of quantitation in the test for pyrogens. "Pyrogens and Fever," Edinburgh & London: Churchill Livingstone, p 207-212.

Clumeck N, Lauwers S, Kahn A, Mommens M, Butzler, JP (1976). Contribution of the *Limulus* test to the diagnosis of endotoxemias and meningitis due to Gram negative bacteria. Nouv Presse Med 6(17):1451.

Cooper JF, Hockstein HD, Seligmann EB (1972). The *Limulus* test for endotoxin (pyrogen) in radiopharmaceuticals and biologicals. Bull Parenter Drug Assoc 26:153.

Elin RJ, Sandberg AL, Rosenstreich DL (1976). Comparison of the *Limulus* activity, mitogenicity and complement reactivity of several bacterial endotoxins and related compounds. National Institute of Allergy and Infectious Diseases, NIH, Bethesda MD 90014.

Elin RJ, Wolff SM (1973). Nonspecificity of the limulus amebocyte lysate test: positive reactions with polynucleotides and proteins. J Infect Dis 128:349.

Federal Register 42, p 57745, Nov. 4, 1977.

Federal Register 43, p 35732, Aug. 11, 1978.

Goldstein JA, Keller LB, Wans WL (1976). *Limulus* Amebocyte Lysate test in neonates. Am J Clin Pathol 66:1012.

McAuley RJ, Ice RD, Curtis EG (1974). *Limulus* test for *in vitro* pyrogen detection. Am J Hosp Pharm 31:688.

Murata H, Kobayashi M, Iio M, Yamada H, Chilba K (1976). Sensitivity of the *Limulus* test and inhibitory factors in the radiopharmaceuticals. Nucl Med 17:1088.

Musiani M, Gatti M, Cevenini R, LaPlace M (1976). Study of the reliability of the *Limulus* Amebocyte Lysate test. Quad Sclavo Diagn 12(2):123.

Newsome PM (1977) Penicillins and the *Limulus* Amebocyte Lysate test for endotoxin. J Pharm Pharmacol 29:704.

Ross S, Rodriguez W, Controni G, Korengold G, Watson S (1975). *Limulus* Lysate test for Gram-negative bacterial meningitis. J Am Med Assoc 233:1366.

Siegel SE, Nachum R, Leimbrock S, Karon M (1976). Detection of bacterial endotoxin in antitumor agents. Cancer Treatment Reports 60(1):9.

Sullivan JD and Watson SW (1975). Inhibitory effect of heparin on the *Limulus* test for endotoxin. J Clin Microbiol 2(2):151.

Suzuki M, Mikami T, Matsumoto T, Suzuki S (1977). Gelation of *Limulus* Lysate by synthetic dextran derivatives. Microbiol Immunol 21:419.

Thye Yin E (1975). Endotoxin, thrombin and the *Limulus* Amebocyte Lysate test. J Lab Clin Med 86:430.

Van Noordwijk J and de Jong Y (1977). Comparison of the *Limulus* Amebocyte Lysate (LAL) test with the rabbit test; false positives and false negatives. Dev Biol Stand 34:39.

Wachtel RE and Tsuji K (1977). Comparison of *Limulus* Amebocyte Lysates and correlation with the U.S. Pharmacopeial pyrogen test. Appl Environ Microbiol 33:1265.

ANTIMICROBIAL DEFENSE MECHANISMS IN LIMULUS POLYPHEMUS

Ronald Nachum

Department of Pathology, C.R. Drew Postgraduate Medical School, 1621 East 120th Street, Los Angeles, California 90059.

INTRODUCTION

Limulus polyphemus blood contains one type of circulating cell, the amebocyte. The Limulus amebocyte contains the entire coagulation mechanism of this marine invertebrate (Levin and Bang, 1968; Young, et al., 1972), and more recently Mürer, et al. (1975) have shown that all of the factors required for clotting of Limulus blood are contained within the cytoplasmic granules. Exposure of Limulus blood to bacterial endotoxin results in amebocyte aggregation, with subsequent cell disruption and degranulation and clot formation. This endotoxin-induced clotting phenomenon has been characterized as a possible defense mechanism, serving to immobilize invading Gram-negative bacteria (Bang, 1956; Levin, 1975), and it has been postulated that the Limulus amebocyte may play an important role in host defense against bacterial invasion and dissemination (Bang, 1956; Levin, 1967; Levin and Bang, 1968; Shirodkar, et al., 1960). Additional evidence for the role of the Limulus amebocyte in host defense was reported by Nachum, et al. (1978), showing that the amebocyte contains an endotoxin inactivating capability. Limited ability of Limulus amebocytes to kill marine bacteria was described by Shirodkar, et al. (1960); however, Pistole and Britko (1978) were unable to demonstrate significant kill of four bacterial species with intact amebocytes. Significant kill of *Escherichia coli* with homologous serum or plasma, was observed.

The present investigation was undertaken to further define the role of the Limulus amebocyte in host defense against bacterial infection. Data presented in this study describes

Pages 513–524, Biomedical Applications of the Horseshoe Crab (Limulidae)

the presence of a potent "broad spectrum" antimicrobial activity in aqueous extracts of L. polyphemus amebocytes.

MATERIALS AND METHODS

Limulus amebocyte lysate (LAL) was obtained as a lyophilized preparation from Associates of Cape Cod, Inc., Woods Hole, Mass. and reconstituted with pyrogen-free water per the Manufacturer's instructions. Methods of LAL preparation are described elsewhere (Levin and Bang, 1968; Sullivan and Watson, 1974).

Bacterial species and strains used as test organisms are listed in Table 1. American Type Culture Collection species were derived from Bactrol Disks (Difco Laboratories, Detroit, Michigan) and clinical isolates from the Microbiology Laboratory, Martin Luther King, Jr., General Hospital, Los Angeles, California. Bacteria for assay were subcultured from an overnight Tryptic Soy Broth (TSB, Difco) to TSB and incubated in a shaking water bath at 35° C for 3-4 hours to obtain log phase cells. Gram-negative bacteria prepared for assay were washed two times in 0.9% NaCl, resuspended in 0.9% NaCl and diluted to a predetermined optical density at a wavelength of 660 nm. Gram-positive bacteria were washed and diluted in 3.0% NaCl containing 0.05% peptone (Difco).

Heat-derived LAL extracts were prepared by heating reconstituted LAL to 60° C for 10 minutes, this procedure serving to inactivate the clotting enzyme (Sullivan and Watson, 1975). Denatured protein was removed by centrifugation in the cold. Dialysis of heat-derived LAL extracts was performed overnight at 5° C against de-ionized water, with stirring, and two changes of liter each.

All assays for antimicrobial activity were done with either heat-derived LAL extracts or with untreated LAL. A control tube was run with each test and consisted of 60° C LAL extracts additionally heated to 100° C for 30 minutes. This preparation was devoid of antimicrobial activity. All assays for antimicrobial activity were run at 37° C.

Viable counts were done by pour plate. All data points were taken in duplicate and the counts averaged. Percent kill was calculated from the initial inoculum size as determined at time "0" from the control tube.

Table 1. Test organisms

Bacterium
Gram-negative bacteria
ESCHERICHIA COLI ATCC 25922
ESCHERICHIA COLI Lab
ESCHERICHIA COLI Mucoid
KLEBSIELLA PNEUMONIAE ATCC 13883
ENTEROBACTER CLOACAE ATCC 23355
SHIGELLA SONNEI
CITROBACTER FREUNDII
SALMONELLA TYPHIMURIUM ATCC 14028
SERRATIA MARCESCENS ATCC 8100
PROTEUS RETTGERI
PROTEUS VULGARIS
PROTEUS MIRABILIS
PROVIDENCIA STUARTII
YERSINIA ENTEROCOLITICA
PSEUDOMONAS AERUGINOSA ATCC 27853
PSEUDOMONAS AERUGINOSA Mucoid
Gram-positive bacteria
STAPHYLOCOCCUS AUREUS ATCC 25933
STAPHYLOCOCCUS EPIDERMIDIS ATCC 12228
ENTEROCOCCUS (Group D Streptococci)

RESULTS

Heat-derived LAL extracts were tested against eight Gram-negative species of the Family Enterobacteriaceae to determine if these extracts were bactericidal (Figure 1). An inoculum size of approximately 2000 cells/ml reaction mixture was used. The heat-derived LAL extracts were found to be bactericidal for some of the species tested. Sensitive organisms were rapidly killed with greater than 99.9% reduction in viable counts observed at 60 minutes incubation time. Several species were resistant, with outgrowth occurring during the two hour test period, although a small reduction in counts was observed after 30 minutes incubation time. <u>Escherichia</u>

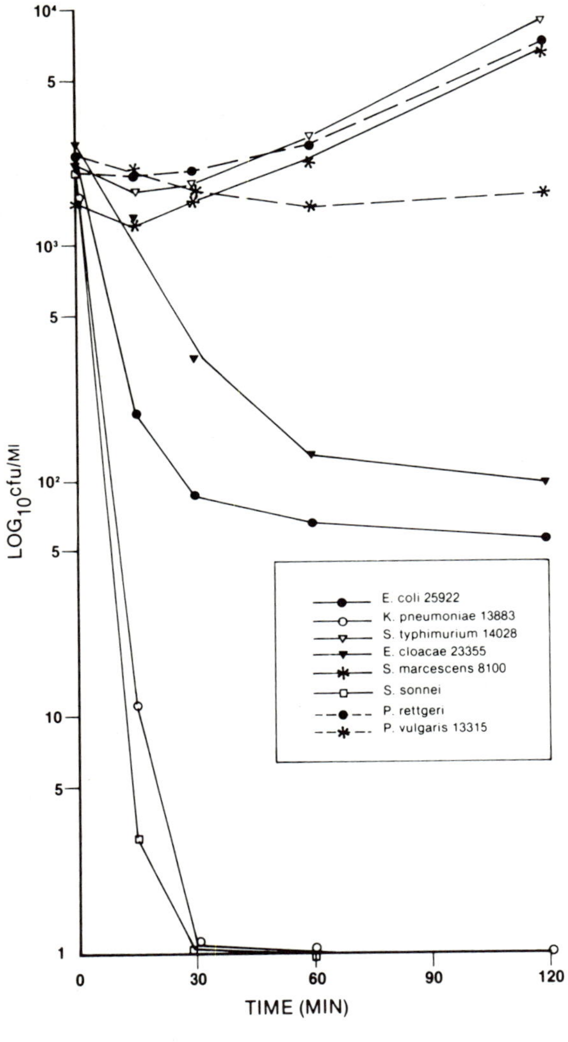

Figure 1. Bactericidal activity of heat-derived LAL extracts on various Gram-negative bacteria.

coli ATCC 25922 and *Enterobacter cloacae* were rapidly killed; however, a small number of viable cells remained at the end of the two hour test, with outgrowth of both species upon prolonged incubation.

To potentiate bactericidal activity heat-derived LAL extracts were prepared to yield a two-fold increase in concentration. This preparation was tested against three susceptible and two resistant species (Table 2). Increasing the concentration of the extracts resulted in a significant decrease ($P=<0.01$) in bactericidal activity against the three susceptible species and no increase in activity against the two resistant species. These data suggested the presence of some inhibitor(s) of bactericidal activity. Dialysis of both single and double strength heat-derived LAL extracts resulted in a significant increase ($P=<0.01$) in bactericidal activity against both sensitive and resistant species (Table 2).

Divalent cation inhibition of endotoxin inactivation with heat-derived LAL extracts was previously described (Nachum, et al., 1978). To determine if cation was similarily responsible for inhibition of bactericidal activity divalent salts were added to the heat-derived extracts (Table 3). Both calcium and magnesium were inhibitory to bactericidal activity when assayed against *E. coli*, with complete inhibition at 0.02 M. The inhibitory effect of $CaCl_2$ was more pronounced when *Proteus rettgeri* was used as test organism, with total inhibition of bactericidal activity at a concentration of 0.005 M.

The antimicrobial activity of heat-extracted, dialyzed LAL was assayed against 16 species and strains of Gram-negative bacteria and three Gram-positive species. The inoculum size for these experiments was increased to 5000 cells/ml unless otherwise indicated. Figures 2, 3, and 4 are presented to show the results of these experiments. All controls increased in viable counts over the four hour incubation time.

Bactericidal activity of the dialyzed extracts on species and strains of the Family *Enterobacteriaceae* are shown in Figure 2. Sensitive species were rapidly killed with a greater than 90% reduction in viable counts occurring in the first 15 to 30 minutes of reaction time. Only *Citrobacter freundii* was observed to increase in numbers at the end of the four hour test period. Of the 14 species and strains of *Enterobacteriaceae* tested, seven were sterilized.

Table 2. Effect of concentration and dialysis on bactericidal activity of heat-derived LAL extracts

Bacterium[a]	N[b]	Percent Kill				P[d]	
		Not Dialyzed		Dialyzed		Strength	Dialysis
		1X[c]	2X[c]	1X	2X	1X vs. 2X	+ or -
E. COLI	3	94.8±4.42	55.7±5.71	92.1±5.63	>99.9	<0.01	<0.01
K. PNEUMONIAE	3	99.9±0.05	92.4±2.02	>99.9	>99.9	<0.01	<0.01
E. CLOACAE	3	94.4±4.12	22.3±3.17	99.9±0.16	>99.9	<0.01	<0.01
S. MARCESCENS	3	0	0	96.1±4.69	>99.9	NS[e]	<0.01
P. RETTGERI	3	0	0	91.5±5.87	>99.9	NS	<0.01

[a] Inoculum size was approximately 2000 cells/ml. Test run at 37° C for two hours. Controls consisted of 100° C LAL extract, concentrated and dialyzed as test.

[b] Number of experiments.

[c] 1X=single strength; 2X=double strength.

[d] Derived by two-way analysis of variance (Winer, B.J., 1971).

[e] Not significant.

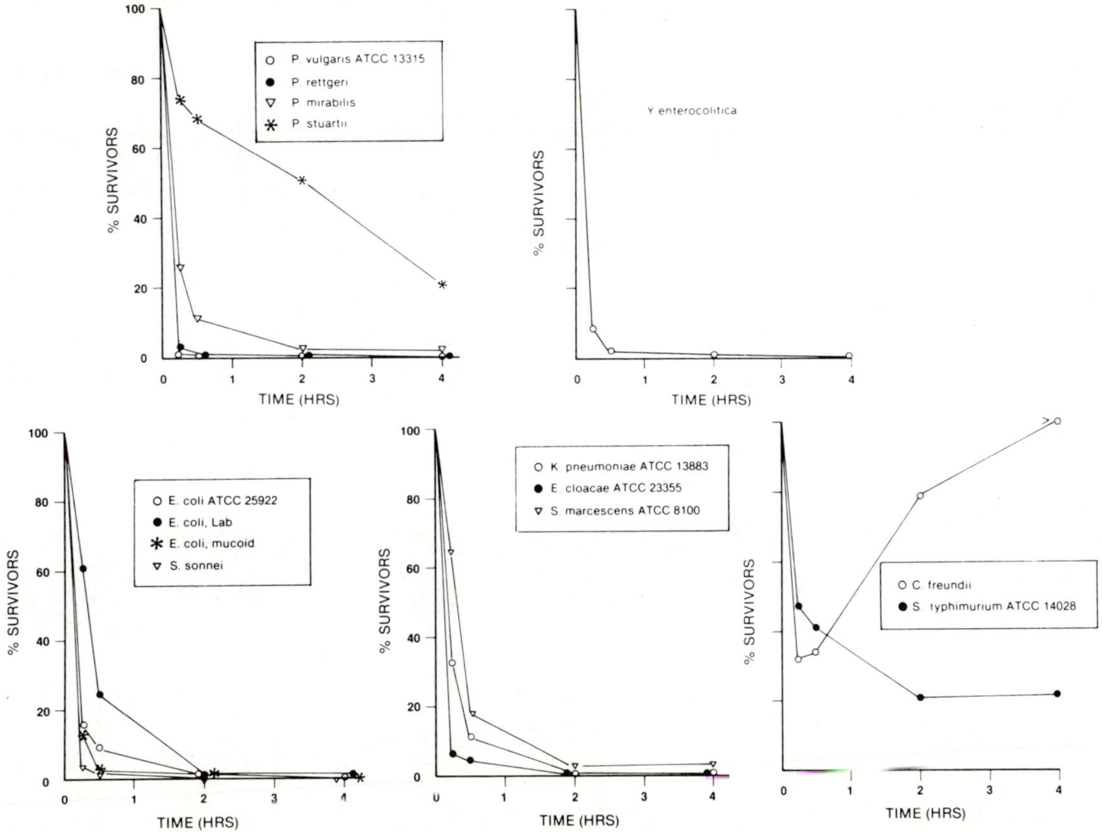

Figure 2. Bactericidal activity of heat-derived, dialyzed LAL extracts against species and strains of the Family <u>Enterobacteriaceae</u>.

Table 3. Effect of divalent salt on bactericidal activity of heat-derived LAL extracts

Salt (M)	$CaCl_2$ Percent Inhibition E. COLI	$CaCl_2$ Percent Inhibition P. RETTGERI	$MgCl_2$ Percent Inhibition E. COLI
None added	0	0	0
0.001	6.1	83.1	0
0.005	24.5	100.0	1.0
0.01	69.5	100.0	28.2
0.02	100.0	100.0	100.0

Both strains of *Pseudomonas aeruginosa* were sensitive to the dialyzed extracts; however, a loss in viable counts in the control tube limited the test to 15 minutes of reaction time (Figure 3).

Antimicrobial activity of dialyzed LAL extracts is shown in Figure 4. *Staphylococcus aureus* was resistant. In contrast, *S. epidermidis* was susceptible. The Enterococci were found to be very sensitive and sterilization of the cell population (10^4 cells/ml) occurred within 30 minutes.

The bactericidal activity of untreated LAL was determined against *E. coli*, *K. pneumoniae*, *E. cloacae*, *S. marcescens*, *C. freundii*, and Enterococci and compared to heat-derived extracts. Both preparations were equally effective against 5/6 bacterial species tested. Untreated LAL was observed to have a greater bactericidal effect on *S. marcescens* (95.9% to 79.6% kill at 60 minutes using an inoculum size of 10^4 cells/ml). Bactericidal titers of untreated LAL, expressed as the reciprocal of the last dilution with a reduction in viable counts of 50%, are shown in Table 4. Although not shown, bactericidal activity of heat-derived LAL extracts was about one-fourth that of untreated LAL.

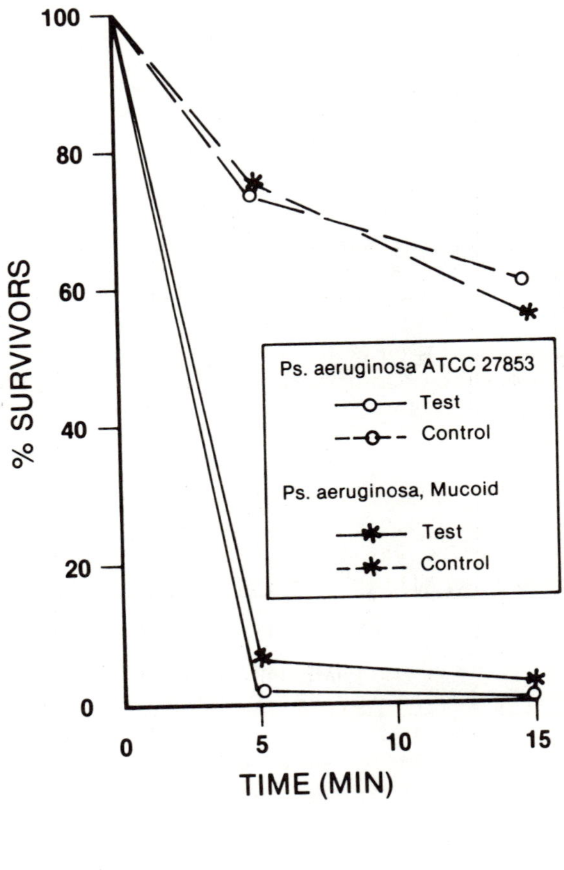

Figure 3. Bactericidal activity of heat-derived, dialyzed LAL extracts for two strains of Pseudomonas aeruginosa. Initial inoculum size was 15 X 10^3 cells/ml reaction mixture.

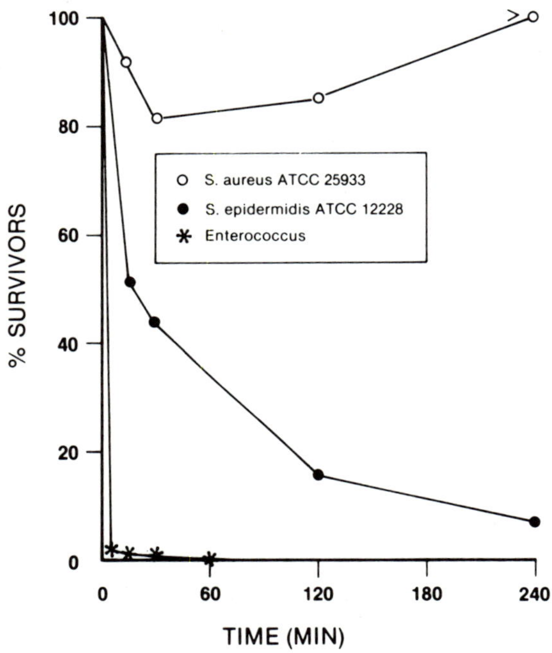

Figure 4. Susceptibility of three species of Gram-positive cocci to heat-derived, dialyzed LAL extracts.

Table 4. Bactericidal activity of Limulus Amebocyte Lysate for various bacterial species

Bacterium	Titer (50% Kill)
ESCHERICHIA COLI	320
KLEBSIELLA PNEUMONIAE	160
ENTEROBACTER CLOACAE	160
SERRATIA MARCESCENS	80
CITROBACTER FREUNDII	20
PROTEUS RETTGERI	0

CONCLUSIONS

A potent antimicrobial substance in Limulus amebocyte lysate prepared from the circulating blood cells of the horseshoe crab, *Limulus polyphemus*, has been described. The present study supports the contention that the *L. polyphemus* amebocyte may play an important role in defense against microbial infection and dissemination, providing a protective mechanism against invading microorganisms.

REFERENCES

Bang FB (1956). A bacterial disease of *Limulus polyphemus*. Bull John Hopkins Hosp 98: 325.

Levin J (1967). Blood coagulation and endotoxin in invertebrates. Fed Proc 26: 1707.

Levin J (1975). Blood coagulation in the horseshoe crab (*Limulus polyphemus*): A model for mammalian coagulation and hemostasis. In "Animal Models of Thrombosis and Hemorrhagic Diseases", DHEW Publication No. (NIH) 76-982, p 87.

Levin J, Bang FB (1968). Clottable protein in *Limulus*: Its localization and kinetics of its coagulation by endotoxin.

Thromb Diath Haemorrh 19: 186.
Mürer EH, Levin J, Holme, R (1975). Isolation and studies of the granules of the amebocytes of *Limulus polyphemus*, the horseshoe crab. J Cell Physiol 86: 553.
Nachum R, Siegel SE, Sullivan JD Jr, Watson SW (1978). Inactivation of endotoxin by *Limulus* amoebocyte lysate. J Invertebr Pathol 32: 51.
Pistole TG, Britko JL (1978). Bactericidal activity of amebocytes from the horseshoe crab, *Limulus polyphemus*. J Invertebr Pathol 31: 376.
Shirodkar, MV, Warwick A, Bang FB (1960). The in vitro reaction of *Limulus* amebocytes to bacteria. Biol Bull 118: 324.
Sullivan JD Jr, Watson SW (1974). Factors affecting the sensitivity of *Limulus* lysate. Appl Microbiol 28: 1023.
Sullivan JD Jr, Watson SW (1975). Purification and properties of the clotting enzyme for *Limulus* lysate. Biochem Biophys Res Comm 66: 848.
Winer BJ (1971). "Statistical Principles in Experimental Design". New York: McGraw-Hill.
Young NS, Levin J, Prendergast RA (1972). An invertebrate coagulation system activated by endotoxin: Evidence for enzymatic mediation. J Clin Invest 51: 1790.

ISOLATION AND CHARACTERIZATION OF A BACTERIAL AGGLUTININ IN THE SERUM OF *LIMULUS POLYPHEMUS*

Kevin J. Gilbride and Thomas G. Pistole
Department of Medicine,
National Jewish Hospital and Research Center,
Denver, CO 80206 and Department of Microbiology,
University of New Hampshire, Durham, NH 03824

INTRODUCTION

The hemolymph of many invertebrates contains glycoproteins which are capable of agglutinating a variety of cell types (Cohen *et al.*, 1975; Finstad *et al.*, 1974; Hall and Rowlands, 1974). These agglutinins demonstrate activity for specific carbohydrate moieties on cell surfaces and may serve as opsonins for the phagocytic removal of foreign cells from the hemolymph (Bang, 1967; Cohen *et al.*, 1972; Cornick and Stewart, 1968; Köhler *et al.*, 1973; McKay *et al.*, 1970; Tyson and Jenkin, 1973). Characterization of agglutinins thus far indicates no physicochemical simularities to vertebrate immunoglobulins; however, their affinity for specific groups of carbohydrates has been well documented (Bang, 1967; Cohen *et al.*, 1972; Köhler *et al.*, 1973; Marchalonis and Edelman, 1968). Since this affinity is directed toward specific groups of carbohydrate ligands, it seems reasonable to consider agglutinins as lectins.

The hemolymph of *L. polyphemus* has served as a model system for studies of agglutinating activity. Noguchi (1903) was first to describe the hemagglutinating activity of this serum. Since that time the hemagglutinin in *L. polyphemus* serum has been investigated by a number of groups (Cohen and Rozenberg, 1974; Finstad *et al.*, 1972; Finstad *et al.*, 1974; Kaplan and Kehoe, 1977; Marchalonis and Edelman, 1968; Nowak and Barondes, 1975; Oppenheim *et al.*, 1974; Roche and Monsigny, 1974). The sialic acid residues on the erythrocyte membrane has been shown to be the sites of attachment for this agglutinin (Cohen *et al.*, 1972; Roelcke *et al.*, 1971;

Pages 525–535, Biomedical Applications of the Horseshoe Crab (Limulidae)

Voigtmann *et al.*, 1971).

In addition to an agglutinin with specificity for N-acetyl neuraminic acid, Cohen *et al.* (1975) have demonstrated in *Limulus* serum a precipitin with anti-galactan activity. In particular, this serum component reacts with O-β-D-galacto-pyranosido(1→3)-D-galactose residues. There thus appears to be good evidence for multiple lectins in *Limulus* serum.

Earlier studies in our laboratory have shown that serum from the horseshoe crab is capable of reacting with a variety of microorganisms (Pistole, 1978). One strain of bacteria, *Staphylococcus aureus* (ATCC 25923), which was shown to be highly reactive in our agglutination assays, was selected for further studies. This paper presents our efforts to isolate and characterize an anti-*S. aureus* lectin from *Limulus* serum.

MATERIALS AND METHODS

Limulus serum.

Sera from horseshoe crabs were obtained, processed and stored as previously described (Pistole, 1978).

Antigen Preparation.

Bacterial cultures, incubated overnight at 37°C in tryp-tosephosphate broth (Difco) with shaking, were washed twice in saline and resuspended in one-tenth the original volume in saline. These suspensions were placed in a 60°C waterbath for 2 hr and subsequently stored at 6-8°C. *S. aureus* (ATCC 25923, Bactrol system, Difco; Wood 46, W.R. Chesbro, University of New Hampshire) and *Salmonella minnesota* (University of New Hampshire stock culture collection) were the organisms used.

Blood from inbred lines of chickens was supplied by R. M. Zsigray (University of New Hampshire). Heparinized blood was washed twice in saline and suspended in the same at a final concentration of 2%.

Agglutination Assays.

The optimal working concentration of each bacterial anti-gen was determined by a preliminary titration of the antigen as previously described (Pistole, 1978). For *S. aureus* the concentration was approximately 1 x 10^8 cells/ml; for *S.*

minnesota, 1 x 10^9 cells/ml.

Protein Estimation

Protein concentrations were determined by the method of Lowry *et al.* (1951) and by the procedure of Bradford (1976, Bio-Rad protein assay). Bovine serum albumin was used as the standard. Protein profiles were also made using absorbance at 280 nm on a Spectronic 600 spectrophotometer (Bausch and Lomb).

Affinity Chromatography

Sepharose 4B (Pharmacia), equilibrated with $CaCl_2$-NaCl buffer (0.01 M $CaCl_2$, 0.154 M NaCl, 0.01 M HEPES, pH 7.5) was used as the ligand. Elution was accomplished by addition of 0.02 M sodium citrate, 0.01 M HEPES, pH 7.5.

Molecular Weight Estimation

The approximate molecular weight of the recovered agglutinin was determined using exclusion chromatography on a Sephadex G-200 (Pharmacia) column. HEPES-buffered citrate (0.02 M sodium citrate, pH 7.5) was used. Three ml fractions were collected at a flow rate of 15 ml/hr.

Polyacrylamide Gel Electrophoresis (PAGE)

Vertical slab PAGE was performed in both the presence and the absence of sodium dodecyl sulfate (SDS) and 2-mercaptoethanol using the procedure of Laemli (1970). A 10 μl sample, consisting of 1.3 μg of protein, was loaded on the gel. The stacking gel consisted of 3%, and the separating gel, of 7% polyacrylamide. Electrophoresis was monitored by used of a tracking dye (bromphenol blue). The gel was fixed and stained according to the procedure of Fairbanks *et al.* (1971).

Scanning Electron Microscopy

Saline suspensions of bacteria were fixed in 0.25% glutaraldehyde for 12 h. (Sigma, Grade A) and post-fixed in 1% osmium tetroxide, (1 h.) then washed extensively in distilled water. The cells were diluted to a final concentration of 1000 organisms/ml. Portions of these suspensions were reacted with whole *Limulus* serum or purified agglutinin at room temperature for 30 min. After repeated washings in distilled water, the cells were placed on aluminum studs, coated with carbon and gold, and examined on an AMR 1000 scanning electron microscope.

RESULTS

Lectin Purification

The lectin was purified by affinity chromatography using Sepharose 4B as the ligand (Fig. 1). Twenty-five ml of whole *Limulus* hemolymph was added to the gel which was then washed with HEPES-buffered $CaCl_2$-NaCl solution. The lectin was eluted by the addition of HEPES-citrate. The eluted protein was shown to have agglutinating activity for *S. aureus* (25923), but did not have activity for *S. minnesota* or chicken erythrocytes. The microtitration results are shown in Table 1. While whole serum agglutinated all three cell types, the eluted fraction reacted only with *S. aureus* cells.

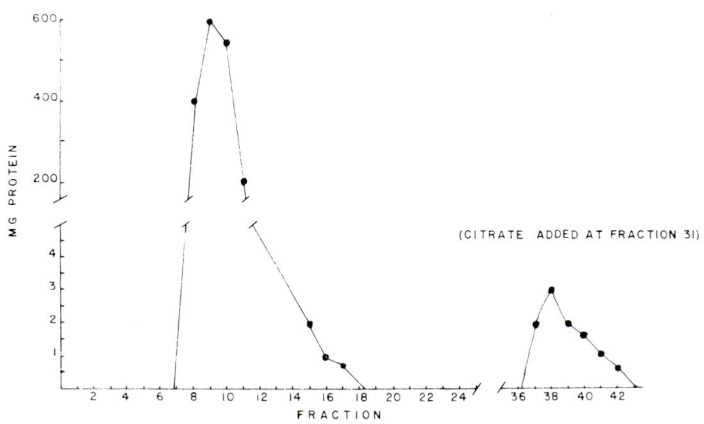

Figure 1. Lectin purification by affinity chromatography. Each fraction contains 10 ml.

TABLE 1
Agglutination Titers of Whole Serum and Fractions Separated by Affinity Chromatography

Antigen	Serum Sample[a]		
	Whole	Fr. 8	Fr. 38
Chicken RBC	12[b]	8	0
S. aureus	9	6	5
S. minnesota	6	0	0

a See Fig. 1
b Log_2 values

Agglutination of *S. aureus* (Wood 46)

Protein A from certain strains of *S. aureus* is known to bind selectively to certain serum glycoproteins, e.g., human IgG subgroups. To determine whether our agglutination results with *S. aureus* were due to protein A reacting with *Limulus* serum components, we prepared a heat-killed antigen from the Wood 46 strain of this organism, which is known to be devoid of protein A. Tests on this antigen revealed that the protein A-deficient strain was capable of being agglutinated by *Limulus* serum; thus, our results are not due to protein A binding of serum proteins.

Polyacrylamide Gel Electrophoresis (PAGE)

The *S. aureus* agglutinin, recovered by affinity procedures, was examined for purity using non-SDS PAGE (Fig. 2). The purified agglutinin (b) appears as a single band in this procedure. Whole *Limulus* serum (a) was similarly analyzed. The third sample (c) represents an eluate from a Sepharose column which had been loaded with an ultracentrifuged concentrate of serum rather than whole serum. Here several bands are discernible.

The purified agglutinin was next subjected to PAGE in the presence of SDS and 2-mercaptoethanol to detect peptide subunits in the whole molecule. As shown in Fig. 3, five bands were obtained which suggested that the overall molecule is composed of heterogeneous subunits.

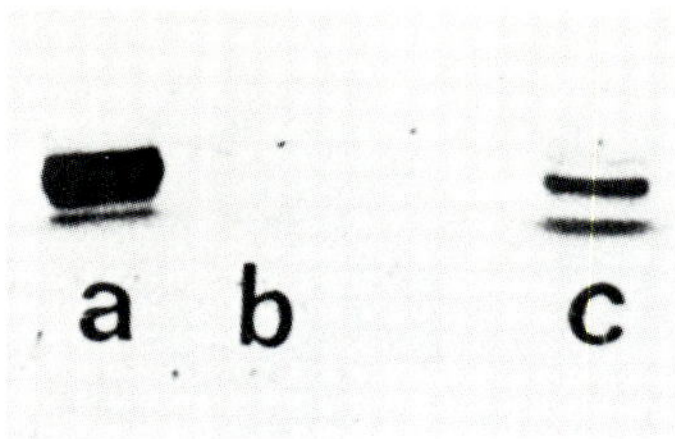

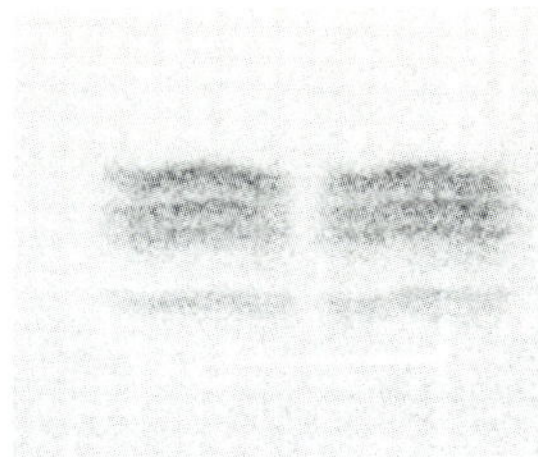

Figure 2. Polyacrylamide gel electrophoresis of whole *Limulus* serum (a), and two eluates from affinity chromatography; (b) as described in the text; (c) using ultracentrifuged serum for the affinity procedure.

Figure 3. SDS-polyacrylamide electrophoresis of the purified lectin.

Molecular Weight Estimation

The purified agglutinin was chromatographed on Sephadex G-200 in the presence of HEPES-citrate for molecular weight determination (Fig. 4). The molecular weight markers were

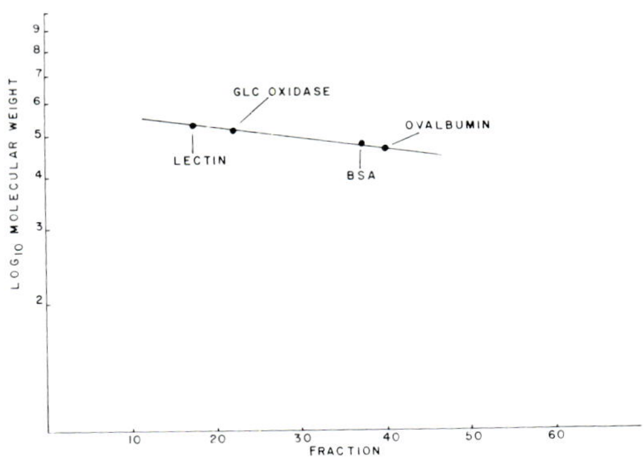

Figure 4. Molecular weight estimation of the purified lectin by gel filtration on Sephadex G-200. See text for key.

(1) ovalbumin, MW 45,000; (2) bovine serum albumin (fraction V), MW 67,000; (3) glucose oxidase, MW 150,000. The purified

lectin (4) was found to have a molecular weight of approximately 200,000.

Scanning Electron Microscopy

S. minnesota and *S. aureus* (25923) cells were incubated with whole *Limulus* serum and agarose-purified agglutinin, then inspected by scanning electron microscopy. The results shown in Fig. 5 indicate that *S. minnesota* cells were agglutinated with whole *Limulus* serum. Fig. 6 is a micrograph of the same organism to which the purified agglutinin was added. No apparent difference between these two figures can be noted, indicating that the lectin does not have agglutinating activity for *S. minnesota*. Fig. 7 represents *S. aureus* (25923) cells to which no whole *Limulus* serum or purified lectin has been added. Normal staphylococcal association is seen; however, this association is greatly different from that seen in Fig. 8 where these cells have been incubated with the purified lectin.

DISCUSSION

A simple method for isolating a carbohydrate-binding protein (lectin) in *Limulus polyphemus* serum has been developed. Since calcium was found to be required for biological activity, elution of the lectin from the affinity column was achieved using a citrate chelator. The lectin was shown to be electrophoretically pure by PAGE, to consist of at least five electrophoretically distinguishable subunits, and to have a molecular weight of approximately 200,000 by gel filtration.

The purification of this lectin resulted in agglutinating activity against *S. aureus* but not *S. minnesota* or chicken erythrocytes. This was revealed by microtitration and scanning electron microscopy. These results indicate the presence of multiple agglutinins in *Limulus* serum.

Since the affinity ligand was Sepharose 4B, the lectin has an apparent affinity for galactose or a galactose-associated molecule. Agarose is a linear polysaccharide consisting of alternating residues of D-galactose and 3,6-anhydro-L-galactose units (Percival, 1970). Recovery of the lectin by the above procedure suggests that this molecule has a moderate affinity for galactose molecules. It is possible

Fig. 5

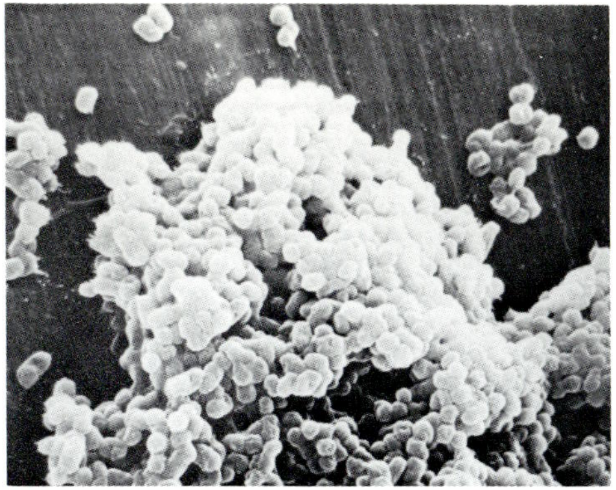

Fig. 6

Figure 5. Scanning electron micrograph of *S. minnesota* cells demonstrating agglutination by whole *Limulus* serum.(Mag. 2400X)
Figure 6. Scanning electron micrograph of *S. minnesota* cells to which agarose-purified agglutinin was added. Control micrographs, with no added serum or serum fraction, appeared identical. (Mag. 6700X)

Fig. 7

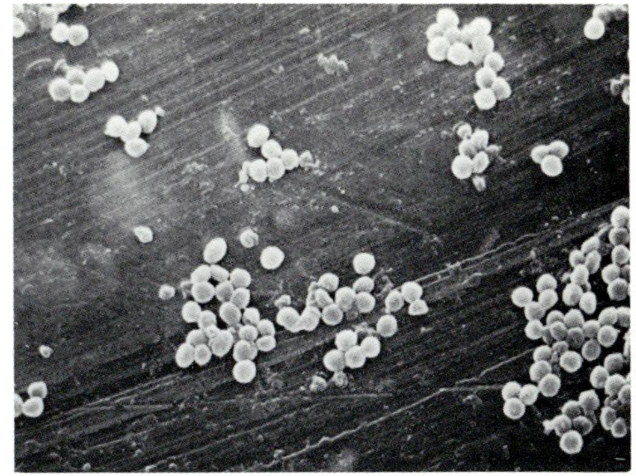

Fig. 8

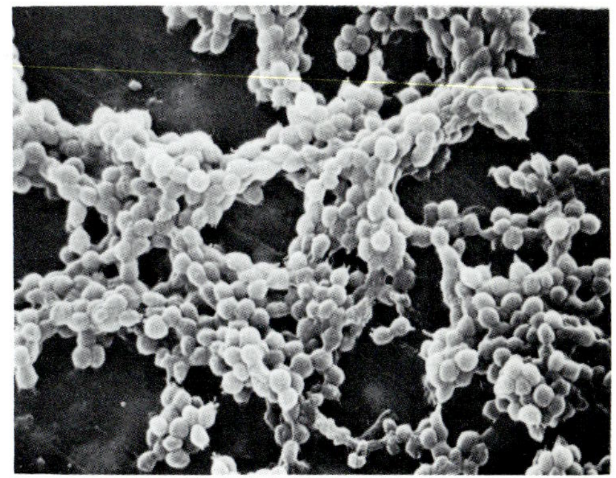

Figure 7. Scanning electron micrograph of *S. aureus* (25923) with no added serum or serum fraction, demonstrating normal association of these cocci. (Mag. 2500X)
Figure 8. Scanning electron micrograph of *S. aureus* (25923) with added agarose-purified agglutinin, demonstraing agglutination. (Mag. 2500 X)

that the lectin may have a stronger avidity for a galactose-associated amino sugar or perhaps a galactose-associated di- or oligosaccharide. Several authors have noted an anti-galactan precipitin in *Limulus* serum (Cohen *et al.*, 1975; Voigtmann *et al.*, 1971). The isolation and characterization of the lectin presented in this paper is supportive of their observations and represents evidence of a lectin with anti-galactose or galactose-associated activity in *L. polyphemus* serum.

ACKNOWLEDGEMENTS

These studies were supported in part by a grant from the National Institutes of Health (AI-12817). We thank Dr. James A. Stewart, Department of Biochemistry, University of New Hampshire, for helpful advice.

REFERENCES

Bang FB (1967). Serologic responses among invertebrates other than insects. Fed Proc 26:1680.

Bradford MM (1976). A rapid and sensitive method for the quantitation of microgram quantities of protein utilizing the principle of protein-dye binding. Anal Biochem 72:248.

Cohen E, Baldo BA, Uhlenbruck G (1975). Anti-galactan precipitans in the hemolymph of *Tridacna maxima* and *Limulus polyphemus*. In Hildemann WH, Benedict AA (eds): "Immunologic Phylogeny," New York: Plenum Press, p 13.

Cohen E, Roberts S, Nordling S, Uhlenbruck G. (1972) Specificity of *Limulus polyphemus* agglutinins for erythrocyte receptor sites common to M and N antigenic determinants. Vox Sang 23:300.

Cohen E, Rozenberg M (1974). Agglutinins of *Limulus polyphemus* (horseshoe crab) and *Birgus latro* (coconut crab). Ann NY Acad Sci 234:28.

Cornick JW, Stewart JE (1968). Interaction of the pathogen *Gaffkya homari* with natural defense mechanisms of *Homarus americanus*. J Fish Res Bd Canada 24:695.

Fairbanks G, Steek TL, Wallach DF (1971). Electrophoretic analysis of the major polypeptides of the human erythrocyte membrane. Biochemistry 10:2606.

Finstad CL, Good RA, Litamn GW (1974). The erythrocyte agglutinin from *Limulus polyphemus* hemolymph: molecular structure and biological function. Ann NY Acad Sci 234:170.

Finstad CL, Litman GW, Finstad J, Good RA (1972). The evolution of the immune response. XIII. The characterization of purified erythrocyte agglutinings from two invertebrate species. J Immunol 108:1704.

Hall JL, Rowlands JM (1974). Heterogeneity of lobster agglutinins. II. Specificity of agglutinin-erythrocyte binding. Biochemistry 13:828.

Kaplan R, Li S, Kehoe JM (1977). Molecular characterization of limulin, a sialic acid binding lectin from the hemolymph of the horseshoe crab, *Limulus polyphemus*. Biochemistry 16:4297.

Köhler W, Prokop O, Kühnemund O (1973). Routine identification of group-C streptococci by means of the albumin gland of the edible snail, *Helix pomatia*. J Med Microbiol 6:127.

Laemli UK (1970). Clevage of structural proteins during assembly of the head of bacteriophage T4. Nature (London) 227: 680.

Lowry OH, Rosebrough NJ, Farr AL, Randall RJ (1951) Protein measurement with the folin phenol reagent. J Biol Chem 193:265.

Marchalonis JJ, Edelman GM (1968). Isolation and characterization of a hemagglutinin from *Limulus polyphemus*. J Molec Biol 32:453.

McKay D, Jenkin C, Rowley D (1970). Immunity in the invertebrates. I. Studies on the naturally occurring hemagglutinins in the fluid from invertebrates. Aust J Exp Biol Med Sci 47:125.

Noguchi H (1903). A study of immunization--haemolysins, agglutinins, precipitins and coagulins in cold-blooded animals. Zbl Bakt Abt Orig 33:353.

Nowak TP, Barondes SH (1975). Agglutinins from *Limulus polyphemus* with formalinized horse erythrocytes as the affinity adsorbent. Biochim Biophys Acta 393:115.

Oppenheim, JD, Nachbar MS, Salton MR, Aull F (1974). Purification of a hemagglutinin from *Limulus polyphemus* by affinity chromatography. Biochim Biophys Acta 58:1127.

Percival E (1970). Algal polysaccharides. In Pigman W, Horton D, Herp A (eds): "The Carbohydrates. Chemistry and Biochemistry", 2nd ed, New York: Academic Press, p 537.

Pistole TG (1978). Broad-spectrum bacterial agglutinating activity in the serum of the horseshoe crab, *Limulus polyphemus*. Develop Comp Immunol 2:65.

Roche AC, Monsigny M (1974). Purification and properties of limulin: a lectin (agglutinin) from hemolymph of *Limulus polyphemus*. Biochim Biophys Acta 371:242.

Roelcke D, Uhlenbruck G, Metaxas MN (1971). Untersuchungen über die serologische Spezitität neuraminsäurehaltiger Rezeptoren auf Humanerythrozyten. Z Immun-Forsch 141:141.

Tyson C, Jenkin CR (1973). The importance of opsonic factors in the removal of bacteria from the circulation of the crayfish (*Parachaeraps bicarinatus*). Aust J Exp Biol Med Sci 51:609.

Voigtmann R, Salfner B, Uhlenbruck G (1971). Studies on broad spectrum agglutinins. IX Specific and unspecific reactions between *Limulus polyphemus* hemolymph and snail extracts with anti-A specificity. Z Immun-Forsch 141:488.

SITES ON THE LIPOPOLYSACCHARIDE MOLECULE REACTIVE WITH *LIMULUS* AGGLUTININS

Homayoun Rostam-Abadi and Thomas G. Pistole

Department of Microbiology
University of New Hampshire
Durham, New Hampshire 03824

INTRODUCTION

The hemolymph of many invertebrate species contains glycoproteins capable of agglutinating a variety of cell types. They function by binding specifically to polysaccharides on the surface of cells, and thus provide a tool for the study of cell surface structures. Although the physiological importance of agglutinins is not yet understood, they may serve in a host-defense capacity similar to that of vertebrate antibodies. They agglutinate a variety of heterologous cells but do not agglutinate homologous cell types. Thus a role of "self recognition" can be suggested.

Recent evidence indicates that cell-free *Limulus polyphemus* hemolymph is capable of agglutinating a variety of microorganisms (Pistole, 1978). Using both direct agglutination, and agglutination-inhibition studies, Pistole (1978) has demonstrated that the deep-core mutants of *Salmonella minnesota* (Re mutant) are capable of removing the agglutinating activity directed against the smooth, wild-type, strain of this species. Preliminary evidence suggests that 2-keto-3-deoxyoctonate (KDO) may act as the immuno-determinant group in the core region of *Enterobacteriaceae*. KDO was initially shown to be a component of the lipopolysaccharide of *Escherichia coli* (Heath *et al.*, 1963) and *Salmonella typhimurium* (Osborn, 1963). KDO occupies the inner most sugar-residue position in the core polysaccharide and provides the

Pages 537–545, Biomedical Applications of the Horseshoe Crab (Limulidae)

point of attachment to the lipid A moiety which is embedded in the outer membrane of Gram-negative *Enterobacteriaceae*. Lipid A is composed of a glucosamine-phosphate backbone to which are attached ester- and amide-linked fatty acids (Galanos *et al*., 1977). KDO and lipid A are present in virtually all strains of *Enterobacteriaceae*, although KDO is absent in some Gram-negative organisms, e.g., *Pseudomonas pseudotuberculosis*, *Bacteroides fragilis*, *Bacteroides melaninogenicus*, and a number of other Gram-negative bacteria (Galanos *et al*., 1977).

We present direct evidence that *Limulus* hemolymph contains precipitins directed against both KDO, and the glucosamine-phosphate backbone of lipid A obtained from the glycoplipid of Rc mutants of *Salmonella minnesota*. Inhibition of precipitation using free KDO provides further evidence in support of the presence of precipitins in *Limulus* hemolymph directed against this sugar residue.

MATERIALS AND METHODS

Bacterial Cultivation

R mutants of *S. minnesota* grown on blood agar plates were checked for homogeneity of growth and purity. Rough colonies were inoculated onto triple sugar iron agar (TSI) and those which showed typical reactions were selected for futher cultivation. *S. minnesota* mutant strains of the following chemotypes were used: Rc, Rd_1, and Re. They were grown in trypticase soy broth in Fernbach flasks at 37°C with constant stirring.

Extraction Procedure

A mixture containing liquid phenol (9 grams dry phenol plus 1.1 ml distilled water), chloroform and petroleum ether (boiling point 40-60°C) in a volume ratio of 2:5:8, respectively, was used. Glycolipid was isolated from the rough strains with this extraction mixture at 4°C using the method of Galanos *et al*. (1969). Each glycolipid was purified by repeated ultracentrifugation at $10^5 g$ for 4 hr. The purified material was lyophilized and stored under vacuum at room temperature until used.

Hydrolysis of Glycolipids and Core with Mild Acid

The lipid A from the respective glycolipids was removed by 1% acetic acid (pH 3.4) at 100°C for 1.5 hr. After removal of lipid A by centrifugation at 3000*g* for 10 min, the water-soluble portion was further purified by ultracentrifugation at $10^5 g$ for 4 hr to remove intact glycolipid. Rc core was hydrolyzed in 2N HCl (100°C, 1 hr) and the hydrolysate was subjected to paper chromatography.

Purification of Rc Core

The supernatant, recovered from the above procedure and containing KDO, KDO-phosphorylethanolamine, and a "heptose fraction", was subjected to high voltage paper electrophoresis. The slowest moving fraction (Dröge *et al.*, 1970), identified as the "heptose fraction", was eluted from the paper, lyophilized, and stored at -15°C.

Hydroxylaminolysis

Fifteen mg of lipid A were treated with 3 ml of 2% hydroxylamine in 4% NaOH in ethanol at 68°C for 6 min, a modification of a technique originally described by Snyder and Stephens (1959). The liberated fatty acids in the form of hydroxymates were in the supernatant fluid. The precipitate was washed exhaustively with small fractions of alkaline hydroxylamine, being recovered each time by centrifugation (10 min, 3000*g*, 10°C). The precipitate was resuspended in a small volume of standard diluent (see below) and extracted three times with chloroform. The chloroform-insoluble fraction was dried in a vacuum desiccator to remove hydroxylamine, and lyophilized after dissolving in standard diluent. Both chloroform-soluble and insoluble fractions were subjected to paper chromatography in 1-propanol-water (3:1), and stained with ammonium silver nitrate and ninhydrin. Glucosamine and chitobiose served as standards.

Liberation of Fatty Acids from the Glycolipid

Three mg of Rd_1 glycolipid were heated at 100°C for 24 hrs in 4N HCl in a sealed tube. The hydrolysates were extracted exhaustively with chloroform-methanol (2:1). The combined chloroform extracts were dried under vacuum.

Modification of Fatty Acids

Fatty acids were methylated in the presence of BCl_3. The methylated fatty acids were concentrated *in vacuo* and the residue was dissolved in methylene chloride for gas-liquid chromatography analysis.

Solubilization of Lipid A

Lipid A prepared from Re glycolipid by acetic acid hydrolysis was checked for absence of KDO and heptose. Ten mg of lipid A in distilled water (5 ml) were solubilized by the addition of triethylamine(25-35 µl). This solution was mixed with a solution of bovine serum albumin (10 mg) in water (5 ml). The resulting mixture was dried in a rotary evaporator (Galanos *et al.*, 1972). The complex was redissolved in standard diluent (0.01 M $CaCl_2$, 0.154 M NaCl , pH 7.4) and lyophilized. Stock solutions were prepared by dissolving the complex in standard diluent.

Authentic Fatty Acids

β-hydroxy myristic acid and other saturated and unsaturated fatty acid methyl esters (1%) used as standards for gas liquid chromatography were obtained from M. Ikawa.

Paper Electrophoresis and Paper Chromatography

Paper electrophoresis was carried out in pyridine-acetic acid-water (100-40-860 v/v/v pH 5.3). Approximately 50 µl of sample was applied to a 6 x 57 cm chromatography-grade paper (Whatman #1), and 45V/cm for 1 1/2 hr were applied. Ascending paper chromatography was performed on Whatman #1 paper in 1-propanol-water (3:1).

Gas-Liquid Chromatography

Gas-liquid chromatography was carried out in a Barber Colman Selectra System series 5000 with flame ionization detector on a column packing of 12% polyethylene glycol succinate on 80/90 mesh Anachrom A (1/4" x 6') at 210°C. The carrier gas was nitrogen.

Colorimetric Determinations

Throughout these studies KDO was identified using the thiobarbituric acid spray method; glucosamine in acid hydrolysates (4N HCl, 24 hr, 100°C) by ninhydrin; and heptose with H_2SO_4/cysteine as described by Nowotony (1969). Phosphate was detected by the spray technique of Hares and Isherwood as described by Nowotny (1969).

Qualitative Precipitin Test (Ring-Test)

Test antigens were suspended in high purity water (Milli-Q system, Millipore; 18 megamho) or standard diluent. Precipitation was carried out in 4-mm-diameter test tubes using 0.2 ml each of Ag and whole *Limulus* hemolymph or fractions obtained after ultracentrifugation of whole serum at $10^5 g$ for 2 hr.

RESULTS

Analysis of Degraded Glycolipids

Glycolipids from various mutants of *S. minnesota* were extracted as described (Galanos *et al.*, 1969). After acid hydrolysis (1% acetic acid, 100°C, 1.5 hr), the degraded glycolipids from the Rc mutant of *S. minnesota* were analyzed for their sugar constituents, using high voltage paper electrophoresis. The hydrolysate consisted of three fractions (Dröge, 1970): the KDO fraction, having the fastest electrophoretic mobility, was free of heptose and glucose and stained intensively with thiobarbituric acid. This fraction represents a mixture of free KDO and KDO which is altered in an unknown way by the action of acetic acid. A second fraction, representing KDO-phosphorylethanolamine, stained with ammonium silver nitrate (Ag^+/OH^-), thiobarbituric acid, and ninhydrin, indicating the presence of ethanolamine bound to KDO. The "heptose fraction" or core of *S. minnesota* chemotype Rc, which exhibited the least electrophoretic mobility, stained with thiobarbituric acid and ammonium silver nitrate (Ag^+/OH^-) reagent, but not with ninhydrin. The presence of heptose was confirmed by analysis of core sugars, and comparing the mobility of residues with authentic heptose after staining with H_2SO_4/cysteine. Purity of this fraction was further investigated using gas-liquid chromatography.

Analysis of the Heptose Fraction

Chromatographic analysis of the "heptose fraction" revealed no fatty acids, and since this fraction did not stain with ninhydrin, glucosamine was also absent from the purified material.

Analysis of Hydroxylamine-treated Lipid A and Isolation of Glucosamine Disaccharide

Chromatographic analysis of the chloroform-insoluble residue indicated that more than 99% of all fatty acids, including amide-linked β-hydroxymyristic acid, were removed. Examination of this same fraction by paper chromatography showed that it had a mobility comparable to the N'N'-diacetylglucosamine, chitobiose, and stained with ninhydrin. Figure 1.

Qualitative Precipitation Analysis

Qualitative analysis of precipitin reactions clearly demonstrated that both Re and Rc glycolipids equally precipitated *Limulus* hemolymph as well as the supernatant fractions of serum obtained by ultracentrifugation (See Table 1). The

TABLE 1

QUANTITATIVE PRECIPITIN TESTS USING WHOLE *LIMULUS* SERUM AND ULTRACENTRIFUGED FRACTIONS

Test substance	Conc. (mg/ml)	Whole serum	Fractions I	Fractions II
Re glycolipid	1	3+	4+	4+
Rc glycolipid	1	3+	4+	4+
Rc core (heptose frac.)	3	3+	4+	4+
NH_2OH-treated Lipid A	3	3+	4+	4+
Chitobiose	2	1+	2+	2+
Melibiose	12	0	0	0
Glucosamine	5	0	0	0
Cellobiose	12	0	0	0
Rc core (mixture)	5	0	0	0
BSA	5	+/-	+/-	+/-
BSA-Lipid A	1	4+	4+	4+
Diluent ($CaCl_2$-NaCl)	0	0	0	0

precipitin reaction appeared to be specific in that neither melibiose nor cellobiose in high concentrations (12 mg/ml) demonstrated any reaction with *Limulus* hemolymph. Chitobiose, a disaccharide of N-acetylglucosamine, which is obtained by partial hydrolysis of chitin, weakly precipitated these same serum fractions. Glucosamine monosaccharide even in high concentration (5 mg/ml) did not precipitate whole serum. Rc core and glucosamine backbone of lipid A precipitated these fractions while the core mixture which contains "heptose fraction", KDO and KDO-PN did not precipitate *Limulus* serum or any of its fractions. Bovine serum albumin (BSA)-solubilized lipid A also formed a strong precipitin line with all fractions tested.

DISCUSSION

Analytical studies using gas-liquid chromatography and paper chromatography have shown that both the "heptose fraction" and unreduced lipid A backbone, glucosamine β(1-6) glucosamine were devoid of fatty acids, and the former also lacked glucosamine since this fraction did not stain with ninhydrin. Analysis of the alkaline hydroxylamine-treated lipid A product (chloroform-insoluble fraction) by gas chromatography clearly demonstrated that the modification of technique of Snyder *et al.* (1959) outlined in the Methods section can be used to isolate the glucosamine backbone of lipid A, free of all contaminating fatty acids, including the β-hydroxymyristic acid amide-linked to the glucosamine disaccharide. In view of the fact that the precipitate obtained after hydroxylaminolysis was further treated with hydroxylamine during centrifugation runs, it appears that longer treatment of lipid A with hydroxylamine can remove amide-linked β-hydroxymyristic acid from lipid A. The product which has a mobility comparable to chitobiose is the glucosamine backbone of lipid A.

Serological analysis of isolated fractions of glycolipid clearly indicates that two sites on the glycolipid react with components of *Limulus* hemolymph. Since both Re and Rc glycolipids precipitate *Limulus* hemolymph equally well, it appears that heptose does not play an important role in the precipitation reaction. In addition, since heptose is a reducing sugar in the Rc core oligosaccharide, it may not react with Limulus agglutinins strongly (Galanos *et al.*, 1977).

Since the "heptose fraction" precipitates *Limulus* hemolymph as well as its fractions very strongly, and glucose and heptose appear not to play a major role in precipitating *Limulus* hemolymph, we suggest that KDO represents the immunodeterminant sugar residue in the tetrasaccharide core of the Rc chemotype. Further evidence for the involvement of KDO is provided by the inability of *Limulus* hemolymph and its fractions to precipitate the core mixture, which consists of "heptose fraction", "KDO-PN" and KDO. We suggest that lack of precipitation here is due to inhibition by free KDO which is present in high concentration in this mixture.

Our studies also demonstrate that the glucosamine backbone of lipid A, freed of its fatty acids is capable of strongly precipitating *Limulus* hemolymph and its fractions. This indicates the presence of multiple agglutinins or precipitins in the hemolymph of *Limulus polyphemus* reactive against glycolipid of *S. minnesota* rough mutants.

In conclusion we have shown that KDO and glucosamine in the glycolipid of the Rc chemotype of *S. minnesota* react with agglutinins in the hemolymph of *Limulus polyphemus*. Our laboratory is in the process of isolating the respective agglutinins.

ACKNOWLEDGEMENTS

These studies were supported in part by grants from the National Institutes of Health (AI-12817) and from the Biomedical Research Support Program, UNH (R-373). We thank Dr. M. Ikawa, Department of Biochemistry, University of New Hampshire, for providing advice and technical assistance.

REFERENCES

Cohen E, Rozenberg M (1974). Agglutinins of *Limulus polyphemus* (horseshoe crab) and *Birgus latro* (coconut crab). Ann NY Acad Sci 234:28.

Dröge W, Lehman V, Lüderitz O, Westphal O (1970). Structural investigations on the 2-keto-3-deoxyoctonate regions of lipopolysaccharides. Eur J Biochem 14:175.

Galanos C, Lüderitz O, Westphal O (1969). A new method for the extraction of R lipopolysaccharides. Eur J Biochem 9:245.

Galanos C, Rietschel ET, Lüderitz O, Westphal O, Kim YB, Watson DW (1972). Biological activities of lipid A complexed with bovine-serum albumin. Eur J Biochem. 31:230.

Galanos C, Lüderitz O, Rietschel ET, Westphal O (1977). Newer aspects of the chemistry and biology of bacterial lipopolysaccharides with special reference to their lipid A component. In Goodwin TW (ed): "International Review of Biochemistry. Biochemistry of Lipids II," Baltimore: University of Park Press, p 239.

Heath EC, Ghalambur MA (1963). 2-keto-deoxy-octonate, a constituent of cell wall lipopolysaccharide preparations obtained from *E. coli*. Biochem Biophys Res Commun 10:340.

Nowotny A (1969). "Basic Exercises in Immunochemistry." New York: Springer-Verlag, p 79.

Osborn MJ (1963). Studies on the Gram-negative cell wall. I. Evidence for the role of 2-keto-3-deoxy-octonate in the lipopolysaccharide of *Salmonella typhimurium*. Proc Natl Acad Sci 50:499.

Pistole TG (1978). Broad-spectrum bacterial agglutinating activity in the serum of the horseshoe crab, *Limulus polyphemus*. Develop Comp Immunol 2:65.

Pistole TG, Britko JL (1978). Bactericidal activity of amoebocytes from the horseshoe crab, *Limulus polyphemus*. J Invert Pathol 31:376.

Rietschel ET, Gottert H, Lüderitz O, Westphal O (1972). Nature and linkage of the fatty acids present in the lipid A component of Salmonella lipopolysaccharides. Eur J. Biochem 28:166.

Snyder F, Stephens N (1959). A simplified spectrophotometric determination of ester groups in lipids. Biochim Biophys Acta 34:244.

BACTERIAL AGGLUTININS FROM *LIMULUS POLYPHEMUS*--AN OVERVIEW

Thomas G. Pistole

Department of Microbiology
University of New Hampshire
Durham, New Hampshire 03824

INTRODUCTION

Two activities associated with *Limulus polyphemus* -- the ability to agglutinate erythrocytes and other mammalian cells, and the ability to detect bacterial lipopolysaccharides-- have been well studied. Serum from this animal is also capable of reacting with bacteria. *Limulus* shares this capability of agglutinating bacterial cells with a variety of invertebrates; however, studies in other animals have been limited in both the number and variety of bacterial species examined. Data collected in our laboratory have shown that *Limulus* serum agglutinates a wide spectrum of bacterial species. This bacterial agglutinating activity is the subject of this report.

EXPERIMENTAL FINDINGS

Early studies in our laboratory used *Salmonella minnesota* wild-type for agglutination studies (Pistole, 1976) and this subsequently was used as a reference strain for studies using other bacterial species (Pistole, 1978). Using heat-killed microorganisms (either 60°C for two hours or 100°C for one hour) and whole *Limulus* serum in standard microtitration assays, I and my associates were able to show the range of reactivity of this serum. Table 1 summarizes our findings with a variety of Gram-positive bacteria. Compared to our test organism, *S. minnesota*, most species tested, with the exception of *Micrococcus luteus*, exhibited significant activity. Thus, five out of six Gram-positive cocci and the single Gram-positive bacillus, *Listeria monocytogenes*,

Pages 547–553, Biomedical Applications of the Horseshoe Crab (Limulidae)

were agglutinated by *Limulus* serum.

TABLE 1

Direct Agglutination Studies.
Activity of *L. polyphemus* Serum against a
Variety of Gram-positive bacteria

Bacterial Species	N[a]	Titer[b]	Percent Activity[c]
Streptococcus pyogenes	10	5.0 ± 0.21	100
Staphylococcus aureus	10	4.9 ± 0.15	98
Streptococcus agalactiae	10	4.6 ± 0.30	92
Streptococcus fecalis	10	3.5 ± 0.26	70
Listeria monocytogenes	10	2.8 ± 0.28	56
Staphylococcus epidermidis	10	2.6 ± 0.26	52
Micrococcus luteus	10	0.6 ± 0.27	12
Salmonella minnesota	10	5.0 ± 0.25	--

[a]Number of serum samples tested
[b]Log2 value ± 1 standard error
[c]Compared to the test antigen system: *S. minnesota*

The results of similar studies using a variety of Gram-negative bacilli are shown in Table 2. Again a range of

TABLE 2

Direct Agglutination Studies
Activity of *L. polyphemus* Serum against a
Variety of Gram-negative Bacteria

Bacterial Species	N[a]	Titer[b]	Percent Activity[c]
Flavobacterium sp. UNH	10	4.5 ± 0.28	80
Salmonella typhimurium UNH	10	4.1 ± 0.27	73
Pseudomonas putida UNH	10	3.3 ± 0.35	59
Pseudomonas aeruginosa UNH	10	1.6 ± 0.16	28
Serratia marcescens UNH	10	0.8 ± 0.14	14
Salmonella minnesota UNH	10	5.6 ± 0.18	--

[a]See explanatory notes in Table 1

activity was found. Although not shown on this slide, most other members of the family *Enterobacteriaceae* which we tested reacted similarly to the *Salmonella typhimurium* strain shown in this Table. Of the organisms listed in Table 2 only *Salmonella* and *Serratia* are members of the *Enterobacteriaceae;*

it is therefore clear from these results that the reactive material or materials present in *Limulus* serum is not specific for this family of bacteria.

The difficulty in making direct comparisons in the above studies led us to carry out a series of adsorption studies. Here the heat-killed organisms were incubated with samples of *Limulus* serum. The mixtures were subsequently centrifuged and the clarified supernatant fluids--the adsorbed sera--were tested for their ability to agglutinate our test organism, *S. minnesota*. Unadsorbed sera were included for comparison. Table 3 summarizes the results of one such set of experiments.

TABLE 3

Agglutinin Adsorption Studies.
Agglutination Activity of *L. polyphemus* Serum, Adsorbed with Various Bacterial Antigens, against *S. minnesota*

Adsorbing Antigen	N[a]	Titer[a]	Percent Activity[a]	p Value[b]
Salmonella typhimurium UNH	5	1.0 ± 0.45	18.9	< 0.01
Salmonella tennessee UNH	5	1.8 ± 0.49	34.0	< 0.01
Escherichia coli ATCC 9637	5	2.3 ± 0.37	43.4	< 0.01
Escherichia coli ATCC 10798	5	2.4 ± 0.37	45.3	< 0.01
Salmonella paratyphi A UNH	5	2.4 ± 0.40	45.3	< 0.01
Escherichia coli ATCC 12435	5	2.8 ± 0.20	52.8	< 0.01
Unadsorbed serum	5	5.3 ± 0.37	--	--

[a]See explanatory notes in Table 1.
[b]Compared to unadsorbed serum.

In this study all the antigens used to absorb the serum samples were, like *S. minnesota*, members of the family *Enterobacteriaceae*, and in each case adsorption with these organisms resulted in a statistically significant drop in the agglutinating activity directed against *S. minnesota*. This strongly supports the idea that the same component in *Limulus* serum is reacting with each organism.

The procedure of agglutinin adsorption was expanded to other bacteria species and the results are presented in Table 4. In this Table are found a number of Gram-positive bacteria used as adsorbing antigens. Serum so adsorbed exhibited no statistically significant drop in activity toward *S. minn-*

esota. Since these Gram-positive organisms had already been

TABLE 4

Agglutinin Adsorption Studies.
Agglutination Activity of *L. polyphemus* Serum,
Adsorbed with Various Bacterial Antigens,
against *S. minnesota*

Adsorbing antigen	N[a]	Mean Titer[a]	Percent Activity[a]	pValue[b]
Gram-positive				
S. epidermidis ATCC 12228	5	3.1 ± 0.18	86.1	> 0.10
S. aureus ATCC 25923	5	3.4 ± 0.16	94.4	> 0.10
L. monocytogenes UNH	5	3.5 ± 0.16	97.2	> 0.10
S. agalactiae UNH	5	3.6 ± 0.22	100.0	> 0.10
S. fecalis ATCC 9790	5	3.6 ± 0.21	100.0	> 0.10
Gram-negative				
S. minnesota UNH	5	1.7 ± 0.15	47.2	< 0.01
S. marcescens ATCC 8100	5	1.9 ± 0.18	52.8	< 0.01
Pasteurella multocida UNH	5	2.0 ± 0.26	55.6	< 0.01
Klebsiella pneumoniae ATCC 23357	5	2.1 ± 0.61	58.3	< 0.01
S. marcescens UNH	5	2.2 ± 0.32	61.1	< 0.01
Proteus vulgaris ATCC 6380	5	2.9 ± 0.18	80.6	> 0.10
P. aeruginosa ATCC 14207	5	3.0 ± 0.21	83.3	> 0.10
P. aeruginosa ATCC 10145	5	3.0 ± 0.16	83.3	> 0.10
K. pneumoniae UNH	5	3.3 ± 0.29	91.7	> 0.10
Unadsorbed serum	5	3.6 ± 0.23	--	--

[a]See explanatory notes in Table 1.
[b]Compared to unadsorbed serum.

shown to be directly agglutinated by *Limulus* serum, this adsorption study provided evidence for two separate agglutinins in *Limulus* serum: one directed toward *S. minnesota* and other Gram-negative bacilli, one reactive with Gram-postive microorganisms. It seems likely that the serum component responsible for agglutinating these Gram-positive bacteria is identical to the galactose-reactive lectin reported by others (Gilbride and Pistole, 1978; Cohen *et al*., 1975); whereas the 2-keto-3-deoxyoctonate (KDO)-reactive lectin reported by Rostam-Abadi and Pistole (1978) is responsible for the agglutination of Gram-negative bacteria.

The lower half of Table 4 indicates that not all Gram-negative bacilli were capable of adsorbing out the agglutinating activity directed against *S. minnesota.* Those organisms which were unreactive in this procedure (*Proteus vulgaris, Pseudomonas* spp., and *Klebsiella pneumoniae* UNH strain) were also poorly reactive in direct agglutination studies, which suggests they are either deficient in or devoid of the appropriate ligand, or that it is at least partially cryptic in these organisms.

DISCUSSION

As we examine invertebrate sera for bacterial agglutinating activity, it is necessary to modify the approach used in studying hemagglutination activity. That *Limulus* serum contains a hemagglutinin whose receptor is N-acetyl neuraminic acid (NANA) has been shown by several investigators (Cohen *et al.*, 1972; Roelcke *et al.*, 1971; Voigtmann *et al.*, 1971). If, however, as seems likely, this same lectin is also responsible for binding to KDO, the question arises: to which of these sites, if either, is the lectin directed naturally. If it is KDO, its role in immunological defense seems likely; if **NANA**, a structure found in relatively few bacterial species (Ng and Dain, 1976), its role is less clear. Furthermore, binding to NANA- and KDO-containing structures may be incidental to the, at present, unknown role of this lectin.

The galactose-reactive lectin appears to be responsible for agglutinating Gram-positive bacteria. Studies in our laboratory suggest that this lectin may also bind to the disaccharide core of the lipid A moiety of lipopolysaccharide (Rostam-Abadi and Pistole, 1978). Conversely this same lectin may also react with the structurally similar chitobiose molecule and may therefore play a role in transport of subunits for chitin formation, a general suggestion made earlier by Cohen (1974). Until the native specificity of these lectins is known, their role *in situ* remains speculative. Studies conducted in our laboratory have shown that *Limulus* serum potentiates the bactericidal activity of amoebocytes with an *Escherichia coli* system (Pistole and Britko, 1978). If the opsonizing factor in serum turns out to be an agglutinin, this should strengthen the idea that these molecules are truly a part of the defense system of *Limulus polyphemus*. It is of interest to note that Hall and Rowlands (1974) have recovered two lectins from the American lobster, *Homarus americanus*, whose specificities--N-acetyl-neuraminic acid and N-acetyl-galactosamine--closely parallel those associated with *Limulus*. By examining the similarities between these two marine invertebrates, we may be able to assess better the role of their lectins.

SUMMARY

In summary, using direct agglutination tests, we have demonstrated a broadly reacting bacterial agglutinating activity in *Limulus* serum. Subsequent agglutinin-adsorption studies indicated the presence of two separate agglutinins: one, reactive with Gram-positive bacteria, which is probably identical to the anti-galactan previously reported; the other, reactive with a variety of Gram-negative bacteria, with apparent affinity for KDO in bacterial lipopolysaccharide. Preliminary evidence suggests the latter is equivalent to the well-documented NANA-reactive lectin. Speculation on the role of these lectins is presented as well as areas for further study.

ACKNOWLEDGEMENTS

These studies were supported in part by a grant (AI-12817) from the National Institutes of Health. Tables 1-4 are reprinted with permission from *Developmental and Comparative Immunology*, Vol. 2, No. 1, Pistole TG. Broad-spectrum bacterial agglutinating activity in the serum of the horseshoe crab, *Limulus polyphemus*. 1978, Pergamon Press, Ltd.

REFERENCES

Cohen E, Roberts SC, Nordling S, Uhlenbruck G (1972). Specificity of *Limulus polyphemus* agglutinins for erythrocyte receptor sites common to M and N antigenic determinants. Vox Sang 23:300.

Cohen E, Rozenberg M, Massaro EJ (1974). Agglutinins of *Limulus polyphemus* (horseshoe crab) and *Birgus latro* (coconut crab). Ann NY Acad Sci 234:28.

Cohen E, Baldo BA, Uhlenbruck G (1975). Anti-galactan precipitins in the hemolymph of *Tridacna maxima* and *Limulus polyphemus*. In Hildemann WH, Benedict AA (eds): "Immunologic Phylogeny", New York: Plenum Press, p 13.

Gilbride KJ, Pistole TG (1978). Isolation and characterization of a bacterial agglutinin (lectin) in the serum of the horseshoe crab, *Limulus polyphemus*. This volume.

Hall JL, Rowlands, Jr DT (1974). Heterogeneity of lobster agglutinins. II. Specificity of agglutinin-erythrocyte binding. Biochemistry 13:828.

Ng S-S, Dain JA (1976). The natural occurrence of sialic acids In Rosenberg A, Schengrund C-A (eds): "Biological Roles of Sialic Acid", New York: Plenum Press, p 59.

Pistole TG (1976). Naturally occurring bacterial agglutinin in the serum of the horseshoe crab, *Limulus polyphemus*. J Invert Pathol 28:153.

Pistole TG (1978) Broad-spectrum bacterial agglutinating activity in the serum of the horseshoe crab, *Limulus polyphemus*. Develop Comp Immunol 2:65.

Pistole TG, Britko JL (1978). Bactericidal activity of amoebocytes from the horseshoe crab, *Limulus polyphemus*. J Invert Pathol 31:376.

Roelcke D, Uhlenbruck G, Metaxas MN (1971). Untersuchungen über die serologische Spezitität neuraminsäurehaltiger Rezeptoren auf Humanerythrozyten. Z Immun-Forsch 141:141.

Rostam-Abadi H, Pistole TG (1978). Sites on the lipopolysaccharide molecule relative with *Limulus* agglutinins. This volume.

Voigtmann R, Salfner B, Uhlenbruck G (1971). Studies on broad spectrum agglutinins IX. Specific and unspecific reactions between *Limulus polyphemus* haemolymph and snail extracts with "anti-A" specificity. Z Immun-Forsch 141: 488.

DISCUSSION

Erwin Neter

State University of New York at Buffalo
Children's Hospital
219 Bryant Street
Buffalo, NY 14222

The organizers of this conference deserve our gratitude, for this meeting brings together much new information on limulus lysate. The lysate has become of great practical value for the detection of minute amounts of endotoxin. Of course, nature did not create the crab for the sole purpose of making it possible for scientists to detect bacterial lipopolysaccharide in specimens from patients and, thus, aid in the diagnosis of endotoxic conditions, including shock, or to detect endotoxin in therapeutically used materials to prevent iatrogenic toxic effects in patients.

This particular session is devoted to a discussion of the agglutinating activities of the lysate, with particular reference to bacteria. Of special interest is the observation that lysate may produce killing of certain microorganisms. The question arises, therefore, whether the agglutinating and bactericidal activities are beneficial to the crab itself. An answer to this question will depend on information as to whether microorganisms that are clumped or killed by the lysate may cause disease of the crab. It will be of great interest to learn also whether the agglutinating and killing activities are due to one and the same factor and whether these reactions are specific in nature. Absorption of the lysate with one particular microbial species or type should clarify this question when tests are made with both absorbing and other microorganisms. It will also be of interest to elucidate the nature of the determinant on the bacterial surface which is operative in these reactions. Finally, the question is asked whether electrolytes are required for the agglutination reaction by limulus lysate as

Pages 555–556, Biomedical Applications of the Horseshoe Crab (Limulidae)

they are for agglutination by bacterial antibodies. So far as the latter phenomenon is concerned, immunoglobulins become attached to the bacterial surface even in the absence of electrolytes, but agglutination occurs only in their presence. Clearly, this will not be the last conference on *Limulus polyphemus*.

DISCUSSION COMMENTARY

Thomas G. Pistole

Department of Microbiology
University of New Hampshire
Durham, New Hampshire 03824

A major outcome of this portion of the Symposium is the realization that, in addition to the well-recognized clotting reaction in the presence of bacterial lipopolysaccharide (LPS), Limulus possesses additional systems reactive with microbes. The antibacterial activity in amebocyte lysates, reported by Nachum, is particularly interesting because of its broad spectrum of activity among the Eubacteriales. This, coupled with the finding that antibacterial activity remains when the amebocyte lysate is treated with mild heat to inactivate the clotting system, indicates a system at least partly distinct from the LPS-activated reaction.

Serum from L. polyphemus is capable of agglutinating a wide variety of bacteria. Agglutinin-adsorption studies, reported by Pistole, suggest the presence of multiple bacterial agglutinins, and this has been confirmed in the studies of Gilbride and Pistole, using a purified agglutinin. This serum-derived protein which agglutinates Staphylococcus aureus appears to have a molecular specificity for galactose-like residues.

A second lectin, which reacts with Gram-negative bacteria, was reported by Rostam-Abadi and Pistole with probable specificity for the 2-keto-3-deoxy-octonate moiety of the LPS molecule. Preliminary evidence suggests that this agglutinin is identical to the serum hemagglutinin with reported specificity for N-acetyl neuraminic acid.

Given the data presented in this session, we are faced with the major question, as yet unanswerable: What is the

Pages 557–558, Biomedical Applications of the Horseshoe Crab (Limulidae)

significance of all this for Limulus? The presence in the hemolymph and its component amebocytes of bacterial recognition molecules as well as demonstrable antibacterial activity provides compelling evidence for an internal defense mechanism in these marine invertebrates. Yet the crucial link of showing agglutinating or antibacterial activity directed toward natural pathogens of Limulus has not yet been made. So, despite the compelling data, we are as yet unable to determine whether the major findings reported here describe a natural defense system or a laboratory curiosity.

A second major question which arises here is the relationship among these various activities. Is the same system responsible for agglutinating bacteria also involved in inhibiting their growth? Are either of these systems linked to the LPS-activation of the clotting system? These questions are as yet unanswered because the findings presented here were, in general, obtained using unpurified material. The logical step now is to isolate the components responsible for each of the activities here reported and determine their ability to mediate the other reported reactions.

It will also be useful to determine the sites on the bacterial surface to which these agglutinating and growth-limiting substances bind. One possible starting point here is to determine the differences in surface structure between those bacterial strains which are highly reactive and those which are minimally so.

Because of its size and availability, and the general information already available on it, Limulus represents an ideal model for further studies in the area of "invertebrate immunology". No doubt this Conference will stimulate efforts in this area.

SOME ASPECTS OF THE BIOMEDICAL SIGNIFICANCE OF CELL-SURFACE SIALIC ACID AND THE HAEMAGGLUTININ OF LIMULUS POLYPHEMUS

G.W.G. Bird

Regional Blood Transfusion Service,

Birmingham, England.

INTRODUCTION

This is a brief annotation on cell-surface sialic acid and the lectin (haemagglutinin) of the haemolymph of the Horseshoe Crab *Limulus polyphemus*.

Lectins are a group of proteins, found in plants and in some invertebrate and lower vertebrate animals, which are able to combine specifically with simple sugars, and are therefore able to react, often specifically, with human red blood corpuscles and with other cells (Bird, 1974).

CELL-MEMBRANE SIALIC ACID

The human red cell membrane consists of seven simple sugars: galactose, mannose, glucose, fucose, N-acetyl-glucosamine, N-acetylgalactosamine and N-acetylneuraminic acid (NANA) (Bird, 1974). NANA, a sialic acid, is an important constituent of many mammalian cell membranes. Its function is not yet fully clear. However, ionised NANA (its α-carboxyl groups) is the principal contributor to the nett negative charge of the cell surface and therefore to the maintenance of the separate state of cells, e.g. red cells, suspended in a fluid medium. NANA is essential in the biosynthesis of several blood group antigens and of the myxovirus receptor and the infectious mononucleosis receptor of sheep and ox erythrocytes. The role of NANA in the ageing and elimination of red cells has been the subject of speculation for many years (Durocher et al, 1975; Jancik and

Pages 559–565, Biomedical Applications of the Horseshoe Crab (Limulidae)

Schauer, 1974; Lauf, 1977; Luner et al, 1977; Seaman et al, 1978). It seems however from the work of Luner et al (1977) and Seaman et al (1978) that although 'old' red cells contain about 10% less NANA than 'young' red cells, the nett number of sialyl residues per unit area of cell surface membrane remains the same, so that it is likely that other factors are responsible for the shorter survival of 'old' cells.

NEURAMINIDASE

The enzyme neuraminidase (EC 3.2.1.18), also known as sialidase or Receptor Destroying Enzyme (RDE) is secreted by various bacteria (table 1) and is an integral part, coded by the virus genome (Drzeniek, 1967), of the myxovirus particle. There are subtle differences between various neuraminidases, but all cleave NANA from the cell surface and expose the T-antigen of Friedenreich (Bird, 1977) in most human body cells and in various body fluids e.g. saliva (Kline and Issitt, 1976) and plasma (Crookston, 1978). Treatment of cells with neuraminidase leads to a decrease of 50% or more of their electron charge density (electrophoretic mobility) concurrent with the release of NANA in the incubation medium. This effect is important in medicine and biology. Tumour cells become more immunogenic after loss of sialic acid. Similar considerations pertain to embryonic or normal lymphoid cells. This subject has been reviewed by Sachtleben *et al* (1973). Curiously, the human red cell membrane itself contains neuraminidase (Bruce Bosmann, 1974); its role in this situation is not yet clear. Neuraminidase inhibitor is normally present in human plasma. Microbial neuraminidase can only cause *in vivo* T-transformation of red cells if present in amounts sufficient to overcome the action of the inhibitor (Bird, 1977). Myxovirus attaches itself to cells by its haemagglutinin and releases itself by its neuraminidase. The possible role of neuraminidase in resistance to influenza has been reviewed in a Lancet Annotation (1972).

LIMULUS POLYPHEMUS

Haemagglutinins in the haemolymph of the Horseshoe Crab, *Limulus polyphemus*, were first described by Noguchi (1903). Although the haemolymph was later used to demonstrate haemagglutinins in the classroom, little scientific interest was taken until Marchalonis (1964) and Cohen et al (1972) revived interest in the haemagglutinin. Cohen et al (1972) demonstrated the NANA-specificity of the haemagglutinin; this was confirmed by Pardoe et al (1972). Red cells, after removal of NANA by neuraminidase, are no longer agglutinated by the haemolymph. Marchalonis and Edelman (1968) isolated the agglutinin and described its physical properties, and its molecular structure and biological function have been studied by Finstad et al (1974).

Cohen et al (1972) have shown that the binding of the Limulus haemagglutinin to human red cells can be blocked by prior coating of the cells with anti-M and anti-N antibodies. Cohen et al (1976) have also shown that human peripheral lymphocytes from patients with chronic lymphocytic leukaemia are clearly more strongly agglutinated by the haemolymph than either normal human peripheral lymphocytes, or those of a line of cultured human T-lymphocytes. The haemolymph was used by Pardoe et al (1972) in studies of Burkitt EB 2 lymphoma cells: the cells were shown to have terminal NANA groups, because they absorbed the Limulus agglutinin. These workers showed that the Limulus agglutinin was a useful tool in the serological or immunochemical detection of terminal NANA of various animal cell surfaces.

Table 1 SOME MICROBES WHICH PRODUCE HIGH LEVELS OF NEURAMINIDASE

Vibrio cholerae

Clostridium perfringens

Diploccocus pneumoniae

Bacteroides fragilis (many strains)

Pasteurella multocida

Corynebacterium diptheriae

Myxoviruses

(For full details see Müller, 1974)

The haemagglutinin, which is Ca•-dependent is a glycoprotein with a molecular weight of 400,000 daltons, consisting of six major subunits, each of which contains three minor subunits (Finstad et al, 1974; Marchalonis and Edelman, 1968).

SIALIC ACID AND HUMAN BLOOD GROUPS

The relationship of NANA to human blood group antigen structure has been studied by Springer and Ansell (1958) and Bird and Wingham (1970a). Besides its important role in M- and N-specificity, sialic acid also forms part of various other antigens (table 2), e.g. the Pr antigens Pr_1, Pr_2 and Pr_3 (but not Pr_a), which are inactivated by neuraminidase, and the Gd antigen which is also inactivated. Pr_{1-3} determinants are inactivated by both neuraminidase and proteases, whereas Gd determinants are inactivated only by neuraminidase (Roelke and Ebert, 1977). The I^F antigen is unaffected by neuraminidase, whereas other I antigens are enhanced (Dzierzkowa-Borodej et al, 1978). The antigen p of the P blood group system is believed to be sialosylparagloboside (Schwarting and Marcus, 1977). Removal of NANA enhances the agglutinability of various receptors besides T; these

Table 2 HUMAN BLOOD GROUPS IN WHICH SIALIC ACID HAS OR APPEARS TO HAVE A STRUCTURAL PART

M
N
Pr_{1-3}
Gd
p

receptors are H, as measured by certain lectins e.g. *Ulex europaeus* (Bird and Wingham, 1970a, 1970b), and the receptors detected by the lectins of *Bauhinia purpurea* and *Moleccella laevis*. Sialic acid-deficient red cells, e.g. T, Tn, En(a-), M^k, M^g, etc., have altered membranes and abnormal serological and electrophoretic properties (Bird, 1974).

CONCLUSION

It is hoped that this highly concentrated summary of some aspects of the biomedical significance of NANA serves to illustrate the importance of the NANA-specific agglutinins of *Limulus polyphemus*.

REFERENCES

Annotation (1972). Neuraminidase and resistance to influenza. Lancet ii:264.

Bird GWG (1974). Plant and other agglutinins in the study of some human erythrocyte membrane anomalies. Ann N Y Acad Sci 234:129.

Bird GWG (1977). Complexity of erythrocyte polyagglutinability. In Mohn JF, Plunkett RW, Cunningham RK, Lambert RM (eds): "Human Blood Groups. 5th Int Convoc Immunol Buffalo NY 1976." Basel: S Karger p 335.

Bird GWG (1978). The application of lectins to some problems in blood group serology. Rev Fr Transf Immuno-hemat 21:103.

Bird GWG, Wingham J (1970a). N-acetylneuraminic (sialic) acid and human blood group antigen structure. Vox Sang 18:240

Bird GWG, Wingham J (1970b). Anti-H from *Cerastium tomentosum* seeds. A comparison with other seed anti-H agglutinins. Vox Sang 19:132.

Bruce Bosmann H (1974). Red cell hydrolases. III. Neuraminidase activity in isolated human erythrocyte plasma membranes. Vox Sang 26:497.

REFERENCES (Cont)

Cohen E, Roberts SC, Nordling S, Uhlenbruck G (1972). Specificity of Limulus polyphemus agglutinins for erythrocyte receptor sites common to M and N antigenic determinants. Vox Sang 23:300.

Cohen E, Minowada J, Pliss M, Pliss L, Blumenson LE (1976). Differentiation of human leukaemic from normal lymphocytes by Limulus serum agglutination. Vox Sang 31:117.

Crookston MC (1978). Antigens common to red blood cells and plasma. Plenary Sessions. Main Lectures XVII Cong Int Soc Haemat and XV Cong Int Soc Bl Transf, Paris, p 51.

Drzeniek R (1967). Differences in splitting capacity of virus and V. cholerae neuraminidase on sialic acid type substrates. Biochem Biophys Res Commun 26:631.

Durocher JR, Payne RC, Conrad ME (1975). Role of sialic acid in erythrocyte survival. Blood 45:11.

Dzierzkowa-Borodej W, Seyfreid H, Lisowska E (1975). Serological classification of anti-I sera. Vox Sang 28:110.

Finstad CL, Good RA, Litman GW (1974). The erythrocyte agglutinin from Limulus polyphemus hemolymph: molecular structure and biological function. Ann N Y Acad Sci 234:170.

Jancik J, Schauer R (1974). Sialic acid - a determinant of the lifetime of rabbit erythrocytes. Hoppe-Selye's Z Physiol Chem 355:395.

Kline WE, Issitt CH (1976). T-substance in body fluids. Abstract S-35 Prog 29th Ann Meeting Amer Assoc Blood Banks, San Francisco, p 126.

Lauf PK (1977). Blood Group antigens and membrane permeability in erythrocytes of man and ruminants. An immunophysiologic approach. In Mohn JF, Plunkett RW, Cunningham RK, Lambert RM (eds): "Human Blood Groups. 5th Int Convoc Immunol Buffalo NY 1976". Basel: S Karger p 383.

Luner SJ, Szklarek D, Knox RJ, Seaman GVF, Josefowicz, Ware BR (1977). Red cell charge is not a function of age. Nature Lond 269:719.

REFERENCES (Cont)

Marchalonis JJ (1964). Natural haemagglutinins from *Limulus polyphemus*. Fed Proc 23:345.

Marchalonis JJ, Edelman GM (1968). Isolation and characterization of a haemagglutinin from *Limulus polyphemus*. J Molec Biol 32:453.

Müller HE (1974). Neuraminidases of bacteria and protozoa and their pathogenic role. Behring Inst Mitt No. 55:34.

Noguchi H (1903). On the multiplicity of the serum haemagglutinins of cold blooded animals. Zentr Bakt Abt I Orig. 34:286.

Pardoe GI, Uhlenbruck G, Bird GWG (1972). Studies on some heterophile receptors of the Burkitt EB 2 lymphoma cell. Immunology 18:73.

Roelke D, Ebert W (1977). Defining cold agglutinin specificities. Arztl Lab 23:500.

Sachtleben P, Gsell R, Mehrishi JN (1973). Neuraminidase and anti-neuraminidase serum: effect on the cell surface properties. Vox Sang 25:519.

Schwarting GA, Marcus DM, Metaxas M (1977). Identification of sialosylparagloboside as erythrocyte receptor for 'anti-p' antibody. Vox Sang 32:257.

Seaman GVF, Knox RJ, Nordt FJ, Zukoski GF (1978). Role of sialic acid (NANA) in red blood cell recognition and elimination *in vivo*. Abstracts, XVII Cong Int Soc Haemat and XV Cong Int Soc Bl Transf Paris, p 284.

Springer GF, Ansell NJ (1958). Inactivation of human erythrocyte agglutinogens M and N by influenza virus and receptor destroying enzyme. Proc nat Acad Sci Wash 44:182.

GANGLIOSIDE-LIMULIN INTERACTIONS

Régine MAGET-DANA, Annie-Claude ROCHE and Michel MONSIGNY
Centre de Biophysique Moléculaire, C.N.R.S.,

45045 ORLEANS Cedex, France

Besides the glycoproteins which are the main glycoconjugates of plasma membranes, cells contain various amounts of glycolipids and gangliosides. Because Limulin selectively binds *N*-acyl neuraminyl glycoconjugates and because the concentration of Limulin required to agglutinate horse red blood cells is much lower than that required to agglutinate human red blood cells, it was noteworthy to point out the composition difference of the red blood cells membrane of these two species. According to Eylar *et al.*(1962) horse red blood cells contain four times less neuraminic acid than human ones. However, 20 p 100 of the neuraminic acids present in horse red blood cells are ganglioside components, while human red blood cells do not contain gangliosides (Yamakawa *et al.*, 1960). Furthermore gangliosides are known to be present in various cells (Wiegandt, 1973) and to be located at the outer half of plasma membrane (Gahmberg and Hakomori, 1973) and to play a significant role as membrane receptors in Cholera toxin binding (Van Heiningen, 1974) and in Sendai virus binding (Haywood, 1974). Gangliosides may be also involved in the cell-cell adhesion and may be affected during malignant transformation (Hakomori, 1973, 1975). So, the demonstration that Limulin was able to selectively bind gangliosides could allow the use of Limulin as tool to study the biological function of gangliosides. Therefore, we became quite interested to study the interactions between Limulin and gangliosides.

Pages 567–578, Biomedical Applications of the Horseshoe Crab (Limulidae)

GANGLIOSIDES

The main ganglioside of horse red blood cells was identified as hematoside (Yamakawa and Suzuki, 1951 ; Yamakawa, 1974) NG1GM$_3$ (Fig. 1).

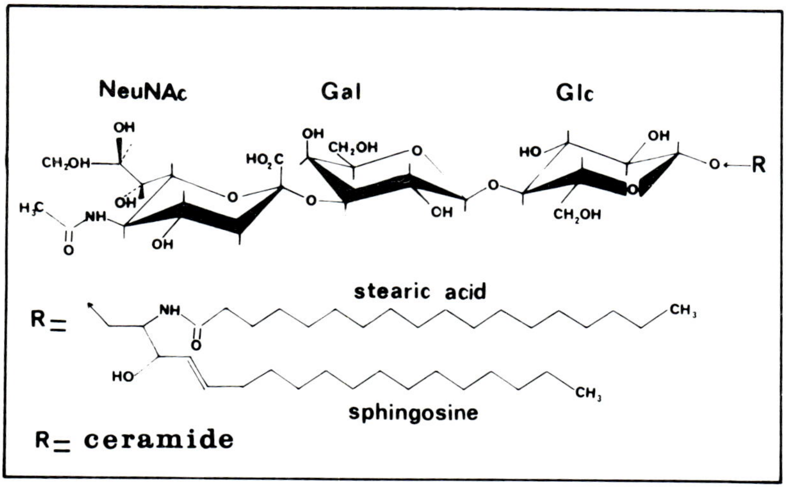

Figure 1 - Structure of the ganglioside GM$_3$.

Fresh horse blood was obtained from a slaughter house. Red blood cell membranes were prepared by the method of Tanner and Boxer (1972). The lipids were extracted from freeze dried membranes with a chloroform-methanol mixture and partitioned in two phases according to Folch *et al.* (1957). The aqueous phase containing gangliosides was dialysed, freeze dried and then treated with 0.2 M NaOH in methanol. Further purification was achieved by precipitation of the crude ganglioside fraction with acetone at - 20°C. An aliquot of the crude ganglioside fraction (about 10 per cent) was labelled by a mild periodate oxidation followed by a reduction with NaB^3H_4 according to Veh *et al.* (1977). The experimental conditions were optimized by the authors to yield a maximum amount of the C_8-aldehyde neuraminic acid analog (Fig. 2). The products obtained behaved as the untreated samples on thin-layer chromatography. So, they were able to be used as radioactive tracers during column chromatography.

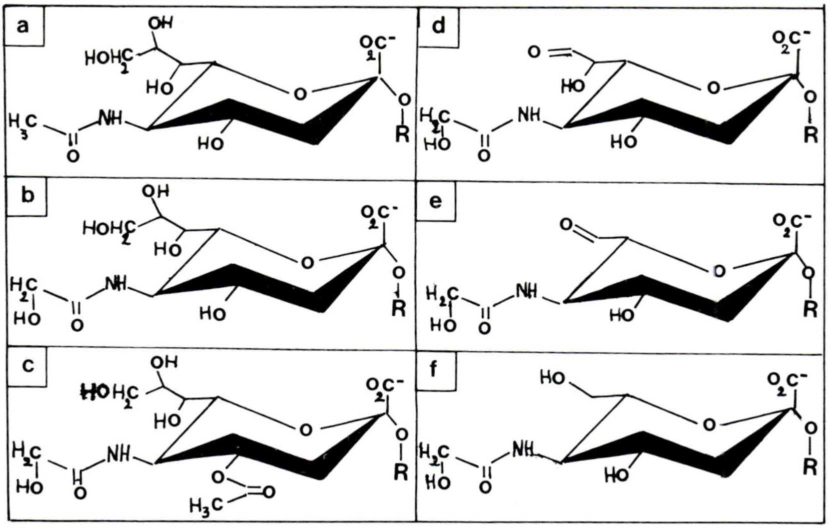

Figure 2 - Structure of chemically modified GM_3.
(a) GM_3, (b) $NGlGM_3$, (c) 4-$OAcNGlGM_3$, (d) C_8-aldehyde-$NGlGM_3$, (e) C_7-aldehyde-$NGlGM_3$, (f) C_7-$NGlGM_3$.

Gangliosides were dissolved in chloroform, applied onto a column packed with silicic acid in petroleum ether-chloroform 1:1 (vol/vol) and eluted stepwise by solvents with very steadily increasing polarity. The major ganglioside $NGlGM_3$ was eluted in a pure form by the solvent mixture : chloroform-methanol-water (60:35:8). Ganglioside analogs of NGl-GM_3 (Fig. 2) were prepared (Veh *et al.*, 1977) and given to us by Dr. R. Veh (Institut für Anatomie, Universität Bochum (RFA)). GM_3 was a gift from Dr. G. Rebel (Centre de Neurochimie, Strasbourg, France). Bovine brain ganglioside mixture was from Sigma.

GANGLIOSIDES EMBEDDED IN LIPID VESICLES

Gangliosides could not be used as inhibitors of the red blood cell agglutination induced by Limulin because they were taken up by plasma membranes of erythrocytes. Indeed, it was found that addition of gangliosides to a suspension of horse red blood cells enhanced the hemagglutination activity of Limulin. That finding was the first evidence of a ganglioside-Limulin interaction. In order to avoid gangliosides take up into red blood cells, gangliosides were incorporated into lipid vesicles. Vesicles were generated by dispersing the dry lipids (lecithin, cholesterol and gangliosides in the molar ratio 1:0.25:0.01 to 0.12 respectively) in 0.1 M NaCl, 0.05 M Tris-HCl buffer pH 8.5. Sonication was carried out at 0° under nitrogen stream for 10 min (Huang, 1969). The lipid dispersion was further centrifuged at 130,000 g for 60 min. The clear supernatant was collected. Vesicles were isolated after gel filtration on a Sepharose 4B column. Vesicles, checked by electron microscopy by Dr J. Schrevel (Poitiers, France), were 25 to 30 nm in diameter. Vesicles were used at the concentration of 0.5 mg of phospholipid per ml.

LIMULIN-GANGLIOSIDE INTERACTIONS IN LIPID VESICLES

The purification and the main properties of Limulin are described elsewhere (Roche and Monsigny, 1974 ; Roche *et al.*, 1975 and Roche *et al.*, this symposium). Two approaches were done to study Limulin-gangliosides interactions : inhibition of erythrocyte agglutination and turbidity (light scattering) measurements.

Inhibition of Erythrocyte Agglutination

Agglutination assays were carried out using a 3 % horse erythrocyte suspension in 0.1 M NaCl, 0.05 M $CaCl_2$ as previously described (Roche and Monsigny, 1974). It is possible to use vesicles containing gangliosides in the same conditions as oligosaccharides. The inhibition activity is given in terms of the ganglioside concentration required to inhibit one agglutination unit and expressed as neuraminic acid concentration. Bovine brain gangliosides, which contain mainly GM_2 and GD_1 with *N*-acetyl neuraminic acid, embedded in vesicles, were not able to inhibit the agglutination activity of Limulin. Conversely, horse erythrocyte gangliosides embedded in vesicles strongly inhibited the effect of Limulin (Table 1).

The activity was dependent on the ganglioside-phospholipid molar ratio, suggesting that the density of the gangliosides on the vesicle surface is a limiting factor. Vesicles without ganglioside had no inhibitory effect.

Table 1 - Inhibition of horse erythrocyte agglutination by vesicles containing gangliosides.

Vesicles Sialic acid/phospholipid molar ratio	Concentration of gangliosides required to inhibit one unit of agglutination (expressed as sialic acid concentration) (μM)
Bovine Brain gangliosides	
0.05	> 2,000
Horse erythrocyte gangliosides	
0.125	50
0.09	50
0.05	130

Optical Studies (Maget-Dana *et al.*, 1977)

Lipid-ganglioside vesicles interacted with Limulin and formed aggregates. The aggregation of vesicles by Limulin suggests that the lectin bears more than one binding site. This aggregation process was studied by monitoring the absorbance (at 500 nm) of the vesicle suspension after addition of Limulin.

Theoretical studies on the scattering of phospholipid dispersions were reported by Chong and Colbow (1976). These studies were based on the Rayleigh-Gans theory on scattering by particles much smaller than the wavelength of the incident light. The turbidity τ of the sample is related to the number N of particles per unit of volume and to the volume V of the particles :

$$\tau = k V^2 N$$

or with $\tau = 2.303$ A, $A = KV^2 N$

That function can be used because the chosen wavelength (500 nm) was 20 times higher than the size of the vesicles 25 to 30 nm.

The binding of Limulin to gangliosides embedded in lipid vesicles induced the formation of aggregates, i.e. an increase in the volume of the particles and an increase in the absorbance of the lipid dispersion. The enhancement of the absorbance versus time was drawn as log $(A_\infty - A_t)/(A_\infty - A_0)$ and was at least bimodal ($A_\infty - A_t$ are the absorbance of the lipid dispersion at times $t = \infty$ and t respectively). A fast process occured in the first 5 minutes which could be related to the diffusion of Limulin towards vesicles. Then a much slower one which appeared to be a pseudo-first order phenomenon. The rate constant K of aggregate formation was calculated from the slope k of the curve : K = k/C, where C is the molar concentration of Limulin (Fig. 3).

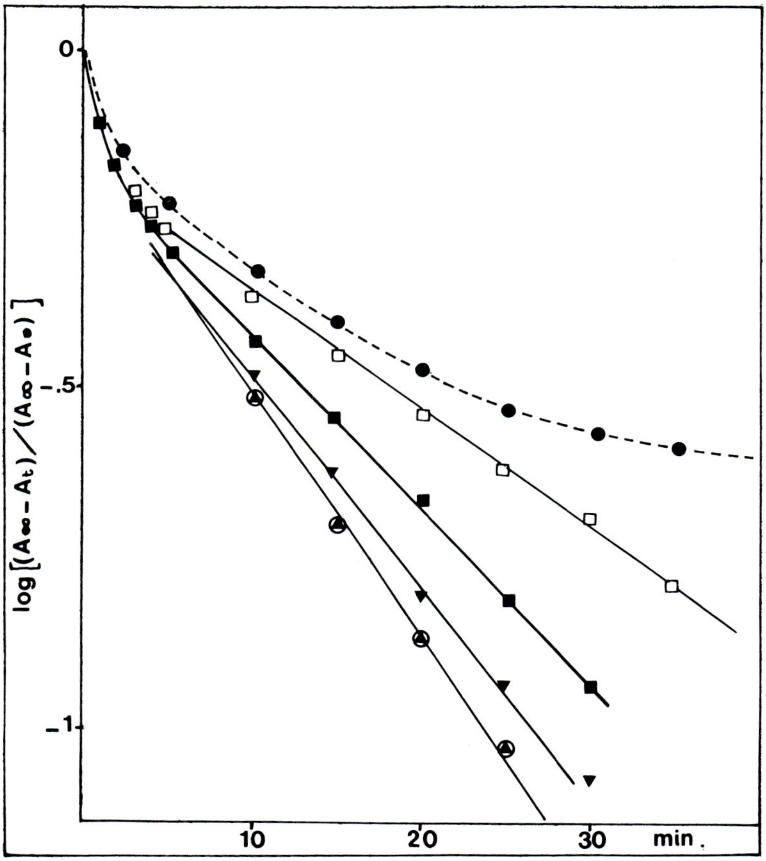

Figure 3 - Kinetics of aggregation of vesicles after addition of Limulin : plot of log $(A_\infty - A_t)/(A_\infty - A_0)$ versus time. Limulin concentration 2.4 10^{-8} M ; ganglioside/lecithin molar ratio : 0.125 (8 10^{-5} M in ganglioside). --●--●-- GM_3 ; —■—■— $NGlGM_3$; —□—□— C_8-aldehyde-$NGlGM_3$; —Ⓐ—Ⓐ— C_7-aldehyde-$NGlGM_3$; —▼—▼— C_7-$NGlGM_3$.

These results contrast with those obtained by Surolia *et al.* (1975) who could detect only one step using *Ricinus communis* agglutinin and GM_1 embedded vesicles.

Binding Specificities

For vesicles containing crude horse erythrocyte gangliosides with a molar ganglioside/phospholipid ratio R = 0.125, K was found to be 0.7 10^4 $mole^{-1}$ sec^{-1}. The interaction of Limulin with vesicles containing pure $NGlGM_3$ (the major ganglioside from horse erythrocytes) followed the same kinetics as with crude horse erythrocyte gangliosides. In similar experimental conditions K was 1.7 10^4 $mole^{-1}$ sec^{-1}. GM_3 analogs with modified sialic acid (Fig. 2) were also tested for their affinity towards Limulin (Maget-Dana *et al.*, 1979b). $NGlGM_3$ analogs with shorter lateral chain of sialic acid (i.e. C_8- and C_7-aldehyde-$NGlGM_3$ and C_7-$NGlGM_3$) interacted with Limulin and the aggregation kinetics were similar to that obtained with pure $NGlGM_3$ (the K values corresponding to the slower process were respectively 1.5 10^4, 2.5 10^4 and 2.2 10^4 $mole^{-1}$ sec^{-1}). GM_3 was also recognised by Limulin but the aggregation process was slow and the kinetics of the aggregation did not fit with a pseudo-first order process. Conversely, 4-O-acetyl- $NGlGM_3$ did not interact at all with Limulin.

In all cases, the maximal increase of turbidity for a given Limulin concentration was dependent on ganglioside concentration (8 10^{-6} M to 8 10^{-5} M related to the molar ratio ganglioside/phospholipid in vesicles). The maximal increase of turbidity for a given ganglioside concentration was dependent on Limulin concentration (2.4 10^{-8} M to 11 10^{-8} M). The non agglutinating fraction ACF-I (see Roche and Monsigny, this symposium) up to 6 10^{-7} M, which has many molecular properties similar to those of the Limulin, except the binding to sialoglycoconjugates (Roche *et al.*, 1975) did not increase the turbidity of the vesicle suspension.

Another lectin, Concanavalin A (up to 2.4 10^{-7} M) did not interact either with lipid vesicles containing gangliosides or with ganglioside-free lipid vesicles. WGA interacted only and in a few extent with the C_7-$NGlGM_3$ analog at a very high lectin concentration (7 10^{-6} M, 250 µg/ml). When aggregation occurred, addition of specific inhibitors reversed the process : ethylene diamine tetracetate which is a strong inhibitor because the binding activity of Limulin requires Ca^{++} (Marchalonis and Edelman, 1968 ; Roche *et al.*, 1975) and bovine submaxillary mucin were able to reverse the aggregation effect of Limulin. This reversibility supports strongly a mechanism of interaction similar to the passive agglutination process and does not fit with a fusion mechanism as described by several authors using high concentrations of lectins (Van der Bosch and Mc Connel, 1975 ; Redwood and Polefka, 1976 ; Boldt *et al.*, 1977). This discrepancy can be related to the concentration of the lectins. Under our experimental conditions, Limulin concentration was low (5 - 50 µg/ml) corresponding to those required to agglutinate human red blood cells and to induce mitosis of mouse splenocytes (Roche *et al.*, 1977). The concentration of ganglioside sialic acids were between 10-100 µM close to the actual concentration of bound sialic acids (40 µM) in a 3 % suspension of horse erythocytes.

DYNAMICAL STUDIES OF INTERACTION BETWEEN LIMULIN AND GANGLIOSIDES EMBEDDED IN VESICLES

Addition of mitogenic lectins on lymphocytes was found to induce a decrease of the microviscosity of lipid membranes (Toyoshima and Osawa, 1975 ; Monsigny *et al.*, 1978). Furthermore, malignant transformation of cells is generally related to an increase of agglutinability by lectins and to a change in the mobility of lectin receptors on the membrane surface (Nicolson, 1976). It was suggested that mobility of membrane receptors was controlled by the microviscosity of the lipid bilayer (Fuchs *et al.*, 1975 ; Shinitzky and Inbar, 1976). So, it was noteworthy to monitor the microviscosity of the lipid upon adding Limulin. The monitoring of the changes of membrane microviscosity can be achieved by measuring the polarization of a fluorescent probe embedded in the bilayer. However because a bilayer is not an isotropic medium (Chen *et al.*, 1977 ; Sené *et al.*, 1978) the polarization of the probe cannot be quantitatively related to the microviscosity of the membrane, but can only be qualitatively related to it.

To point out an expected change in membrane fluidity induced by Limulin binding to gangliosides, the fluorescent probe (1,6-diphenyl-1,3,5-hexatriene, DPH) was embedded in lipid vesicles (Roche *et al.*, 1978). The fluorescence polarization P of DPH was monitored in the absence and in the presence of Limulin. As soon as the lectin was added, we observed a small decrease of P (2 to 3 per cent of the initial value). To know the actual effect of the lectin on the P values, light-scattering depolarization of fluorescence must be taken into account. So, P values were extrapolated to zero absorbance. On these bases it appeared that there was no significant change with fluorescence polarization related to a modification of the dynamical properties of the bilayer. This conclusion is also supported by the results of pulse fluorimetry experiments conducted using a monoexponential analysis : the average fluorescence life time of DPH was found to be insensitive to the aggregation induced by adding Limulin.

Influence of the Lecithin Nature and of the Physical State of the Lipid Bilayer

Interactions between a lectin and gangliosides embedded in the vesicle bilayers could be on the dependence of the nature of the fatty acid moieties, therefore studies were undertaken to detect any variation of the interaction process and of polarization induced by lectin binding, below, at the level of or above the phase transition temperature. When using egg lecithin vesicles, no phase transition can be expected in the 4° to 45°C temperature range because their transition temperature is lower than 0°C. Dimyristoyl lecithin vesicles (transition temperature : 24°C) containing horse erythrocyte gangliosides were aggregated by Limulin in a way very similar to that of egg lecithin ones. The turbidity increase was the same below and above the transition temperature and the change in the depolarization of DPH was quite similar whatever the temperature is (Fig. 4). So, the aggregation process and the polarization of the fluorescence probe could not be related to the physical state of the bilayer. These results do not agree with those of Alving and Richards (1977) who found that the activities of anti galactosyl ceramide antibodies were depending on the composition in vesicle fatty acid. This discrepancy could be attributed to specific properties of the antibody or more likely to the fact that Alving and Richards (1977) were using a glycolipid containing only one sugar while gangliosides contain at least three sugars ; the monosaccharide can be hidden in the polar head of Lecithin while the sugar moiety of gangliosides are too large to be completely hidden.

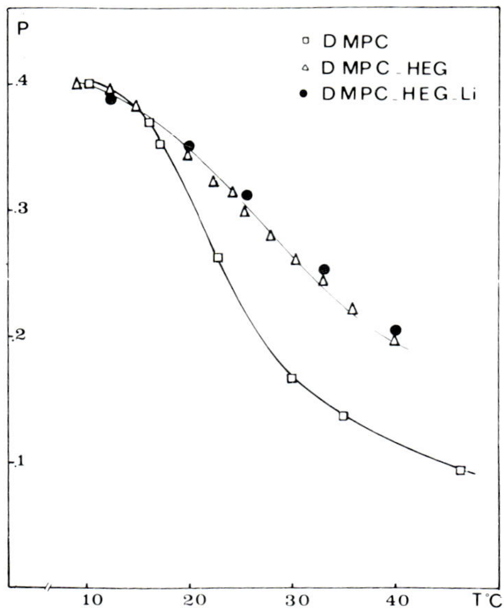

Figure 4 - Changes of fluorescence polarization of DPH embedded in dimyristoyl lecithin vesicles, related to temperature. —□—□— DMPC vesicles ; —△—△— DMPC vesicles containing HEG (ganglioside/lecithin molar ratio 0.130) ; —●—●— DMPC vesicles containing HEG, aggregated by Limulin (42 μg/ml - 12 10^{-8} M).

CONCLUSION

From these studies it can be concluded that Limulin mainly binds *N*-glycolyl neuraminic acid containing gangliosides. The hydroxyl group 4 of neuraminic acid has to be free for Neu-4-OAc-NGlGM_3 do not interact at all with Limulin. Furthermore, the hydroxymethyl groups (C_8 and C_9) of the neuraminic acid lateral chain are not involved in the binding process. So far, there is no evidence for a dynamic change of the bilayer containing ganglioside upon binding Limulin.

ABBREVIATIONS

NeuNAc : *N*-acetyl neuraminic acid ; NeuNGl : *N*-glycoly neuraminic acid ; GM_3 : NeuNAc-α-2-3-Gal-β-1-4-Glc-β-ceramide ; NGlGM_3 : GM_3 containing NeuNGl instead of NeuNAc ; HEG : Horse erythrocyte gangliosides ; DMPC : dimyristoyl phosphatidyl choline ; ACF-I : unadsorbed material on the affinity column during the last step of the purification procedure of Limulin (see Roche and Monsigny, this symposium).

REFERENCES

Alving CR and Richards RL (1977). Immunochemistry 14:373.
Andrich MP and Vanderkooi JM (1976). Biochemistry 15:1257.
Boldt DH, Speckart SF, Richards RL and Alving CR (1977). Biochem Biophys Res Commun 74:208.
Chen LA, Dale RE, Roth S and Brand L (1977). J Biol Chem 252:2163.
Chong CS and Colbow K (1976). Biochim Biophys Acta 436:260.
Eylar EH, Madoff MA, Brody OV and Oncley JL (1962). J Biol Chem 237:1992
Folch J, Lees M and Sloane-Stanley GH (1957). J Biol Chem 226:497.
Fuchs P, Parola A, Robbins PW and Blout ER (1975). Proc Natl Acad Sci USA 72:3351.
Gahmberg CG and Hakomori S (1973). J Biol Chem 12:4311.
Hakomori S (1973). Adv Cancer Res 18:265.
Hakomori S (1975). Biochim Biophys Acta 417:55.
Haywood AM (1974). J Mol Biol 83:427.
Huang CH (1969). Biochemistry 8:344.
Klenk E and Lauenstein K (1951). Z Physiol Chem 288:220.
Maget-Dana R, Roche AC and Monsigny M (1977). FEBS Lett 79:305.
Maget-Dana R, Strecker G and Monsigny M (1979). (In preparation).
Maget-Dana R, Veh R, Roche AC, Schauer R and Monsigry M (1979). (In preparation).
Marchalonis JJ and Edelman GM (1968). J Mol Biol 32:453.
Monsigny M, Roche AC and Kiéda C (1978). *In* Commission of the European Communities (ed), "Structure and function of biological membranes. Molecular aspects". Commission of the European Communities, Brussels p 161.
Nicolson GL (1976). Biochim. Biophys Acta 458:1.
Redwood WR and Polefka TG (1976). Biochim Biophys Acta 455: 631.
Roche AC and Monsigny M (1974). Biochim Biophys Acta 371:242.
Roche AC, Schauer R and Monsigny M (1975). FEBS Lett 57:245.
Roche AC, Perrodon Y, Halpern B and Monsigny M (1977). Eur J Immunol 7:263.
Roche AC, Maget-Dana R, Obrenovitch A, Hildenbrand K, Nicolau C and Monsigny M (1978) FEBS Lett 93:91.
Roche AC and Monsigny M (1978). (This symposium).
Sené C, Genest D, Obrénovitch A, Wahl P and Monsigny M (1978). FEBS Lett 88:181.
Shinitzky M and Inbar M (1976) Biochim Biophys Acta 433:133.
Tanner MJA and Boxer DH (1972). Biochem J 129:333.
Toyoshima S and Osawa T (1975). J Biol Chem 250:1655.

Van der Bosch J and Mc Connel M (1975). Proc Natl Acad Sci USA 71:4409.
Van Heiningen WE (1974). Nature 249:415.
Veh RW, Corfield AP, Sander M and Schauer R (1977). Biochim Biophys Acta 486:145.
Wiegandt H (1973). Hoppe-Seyler's Z Physiol Chem 354:1049.
Yamakawa T and Suzuki S (1951). J Biochem 38:199.
Yamakawa T, Irie R and Iwanaga M (1960). J Biochem 48:490.
Yamakawa T (1974). Yoshikawa H and Rapoport SM (ed) "Cellular and molecular biology of erythrocytes" Univ Tokyo Press, Tokyo p 15.

REACTIVITY OF LIMULUS POLYPHEMUS HEMOLYMPH WITH D-GLUCURONIC ACID CONTAINING GLYCOSUBSTANCES

Peter Vaith, Gerd Uhlenbruck, Werner E.G. Müller *, and Elias Cohen **
Medical University Clinic Cologne, Dept. Immunobiology, Kerpener Str. 15, D-5000 Köln 41 (GFR). *Institute of Physiological Chemistry, University of Mainz (GFR).
**Roswell Park Memorial Institute, Buffalo (USA).

Two major properties exhibited by Limulus polyphemus hemolymph (LH) are well known for many years: first, the hemagglutinating activity (Noguchi 1903; Marchalonis 1964; Cohen et al., 1965), which was later shown to be directed against terminal neuraminyl groups (Cohen 1968; Pardoe et al., 1970; Cohen et al., 1972); and second, the amoebocyte lysate, which detects bacterial endotoxins (Levin and Bang, 1964). Since the first isolation and characterization of the Limulus hemagglutinin (Marchalonis and Edelman, 1968), a series of papers on the purification and properties of this lectin, later termed Limulin (Roche and Monsigny, 1974), have been published (Finstad et al, 1972; Oppenheim et al., 1974; Nowak and Barondes, 1975; Sela et al., 1975; Kaplan et al., 1977).

Besides these studies concerned with Limulin some additional properties have been detected in Limulus serum. In 1971, our group reported on precipitin reactions between a snail galactan and LH; an activity, which seemed to be independent from Limulin (Voigtmann et al., 1971). Later on, Thomas Pistole discovered the property of Limulus serum to agglutinate and to kill a variety of bacteria (Pistole 1976 ; Pistole and Furman, 1976; Pistole 1978); the relationship of these activities to Limulin as well as to the endotoxin detecting principle was not clear.

Pages 579–587, Biomedical Applications of the Horseshoe Crab (Limulidae)

The results, which we want to present here, describe the reactivity of LH with certain D-glucuronic acid (GluA) containing glycosubstances. These findings have several implications: first, they lead us to a revised interpretation of our previous results obtained with the galactan from the snail *Helix pomatia* (Voigtmann et al., 1971); second, our data let us hypothesize that *Limulin* exhibits a structural and functional subunit heterogeneity; and third, some reactions reported to occur between *Limulus* serum and certain bacteria being devoid of sialic acids (Pistole 1978) may be partially explained by our findings.

MATERIALS AND METHODS.

D-glucuronic acid was obtained from Serva, Ltd., Heidelberg, GFR, and D-galacturonic acid from E. Merck, Darmstadt, GFR.

The origin and preparation of snail galactans from *Helix pomatia*, *Achatina fulica* and *Lymnaea stagnalis* has been given elsewhere (Voigtmann et al., 1971; Baldo and Uhlenbruck, 1975; Uhlenbruck et al., 1977). Bovine pneumogalactan (lot no. 113 065, preparation RO 2-2760) was a generous gift from Hoffmann-La Roche, GFR and USA. Sponge polysaccharides were prepared by phenol-saline extraction (90% phenol, v:v) of dried sponge tissue. The animals: *Geodia cydonium*, *Hippospongia communis* (Lam.) and *Cacospongia scalaris* (Schmidt) were collected in the vicinity of Rovinj (Yugoslavia). *Klebsiella* K16 and K51 capsular polysaccharides were kindly donated from Dr. A.K. Chakraborty (on leave from MPI Freiburg, GFR). *Erwinia amylovora* polysaccharides are a generous gift of Prof. R.N. Goodman (University of Missouri, USA).

Limulus polyphemus hemolymph was obtained as described by Cohen, 1968.

Hydrolysis of polysaccharides was performed as described by Voigtmann et al., 1971. The neutralized hydrolysates were separated by paper and thin layer chromatography (tlc). Descending paper chromatography was done on Whatman no. 1 paper in n-butanol/pyridine/water (6/4/3). Sugars were detected with the alkaline silver reagent. Silica gel 60 plates with a layer thickness of 0.25 mm (E. Merck) were employed for tlc. The same solvent system as above in the ratios (8/4/3) was used. The spots were developed with a ß-naphthylamine spray after Percheron (Sherma and Zweig, 1971).

Ouchterlony experiments were performed in an agar containing 2% agarose (Serva) in Tris-NaCl buffer plus calcium (0.1 M NaCl, 0.01 M $CaCl_2$, 0.05 M Tris-HCl, pH 8.5).

Hemagglutination inhibition tests were slightly modified from a standard procedure (Vaith et al., 1978). All acidic sugars used as inhibitors were dissolved in Tris-NaCl buffer with or without calcium, respectively, and brought to pH 7.2 with 0.1 M NaOH. A suspension (2%) of human red blood cells (blood group 0) in calcium containing Tris-NaCl buffer was prepared as reported by Nowak and Barondes, 1975.

RESULTS AND DISCUSSION.

The starting point for our investigations was the observation that a proteoglycan complex from the marine sponge *Geodia cydonium* (Müller et al., 1976) inhibited the LH-mediated hemagglutination of human red blood cells, though we could not trace sialic acid in this material (Vaith et al., 1978). Sugar analysis revealed that three major monosaccharides constituted almost all of this sponge carbohydrate, namely D-galactose (Gal), D-glucose (Glu) and D-glucuronic acid (GluA). We first believed that the anti-Gal activity in LH, previously supposed by us (Voigtmann et al., 1971), would be involved in this inhibition reaction. Negative results with many different anti-Gal lectins (Vaith et al., 1978), however, and other considerations made this interpretation unlikely. From biological experiments performed with sponge cells we obtained several lines of evidence that GluA in that proteoglycan complex existed in a terminal position (Vaith et al., 1978; Müller et al., 1978), which suggested an interaction of LH with this acidic sugar.

Independent from these findings we made another interesting observation: When LH was filled into one hole of an agar plate and no reaction partner was added, a concentric precipitin ring, which had diffused from LH into the agar, could be seen; LH was appearently able to precipitate with agarose (Fig. 1). This picture is confronted with the pattern obtained, when the snail galactan from Helix pomatia precipitates with LH (Fig. 2).

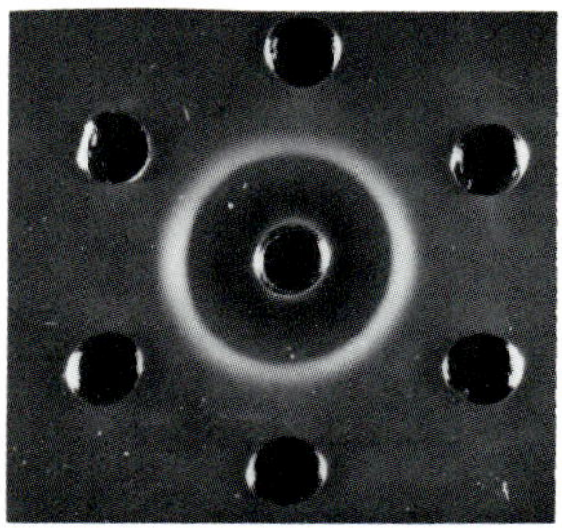

Fig. 1

Center: LH
1-6 : empty

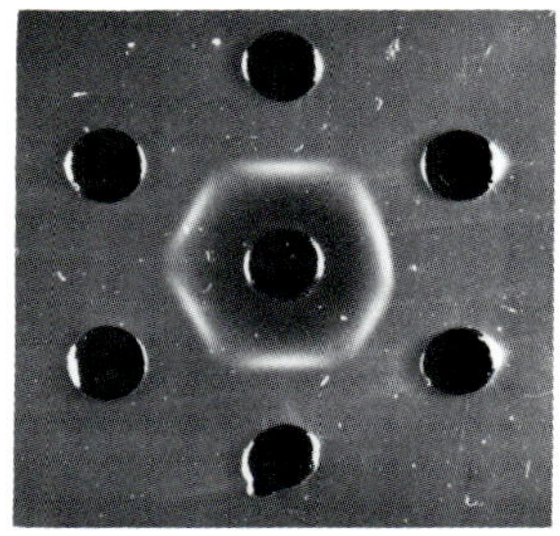

Fig. 2

Center: LH
1-6 : HP-galactan

According to our suggestion derived from the sponge experiments we were not surprised to detect GluA in hydrolyzed samples of agarose as well as Helix pomatia galactan (HP-galactan).

Additional information that GluA is most likely the determinant sugar recognized by LH came from our experiments performed with bacterial polysaccharides: Two capsular polysaccharides from the Gram-negative pathogen Klebsiella K16 and K51 were tested against LH in Ouchterlony experiments; while K51 gave a sharp precipitin line with LH, no reaction could be seen with K16. Both polysaccharides do contain GluA; in K51, however, this sugar is linked to the terminal side chain, whilst in K16 GluA exists in a rather cryptic form inside the main chain (A.K. Chakraborty, personal communication). The same holds true for the strong precipitation seen between LH and a capsular polysaccharide from <u>Erwinia amylovora</u>, a plant pathogen. This polysaccharide is comprised of alternating Gal and Glu residues forming a backbone to which GluA residues are linked as short branches (R.N. Goodman, personal communication). Such an exposed position of GluA seems thus to be a prerequisite for the reaction with LH. Negative reactions with chondroitin sulphate (Serva) and heparin sulfate (Schuchardt) are consistent with this interpretation. Accordingly, we want to predict that the polysaccharides, listed in Table 1, contain such serologically available GluA residues.

TABLE 1

Certain Acidic Glycosubstances Reacting With LH.

Origin of Material	Presence of GluA
Agarose	+
Helix pomatia (HP-galactan)	+
Achatina fulica	nt
Lymnaea stagnalis	nt
Beef lung (Pneumogalactan)	nt
Erwinia amylovora	+
Klebsiella K51	+
Geodia cydonium	+
Hippospongia communis	+
Cacospongia scalaris	+

nt= not tested. Assay for GluA: see 'Methods'.

Whilst our results presented here are strongly in favour of an anti-GluA activity in LH, the molecular nature of the active principle in LH is unknown. We have good reasons to believe, however, that this activity is associated with the anti-neuraminyl (anti-NANA) lectin, <u>Limulin</u>. For example, those polysaccharides which precipitated with LH also inhibited strongly the agglutination of human red blood cells (data not shown); thus an inhibition of the hemagglutinin i.e. Limulin must have occured. Further evidence can be deduced from previous studies of another group (Nowak and Barondes, 1975); they originally reported that agglutination of formalinized horse erythrocytes by purified Limulin could be inhibited not only by NANA, but also by GluA in similar concentrations. Interestingly, D-galacturonic acid (GalA) had no effect up to 100 mM. Unspecific inhibition by GluA, which may be thought to occur by simple charge effects or by binding calcium ions, are very unlikely, as already shown by those authors (Nowak and Barondes, 1975). The polysaccharide solutions we used were likewise carefully buffered in Tris-NaCl buffer containing 20 mM calcium chloride. A further similarity between the anti-GluA principle and Limulin can be seen in the finding that the precipitation of LH with agarose was impeded by 0.1 M EDTA; thus divalent cations seem to be necessary for the activity.

The appearent dual specificity of Limulin against NANA and GluA residues is somewhat surprising, if one assumes an identical combining site for both sugars; especially as GalA was reported to be non-reactive (Nowak and Barondes, 1975). The latter molecule bears much more structural similarity to GluA than GluA to NANA. An associated reactivity against GluA and NANA can best be explained by the assumption of a subunit heterogeneity of Limulin. Certain degrees of electrophoretical (Marchalonis and Edelman, 1968;Roche and Monsigny, 1974) as well as immunological (Kaplan et al., 1977) heterogeneities among the 1.5 s subunits of Limulin have been observed; these heterogeneities have been explained by different conformational states or degrees of polymerization of the same subunit (Marchalonis and Edelman, 1968), by the assumption of "isolectins" with the same specificity (Roche and Monsigny, 1974) or by methodological reasons (Kaplan et al., 1977). If our hypothesis is correct, the structural heterogeneities should have a functional counterpart, in that different types of subunits would be able to recognize different acidic sugars, i.e. GluA and

NANA.

A further consequence of our findings will briefly be mentioned here. Pistole reported in his recent paper (Pistole, 1978) on a broad spectrum bacterial agglutinating activity in *Limulus* serum. Whilst he supposed a possible cross reactivity of *Limulin* with 2-keto-3-deoxy-octonate (KDO) in Re mutants of *Salmonella minnesota*, the binding site on Gram-positive bacteria was unclear. If a serological activity of GluA determinants on these bacteria could be confirmed, the agglutination reactions with *Limulus* serum could obviously be explained by the anti-GluA activity described in this paper.

REFERENCES.

Baldo BA and Uhlenbruck G (1975). Quantitative precipitin studies on the specificity of an extract from Tridacna maxima (Röding). Carbohydr Res 40: 143.

Cohen E (1968). Immunologic observations on the agglutinins of the hemolymph of Limulus polyphemus and Birgus latro. Trans NY Acad Sci 20: 427.

Cohen E, Roberts SC, Nordling S, and Uhlenbruck G (1972). Specificity of Limulus polyphemus agglutinins for erythrocyte receptor sites common to M and N antigenic determinants. Vox Sang 23: 300.

Cohen E, Rowe AW, and Wissler FC (1965). Heteroagglutinins of the horseshoe crab, Limulus polyphemus. Life Sciences 4: 2009.

Finstad CL, Litman GW, Finstad J, and Good RA(1972) The evolution of the immune response. XIII. The characterization of purified erythrocyte agglutinins from two invertebrate species. J Immunol 108: 1704.

Kaplan R, Li St, and Kehoe JM (1977). Molecular characterization of Limulin, a sialic acid binding lectin from the hemolymph of the horseshoe crab, Limulus polyphemus. Biochemistry 16: 4297.

Levin J and Bang FH (1964). The role of endotoxin in the extracellular coagulation of Limulus blood.

Bull John Hopkins Hosp 115: 337.
Marchalonis JJ (1964). Natural hemagglutinin from Limulus polyphemus. Fed Proc 23: 1468.
Marchalonis JJ and Edelman GM (1968). Isolation and characterization of a hemagglutinin from Limulus polyphemus. J Mol Biol 32: 453.
Müller WEG, Müller I, Zahn RK and Kurelec B (1976). Species-specific aggregation factor in sponges. VI. Aggregation receptor from the cell surface. J Cell Sci 21: 227.
Müller WEG, Zahn RK, Kurelec B, Uhlenbruck G, Vaith P and Müller I (1978). Aggregation of sponge cells. XVIII. Glycosyltransferases associated with the aggregation factor. Hoppe-Seyler's Z Physiol Chem 359: 529.
Noguchi H (1903). On the multiplicity of the serum haemagglutinins of cold-blooded animals. Zentr Bakt Abt 1 Orig 34: 286.
Nowak TP and Barondes SH (1975). Agglutinin from Limulus polyphemus. Purification with formalinized horse erythrocytes as the affinity adsorbent. Biochim Biophys Acta 393: 115.
Oppenheim JD, Nachbar MS, Salton MRJ and Aull F (1974). Purification of a hemagglutinin from Limulus polyphemus by affinity chromatography. Biochim Biophys Res Comm 58: 1127.
Pardoe GI, Uhlenbruck G and Bird GWG (1970). Studies on some heterophile receptors of the Burkitt EB2 lymphoma cell. Immunology 18: 73.
Pistole TG (1976). Natually occuring bacterial agglutinin in the serum of the horseshoe crab, Limulus polyphemus. J Invert Pathol 28: 153.
Pistole TG (1978). Broad-spectrum bacterial agglutinating activity in the serum of the horseshoe crab, Limulus polyphemus. Develop Comp Immunol 2: 65.
Pistole TG and Furman RM (1976). Serum bactericidal activity in the horseshoe crab, Limulus polyphemus. Infection and Immunity 14: 888.
Roche AC and Monsigny M (1974). Purification and properties of Limulin: A lectin (agglutinin) from hemolymph of Limulus polyphemus. Biochim Biophys Acta 371: 242.
Sela BA, Wang JL and Edelman GM (1975). Isolation of lectins of different specificities on a single affinity adsorbent. J Biol Chem 250: 7535.

Sherma J and Zweig G (eds) (1971). "Paper Chromatography and Electrophoresis". Vol II. New York: Academic Press, p. 152.

Uhlenbruck G, Steinhausen G and Palatnik M (1977). Similarity of glycoproteo-galactans in the albumin glands from Achatina and Borus snails. Comp Biochem Physiol 57B: 335.

Vaith P, Müller WEG and Uhlenbruck G (1978). On the role of D-glucuronic acid in the aggregation of cells from the marine sponge Geodia cydonium. Develop Comp Immunol (submitted).

Voigtmann R, Salfner B and Uhlenbruck G (1971). Studies on broad spectrum agglutinins. IX. Specific and unspecific reactions between Limulus polyphemus haemolymph and snail extracts with "anti-A" specificity. Z Immun Forsch 141: 488.

ACKNOWLEDGEMENT

Thanks are given to Miss Elisabeth Schieb for the preparation and sugar analyses of sponge polysaccharides. The technical assistence of Mrs Gisela Holz is gratefully acknowledged. Part of this work was supported by the Deutsche Forschungsgemeinschaft (SFB 68/C5).

DIFFERENTIATION OF HUMAN LEUKEMIC FROM NORMAL LYMPHOCYTES BY PURIFIED LIMULUS AGGLUTININ

E. Cohen, L.E. Blumenson, M. Pliss, and J. Minowada.
Departments of Laboratory Medicine and Immunology and Immunochemistry Research and Biostatistics, Roswell Park Memorial Institute, Buffalo, N.Y. 14263

Agglutinin(s) for human erythrocytes and lymphocytes in the serum of the arachnoid *Limulus polyphemus* (horseshoe crab) have been described from this laboratory (Cohen, 1968; Cohen, Rozenberg and Massaro, 1974; Cohen, Minowada, Pliss, Pliss and Blumenson, 1976). The M and N blood group N-acetyl-neuraminyl (NANA) receptor sites were studied by use of *Limulus* serum agglutinin (Cohen, Roberts, Nordling and Uhlenbruck 1972). The agglutination of a variety of normal and abnormal human cells by *Limulus* serum has been reported by Pardoe, Uhlenbruck and Bird, 1970). The serum agglutinin was evaluated as to its ability to differentiate normal, leukemic and B and T human peripheral lymphocytes (Cohen, Blumenson, Pliss, Minowada and Fitzpatrick, 1976 and 1978). Elegant studies of the Japanese horseshoe crab *Tachypleus* lectins that agglutinate mammalian erythrocytes were reported (Shimizu, Ito and Niwa, 1977) with 4 purified lectins fractionated from the hemolymph. The Japanese workers demonstrated that the lectins were heterogenous in activation by Ca Cl_2, as well as specificity. Some showed different sugar specificities than *Limulus* agglutinins. More details of other studies by Shimizu and his associates are reported elsewhere in this proceedings. *Limulus* purified agglutinin (lectin) was reported to be inhibited by both NANA and D-glucaronic acid (Nowak and Barondes, 1975).

Pages 589–600, Biomedical Applications of the Horseshoe Crab (Limulidae)

Sophisticated separation procedures have been reported by a number of other workers, following the isolation of Limulus lectin (Marchalonis and Edelman, 1968) by starch gel electrophoresis. Absorption and elution from formalin-fixed erythrocytes (Nowak and Barondes, 1975); DEAE-Sephadex chromatography (Roche and Monsigny, 1974); BSM-Sephorose chromatography (Oppenheim, Nachbar, Salton and Aull, 1974), (Roche, Schauer and Monsigny, 1975). A Limulus lectin that agglutinated only horse erythrocytes was isolated (Sela, Wang and Edelman, 1975).

The present work reported from Roswell Park Memorial Institute is based on experiments with commercial purified Limulus agglutinin, on human peripheral and cultured lymphocytes.

MATERIALS AND METHODS.

Limulin (W) was obtained from the Worthington Chemical Corporation through Mr. Joseph Teller. It had been purified by the method of Roche and Monsigny 1974.

Human peripheral lymphocytes of chronic lymphocytic leukemia (CLL) and of chronic myelogenous leukemia (CML), were isolated from heparinized (CML) blood samples obtained through the courtesy of Drs. Tin Han and German Gomez, Medicine B Service, R.P.M.I. Acute lymphocytic leukemia (ALL) samples were obtained through the courtesy of Dr. Donald Higby and Edward Henderson and the Staff of Medicine A. Cell harvest was by a density gradient method (Amos, et al 1970).

Cultured lymphoblastoid cell lines of B cells (B411-4) and of T cells (MOLT-4F) had been maintained in the laboratory of Dr. June Minowada, using RPMI medium 1640 supplemented with 10% (v/v) head-inactivated fetal calf serum at 37° C. Their growth was logarithmic with an exponential increase in cell number. The B411-4 line was originated from the peripheral blood lymphocytes of a healthy donor and the MOLT-4F line was originated from a patient who relapsed of this acute lymphoblastic leukemia. The B411 line represents normal B lymphoblastoid cell and the MOLT-4F line represents leukemic T-lymphoid cell. The details of the properties of the cell have been reported previously (Minowada, Chnuma and Moore, 1972).

Human peripheral normal lymphocytes were isolated from heparinized blood samples of healthy volunteer plasmapheresis donors, RPMI Blood Bank through the courtesy of Mrs. Grace PoChedley, RN and her staff. Harvest of lymphocytes was by the same method as the chronic lymphocytic leukemia (CLL) cells.

Microagglutination was performed in plastic (Falcon #3034) trays. Limulin was diluted in Hanks BSS (1X) without calcium and magnesium and 5 lambdas was added to each well. Cell suspensions of lymphocytes (4000-6000 mm^3) in volumes of 5 lambdas were added to each well. The tray was rotated on a Yankee rotator, at 120-160 revolutions per 5 minutes, at room temperature, 22-25° C. Incubation was at room temperature for 25 minutes. Agglutination was read and scored from 1 to 4 plus, in the serial dilutions of the purified lectin.

STATISTICAL ANALYSES.

In accord with previous non-parametric statistical analysis reported (Cohen, Minowada, Pliss, Pliss and Blumenson, 1976), the logarithms of reciprocals of titers proved most advantageous to use (Siegel, 1956).

RESULTS AND DISCUSSION.

The mean of the logarithm of the reciprocal titers and mean scores are shown for each cell type with and without the addition of calcium in Tables 1 and 2. Not all samples were matched with and without calcium, so the number of observations will differ between the tables to be presented.

Table 1. LIMULIN (W) AGGLUTINATION OF LYMPHOCYTES MEAN LOG_2 TITER WITHOUT CA++

LOG_2 TITER

CELLS	N	$\bar{X}$	SD	SE
NORMAL	131	4.07	2.206	.193
CLL	25	6.10	1.911	.382
CML	20	5.17	1.679	.375
ALL	5	5.32	3.054	1.366
AML	8	3.49	3.111	1.099
T	2	6.14	.707	.500
B	15	6.89	1.442	.372

Table 1A. LIMULIN (W) AGGLUTINATION OF LYMPHOCYTES MEAN LOG_2 TITER WITH CA++

CELLS	N	$\bar{X}$	SD	SE
NORMAL	109	4.46	2.344	.225
CLL	20	6.09	1.726	.386
CML	17	5.32	1.902	.461
ALL	4	6.35	1.404	.702
AML	8	6.85	1.405	.497
T	2	5.48	3.056	2.161
B	2	5.89	3.627	2.565

Table 2. LIMULIN (W) AGGLUTINATION OF LYMPHOCYTES MEAN SCORE WITHOUT CA++

CELLS	N	$\bar{X}$	SD	SE
NORMAL	131	1.18	1.166	.102
CLL	25	3.64	2.543	.509
CML	20	2.05	1.224	.274
ALL	5	1.60	.742	.332
AML	8	1.44	.942	.333
T	2	.75	.354	.250
B	15	3.57	1.556	.402

Mean Log_2 titer = Mean of Log_2 of reciprocal of titer.

Table 2A. LIMULIN (W) AGGLUTINATION OF LYMPHOCYTES MEAN SCORE WITH CA++

CELLS	N	$\bar{X}$	SD	SE
NORMAL	109	2.10	1.299	.124
CLL	20	4.43	2.336	.522
CML	17	2.41	1.093	.265
ALL	4	2.63	1.600	.800
AML	8	1.56	1.568	.555
T	2	2.00	1.414	1.000
B	2	4.75	5.303	3.750

As can be seen from Tables 3 and 4 respectively, the addition of calcium increased the mean logarithm reciprocal titer and the mean score in all cell types except for the mean titer with the CLL and the T cells. Neither of these exceptions was statistically significant. There was a significant difference for the normal cells and AML cells ($p<0.05$). The increased score was also significant for the normal cells ($p<0.001$) and AML cells ($p<0.05$).

Table 3. LIMULIN (W) AGGLUTINATION OF LYMPHOCYTES COMPARISON OF MATCHED SAMPLES WITH CA++ -- WITHOUT CA++ LOG_2 RECIPROCAL TITER

CELLS	N	$\bar{D}$	$S_{\bar{D}}$	t
NORMAL	107	.65	.29	2.22*
CLL	19	-.22	.45	-.49
CML	16	.34	.42	.81
ALL	3	1.36	1.63	.84
AML	7	3.35	1.00	3.35*
T	1	-.66	1.67	-.39
B	1	.90	.90	1.00

* - Significant at 0.05 Level

*** - Significant at 0.001 Level

$\bar{D}$ - Difference of Means

$S_{\bar{D}}$ - Standard Error of Difference of Means

$t = \bar{D}/S_{\bar{D}}$

Table 4. LIMULIN (W) AGGLUTINATION OF LYMPHOCYTES COMPARISON OF MATCHED SAMPLES WITHOUT CA++

		SCORE		
CELLS	N	$\bar{D}$	$S_{\bar{D}}$	t
NORMAL	107	.48	.13	.370***
CLL	19	.65	.56	1.15
CML	16	.44	.36	1.23
ALL	3	1.13	.85	1.32
AML	7	1.13	.45	2.49*
T	1	1.25	1.25	1.00
B	1	2.50	3.00	.83

* - Significant at 0.05 Level
*** - Significant at 0.001 Level
$\bar{D}$ - Difference of Means
$S_{\bar{D}}$ - Standard Error of Difference of Means

Comparison of agglutination by Limulus agglutinin among groups: The t-test comparisons of the average scores between every pair of cell types, both without and with calcium is shown in Tables 5 and 6.

Table 5. PAIRWISE COMPARISON OF AGGLUTINATION OF HUMAN LYMPHOCYTES BY LIMULIN (W) OF LOG_2 RECIPROCAL TITERS

CELLS	WITHOUT CA++ DIFF. OF MEANS	WITH CA++ DIFF. OF MEANS
NORMAL-CLL	-2.03***	-1.63**
NORMAL-ALL	-1.25	-1.89
NORMAL-CML	-1.10*	- .86
NORMAL-AML	.58	-2.39**
NORMAL-B	-2.82***	-1.43
NORMAL-T	-2.07	-1.02

Table 5. (Con't)

CELLS	WITHOUT CA++ DIFF. OF MEANS	WITH CA++ DIFF. OF MEANS
CLL-ALL	.78	-..26
CLL-CML	.93	.77
CLL-AML	2.61**	- .76
CLL-B	- .79	.20
CLL-T	- .04	.61

* - t Test Significant at 0.05 Level
** - t Test Significant at 0.01 Level
*** - t Test Significant at 0.001 Level

Table 5A. PAIRWISE COMPARISON OF AGGLUTINATION OF HUMAN LYMPHOCYTES BY LIMULIN (W) OF LOG_2 RECIPROCAL TITERS

CELLS	WITHOUT CA++ DIFF. OF MEANS	WITH CA++ DIFF. OF MEANS
ALL-CML	1.03	.15
ALL-AML	1.83	- .50
ALL-B	-1.57	.46
ALL-T	- .83	.86
CML-AML	1.68	-1.53
CML-B	-1.72**	- .67
CML-T	- .97	- .16
AML-B	-3.40**	.96
AML-T	-2.66	1.36
T-B	.74	.40

* - t Test significant at 0.05 Level
** - t Test Significant at 0.01 Level
*** - t Test Significant at 0.001 Level

Table 6. PAIRWISE COMPARISON OF AGGLUTINATION OF HUMAN LYMPHOCYTES BY LIMULIN (W) OF SCORES

CELLS	WITHOUT CA++ DIFF. OF MEANS	WITH CA++ DIFF. OF MEANS
NORMAL - CLL	-1.83***	-2.33***
NORMAL-ALL	.21	- .53
NORMAL-CML	- .24	- .31
NORMAL-AML	.37	- .46
NORMAL-B	-1.76***	-2.65**
NORMAL-T	1.06	.10
CLL-ALL	2.04	1.80
CLL-CML	1.59***	2.02***
CLL-AML	2.20***	1.87***
CLL-B	.07	- .32
CLL-T	2.89***	2.43**

* - t Test Significant at 0.05 Level
** - t Test Significant at 0.01 Level
*** - t Test Significant at 0.001 Level

Table 6A. PAIRWISE COMPARISON OF AGGLUTINATION OF HUMAN LYMPHOCYTES BY LIMULIN (W) OF SCORES

CELLS	WITHOUT CA++ DIFF. OF MEANS	WITH CA++ DIFF. OF MEANS
ALL-CML	- .45	.22
ALL-AML	.16	.06
ALL-B	-1.97*	-2.13
ALL-T	.85	.63
CML-AML	.61	- .15
CML-B	1.30**	.41
CML-T	-1.52	-2.34*
AML-B	-2.13**	-2.19
AML-T	.69	.56
T-B	2.82*	2.75

Table 6A. (Con't)

* - t Test Significant at 0.05 Level
** - t Test Significant at 0.01 Level
*** - t Test Significant at 0.001 Level

The strongest and most consistent difference is between normal and CLL cells, with CLL cells having a significantly larger mean titer ($p<0.001$ without calcium, $p<0.01$ with calcium). There was a significantly larger mean score ($p<0.001$ both with and without calcium).

The most consistent comparisons seem to be those between normal cells and the other six cell types. Normal cells gave lower average titers and lower average scores in 19 of the 24 possible comparisons with the other six groups both with and without calcium. Of the 19 comparisons, 9 were statistically significant, while none of the 5 reversals were significant. On the other hand, CLL cells have higher average titers and higher average scores in 19 of the 24 possible comparisons, with 11 of the 19 comparisons being statistically significant and none of the 5 reversals significant.

In addition to the comparisons involving normal and CLL cells, the following comparisons between the other cell types were statistically significant: B cells had higher average titers without calcium than both CML ($P<0.01$) and AML ($p<0.01$). Also, without calcium, B cells gave higher average scores than ALL cells ($p<0.05$) and AML cells ($p<0.05$). Finally, with the addition of calcium, T had a significantly higher score than CML ($p<0.05$).

To sum up, the major conclusion from our data are as follows:

The addition of calcium has not increased the discriminatory ability of the Limulus purified agglutinin test as judged by these data.

The strongest conclusion is that CLL lymphocytes give higher titers and scores than normal cells, both with and without calcium.

Other conclusions are:
The addition of calcium generally lead to higher titers and scores. This kind of observation was reported earlier for agglutination of erythrocytes by Limulus serum (Cohen, 1968).

Generally, normal lymphocytes give lower titers and scores than the other six groups of cells.

Generally, CLL lymphocytes give higher titers and scores than the other six groups of cells.

The data of this study confirm the demonstration of the significant difference between the agglutinability of B versus T lymphoblastoid cells. However, Limulus agglutinins (lectins) may be heterogenous, as to NANA and sugar specificity, as shown by others for Tachypleus (Shimizu, Ito and Niwa, 1977). If this can be demonstrated by subsequent studies, then certain agglutinin (lectin) fractions may reveal more discriminatory ability for B versus T lymphocytes than do the others. An investigation of fractions of Limulus agglutinins (lectin) appears warranted, for biomedical use for identification and possible separation of human lymphocyte subpopulations, monocytes and other hematopoietic cell types. Recent studies (Vaith, Uhlenbruck, Muller, 1978) and (Monsigny, 1978), are reported elsewhere in this proceedings. The over-all implications appear to be that Limulus polyphemus hemolymph serum contains a variety of useful "cell probe" reagents. Further purification, characterization and application of Limulus agglutinins (lectins) is now in progress. A rich legacy based on experiments by an early Woods Hole Marine Biological Laboratory investigator (Noguchi, 1903) has been inherited.

REFERENCES

Amos DB, Bashir H, Boyle W, MacQueen M, Tilikainen A (1970). A simple microcytotoxicity test. Transpl 7: 220.
Cohen E (1968). Immunologic observations of the agglutinins of the hemolymph of Limulus polyphemus

and Birgus latro. Trans. NY Acad Sci 30: 427.
Cohen E, Blumenson LE, Pliss M, Minowada J, Fitzpatrick JE (1975). Differentiation of human leukemic from normal lymphocytes by purified Limulus agglutinin. Exper Hemat 5: 64.
Cohen E, Blumenson LE, Pliss M, Minowada J, and Fitzpatrick JE (1978). Feder Proc 37: 1549.
Cohen E, Minowada J, Pliss L, and Blumenson LE (1976). Differentiation of Human Leukemic from Normal Lymphocytes by Limulus Serum Agglutination. Vox Sang. 31: 117-123.
Cohen E, Roberts SC, Nordling S, Uhlenbruck G (1972). Specificity of Limulus polyphemus agglutinins for erythrocyte receptor sites commen to M and N antigenic determinants. Vox Sang 23: 300.
Cohen E, Rozenberg M, Massaro EJ (1974). Agglutinins of Limulus polyphemus (horseshoe crab) and Birgus latro (coconut crab). Ann NY Acad Sci 234: 28.
Marchalonis JJ, Edelman GM (1968) Isolation and characterization of a hemagglutinin from Limulus polyphemus. J Mol Biol 32(2): 453.
Minowada J, Ohnuma T, Moore GE (1972). Rosette forming human lymphoid cell lines. Establishment and evidence for origin of thymus derived lymphocytes J Nat Can Inst 49: 891.
Noguchi H (1903) On the multiplicity of the serum haemagglutinins of cold-blooded animals. Zentr Bakt Abt I orig 34: 286.
Nowak TP, Barondes SH (1975) Agglutinin from Limulus polyphemus. Purification with formalinized horse erythrocytes as the affinity absorbent. Biochem Biophys Acta 393: 115.
Oppenheim JD, Nachbar MS, Salton MRJ, Aull F (1974). Purification of a hemagglutinin from Limulus by affinity chromatography. Biochem Biophys Res Com 58: 1127.
Pardoe GI, Uhlenbruck G, Bird GWG (1970). Studies in some heterophile receptors of the Burkit EB2 lymphoma cell. Immunology 18: 73.
Roche AC, Monsigny M (1974). Purification and properties of limulin: A lectin (agglutinin) from hemolymph of Limulus polyphemus. Biochim Biophys Acta 371: 242

Roche AC, Schauer R, Monsigny M (1975). Protein sugar interactions. Purification by affinity chromatography of Limulin, an N-acyl neuramyl binding protein. FEBS Letters 57: 245.

Sela BA, Wang JL, Edelman GM (1975). Isolation of lectins of different specificities on a single affinity absorbent. J Biol Chem 250: 7535

Shimizu S, Ito M, Niwa M (1977). Lectins in the hemolymph of Japanese Horseshoe Crab, Tachypleus tridentatus. Biochem Biophys Acta 500: 71.

Siegel S (1956). Non-parametric statistics. McGraw-Hill, NY, 213.

ACKNOWLEDGEMENTS

The gracious help of Drs. Tin Han, Donald Higby, Edward Henderson (Chief Med. A) and Staff, J. Sokal (Chief, Med. B), G. Gomez and Staff facilitated acquisition of human leukemic blood samples. Dr. J.E. Fitzpatrick (Chief, Dept. Lab. Medicine) provided facilities for performance of the study. Plasmapheresis donor room nurses, Mrs. G. PoChedley, Mrs. B. McGaw and Mrs. P. Pogodzinski obtained normal blood samples for this study. Our appreciation is extended to all the voluntary donors of blood samples. The study was supported in part by a USPHS grant CA-14413 NCI and by the Juliette and Israel Cohen Memorial Research Fund.

DISCUSSION TO SECTION V

Dr. G.W.G. Bird

Red cell N-acetylneuraminic (sialic) acid occurs in the N-acetyl or N-glycosyl forms. M. Monsigny was quite right to draw attention to this fact which might explain why horse erythrocytes, in which sialic acid is in the N-glycosyl form is a better inhibitor of Limulin that human erythrocytes or brain gangliosides in which it is in the N-acetyl form.

The possibility that the haemagglutinin of the Horseshoe "Crab" has specificity for sugars other than sialic acid is interesting. Dr. Pistole has found evidence of galactose-specificity. Dr. Vaith whose group believed that it also has galactose-specificity, has changed his mind and now believes it has glucuronic acid-specificity. One wonders what glucuronic acid has to do with the hunan red cell surface. This brings to mind the work of Martin and Sachs in Kiel who successively treated human red cells with proteases and B-glucuronidase and exposed a hitherto undescribed red cell cryptantigen, (Blut, 26, 277, 1973).

Dr. Vaith has suggested that it is possible that the B-glucuronidase was not a pure enzyme. This may be correct. It is not easy to get a completely pure enzyme. However, the question of <u>Limulus</u> haemagglutinin specificities other than for sialic acid must be reinvestigated.

Pages 601–602, Biomedical Applications of the Horseshoe Crab (Limulidae)

Dr. Waxdal has pointed out that in cancer cells there may be movement of the cell membrane with clustering of receptor sites. The fluid mosaic hypothesis of Singer and Nicolson (Science, 175, 720, 1972) is relevant not only to cancer cells but also to cells treated with enzymes. At least as far as neuraminidase is concerned there is a difference between red blood cells and tumour cells. Although red blood cells are alive in the sense that they are metabolically active, they are dead in the sense that they have no nucleus and cannot regenerate lost membrane components. If they are treated with neuraminidase they stay devoid of sialic acid. When EB2 lymphoma cells are treated with neuraminidase they regenerate their sialic acid quickly.

LIMULIN (*LIMULUS POLYPHEMUS* LECTIN). ISOLATION, PHYSICOCHEMICAL PROPERTIES, SUGAR SPECIFICITY AND MITOGENIC ACTIVITY.

Annie-Claude ROCHE and Michel MONSIGNY

Centre de Biophysique Moléculaire, C.N.R.S.,

45045 ORLEANS Cedex, France.

The presence of an agglutinin in the hemolymph of *Limulus polyphemus* was first reported by Noguchi (1903). The first specificity studies of that lectin were conducted by Marchalonis (1964) and by Cohen (1965, 1968) in the sixties, leading to the conclusion that Limulin binds *N*-acetyl neuraminic acid (Cohen, 1968). The first isolation and characterization were described using preparative electrophoresis by Marchalonis and Edelman (1968) and, ion exchange and gel filtration chromatography in our laboratory (Roche and Monsigny, 1974), but in both cases the agglutination activity was relatively low. More recently, several procedures using affinity chromatography were proposed (Oppenheim *et al.*, 1974 ; Roche *et al.*, 1975 ; Nowak and Barondes, 1975 ; Sela *et al.*, 1975). In the present report, we describe the isolation of Limulin by affinity chromatography on immobilized submaxillary mucin (Roche *et al.*, 1975) leading to a quite pure protein with a very high agglutinating activity. Some of the main physicochemical and biological properties will be also described.

ISOLATION

Frozen hemolymph of *Limulus polyphemus* (obtained from Marine Biological Laboratory, Woods Hole, Mass., USA) was thawed and centrifuged (1,200 g, 30 mn). The supernatant was dialysed against 0.1 M NaCl, 0.01 M $CaCl_2$, 0.05 M Tris-HCl buffer, pH 8.5 and then applied onto a DEAE-Sephadex A 50 column (Pharmacia, Uppsala, Sweden) or a DEAE-Ultrogel AcA 22 column (Pharmindustrie, IBF-Réactifs, Clichy, France)

Pages 603–616, Biomedical Applications of the Horseshoe Crab (Limulidae)

equilibrated in the same buffer. The hemocyanin (blue protein) was adsorbed on the ion exchange gel, while the agglutinating material was slightly retarded (Roche and Monsigny, 1974). The active material was applied directly onto a column of submaxillary mucin-substituted Ultrogel AcA 22 (Pharmindustrie) or Sepharose 4B (Pharmacia) equilibrated in the above buffer (Roche *et al.*, 1975). A large part of the material passed through the column and was devoid of agglutinating activity (affinity chromatography fraction I : ACF-I). A second fraction (ACF-II) was eluted using a high ionic strength buffer (1 M NaCl, 0.01 M $CaCl_2$, 0.05 M Tris-HCl buffer, pH 8.5) ; the ACF-II was also devoid of agglutinating activity. The active material ACF-III was eluted with a calcium-free buffer (1 M NaCl, 0.05 M Tris-HCl, pH 8.5). To this fraction, $CaCl_2$ was added to reach a 0.01 M concentration. The ACF-III was then concentrated by affinity chromatography (Limulin) using a small column. The results are summarized in Table 1. Alternatively, the procedure previously described (Roche *et al.*, 1975) which leads to a quite similar result can be used.

Table 1 - Purification of Limulin

Fraction	Volume ml	Protein content (a) mg	Total activity (b) $\times 10^{-3}$	Specific activity u/mg
Dialysed hemolymph	100	17,000	25	1.5
DEAE-Ultrogel (c)	500	1,450	15	10
ACF-III (d)	20	15	320	21,000
Limulin (e)	5	0.73	320	440,000

(a) The protein content was estimated assuming $A^{1\%}_{280\ nm}$ of 10.
(b) Agglutination activity using horse erythrocytes 3 % in the presence of 0.01 M $CaCl_2$.
(c) 3 x 40 cm column
(d) 4 x 4 cm column
(e) 1 x 4 cm column.

The purification factor (250,000) is much higher than that obtained using other purification procedures : 20 (Marchalonis and Edelman, 1968), 60 (Roche and Monsigny, 1974) or 1,700 (Oppenheim *et al.*, 1974). The yield calculated on the basis of agglutinating activity is higher than 100 p 100, because inhibitors are eliminated during the purification procedure. The amount of pure protein (Limulin) obtained from 1,000 ml of hemolymph is about 6 mg, the minimal concentration required to agglutinate horse red blood cells (3 % suspension in 0.15 M NaCl, 0.01 M $CaCl_2$) is lower than 1 ng/ml. Using the procedure described by Oppenheim *et al.*, 1974, we obtained a preparation corresponding to a mixture of ACF-II and ACF-III. Using affinity chromatography on fetuin substituted Sepharose (Sela *et al.*, 1975), the yield - based on total activity- and the specific activity were very low. Using affinity chromatography on formalinized horse erythrocytes, elution with NeuNAc, and subsequent fractionation on Sephadex G200, the procedure (Nowak and Barondes, 1975) led to a very active preparation, but the yield -based on total activity- was low. Using polyacrylamide gel electrophoresis in sodium dodecylsulfate, it was shown that the protein obtained from the DEAE-Sephadex column migrated as a single band (Roche and Monsigny, 1974). But if the purity of the lectin is monitored by crossed immunoelectrophoresis using rabbit antisera against the active fraction, the same fraction appeared quite heterogeneous (Roche *et al.*, 1977). Limulin (ACF-III) eluted with a calcium free buffer from an affinity chromatography on immobilized BSM was the only fraction which appeared pure (Roche *et al.*, 1977). The absence of glycosylhydrolase activity was checked using various *p*-nitrophenyl glycosides.

AMINO ACID COMPOSITION (Table 2)

Limulin is an acidic protein, its isoelectric point of 5 is consistent with the relative preponderance of acidic amino acids residues (Roche, 1978). The content of carbohydrate is less than 4 % with 5 glucosamine per 1,000 residues. Similar results were obtained by Nowak and Barondes (1975).

SPECTROSCOPIC ANALYSIS

Absorbance

The absorbance spectrum showed a maximum at 278 nm, a shoulder at 290 nm and a minimum at 250 nm. After freeze-drying or concentration the shoulder may disappear and the maximum may be displaced indicating an aggregation process.

Table 2 - Amino acid composition (a) of Limulin.

Amino acids	Residues per monomer	Amino acids	Residues per monomer
Lysine	10	Glycine	25
Histidine	6	Alanine	13
Arginine	3	Valine	16
Cystin/2	1	Methionine	9
Aspartic acid	30	Isoleucine	17
Threonine	15	Leucine	22
Serine	22	Tyrosine	9
Glutamic acid	32	Phenylalanine	15
Proline	5	Tryptophane (b)	5

(a) Average values obtained after HCl 5.6 N hydrolysis (24 h, 48 h and 72 h) in duplicate and determinated with an amino acid analyser Beckman 120 C.
(b) Determinated by spectrophotometric analysis.

Fluorescence

The excitation spectrum showed a maximum at 280 nm and a shoulder at 290 nm. The emission spectrum, obtained using an excitation wavelength between 280 and 295 nm, showed a maximum at 333 nm, in agreement with the fluorescence of tryptophan residues in a slightly polar environment.

Circular dichroism

In order to assess the nature and the extent of secondary structure in the lectin of *Limulus polyphemus* and the existence of a correlation between the Ca^{++} binding capacity of the lectin and the conformation of the protein, the circular dichroism of the protein was undertaken (Roche *et al.*, 1978). The CD spectrum obtained from the non agglutinating fraction (ACF-I) was quite different of that of Limulin (ACF-III), the active protein. The shape of the spectrum of the inactive proteins (ACF-I) was very similar to that obtained by Finstad *et al.* (1972), (Fig. 1).

The CD spectrum of the active fraction (Limulin, ACF-III) in the near ultraviolet had a shape similar to that of

ACF-I, but with a lower intensity ; a small contribution of the disulfide chromophores seems likely. The CD spectrum in the far UV had a very low intensity, perhaps the lowest, so far reported, for a protein and that explains why we were not able to calculate the fractions of the various conformations involved in the secondary structure. The conformation was shown to be sensitive to freezing. The change was almost irreversible, however, the agglutinating activity was only slightly lowered.

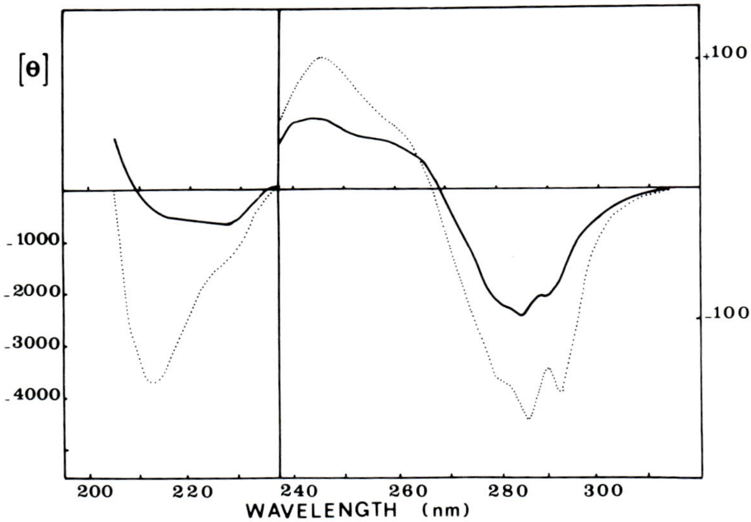

Figure 1 - The far and near ultraviolet circular dichroism spectra of ACF-I and of Limulin at pH 8.5 (0.01 M $CaCl_2$, 0.1 M NaCl 0.05 M Tris-HCl). (·····) ACF-I ; (———) Limulin. (From Roche, Maurizot and Monsigny, 1978).

Removal of calcium abolished the agglutinating activity (Marchalonis and Edelman, 1968) and the adsorption of the lectin on immobilized mucin (Roche *et al.*, 1975). The removal of calcium induced an appreciable change in the far ultraviolet CD and therefore a partial conformational change of the protein. However because this change was very limited, the environment of the aromatic amino acids did not vary substantially.

MOLECULAR WEIGHT AND SUBUNITS STRUCTURE

Limulin sedimented in a single symmetrical boundary, with no detectable heterogeneity (sedimentation coefficient : $s^0_{20,w}$ 13.9). The molecular weight calculated from the diffusion constant and the sedimentation coefficient was about 340,000. In polyacrylamide gel electrophoresis in the presence of sodium dodecylsulfate and of mercaptoethanol, the average subunit molecular weight was 29,000 ± 1,000. On these bases, Limulin could be constituted of 12 subunits.

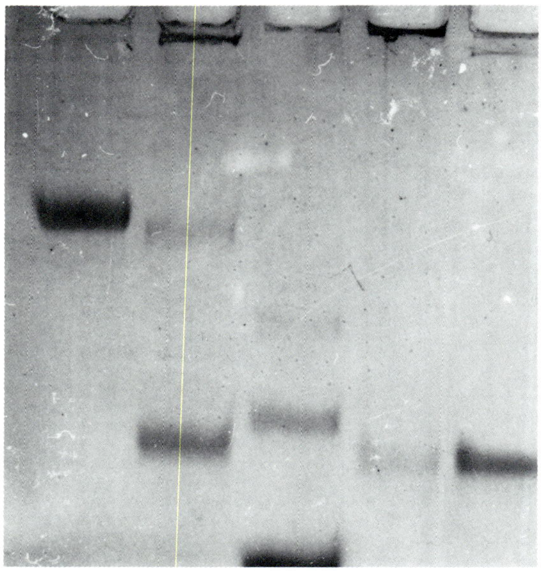

Figure 2 - Cross linking of subunits with dimethylsuberimidate polyacrylamide gel electrophoresis in the presence of 1 % sodium dodecylsulfate and 0.1 % mercaptoethanol. BSA : Bovine Serum Albumin ; Li : Control Limulin. WGA-Scl : Cross linked WGA ; the lower band corresponds to the monomer (m.w. 18,000), the other bands to the dimer, trimer, etc... Li-Scl: Cross-linked Limulin ; the lower band corresponds to the monomer (m.w. 29,000), the upper band to a dimer (m.w. 60,000).

In order to determine the arrangements of the Limulin subunits, diimidoesters were used as cross linking agents (Davies and Stark, 1970). The best results were obtained using dimethylsuberimidate. As shown by polyacrylamide gel electrophoresis in the presence of mercaptoethanol and sodium dodecylsulfate (Fig. 2), two species were obtained corresponding to a monomer and a dimer (molecular weights of 29,000 and 60,000, respectively).

The absence of higher molecular weight species suggests that the native dodecameric protein could be built as an hexameric association of dimers. This proposal is also in agreement with the electron microscopy study (unpublished results) using ultrathin carbon film and uranyl formiate staining. Molecules appeared as discs with a central hole, their diameter is about 12 nm. A similar image was obtained when a Ca^{++}-free buffered solution of Limulin was used. However after freeze-drying, aggregated forms were present.

SPECIFICITY

Horse and cat erythrocytes were agglutinated by a very low concentration of Limulin, 1ng/ml. The specificity of Limulin was determined using oligosaccharides, glycopeptides or glycoproteins as inhibitors of erythroagglutination. The results are summarized in the Table 3.

Free *N*-acetyl neuraminic acid and free *N*-glycolyl neuraminic acid inhibited the agglutination at high concentration 200 mM. 2-Deoxy-2,3 dehydro-*N*-acetyl neuraminic acid which is a strong inhibitor of several neuraminidases inhibited more strongly the agglutinating activity of Limulin. 2-*O*-methyl-β-neuraminic acid (with a free amino group) had not effect at all but the 2-*O*-methyl-α- or β-*N*-acetyl neuraminic acid were inhibitors at lower concentration than unsubstituted *N*-acetyl neuraminic acid. The acyl group was necessary to the recognition by Limulin, and the osidic substitution increased the affinity. *N*-acetyl hexosamines were not inhibitors in contrast with the lectins isolated by Shimizu *et al.* (1978) from the japonese horseshoe crab (*Tachypleus tridentatus*). Amongst the oligosaccharides tested, *N*-acetyl neuraminyl-α-2-3-(6)-*N*-acetyl galactosamine and some of its derivatives are the only ones which inhibited strongly the agglutination by Limulin. *N*-acetyl neuraminyl α-2-3-(6)-galactose and glycoconjugates containing this type of oligosaccharide were not inhibitors.

Table 3 - Sugar specificity of Limulin.

Compounds	Concentration mM (a)
NeuNAc	200
NeuNGl	200
2-deoxy 2,3-dehydro-*N*-acetyl neuraminic acid (d)	40
2-*O*-methyl-β-neuraminic acid (b)	* 300
2-*O*-methyl *N*-acetyl α-neuraminic acid (b)	10
2-*O*-methyl *N*-acetyl β-neuraminic acid (b)	20
GlcNAc	* 55
GlcNAc-β-1-4-GlcNAc	* 200
NeuNAc-α-2-3(6)-Gal-β-1-4-Glc (Sigma)	* 46
NeuNAc-α-2-6-Gal-GlcNAc NeuNAc-α-2-6-Gal-GlcNAc (Man)$_3$GlcNAc (c)	* 20
NeuNAc-α-2-6-GalNAc (d)	6
NeuNGl-α-2-6-GalNAc (d)	3
NeuNAc-α-2-6-GalNAc itol (d)	6
NeuNGl-α-2-6-GalNAc itol (d)	2
NeuNAc-α-2-6-Gal-β-1-3-GlcNAc-Gal-Glc (c)	* 2
NeuNAc-α-2-6-Gal-β-1-3 NeuNAc-α-2-6-GlcNAc-Gal-Glc (c)	* 2.4
Bovine submaxillary mucin (BSM)	0.002
Equine submaxillary mucin	0.3
Ovine submaxillary mucin	0.1
Porcine submaxillary mucin	0.03
Desialylated BSM	* 0.5
Fetuin (Sigma)	0.01

* No inhibition up to that concentration.
(a) The concentration of compounds is expressed as the sialic acid concentration required to inhibit one agglutination unit (Roche and Monsigny, 1974).
(b) Gift from R. Schauer (Kiel, G.F.R.)
(c) Gift from J. Montreuil and G. Strecker (Lille, France)
(d) Gift from R. Veh (Bochum, G.F.R.).

Amongst all the sialoglycoproteins tested, bovine submaxillary mucin was the best inhibitor ; a concentration (expressed as sialic acid) as low as 0.002 mM inhibits erythroagglutination. The inhibitory effect of BSM is abolished after desialylation (H_2SO_4 0.1N, 80°, 1 h). The other mucins (ovine, porcine, equine) had different inhibitor effects, this could be related to the type of sialic acid. Equine submaxillary mucin contains a large proportion of *N*-acetyl 4-*O*-acetyl neuraminic acid which seemed to lower the affinity (See also Maget-Dana *et al.*, 1978). As in the case of many other lectins, the affinity of the best free oligosaccharide ligands was about two orders of magnitude lower than that of a polyligand, such as protein substituted by several oligosaccharidesor as cell surface glycoproteins. This large differences can be related to the avidity phenomenon (Monsigny *et al.*, 1978). In conclusion, Limulin appeared to recognize specifically glycoproteins containing *N*-acetyl neuraminyl-α-2-3(6)-*N*-acetyl galactosamine.

VISUALIZATION OF GLYCOCONJUGATES CONTAINING *N*-ACYL NEURAMINIC ACID

Limulin, as other lectins (Monsigny *et al.*, 1978) can be easily used to visualize glycoconjugates at the cell surface level, either by fluorescence microscopy or by electron microscopy. Limulin was treated with fluoresceinyl- or tetramethyl rhodaminyl- isothiocyanate in an amine-free buffer, containing 0.01 M $CaCl_2$; the fluorescent lectin was then purified by affinity chromatography as described above. Using fluoresceine-substituted Limulin, sialoglycoconjugates of the cell surface of various cells were easily visualized (as mouse leukemic L 1210 cells, BHK 21 cells, lymphocytes, cells of axial organ from *Asterias rubens*...). By incubating mouse splenocytes at 37°, cap formation was detected as with several other lectins. A Limulin-horse radish peroxidase (Limulin-HRP) conjugate was prepared using glutaraldehyde as cross-linking agent (Avrameas, 1969) and purified by gel filtration chromatography. The Limulin-HRP conjugate was used to visualize sialoglycoconjugates on human bronchial mucosa sections (Mazucca *et al.*, 1977). Goblet cells were selectively stained at their apical part. The whole cytoplasm of the serous cells (Gianuzzi's crescents of submucosal glands) was stained. The serous cells were mainly stained at their apical pole. On the opposite, mucous cells which are known to contain sialoglycoconjugates, and are stained with the Alcian blue and periodic acid-schiff-reagents, were not stained with the Limulin-HRP conjugate. The selective staining with the Limulin-HRP conjugate is consistent with the high specificity of Limulin towards *N*-acyl neuraminyl-α-GalNAc disaccharide. However, according to the large size of the lectin, we cannot exclude that the Limulin-HRP conjugate could not reach some sialoglycoconjugates.

MITOGENIC EFFECT OF LIMULIN

The ability of the different fractions obtained from the affinity column to induce blastogenesis *in vitro* was assayed using human blood peripheral lymphocytes (Roche *et al.*, 1977) or mouse lymphocytes (from blood, spleen, thymus, mesenteric ganglions). The ACF-I and the ACF-II had no blastogenic activity up to 200 µg of protein per ml. The *Limulus* lectin (Limulin) purified on Sepharose fetuin column (Sela *et al.*, 1975) which had a low agglutination activity was not mitogenic. However, the pure lectin, Limulin, stimulated human blood lymphocytes from all the donors, with an optimal dose between 10 and 30 µg/ml (Fig. 3). 40-60 % of the cells were susceptible to blastic transformation.

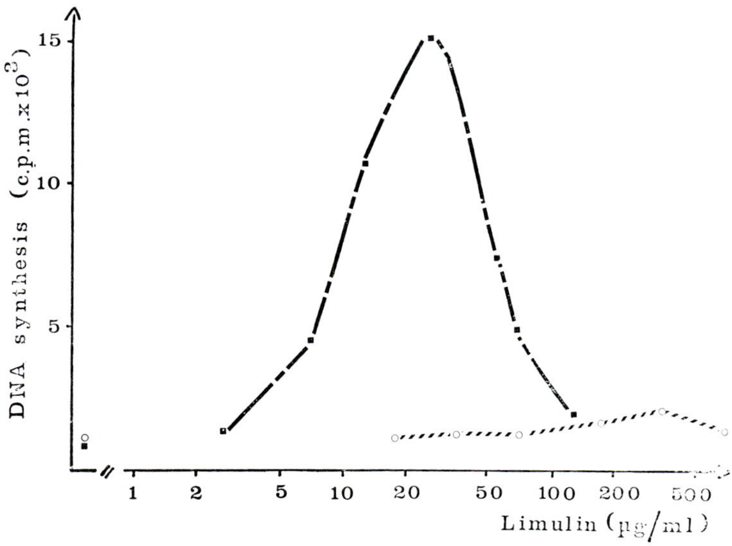

Figure 3 - Typical dose response curve of human peripheral lymphocytes culture (2.5 x 10^5 cells) stimulated with Limulin (- — -) and with ACF-I (······). ^3H thymidine incorporation after 48 h culture.

The mitogenic activity of Limulin could be inhibited when lymphocytes were incubated in the presence of an optimal dose of Limulin and of various amounts of bovine submaxillary mucin. The desialylated BSM was not able to inhibit the mitogenic activity of the lectin. The mitogenic activity of

PHA, used as a control was not inhibited either with BSM or with desialylated-BSM (Table 4). These findings suggest that the triggering proteins of lymphocyte cell surface are sialoglycoproteins and that these glycoproteins contain the *N*-acyl neuraminyl-α-GalNAc disaccharides.

Table 4 - Stimulation of human peripheral lymphocytes by Limulin evaluated by ^{3}H thymidine incorporation in 48 h culture (2.5 x 10^5 cells) in triplicate. Data shown are average ; variations were less than 10 %.

Mucin µg/ml	^{3}H-thymidine cpm	Lectin	Neuraminidase	^{3}H-thymidine cpm
0	500	-	0	400
0	10,500		0	10,000
N60	900	Limulin	1.5 u/ml	2,800
D60	8,700	18 µg/ml		
0	6,700	PHA	0	5,800
N60	5,600	16 µg/ml	1.5 u/ml	5,200
D60	7,100			

N : Native bovine submaxillary mucin,
D : Desialylated bovine submaxillary mucin.

The mitogenic effect of Limulin was very strong on mouse blood lymphocyte, with an optimal dose at the level of 10 µg/ml. Spleen cells and ganglion cells were also stimulated but thymocytes were not stimulated at all. We have also established that Limulin stimulated thymus-dependent lymphocytes (T) but had no mitogenic effect on thyms-independent lymphocytes (B). Two subpopulations of Balb C splenocytes were isolated according to Trizio and Cudkowicz (1974) the non adherent cells with about 80 % of T-cells and the adherent cells with 4 % of T-cells ; T-cells were detected by specific lysis using anti θ antibody and complement. The non-adherent splenocytes as the unfractionated splenocytes were activated by Limulin as by PHA. On the opposite, the adherent splenocytes of Balb C mouse and the unfractionated splenocytes of nude mouse were not activated by Limulin nor by PHA, but were activated by LPS (Table 5). Therefore, Limulin, as PHA and Concanavalin A is a T-lymphocyte specific mitogen. Although, Limulin has a very high molecular weight and is a dodecameric protein, it is devoid of mitogenic activity on B-lymphocytes. These findings suggest either that only some of the twelve subunits have an active binding site or that a poly-binding site T-mitogen is not a B-mitogen

Table 5 - Stimulation of mouse lymphocytes by Limulin evaluation by ^{3}H-thymidine incorporation in 48 h culture (2.5 x 10^5 cells) in triplicate. Data shown are average ; variations were less than 10 %.

	Balb C mouse splenocytes			Nude mouse
Mitogen	unfractionated	non adherent	adherent	splenocytes
-	600	400	2,000	3,600
Limulin (a)	17,000	20,600	4,000	3,200
PHA (b)	33,600	31,600	4,500	3,000
LPS (c)	17,500	5,500	26,500	21,000

(a) 15 µg/ml ; (b) 50 µg/ml ; (c) 500 µg/ml.

ACKNOWLEDGEMENTS

We wish to thank Mrs Y. Perrodon and Mrs M.L. Welter for excellent technical assistance. This work was partly supported by a grant 75.4.074.3 from Institut National de la Santé et de la Recherche Médicale and a grant 76.2302 from Centre National de la Recherche Scientifique. A.C. Roche is "Chargée de Recherche" INSERM.

REFERENCES

Avrameas S (1969). Coupling of enzymes to proteins with glutaraldehyde. Use of the glycoconjugates for the detection of antigens and antibodies. Immunochem 6:43.

Cohen E (1968). Immunologic observations on the agglutinin of the hemolymph of *Limulus polyphemus* and *Birgus latro*. Trans N Y Acad Sci 20:427.

Cohen E, Rowe AW and Wissler FC (1965). Heteroagglutinins of the horseshoe crab *Limulus polyphemus*. Life Sciences 4:2009.

Davies GE and Stark GR (1970). Use of dimethylsuberimidate, a cross-linking reagent in studying the subunit structure of oligomeric proteins. Proc Nat Acad Sci 66:651.

Finstad CL, Good RA and Litman GW (1972). The evolution of the immune response. The characterization of purified erythrocyte agglutinins from two invertebrate species. J Immunol 108:1704.

Maget-Dana R, Roche AC and Monsigny M (1978). Ganglioside-Limulin interactions. This conference.

Marchalonis JJ (1964). Natural hemagglutinin from *Limulus polyphemus*. Fed Proc 23:1468.

Marchalonis JJ and Edelman GM (1968). Isolation and characterization of a natural hemagglutinin from *Limulus polyphemus*. J Mol Biol 32:453.

Mazucca M, Roche AC, Lhermitte M and Roussel P (1977). *Limulus polyphemus* lectin sites in human bronchial mucosa. J Histochem Cytochem 25:470.

Monsigny M, Roche AC and Kieda C (1978). Lectins as tools to study cell surface membranes. *In* Commission European Communities (ed) : "Structure and functions of biological membranes. Molecular aspects". Commission of European Communities, Brussels.p 161.

Noguchi H (1903). On the multiplicity of the serum haemagglutinin of cold blooded animals. Zentr Bakt Abt 1 Orig 34:286.

Nowak TP and Barondes SH (1975). Agglutinin from *Limulus polyphemus*. Purification with formalinized horse erythrocytes as the affinity adsorbent. Biochim Biophys Acta 393:115.

Oppenheim JD, Nachbar MS, Salton MRJ and Aull F (1974). Purification of a hemagglutinin from *Limulus polyphemus* by affinity chromatography. Biochem Biophys Res Commun 58: 1127.

Roche AC (1978). Interactions glycanne-protéine : isolement et propriétés de la Limuline. Thèse de Doctorat ès-Sciences, Orléans, France.

Roche AC, Maurizot JC and Monsigny M (1978). Circular dichroism of Limulin : *Limulus polyphemus* lectin. FEBS Letters 91:233.

Roche AC and Monsigny M (1974). Purification and properties of Limulin : a lectin (agglutinin) from hemolymph of *Limulus polyphemus*. Biochim Biophys Acta 371:242.

Roche AC, Perrodon Y, Halpern B and Monsigny M (1977). Limulin (*Limulus polyphemus* lectin) : mitogenic effect on human peripheral lymphocytes. Eur J Immunol 7:263.

Roche AC, Schauer R and Monsigny M (1975). Protein sugar interactions. Purification by affinity chromatography of Limulin, an *N*-acyl neuraminyl-binding protein. FEBS Letters 57:245.

Sela BA, Wang JL and Edelman GM (1975). Isolation of lectins of different specificities on a single affinity adsorbent. J Biol Chem 250:7535.

Shimizu S, Ito M and Niwa M (1977). Lectins in the hemolymph of japonese horseshoe crab, *Tachypleus tridentatus*. Biochim Biophys Acta 500:71.

Trizio D and Cudkowicz G (1974). Separation of T and B lymphocytes by nylon wool columns : Evaluation of efficacy by functional assays *in vivo*. J Immunol 113:1093.

FUNCTIONAL IMPLICATIONS OF THE COVALENT STRUCTURE OF LIMULIN: AN OVERVIEW

J. Michael Kehoe,[1] Ruth Kaplan,[2] Steven S.-L. Li[3]

[1]Northeastern Ohio Universities College of Medicine
Rootstown, Ohio 44272

[2]George Mason University, Fairfax, Virginia 22030

[3]NIEHS, NIH, Research Triangle Park, North
Carolina 27709

A. INTRODUCTION

The covalent structure of a protein is of interest for several reasons. In the first place, knowledge of the amino acid sequence provides the best test of a given protein's relationship to other proteins. Other parameters including molecular weight estimations, behavior on polyacrylamide gels, and comparable measurements may be suggestive but cannot be definitive in this respect. Sequence analysis also provides some consideration of the degree of relatedness among proteins that do show sequence homologies. Sequence data may thus provide clues regarding the origins of particular proteins and even some idea of the point in evolutionary time when the protein appeared.

Knowledge of the covalent structure of a protein is also a prerequisite to the full understanding of the exact nature of the various functional sites the molecule possesses. This is especially true with respect to any specific ligand binding sites a given protein manifests. Here, a comparative approach involving a number of similar, but distinct, ligand binding proteins is often productive. Such has been the case with immunoglobulins (Kehoe and Capra, 1978) where the comparative analyses of the variable region sequences of a number of antibodies has provided useful insights into the binding properties of this family of proteins. Finally,

Pages 617–623, Biomedical Applications of the Horseshoe Crab (Limulidae)

complete amino acid sequence data is a *sine qua non* for the interpretation of X-ray diffraction data needed for the establishment of the tertiary structure of proteins at high resolution.

It was with these concepts in mind that we initiated the determination of the amino acid sequence of limulin, the sialic acid binding agglutinin found in the hemolymph of *Limulus polyphemus*.

B. MATERIALS AND METHODS

Limulin was isolated from the hemolymph of the horseshoe crab (*Limulus polyphemus*) essentially by the method of Roche and Monsigny (1974). The purified subunit product (Marchalonis and Edelman, 1968) showed more than one band in sodium-dodecyl sulfate-polyacrylamide gel electrophoresis.

Amino acid sequence analysis using the Edelman degradation was carried out by automated methods using a Beckman 890 sequencer. The fragmentation of the protein and the isolation of the resulting subfragments have been described (Kaplan, Li, and Kehoe, 1977). Phenylthiohydantoin residues of the liberated amino acids were identified by gas chromatography (Pisano *et al*., 1972) or thin layer chromatography (Summers *et al*., 1973).

C. RESULTS

The amino acid sequence of the first fifty residues of limulin is shown in Figure 1. This sequence is displayed with a human kappa light chain as a representative immunoglobulin and with the available sequence of an influenza virus hemagglutinin as another representative sialic acid binding protein. As indicated, no homologies are apparent between limulin and either of these other proteins.

D. DISCUSSION

The first observation of note concerning the primary structure analysis of limulin was that the protein yielded a unique sequence with no indication of heterogeneity, either at the amino-terminus or in the interior of the

HUMAN KAPPA CHAIN

10 ... 20

Asp Ile Gln Met Thr Gln Ser Pro Ser Thr Leu Ser Ala Ser Val Gly Asp Arg Val Thr Ile Thr Cys Arg Ala

LIMULIN

10 ... 20

Leu Glu Glu Gly Glu Ile Thr Ser Lys Val Lys Phe Pro Pro Ser Ser Ser Pro Ser Phe Pro Arg Leu Val Met

HUMAN INFLUENZA VIRUS HEMAGGLUTININ

10 ... 20

Gly Leu Phe Gly Ala Ile Ala Gly Phe Ile Glu Asn Gly Trp Glu Gly Leu Ile Asp Gly ? Tyr Gly Tyr

HUMAN KAPPA CHAIN

30 ... 40 ... 50

Ser Gln Ser Ile Asn Thr Trp Leu Ala Trp Tyr Gln Gln Lys Pro Gly Lys Ala Pro Lys Leu Leu Met Tyr Lys

LIMULIN

30 ... 40 ... 50

Val Gly Thr Leu Pro Asp Leu Gln Glu Ile Thr Leu Cys Tyr Trp Phe Lys Val Asn Gln Leu Lys Gly Thr Leu

Figure 1. Primary structure of the amino-terminal segment of limulin compared with a human kappa chain (protein Eu, Edelman *et al.*, 1969) and a hemagglutinin isolated from human influenza virus (Bucher *et al.*, 1976). The limulin sequence is from Kaplan *et al.*, (1977) and is reproduced with the permission of the American Chemical Society.

protein sequenced to date. This is true in spite of the fact that the isolated protein exhibits a mild degree of heterogeneity on polyacrylamide gels. Careful determination of the yield of the automated Edman degradation showed that greater than 90% of the protein available was being sequenced. There is consequently little likelihood that a significant portion of the isolated protein contains a blocked amino-terminus, another possible source of heterogeneity.

The presence of a definite, and unique, sequence also assumes importance because of the demonstration by Roche and Monsigny (Roche and Monsigny, 1977: Monsigny, this conference) that not all of the protein isolated by the chromatographic procedures mentioned above bind to an affinity column. These workers have in fact described three distinct fractions of this protein, as determined by affinity chromatography. The results of the sequence studies would imply that this variation is due to some loss of specific ligand binding capacity by a certain proportion of the molecules rather than to heterogeneity in the primary structure of the molecules in the limulin pool. Perhaps a large proportion of the isolated limulin picks up sialic acid containing moieties during the isolation process, an occurrence that could have a marked effect on the affinity column behavior of the preparation. The distinct bands seen on polyacrylamide gel electrophoresis are, however, most likely due to variations in the carbohydrate content of the glycoprotein itself.

The sequence analysis also indicates that this invertebrate lectin bears no relationship to the immunoglobulin family of proteins as known in higher vertebrates. As pointed out previously, knowledge of the primary structure is essential to reach definitive conclusions regarding the presence or absence of such relationships. This data is thus consistent with most results reported to date from a variety of systems indicating that immunoglobulin progenitor molecules are not to be found in invertebrates. This reinforces the view that the adaptive immune response as known in higher mammals is an evolutionary development of vertebrates alone.

It is not currently possible to extensively compare the primary structure of this protein with numerous other plant or animal lectins because very little information on such lectins is available. The complete primary structure of conconavalin A is known, however (Edelman *et al*., 1972), and a comparison of the established structure of limulin with this sequence shows no apparent relationship. Similarly, no homologies are apparent with selected stretches of a lentil lectin (Foriers *et al*., 1978). It is conceivable that clear relationships among lectins will be apparent when more primary structure data is available. Such information is especially needed to ascertain whether comparable active sites are to be found in these diverse proteins that all possess anti-carbohydrate binding activities.

The sequence data also points up some potentially important functional features of limulin. The region from residue 13 to 19 (See Figure 1) is a stretch particularly rich in proline and serine residues. It has been pointed out that such regions are often points of attachment for carbohydrate in glycoproteins (Spiro, 1973). Carbohydrate may well be attached to the protein component of limulin at this point.

Some difficulties have been observed in iodinating intact limulin. To the extent that the presence of tyrosine residues might be responsible, the sequence and compositional data obtained may be informative. The intact molecule of 163 amino acids appears to contain two tyrosine residues, one at position 39 and the second at position 61. The tyrosine at position 39 is immediately adjacent to a half-cystine residue that is part of a disulphide bond that binds this region of the molecule with the carboxy terminal segment. This tyrosine may consequently be deeply buried within the molecule's tertiary structure and be relatively inaccessible to the iodination process. It will, of course, be necessary to determine the tertiary structure of the molecule by X-ray diffraction to verify this hypothesis.

Current work on completing the determination of the covalent structure of this molecule involves cleavage by trypsin after citraconylation of the intact molecule. Compositional and sequence analyses have suggested the presence of only 3-4 arginines in the intact subunit. Thus, trypsin digestion following citraconylation should yield 4 or 5 fragments highly complementary for sequence analysis to the cyanogen bromide fragments prepared previously. Preliminary gel analysis of such fragments is encouraging in this respect.

E. SUMMARY

Amino acid sequence analysis of the subunit of limulin, the sialic acid binding lectin isolated from the horseshoe crab (Limulus polyphemus), has been carried out. The studies showed that this lectin is composed of a unique polypeptide chain that shows no amino-terminal or internal micro-heterogeneity. It contains one intrachain disulphide bond. No unusual amino acids were detected. No homologies are apparent between limulin and the immunoglobulin family of higher mammals or the limited number of other plant or animal lectins that have been sequenced to date. It is not yet clear what precise region of this molecule bears the sialic acid binding site, although the exact location of certain functionally important residues such as tyrosine has been determined. The carboxy-terminal segment of the molecule shows a marked tendency to self-aggregate, a phenomenon that could be related to the association of the identical subunits *in vivo*.

REFERENCES

Bucher DJ, Li S-L, Kehoe JM, and Kilbourne ED (1976). Chromatographic isolation of the hemagglutinin polypeptides from influenza virus vaccine and determination of their amino terminal sequences. Proc Nat Acad Sci US 73:238.

Edelman GM, Cunningham BA, Gall WE, Gottlieb PD, Rutishauser U, and Waxdal MJ (1969). The covalent structure of an entire γG immunoglobulin molecule. Proc Nat Acad Sci US 63:78.

Edelman GM, Cunningham BA, Roche GN, Jr, Becker JW, Waxdal MJ, and Wang JL (1972). The covalent and three-dimensional structure of concanavalin A. Proc Nat Acad Sci US 69:2580.

Foriers A, deNeve R, Kanarek L, and Strosberg AD (1978). Common ancestor for concanavalin A and lentil lectin? Proc Nat Acad Sci US 75:1136

Kaplan R, Li S-L, and Kehoe JM (1977). Molecular characterization of limulin, a sialic acid binding lectin from the hemolymph of the Horseshoe Crab, *Limulus polyphemus*. Biochemistry 16:4297.

Kehoe JM and Capra JD (1978). Patterns of sequence variability in immunoglobulin variable regions: functional, evolutionary and genetic implications. In Good RA and

Day SB (eds): "Comprehensive Immunology," New York: Plenum Medical Book Co., Vol. 5, p. 273.
Marchalonis J and Edelman GM (1968). Isolation and characterization of a hemagglutinin from Limulus polyphemus. J Mol Biol 32:453.
Roche A-C and Monsigny M (1974). Purification and properties of limulin: a lectin (agglutinin) from hemolymph of Limulus polyphemus. Biochim Biophys Acta 371:242.
Roche A-C, Perrodon Y, Halpern B, and Monsigny M (1977). Limulin (Limulus polyphemus lectin): mitogenic effect in human peripheral lymphocytes. Eur J Immunol 7:263.
Spiro RG (1973). Glycoproteins. Adv Prot Chem 27:349.

PURIFICATION AND PROPERTIES OF LECTINS FROM THE JAPANESE HORSESHOE CRAB, TACHYPLEUS TRIDENTATUS

Satoru Shimizu, Mie Ito, Noriko Takahashi* and Makoto Niwa**
Aichi Cancer Center Research Institute, Nagoya,
*Nagoya City University School of Medicine, Nagoya,
and **Osaka City University Medical School, Osaka,
Japan

The body fluid of some invertebrates has been reported to contain lectins that has an affinity for sialic acid, and the heterogeneity of the lectins has often been shown biochemically by isolating the lectins of different specificity (Hall and Rowlands 1974, Hardy et al 1977). From the hemolymph of the American horseshoe crab, *Limulus polyphemus*, a lectin that has a specificity for sialic acid has been purified by a variety of methods (Marchalonis and Edelman 1968, Roche and Monsigny 1974, Oppenheim et al 1974, Nowak and Barondes 1975, Sera et al 1975). However, the heterogeneity of the lectin in the American horseshoe crab has not been shown, although cross-absorption studies with the various erythrocytes suggested the heterogeneity of the lectin in the hemolymph (Noguchi 1903, Finstad et al 1972). The hemolymph of the Japanese horseshoe crab, *Tachypleus tridentatus*, contains lectins that strongly agglutinate mammalian erythrocytes. A previous study on the isolation of the lectins by affinity chromatography on a column of bovine submaxillary gland mucin-conjugated Sepharose showed that the lectins in the hemolymph were heterogeneous and that it contained one major and three minor lectins (Shimizu et al 1977). The specificity of these lectins for sugar, suggested by the inhibition study of agglutination, was different from that of the lectin from the American horseshoe crab which is believed to be specific for sialic acid; *Tachypleus* lectins were commonly inhibited by the sugars containing the N-acetylamino group such as sialic acid, N-acetyl-D-glucosamine (GlcNAc) and N-acetyl-D-galactosamine (GalNAc). The major lectin had unique properties compared with the other lectins. It did not agglutinate porcine erythrocytes and

Pages 625–639, Biomedical Applications of the Horseshoe Crab (Limulidae)

did not require calcium ion for agglutination. The present study intended to purify the lectin and to characterize the molecular structure and the specificity of the lectin in order to confirm the heterogeneity of the lectins in the hemolymph. In the course of the purification,a new type of lectin was found to exist in the hemolymph. The lectin had a high affinity for a GalNAc-Sepharose column and was eluted by a high concentration of GlcNAc.

MATERIALS AND METHODS

Hemolymph of *T.tridentatus* was collected according to the method of Nakamura et al (1976),and stored at -20°C. A hemagglutinating assay was performed using microtiter V-plate and human O erythrocyte unless otherwise specified,and a 2-fold serial dilution of the samples was made in 25 µl of Tris-buffered saline(50 mM Tris-HCl,pH 7.0-150 mM NaCl). As the major lectin did not require calcium ion for agglutination,it was omitted from the diluent. Laemmli's buffer system(Laemmli 1970) was applied for polyacrylamide gel electrophoresis(PAGE) in the presence and absence of sodium dodecylsulphate(SDS). A 5 % gel containing 50 mM GlcNac was used for PAGE of intact lectins and 0.1 % SDS was added to the 7.5 % gel for PAGE of subunit polypeptides,the lectins were treated with 2 % SDS at 37°C for 1 h in the presence or absence of 40 mM dithiothreitol. The gels were fixed and stained with Coomassie blue by the method of Fairbanks et al(1971). The preparation of anti-crude lectin serum in rabbits was previously described (Shimizu et al 1977). Anti-serum to purified lectin was prepared by injecting the lectin mixed with Freund complete adjuvant intradermally to the rabbit two times at 10 days interval. The rabbit was bled 2 weeks after the last injection. Immunodiffusion tests were performed in 1.5 % agar gel containing 50 mM GlcNac and 20 mM EDTA. Amino acid analysis was performed by the method of Spackman et al(1953). The cysteine content was determined after converting it to S-sulphocysteine by the method of Liu and Inglis(1972). N-terminal amino acid was determined by the Edman's method modified by Iwanaga et al(1969). GalNAc-Sepharose was prepared by conjugating GalNAc(50 mg) to epoxy-activated Sepharose 6 B(1 g) according to the instruction mannual "Affinity Chromatography" of Pharmacia Fine Chemicals. *Limulus polyphemus* agglutinin, wheat germ lectin,peanut lectin and *Phaseolus vulgaris* agglutinin were purchased from E-Y Laboratories(San Mateo,California,USA),and neuraminidase(*Arthrobacter ureafaciens)* from

Nakarai Chemicls(Osaka,Japan).

RESULTS

Purification

Eighty ml of the hemolymph was added to 40 ml of GalNAc-Sepharose washed with Tris buffered saline and the mixture was stirred for 30 min in a cold room. After separated on a glass filter,GalNAc-Sepharose was poured into a column and washed with Tris-buffered saline until no UV absorption was obserbed in the eluate. Lectins were eluted from the column by the 3 steps of buffer systems described previously and finally with 500 mM GlcNAc in the buffer. Fig 1 shows the elution profile of the proteins and the agglutinating activities.

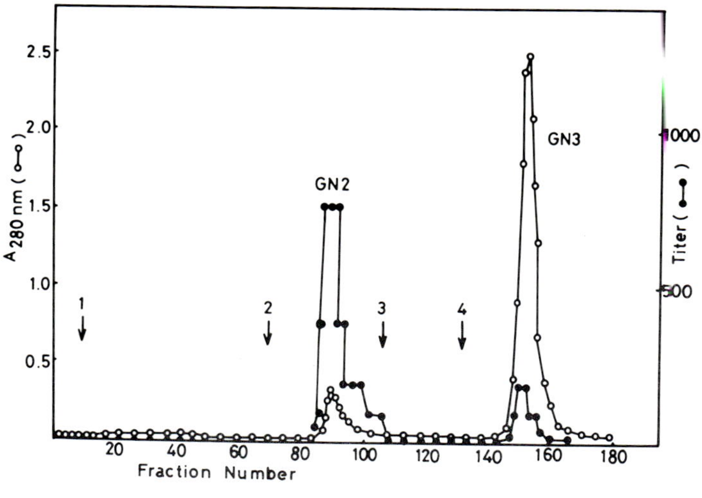

Fig 1. Purification of *Tachypleus* lectins on a GalNAc-Sepharose column. From GalNAc-Sepharose that was treated with the hemolymph,lectins were eluted by four buffer systems. Change in eluent are indicated by the arrows. Two ml fractions were collected. All fractions were monitored for absorbance at 280 nm and agglutinating activity. Buffer system 1,0.1 M Na-borate,pH 8.5; 2,50 mM GlcNAc in Tris-buffered saline; 3, 1.0 M NaCl in Tris-HCl,pH 7.0; 4, 0.5 M GlcNAc in Tris-buffered saline.

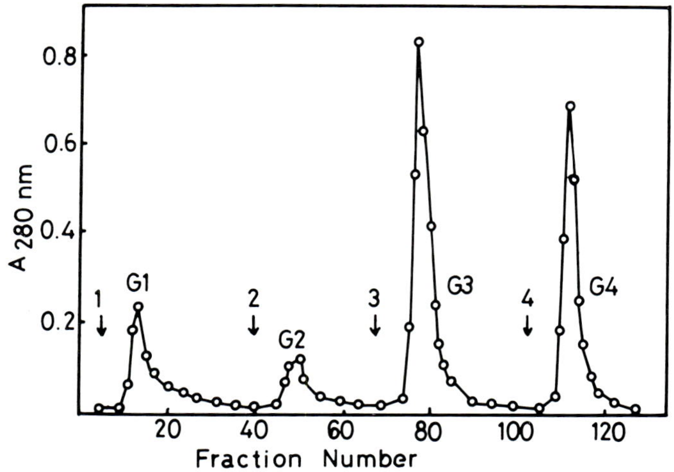

Fig 2. Rechromatography of the GN2 lectin on a GalNAc-Sepharose column. The GN2 lectin(5.3 mg) was passed through a column of GalNAc-Sepharose(0.9x12 cm) equilibrated with Tris-buffered saline and the column was washed throughly with the buffer. Lectins were eluted by the stepwise elution of increasing concentrations of GlcNAc in the buffer. One ml fractions were collected. Changes in eluent are indicated by the arrows. 1. 5 mM; 2, 10 mM; 3, 50 mM; 4, 500 mM GlcNAc in Tris-buffered saline.

Table 1. Results of Purification

	Volume ml	Protein mg/ml	Unit	Specific Activity	Yield %
Hemolymph	80.0	73.6	768	10.4	100.0
GN2	26.3	0.053	384	7245	16.4
GN3	26.0	0.119	96	807	4.1
GN2	92.0	0.058	384	6621	100.0
G1	13.2	0.037	768	20757	28.7
G2	11.8	0.025	768	30720	25.7
G3	12.0	0.160	768	4800	26.1
G4	7.6	0.213	384	1803	8.3

Unit of agglutination per ml was expressed as titer^{-1} (See text). Specific activity : unit/mg protein.

At the elution steps of 1 and 3 where borate buffer and 1 M NaCl were used,respectively,no agglutinating activity was eluted. Agglutinating activities were detected in 50 mM and 500 mM GlcNAc elution steps (GN2 and GN3). GN2 corresponds to the major lectin(M3 lectin in previous report,Shimizu et al, 1977),and GN3 is a newly detected lectin. Both lectins, after being dialyzed against Tris-buffered saline,were submitted to rechromatography on a small column of GalNAc-Sepharose. GN3 was eluted as a broad but single peak in rechromatography when eluted with 500 mM GlcNAc after the column was washed with 250 mM GlcNAc. GN2 was divided into 4 fractions (G1,G2,G3 and G4) by the stepwise elution of increasing concentrations of GlcNAc as shown in Fig 2. Results of purification are shown in Table 1. One unit of agglutination per ml was expressed as titer^{-1},which means the reciprocal of the highest dilution of sample which gave positive agglutination with human O erythrocytes.

Molecular Structure

Fig 3 shows the protein patterns of the lectins detected by PAGE. GN3 gave a broad but single band,but GN2 had several bands(Fig 3A). Of the multiple bands of GN2,characteristic bands were condensed in each of G1,G2,G3 and G4,but none of these fractions showed a single band. In G1 and G2,three bands were commonly contained, but the relative quantity of the bands in the fractions were different. In GN3, at least 7 bands were distinguished, and a band having the largest mobility was a major protein component. G4 contained almost all of the protein bands in GN3 as a minor components, and the major component of GN4 were the bands showing much smaller mobility that were absent in GN3.

In the presence of SDS, proteins in these fractions were dissociated into smaller polypeptides(Fig 3B). Both G1 and G2 gave 2 bands of molecular weights of 73,000 and 65,000. In G1, the larger polypeptide was the major component,and in G2, the smaller polypeptide was predominant. G3 and G4 had a single band in SDS-PAGE respectively. The molecular weights of the SDS-dissociated polypeptides were estimated as 65,000 for G3 and 67,000 for G4. GN3 was dissociated into 2 main and one minor polypeptides and the molecular weight of these polypeptides were 25,000,23,000 and 21,000 respectively.

Treatment of the lectins with dithiothreitol in the

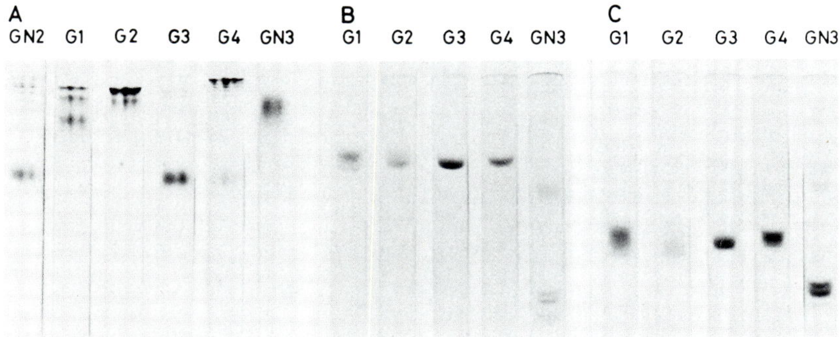

Fig 3. Polyacrylamide gel electrophoresis of *Tachypleus* lectins. A. Ten μg of each lectin was loaded on a 5 % gel. Electrophoresis was performed in the absence of SDS. B and C. Each lectin was incubated with 2 % SDS in the presence (C) or absence (B) of 40 mM dithiothreitol at 37° C for 1 h, and loaded on 7.5 % gel. Electrophoresis was performed in the presence of 0.1 % SDS.

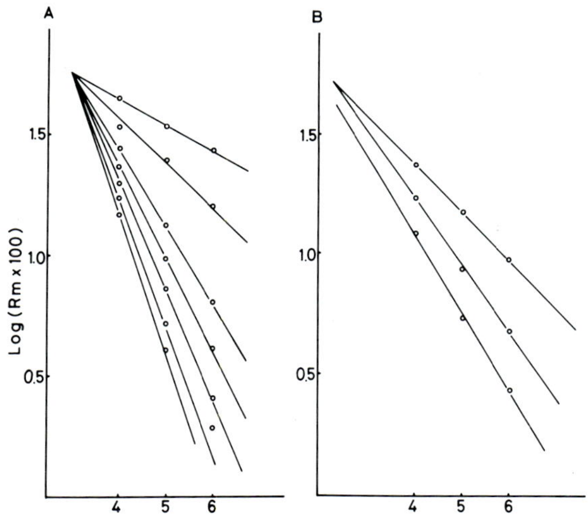

Fig 4. Effects of different gel concentrations on the mobilities of the components of lectins. A, G3 lectin; B, G4 lectin.

presence of SDS led to the reduction of the molecular weight of SDS-polypeptides into smaller subunit polypeptide except in the case of GN3 in which no change of PAGE pattern was observed,suggesting that interchain disulphide linkages were involved in the SDS-polypeptides of G1,G2,G3 and G4,and not in the SDS-polypeptide of GN3 (Fig 3C). Both G1 and G2 gave three bands of molecular weight of 38,000,35,000 and 32,000. In G1,a band of molecular weight of 38,000 was the most intensely stained and in G2, a band of molecular weight of 32,000 was the main component. G3 gave only one band of molecular weight of 34,000,and G4 had a band of molecular weight of 36,000 and a small amount of the component of molecular weight of 34,000. These results of PAGE indicate that the GN3 lectin differs in the molecular structure from the GN2 lectin which contained "isolectins" of G1,G2,G3 and G4, and that these isolectins could be divided into two groups: the G1 group including G1 and G2 lectins, and the G3 group including G3 and G4 lectins.

In order to estimate the molecular weight of intact lectins,PAGE of intact lectins was performed in the gel con-concentrations of 4,5 and 6 %, and the logarithmic values of the relative mobility of each band were plotted against gel concentrations (Fig 4). The plots from seven bands of G3 lectin gave nonparallel lines and when extrapolated these lines crossed at the same point. Three lines from the plots of G1 bands had different crossing points. According to Hedrick and Smith(1968),the slope of the line has linear relation to the molecular weight of protein, and the lines from proteins having the same charge but different molecular weight have the same crossing point. The results obtained indicate that intact G3 might be aggregates consisted of protein subunits of various molecular weight with the same charge,whereas the subunits of G1 might differ in both their size and charge. The molecular weight of the smallest aggregate size of undissociated G1 and G3 estimated from the standard line obtained from the slopes of ferritin and aldolase were molecular weight of 480,000 and 180,000,respectively. The molecular weight of GN3 protein,estimated by using a low angle light scattering apparatus,Chromatix KMX-6,was found to be as large as 540,000.

Antigenicity

Precipitin lines of the lectins against anti-serum to

lectins are shown in Fig 5. Fig 5A shows the results of anti-crude lectin serum. The G1 group gave one common precipitin line,whereas the G3 group had two common lines. These lines fused with the corresponding lines against hemolymph,but the line against the G1 group did not fuse with the lines against the G3 group. The GN3 lectin gave no line against anti-crude lectin serum. It is especially interesting that *Limulus* lectin did not make any precipitin line against anti-*Tachypleus* crude lectin serum,suggesting considerable differences in antigenicity of lectins from the both species. Such difference in antigenicity was obserbed in the case of coagulogens from *Tachypleus tridentatus* and *Limulus polyphemus*. Fig 5B and 5C illustrate the lines of each fractions against anti-G1 as well as anti-G3 sera,respectively. These antisera gave precipitin lines not only against the respective lectin fraction to which antiserum was generated, but also against other lectin fractions. Both anti-G1 and anti-G3 sera had one precipitin line against the G1 group and two lines against the G3 group. The lines against the G1 group and the G3 group did not fuse with each other. This suggests that lectins in the G1 group have a different antigenic determinant from that of the G3 group.

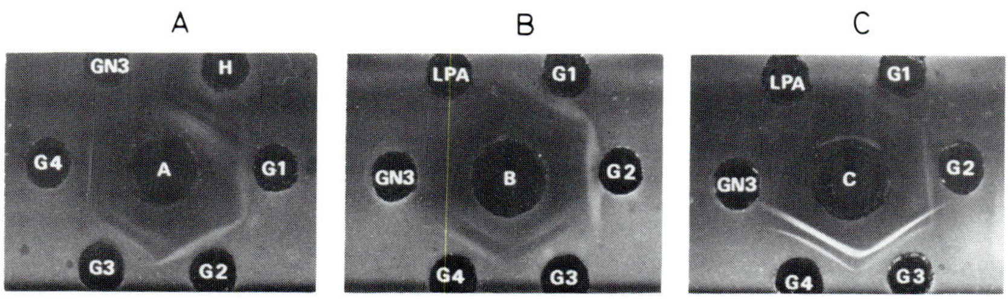

Fig 5. Immunodiffusion tests of *Tachypleus* lectins. A, rabbit anti-crude lectin serum; B, rabbit anti-G1 serum; C, rabbit anti-G3 serum; H, hemolymph; LPA, Limulus lectin.

In order to check whether these antisera contained antibodies to all of the subunit polypeptides of these lectins, the reaction of antisera against SDS polypeptides was tested. As shown in Fig 6, when SDS-polypeptides were used as the antigens,anti-G3 serum gave one common precipitin line against only the G3 group lectins (Fig 6B), and anti-G1 serum had two diffused lines against the G1 group lectins and one line against G3 (Fig 6A). The line against G3 fused with the

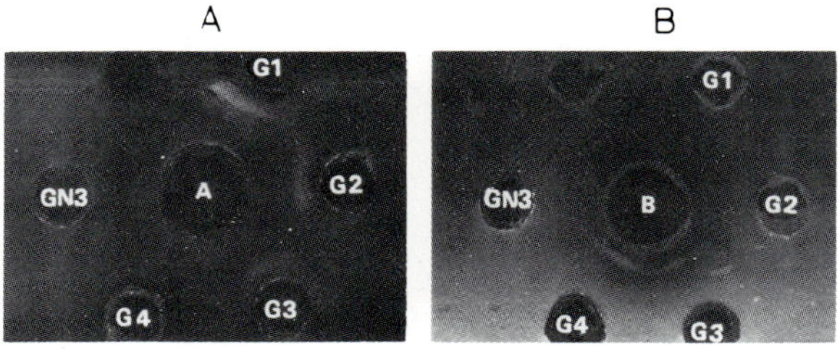

Fig 6. Immunodiffusion tests of SDS-polypeptides of *Tachypleus* lectins. Each lectin was treated with 0.1 % SDS and used as an antigen. A, anti-G1 serum; B, anti-G3 serum.

line against the G1 group. These results may arise from any slight contamination of the fractions,but it might be also possible that the antigenic expressions of the intact lectins would be quite different from those of their subunits.

Amino Acid Compositions

Table 2 shows the amino acid compositions of G1,G2,G3, G4 and GN3 lectins. All of the lectins had a relatively high content of Asp,Glu and Gly,but G1 group,G3 group and GN3 lectins each showed a slightly different amino acid compositions. In GN3 increase in the content of Val,Ser and Pro and decrease in the content of Tyr,Thr and Met were observed. The amino acid compositions of these *Tachypleus* lectins differed from that of the *Limulus* lectin; decrease in the content of Tyr and increase in the content of Leu and Ser were distinct in the latter(Marchalonis and Edelman 1968, Finstad et al 1977,Roche and Monsigny 1974,Nowak and Barondes 1975,Kaplan et al 1977). The amino-terminal sequence of the G3 and GN3 lectins were Pro-Ile-Pro- and Gln-Try-Pro, respectively, and these amino-terminal sequences were quite different from the amino-terminal sequence (Leu-Glu-Glu) of the *Limulus* lectin previously reported(Kaplan et al 1977). These results,though still being fragmentary, may provide a very interesting paradigm of molecular evolution and furnish useful hints on the phylogeny of the Asian and American species of the horseshoe crab. Comparative studies on the chemical and immunological properties of lectins from four species of the horseshoe crab,*Tachypleus tridentatus*, *Tachypleus gigas*,*Carcinoscorpius rotundicauda* and *Limulus polyphemus* are now undertaken.

Table 2. Amino Acid Compositions of *Tachypleus* Lectins.

Amino acid	*Tachypleus tridentatus* G1	G2	G3	G4	GN3
Asp	16.56	14.42	12.43	11.86	13.56
Thr	5.28	5.56	6.29	6.58	3.71
Ser	6.03	6.25	5.89	6.09	8.15
Glu	8.99	9.36	9.63	10.17	8.33
Pro	3.31	3.59	3.74	3.98	5.58
Gly	10.18	10.49	9.01	9.02	11.45
Ala	3.78	3.80	3.81	3.75	3.40
Cys/2	1.98	2.03	1.55	1.83	2.23
Val	3.59	3.78	4.71	3.80	10.25
Met	1.84	1.93	2.49	2.38	1.01
Ile	5.48	5.90	7.21	5.32	6.53
Leu	6.43	6.49	6.08	7.39	5.77
Tyr	7.57	7.70	5.03	5.83	2.52
Phe	3.92	3.86	5.13	4.51	3.65
Lys	7.24	6.22	7.77	8.37	7.00
His	3.61	3.83	4.86	5.06	2.99
Arg	4.23	4.80	4.40	4.07	3.88

Comparative Agglutination Titers for Various Erythrocytes.

Table 3 shows the comparative agglutination titers of each lectin for erythrocytes from various mammalian species plotted against the titer for human O erythrocytes,taken as 512. The relative activity of each lectin was distinct. The G1 group of lectins had a high activity for human O erythrocytes,and the G3 group of lectins and GN3 lectin gave high titers for horse and mouse erythrocytes. All of the lectin fractions did not agglutinate porcine and bovine erythrocytes.

Inhibition of Agglutination

As was suggested from the purification procedure,

Table 3. Comparative Agglutination Test

Erythrocytes	Titer G1	G2	G3	G4	GN3
Human O	512	512	512	512	512
" A	128	128	256	512	512
" B	128	128	256	512	512
Horse	64	128	2048	4096	8192
Rat	128	256	256	1024	2048
Mouse	64	64	1024	2048	4096
Rabbit	8	16	128	256	512
Cattle	0	0	0	0	0
Pig	0	0	0	0	0

agglutination by these lectins was inhibited by N-acetylhexosamine(Fig 7). Free N-acetylneuraminic acid(NANA) also inhibited the reaction,but its inhibitory ability was the same order as N-acetylhexosamine. The concentrations of these N-acetylamino sugars required for the inhibition of agglutination were slightly different among these lectins. In addition to the N-acetylamino sugars, the agglutination was inhibited by the compounds containing N-acetylamino group such as N-acetylaminoethanol,N-acetylalanine and N-acetylglycine (Fig 8). Five to 25 mM of these compounds was sufficient to inhibit the reaction and this concentration was not so high when compared to the concentration of N-acetylamino sugars required to exhibit their inhibition.

Table 4. Effects of Neuraminidase Treatment of Human Erythrocytes on Agglutination by various lectins. Figures represent agglutination titer.

Lectins		Intact	Neuraminidase treated
Tachypleus tridentatus	G1	2048	0
	G2	2048	0
	G3	2048	0
	G4	128	0
	GN3	16	0
Peanut Lectin		0	2048
Wheat Germ Lectin		64	256
Phaseolus vulgaris Agglutinin		256	1024

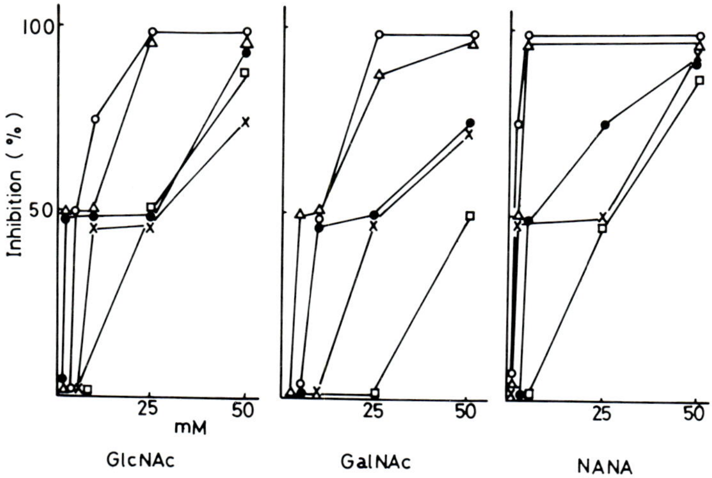

Fig 7. Inhibition of agglutination by N-acetylamino sugars. Agglutination titer was determined in the presence of a inhibitor,and a one-step reduction of titer was interpreted as indicating 50 % inhibition of agglutination. –○–,G1 ; –△–,G2; –×–,G3; –●–,G4; –□–,GN3.

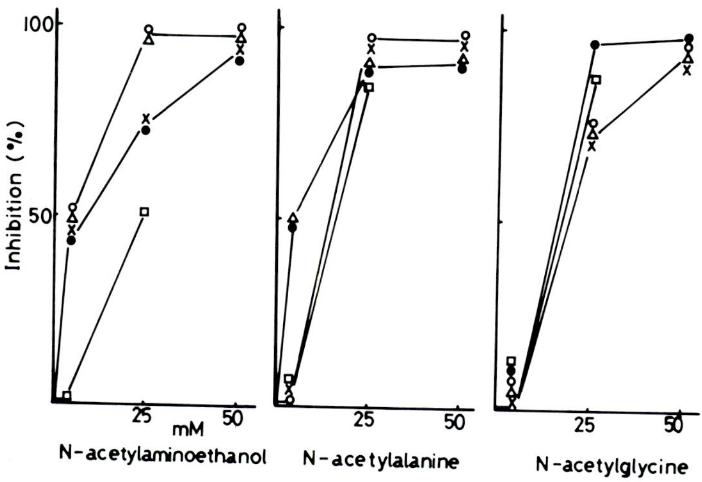

Fig 8. Inhibition of agglutination by N-acetylamino compound –○–,G1; –△–,G2; –×–,G3; –●–,G4; –□–,GN3.

Submaxillary gland mucins were poerful inhibitors of the agglutination,and the sialic acid concentration of mucins required for inhibition was about 1,000 times lower than the inhibitory concentration of the N-acetylamino compounds.

Table 4 shows the effect of the treatment of human O erythrocytes with neuraminidase on the agglutination by various lectins. Neuraminidase treatment changed the susceptibility of the cells to the lectins. Change in the agglutination titer of peanut lectin from 0 to 2048 indicated that sialic acids were effectively removed from the cell surface and that the terminal galactose residues were newly exposed to which peanut lectin could bind. Slight increase in the titers were observed in the agglutination by wheat germ lectin and *Phaseolus vulgaris* agglutinin which were specific for GlcNac and GalNAc,respectively. On the contrary, all of the *Tachypleus* lectins did not agglutinate neuraminidase treated erythrocytes.

DISCUSSION AND CONCLUSION

The present study on the purification of the lectins in the hemolymph of Japanese horseshoe crab intended to confirm the suggestion proposed previously (Shimizu,1977) that the hemolymph of Japanese horseshoe crab contains a variety of lectins with the almost same specificity. In the present study two types of lectins were purified by affinity chromatography on a GalNAc-Sepharose column, and the differences in subunit structure,antigenicity,amino acid compositions and agglutinating activity were demonstrated in these lectins. One lectin corresponding to the major lectin in the hemolymph contained four "isolectins",which could be divided into two groups of lectins;one strongly agglutinated human O erythrocytes and the other had a high affinity for horse and mouse erythrocytes. These results indicate that the lectins in the hemolymph are heterogeneous or polymorphic, although the possibility that this heterogeneity could arise artificially in the course of the purification procedures is not completely excluded.

The second conclusion is that the specificity of these lectins suggested by the inhibition studies of agglutination is similar among these lectins,and that they are commonly inhibited by the compounds containing the N-acetylamino group.

No lectin was reported,agglutination by which was inhibited by the compounds other than sugars,and this suggests that *Tachtpleus* lectins might have a primary binding specificity for N-acetylamino groups of sugars on the cell surface. On the other hand,neuraminidase treatment of erythrocytes completely abolished the susceptibility of the cells to the agglutination by the lectins,suggesting the importance of sialic acid as receptors on the cell surface. However, we are not yet sure whether all of the N-acetylamino groups and sialic acids on the cell surface could be the receptor sites for the lectins. The difference in the comparative agglutination titers among these lectins might be some reflection of a slight difference in the specificity and affinity,and this suggests that some structure other than N-acetylamino group and sialic acid might be involved in the receptor sites of these lectins. Studies on the recptors of erythrocytes are on going.

Acknowledgments

The authors are indebted to Dr.M.Yoshida,Marine Laboratory of Okayama University and Dr.H.Nishii,Kasaoka,Okayama Prefecture, for their support in the collection of the horseshoe crabs.

REFERENCES

Fairbanks g,Steck TL,Wallach DFH (1971). Electrophoretic analysis of the major polypeptides of the human erythrocyte membrane. Biochemistry 10:2606.

Finstad CL,LitmanWG,Finstad J,Good RA (1972). The evolution of the immune response. XIII. The characterization of purified erythrocyte agglutinins from two invertebrate species. J Immunol 108:1704

Hall JL,Rowlands JrDT (1974). Heterogeneity of lobster agglutinins I. Purification and physicochemical characterization. Biochemistry 13:821

Hardy SW,Grant PT,Fletcher TC (1977). A haemagglutinin in the tissue fluid of the Pacific oyster,*Crassostrea gigas*, with specificty for sialic acid residues in glycoproteins. Experientia 33:767

Hedrick JL,Smith AJ (1968). Size and charge isomer separation and estimation of molecular weight of protein by gel electrophoresis. Arch Biochem Biophs 126:155

Iwanaga S,Wallén P,Gröndahl NJ,Henschen A,Blombäck B (1969). On the primary structure of human fibrinogen. Isolation and characterization of N-terminal fragment from plasmic digests. Eur J Biochem 8:189

Kaplan S,Steven S-L,Kehoe JM (1977). Molecular characterization of limulin,a sialic acid binding lectin from the hemolymph of the horseshoe crab,*Limulus polyphemus*. Biochemistry 16:4297.

Laemmli UK (1970).Cleavage of structural proteins during the assembly of the head of bacteriophage T4. Nature 227:680

Liu T-Y,Inglis S (1972). Determination of cystine and cysteine as S-sulfocysteine. In Hirs CH,Timasheff SN (eds) :"Methods in Enzymology",New York :Academic Press,vol 25,p55.

Marchalonis JJ,Edelman GM (1968). Isolation and characterization of a hemagglutinin from *Limulus polyphemus*. J Mol Biol 32:453.

Nakamura S,Iwanaga S,Harada T,Niwa M (1976). A clottable protein(Coagulogen) from amoebocyte lysate of Japanese horseshoe crab(*Tachypleus tridentatus*). Its isolation and biochemical properties. J Biochem 80:1011.

Noguchi H (1903). On the multiplicity of the serum hemagglutinins of the cold-blooded animals. Zentr Bakt Abt I Orig 34:286.

Nowak TP,Barondes SH (1975). Agglutinin from *Limulus polyphemus*.Purification with formalinized horse erythrocytes as the affinity adsorbent. Biochim Biophys Acta 393:115.

Oppenheim JD,Nachbar MS,Salton MRJ,Aull F (1974).Purification of a hemagglutinin from *Limulus polyphemus* by affinity chromatography. Biochem Biophys Res Commun 58:1127.

Roche A-C,Monsigny M (1974). Purification and properties of limulin:a lectin(agglutinin)from hemolymph of *Limulus polyphemus*. Biochim Biophys Acta 371:242.

Roche A-C,Schauer R,Monsigny M (1975). Protein-sugar interactions.Purification by affinity chromatography of limulin, an N-acylneuraminyl-binding protein. FEBS Lett 57:245.

Sela B-A,Wang JL,Edelman GM (1975).Isolation of lectins of different specificities on a single affinity adsorbent. J Biol Chem 250:7535.

Shimizu S,Ito M,Niwa M (1977).Lectins in the hemolymph of Japanese horseshoe crab,*Tachypleus tridentatus*. Biochim Biophys Acta 500:71.

Spackman DH,Stein WH,Moore S (1953). Automated recording apparatus for use in the chromatography of amino acids. Anal Chem 30:1190.

AFFINITY CHROMATOGRAPHY WITH AGAROSE-LECTIN DERIVATIVES: SEPARATION OF HUMAN GLYCOPROTEINS AND APPLICATION TO ERYTHROPOIETIN PURIFICATION

J.L. Spivak and M.D. Hollenberg

Johns Hopkins University School of Medicine
Baltimore, Maryland

Erythropoiesis is regulated by the hormone erythropoietin, a sialoglycoprotein produced in the kidneys (Jacobson et al., 1957) and to a minor extent in the liver (Fried, 1972). Whether erythropoietin is secreted in an active form (Erslev, 1974) or as a precursor that must be activated by a plasma factor (Libbin et al., 1974; Peschle and Condorelli, 1975) is unsettled. There appears to be no storage pool of the hormone because it cannot be isolated from the kidney in significant quantities (Contrera et al., 1965; Sherwood and Goldwasser, 1978) and because inhibitors of RNA and protein synthesis block its secretion in response to a physiologic stimulus (Schooley and Mahlman, 1972). The concentration of erythropoietin in normal plasma has been estimated to be 0.02 units/ml (Alexanian, 1973) or approximately 0.3 ng/ml. Less than 10% of the hormone is excreted in the urine each day (Alexanian, 1969); the liver is probably the major site of degradation (Goldwasser et al., 1974).

Erythropoietin increases the number of cells capable of differentiating into erythrocytes (Reissmann and Samorapoompichit, 1970), triggers their differentiation (Gurney et al., 1962), augments hemoglobin synthesis in developing erythroblasts (Glass et al., 1975) and facilitates the early release of newly formed red cells from the marrow (Papayannopoulou and Finch, 1972). Owing

Pages 641–653, Biomedical Applications of the Horseshoe Crab (Limulidae)

to the lack of suitably pure preparations of the hormone, the mechanism by which erythropoietin activates its target cells is unknown. Efforts to purify the hormone have been hindered by lack of a body fluid or tissue rich in erythropoietin. Recently, Miyake, Kung, and Goldwasser (1977) purified human urine erythropoietin to homogeneity using a conventional multistep procedure. However, twenty-five hundred and fifty liters of urine were required for the purification procedure which yielded approximately 10 mg of hormone. The need for such large quantities of starting material when conventional techniques for separating proteins on the basis of size and charge are employed, is in part due to the presence in urine of at least twelve glycoproteins with a molecular weight similar to erythropoietin and seven with a similar isolectric point. When adequate quantities of crude starting material are available, the protocol established by Miyake et al., should serve as a model for the purification of erythropoietin from the urine. Until then, alternative methods must be sought for the isolation of erythropoietin from the relatively restricted sources of crude hormone currently available. Since erythropoietin is a glycoprotein, we have employed affinity chromatography using agarose-bound lectins to isolate the hormone from other proteins in crude extracts of human urine (Spivak et al., 1977; Spivak et al., 1978). This report reviews our experience with this technique.

MATERIALS AND METHODS

Crude erythropoietin-rich extracts of human urine (40-55 units/mg protein) were obtained from the National Heart and Lung Institute. Bioassay of hormone preparations and column fractions was performed in vivo using exhypoxic polycythemic mice (Spivak, 1976). The second International Reference Preparation of human urine erythropoietin obtained from the World Health Organization was used as the standard for the bioassay.

Affinity Adsorbents

Ricinus communis 120 (RCA-120), peanut agglutinin (PA), soybean agglutinin (SA), concanavalin A (Con-A), phytohemagglutinin (PHA-P and L-PHA), *Lens culinaris* A (LCA) and B (LCB) and anti-A and anti-H lectins were purchased commercially. Wheat germ agglutinin (WGA) and phytohemogglutinin (E-form) were provided by Dr. J. H. Shaper. Hemolymph from *Limulus polyphemus* (provided by Dr. J. Levin) was employed to obtain limulin by the technique of Roche, Schauer and Monsigny (1975). Individual lectins were coupled to beaded agarose (Sepharose 4B or CL-4B, 2-10 mg lectin/ml of agarose) after cyanogen bromide activation (March et al., 1974). Coupling efficiency varied from 50 to 70%.

Chromatography

Chromatography was performed at 4°C with disposable glass pipets, 0.5 X 9 cm, plugged with glass wool or 5 ml plastic syringes stoppered with sintered glass discs as columns. The columns which contained approximately 2 ml of affinity adsorbent were equilibrated with phosphate-buffered saline (PBS) at pH 7.4. The buffer was allowed to flow by gravity at a rate of 12 ml per hour and 1 ml fractions were collected. Column fractions were dialyzed against PBS before quantitation of protein (Lowry et al., 1951) and bioassay for erythropoietin.

RESULTS

The sugar content of human erythropoietin has not been unequivocally established. Consequently, lectins were chosen for study not only on the basis of their affinity for sugars present in erythropoietin but also on the basis of their affinity for other sugars known to occur in human glycoproteins. To date, we have examined the ability of 13 lectin-agarose derivatives to bind erythropoietin (Spivak et al., 1978). The results of these studies are shown in Table 1. Five of the 13 lectins (LCA, LCB, L-PHA, PHA-P and WGA) bound hormone which could be

TABLE 1

BINDING OF HUMAN URINE ERYTHROPOIETIN BY AGAROSE-BOUND LECTINS

LECTIN	ELUTING AGENT	BIOASSAY FOR ERYTHROPOIETIN	
		VOID VOLUME	ELUATE FRACTIONS
LIMULIN	NANA	+	-
RCA-120	Gal	+	-
PA	Gal	+	-
SA	GalNAc	+	-
ANTI-A	GalNAc	+	-
ANTI-H	Fuc	+	-
CON-A	Man	+	-
LCA	Man	+	+
LCB	Man	+	+
PHA			
L-Form	GalNAc	+	+
P-Form	GalNAc	-	+
E-Form	Guanidine HCl	-	+
WGA	GlcNAc	-	+

Approximately 5-15 units of erythropoietin were applied to each agarose-lectin column. (NANA), N-acetylneuraminic acid; (GalNAc), N-acetylgalactosamine; (Gal), Galactose; (Man), α methyl Mannoside, (GlcNAc), N-acetylglucosamine; (Fuc), Fucose. For lectin abbreviations, see MATERIALS AND METHODS.

eluted by a sugar specific for the particular lectin. However, only PHA and WGA bound the hormone quantitatively.

Binding of erythropoietin by PHA varied with the form of the lectin employed. The leukoagglutinating isolectin, L-PHA, failed to bind erythropoietin quantitatively while the erythroagglutinating isolectin, E-PHA, had such a high affinity for erythropoietin that it could not be released by 0.5M GalNAc, 0.2M NaSCN or 1M NaCl. The hormone was released from the E isolectin by 4M guanidine hydrochloride at pH 7.0. PHA-P, a partially purified extract of Phaseolus vulgaris which contains both the E and L isolectins, also bound erythropoietin quantitatively. Erythropoietin was partially released from PHA-P by 0.1M GalNAc and completely by 4M guanidine HCl. The erythropoietin which bound to L-PHA could also be eluted in the same fashion.

Failure of erythropoietin to bind quantitatively to L-PHA-agarose was not due to molecular heterogeneity of the hormone preparation. Erythropoietin which passed unretarded through one L-PHA agarose column was still capable of being partially bound when applied to a fresh L-PHA-agarose column. Furthermore, when an equivalent amount of E-PHA covalently coupled to agarose was employed, all of the hormone was adsorbed by that isolectin. Since both GalNAc and NaCl can release erythropoietin from L-PHA but not E-PHA, it appears that the L isolectin has a lower affinity for the hormone.

The data in Table 1 also indicate that the binding of erythropoietin by a particular lectin-agarose derivative was not simply a function of that lectin's specificity for a particular monosaccharide. Limulin, which was capable of binding to bovine submaxillary mucin (Roche and Monsigny, 1974), failed to bind erythropoietin, although the hormone has abundant sialic acid residues. Furthermore, although Con A, LCA and LCB all bind mannose, only the latter two lectins bound erythropoietin. The reason for the differences in hormone binding by these three lectins which share the same sugar specificities has not been established.

In our initial studies (Spivak et al., 1977) we observed that only 40% of the erythropoietin bound to WGA-agarose was released when GlcNAc was used as the eluting sugar. Since NANA is also known to bind to WGA we measured the release of erythropoietin from WGA-agarose following the sequential application of the two sugars to the column (Spivak et al., 1978). As shown in Table 2,

TABLE 2

SEQUENTIAL ELUTION OF HUMAN URINE ERYTHROPOIETIN FROM WGA-AGAROSE

ELUTING SUGAR		UNITS OF HORMONE ELUTED
(A)	GLcNAc	7.5
	NANA	2.5
(B)	NANA	2.0
	GLcNAc	10.0

(A) 19 units of erythropoietin were initially applied to the column. (B) 55 units of erythropoietin were initially applied to the column.

these sugars, whether applied alone or sequentially, are unable to release completely erythropoietin bound to WGA-agarose. Since polymers of GlcNAc have a higher affinity for WGA than GlcNAc (Goldstein et al., 1975), we used N,N-diacetylchitobiose as an eluting agent. When this disaccharide was employed, all of the hormone was recovered from the column (Table 3).

Since erythropoietin was bound quantitatively by two different lectins, PHA and WGA, we sought to determine whether they bound erythropoietin in

TABLE 3

ELUTION OF HUMAN URINE ERYTHROPOIETIN FROM AGAROSE-BOUND WGA

COLUMN FRACTION	RECOVERY (Units)	SPECIFIC ACTIVITY (Units/mg Protein)
VOID VOLUME	0	-
N,N DIACETYL CHITOBIOSE ELUATE	40	181
GUANIDINE HCl ELUATE	0.8	0.4

20 units of erythropoietin (0.5 mg, 40 u/mg protein) were applied to a 0.5 x 9.0 cm column. The increase in recovery of hormone probably represents the removal of an inhibitor during chromatography.

the same fashion (Spivak et al., 1978). As shown in Table 4, only NANA was capable of eluting erythropoietin from both PHA and WGA, and then only partially. N,N-diacetylchitobiose quantitatively

TABLE 4

ELUTION OF HUMAN URINE ERYTHROPOIETIN FROM AGAROSE-BOUND WGA AND PHA

	RECOVERY OF BOUND HORMONE (%)	
ELUTING AGENT	WGA	PHA
N-ACETYLGLUCOSAMINE (0.1M)	40	0
N-ACETYLGALACTOSAMINE (0.1M)	0	25
N-ACETYLNEURAMINIC ACID (0.1M)	39	35
N,N DIACETYLCHITOBIOSE (0.01M)	75	0
GUANIDINE HCl (4M, pH 7.0)	50	60
SODIUM THIOCYANATE (0.2-0.5M)	0	0

released erythropoietin from WGA-agarose but not from PHA-agarose. Thus, it appears that the two lectins bind the hormone principally through different sugar determinants.

Although affinity chromatography with lectin-agarose derivatives was capable of isolating erythropoietin from over 90% of the proteins in crude urine extracts, it was not capable of completely purifying the hormone (Spivak et al., 1977). Consequently, we have initiated studies combining both conventional techniques for separation of proteins on the basis of size and charge and lectin-agarose chromatography. The results of one such study are shown in Table 5. While purification to homogeneity has not been achieved, it is noteworthy that

TABLE 5

PURIFICATION OF HUMAN URINE ERYTHROPOIETIN

PURIFICATION STEP	PROTEIN (mg)	ACTIVITY (Units)	SPECIFIC ACTIVITY (Units/mg)	RECOVERY (%)	PURIFICATION FACTOR
	10	550	55	-	-
SEPHADEX G-100	3.28	780	237	141*	4.3
DEAE-AGAROSE	1.0	288	288	52	5.2
CON-A/ANTI-ALBUMIN AGAROSE	0.8	176	220	32	4.0
WGA-AGAROSE	0.07	274	3914	49	71

*Recovery of greater than 100% is due to the removal of an inhibitor during gel filtration chromatography.

substantial purification has been obtained even though the quantity of crude starting material was extremely small.

DISCUSSION

The isolation and purification of erythropoietin from plasma or urine has been difficult owing to the minute concentration of hormone in such fluids. Since many glycoproteins in urine or plasma have a molecular weight or isolectric point similar to erythropoietin, it is not surprising that conventional techniques for separating the hormone from other proteins on the basis of size or charge require generous quantities of starting material. The quantity of crude starting material currently available to most investigators is, however, limited. Since many conventional separation procedures are too cumbersome or inefficient to employ on a small scale, lectin-agarose chromatography is an attractive alternative. Our studies and those of Sieber (1977) have identified several lectins which are likely to be of use in the purification of erythropoietin.

From the data presented, it is apparent that binding of erythropoietin to the lectins studied was complex. For example, limulin which binds to NANA failed to bind erythropoietin while WGA did. It is noteworthy, however, that limulin has a low affinity for NANA, as compared with other derivatives of neuraminic acid, an even lower affinity for GlcNAc and does not bind at all to N,N-diacetylchitobiose (Roche and Monsigny, 1974). Furthermore, limulin appears to bind more avidly to oligosaccharides containing polymers of neuraminic acid than to a single residue (Roche et al., 1975). WGA, on the other hand, while it binds NANA has a much higher affinity for GlcNAc(which is present in erythropoietin) and polymers of that hexosamine (Goldstein et al,1975). PHA has been shown to bind avidly to oligosaccharides which contain a terminal sialic acid residue, a penultimate galactose and a core structure containing mannose (Kornfeld and Kornfeld, 1970), features which are also common to erythropoietin. Thus, when the affinity of limulin, WGA and PHA for certain disaccharides or oligosaccharides is considered, their ability to bind erythropoietin or the lack of it can be explained.

Although lectin-agarose chromatography appears to be a valuable technique for the isolation of erythropoietin from crude urine extracts, we have recently become aware of a situation in which it cannot be employed. Erythropoietin in crude urine extracts may be unstable owing to the presence of proteolytic enzymes and sialidase (Chiba et al., 1972). Consequently, phenol extraction of such extracts has been advocated as means of eliminating the influence of these enzymes. However, erythropoietin which has been subjected to phenol extraction is altered to the extent that it not only does not migrate as expected during gel filtration on Sephadex G-100, but also it fails in whole or in part to bind to either WGA-agarose or PHA-agarose. Since lectin-agarose affinity chromatography serves to remove the bulk of the proteolytic activity from crude urine extracts, phenol extraction is neither necessary or helpful when lectin-agarose chromatography employing WGA-agarose or PHA-agarose is planned.

REFERENCES

Alexanian R (1969). Correlations between erythropoietin production and excretion. J Lab Clin Med 74:614.

Alexanian R (1973). Erythropoietin excretion in bone marrow failure and hemolytic anemia. J Lab Clin Med 82:438.

Chiba S, Kung C K-H, Goldwasser E (1972). Stabilization of urinary erythropoietin. Biochem Biophys Res Comm 47:1372.

Contrera JF, Camiscoli JF, Weintraub AH, Gordon AS (1965). Extraction of erythropoietin from kidneys of hypoxic and phenylhydrazine-treated rats. Blood 25:809.

Erslev AJ (1974). In vitro production of erythropoietin by kidneys perfused with a serum-free solution. Blood 44:77.

Fried W (1972). The liver as a source of extrarenal erythropoietin. Blood 40:671.
Glass J, Lavidor LM, Robinson SH (1975). Studies of murine erythroid cell development. J Cell Biol 65:298.
Goldstein JJ, Hammarstrom S, Sundblad G (1975). Precipitation and carbohydrate binding specificity studies on wheat germ agglutinin. Biochim Biophys Acta 405:53.
Goldwasser E, Kung C K-H, Eliason J (1974). On the mechanism of erythropoietin-induced differentiation. XIII. The role of sialic acid in erythropoietin action. J Biol Chem 249:4202.
Gurney CW, Lajtha LG, Oliver R (1962). A method for investigation of stem cell kinetics. Brit J Haemat 8:461.
Jacobson LO, Goldwasser E, Fried W, Plzak L (1957). Role of the kidney in erythropoiesis. Nature 179:633.
Kornfeld R, Kornfeld S (1970). The structure of a phytohemagglutinin receptor site from human erythrocytes. J Biol Cehm 245:2536.
Libbin RM, Person P, Gordon AS (1974). Renal lysosomes: Role in biogenesis of erythropoietin. Science 185:1174.
Lowry OH, Rosebrough NJ, Farr AL, Randall RJ (1951). Protein measurement with the folin phenol reagent. J Biol Chem 193:265.
March SC, Parikh I, Cuatrecasas P (1974). A simplified method for cyanogen bromide activation of agarose for affinity chromatography. Anal Biochem 60:149.
Miyake T, Kung C K-H, Goldwasser E (1977). Purification of human erythropoietin. J Biol Chem 252:5558.
Papayannopoulou T, Finch C (1972). On the in vivo action of erythropoietin: A quantitative analysis. J Clin Invest 51:1179.
Peschle C, Condorelli M (1975). Biogenesis of erythropoietin: Evidence of pro-erythropoietin in a subcellular fraction of kidney. Science 190:910.
Reissmann KR, Samorapoompichit S (1970). Effect of erythropoietin on proliferation of erythroid stem cells in the absence of transplantable colony-forming units. Blood 36:287.

Roche A-C, Monsigny M (1974). Purification and properties of limulin: A lectin (agglutinin) from hemolymph of Limulus polyphemus. Biochim Biophys Acta 371:242.

Roche A-C, Schauer R, Monsigny M (1975). Protein-sugar interactions. Purification by affinity chromatography of limulin, an N-acylneuraminidyl-binding protein. FEBS Lett 57:245.

Schooley JC, Mahlman LJ (1972). Evidence for the de novo synthesis of erythropoietin in hypoxic rats. Blood 90:662.

Sherwood JB, Goldwasser E (1978). Extraction of erythropoietin from normal kidneys. Endocrinology 103:866.

Sieber F (1977). Chromatography of human urinary erythropoietin and granulocyte colony-stimulating factor on insolubilized phytohemagglutinin. Biochim Biophys Acta 496:146.

Spivak JL (1976). Effect of erythropoietin on chromosomal protein synthesis. Blood 47:581.

Spivak JL, Small D, Hollenberg MD (1977). Erythropoietin: Isolation by affinity chromatography with lectin-agarose derivatives. Proc Natl Acad Sci USA 74:4633.

Spivak JL, Small D, Shaper JH, Hollenberg MD (1978). Use of immobilized lectins and other ligands for the partial purification of erythropoietin. Blood, in press.

INTERACTION OF LIMULIN WITH MEMBRANE GLYCOPROTEINS

Mary A. Fletcher, Timothy M. Lo and Karen E. Caldwell
University of Miami School of Medicine

Miami, Florida, 33132

Sialic acid-containing glycoconjugates are present in every mammalian cell membrane thus far examined. These sialic acid residues occupy a terminal position in oligosaccaharide chains and thus tend to have an external position in the membrane (Gottschalk, 1972). Limulin, with its well documented affinity for sialic acid residues (Cohen, 1968) should be very useful in the study and purification of membrane glycoproteins, although it has not been widely exploited for this purpose. In the present study we examined the reactivity of Limulin with highly purified sialoglycoproteins from the erythrocyte membranes of five mammalian species.

MATERIALS AND METHODS

Extraction and Purification of Glycoproteins

The extraction of glycoprotein from the hemoglobin-free stroma of horse, sheep, goat and human erythrocytes, by a modification of the method of Schwartzweiss and Tomscik (1948), is described in detail elsewhere. (Fletcher and Woolfolk, 1971; Fletcher, Lo and Graves, 1976). In brief, the freeze-dried stroma were extracted sequentially, under reflux, with acetone, absolute ethanol, and 75% ethanol. The 75% ethanol extract was concentrated to a small volume by vacuum distillation, dialyzed against deionized water, and freeze-dried.

Pages 655–664, Biomedical Applications of the Horseshoe Crab (Limulidae)

In addition to the material prepared by hot aqueous ethanol extraction described above, glycoprotein was extracted from both human and bovine erythrocyte stroma by the lithium diiodosalicylate (LIS)-phenol-water extraction method of Marchesi and Andrews (1971).

Purification of the crude extracts was achieved by several steps. The 75% ethanol extract was dissolved in H_2O to a concentration of 2%, and 9 volumes of ethanol added. After 4 hrs at 4°C, the 90% ethanol insoluble material was collected by centrifugation. The pellet was dissolved in 0.02 M sodium citrate, pH 4.1, and the solution was placed on a phosphocellulose column equilibrated and eluted with the same buffer. The column effluent was monitored for sialic acid, hexose, and O.D. 280 nm. The glycoprotein fraction emerged unretarded from the column and was dialyzed against several changes of distilled H_2O and freeze-dried. This fraction was dissolved in distilled H_2O (2 mg/ml) and shaken for 30 min. at 0°C with 5 volumes of ether: ethanol (4:1 v/v). The aqueous phase was freeze-dried, redissolved in H_2O and extracted once more with ether:ethanol. The re-extracted aqueous phase was freeze-dried, then dissolved in H_2O (10mg/ml). Nine volumes of chloroform:methanol (2:1 v/v) were added. The mixture was shaken for 30 min. at room temperature, then centrifuged. The aqueous layer was freeze-dried.

Purification of Limulin

The lectin was purified by a modification of the method of Roche and Monsigny (1974). The hemolymph (2000 ml) was obtained from 15 adult Limulus polyphemus. After 24 hrs at 4°C the "serum" was separated from the clotted cellular material by filtering through cheese cloth and low speed centrifugation. The serum was then separated from the hemocyanin by centrifugation at 100,000 g for 5 hrs. The clear, colorless supernatant was dialysed against 0.1 M NaCl, 0.01M$CaCl_2$, 0.05 M TrisHCl buffer, pH 8.5. The lectin was purified by application to a DEAE-Sephadex A-50 column equilibrated with the above buffer. The peak of lectin activity was concentrated to 2.6 mg/ml protein using an Amicon P-10 membrane. Hemagglutination and hemagglutination inhibition tests using horse erythrocytes were carried out as described earlier (Fletcher and Woolfolk, 1971).

Preparation of ^{131} I Labeled Limulin

Purified Limulin was radioiodinated by both the Chloramine T (Greenwood, Hunter, and Glover, 1963), and lactoperoxidase methods (Marchalonis, 1969). Labelled Limulin was separated from free iodine by passage over a Sephadex G-25 column. Incorporation of iodine in the Limulin was 50-80%.

Analytical Methods

Sialic acids were determined by the alkaline-Ehrlich method and by the thiobarbituric acid method, both as described by Aminoff (1961), after hydrolysis in 0.1N H_2SO_4 at 80°C for 1 hr. N-acetylneuraminic acid (grade III, Sigma) served as standard. Neutral sugars and amino sugars were determined as alditol acetates by gas-liquid chromatography as described previously, (Fletcher, Lo and Graves, 1976). Individual amino acids were determined after hydrolysis by 6N HCl at 110°C in sealed, evacuated tubes for 24 hrs using a Durrum model D-500 analyzer.

Gel electrophoresis in 7.5% polyacrylamide containing 0.1% sodium dodecyl sulfate (SDS), 0.01 M phosphate, pH 7.0 was done as previously described, (Fletcher and Woolfolk, 1972).The samples were preincubated in 1% SDS, 0.1% mercaptoethanol, 0.01 M phosphate buffer, pH 7.0 at 37°C for 30 min. The gels were stained with periodic acid Schiff (PAS) reagent and by Coomassie brilliant blue (CB).

RESULTS AND DISCUSSION

The compositions of the glycoproteins extracted from human, horse, sheep and goat erythrocyte stroma with hot, aqueous ethanol are given in Table 1. It has been previously shown that these glycoproteins correspond to the major PAS-staining bands in erythrocytes from these four species. (Fletcher and Woolfolk, 1972; Fletcher, Lo, and Graves, 1976.) For each glycoprotein the ratio of polypeptide to carbohydrate was 1:1 and sialic acid was the most abundant sugar (Table 1).

In contrast when bovine stroma was extracted with hot, aqueous ethanol a minor 25,500 MW_R glycoprotein was extracted

TABLE 1
Composition of Erythrocyte Glycoproteins Extracted From the Membrane with Hot Aqueous Ethanol

Constituent	Horse	Sheep	Goat	Human
	Moles/10,000 g			
SA[a]	6.96	7.31	8.23	7.64
Gal	7.56	5.62	5.73	6.10
Man	0.89	0.32	0.26	0.85
Fuc	----	----	----	0.74
GalNAc	6.92	3.50	4.24	3.93
GlcNAc	1.07	2.92	2.68	2.11
Polypeptide[b] (%)	49.0	53.0	49.0	53.0

[a]N-acetyl-and N-glycolylneuraminic acids.
[b]Calculated from amino acid composition.

which has only 30% carbohydrate (Table 2). The major glycoproteins of bovine erythrocytes were extracted with LIS-phenol. This glycoprotein extract contained 80% carbohydrate and was heterogeneous on SDS polyacrylamide gel electrophoresis (Table 3).

TABLE 2
Composition of Bovine Erythrocyte Glycoproteins[a]

Constituent	Hot Aqueous Ethanol Extract	LIS-phenol Extract
SA[b]	3.03	6.75
Gal	3.93	18.24
Man	0.41	-----
GalNAc	1.28	2.66
GlcNAc	1.42	9.77
Polypeptide[c] (%)	77	19

[a]These preparations of glycoprotein were made from aliquots of stroma prepared from a single bovine.
[b]N-glycolylneuraminic acid.
[c]Calculated from amino acid composition.

We have tested the activity of these glycoproteins with purified Limulin. Table 4 shows hemagglutination inhibition studies. There was considerable variation in the amount of inhibition seen for the seven glycoprotein preparations tested. The most active inhibitor was the major glycoprotein complex of the bovine erythrocyte extracted from the stroma with LIS-phenol. The least active were the two preparations of human erythrocyte glycoprotein. Obviously the ability to inhibit the agglutination of horse erythrocytes by Limulin is not strictly a function of sialic acid content. However, treatment of each of these glycoproteins with Vibrio Cholera neuraminidase (Behrihgwerke) at 37°C for 24 hours at 200 units/mg glycoprotein destroyed activity. It should be noted that each of these glycoproteins was tested both before and after extraction with chloroform: methanol (2:1), and ethanol:ether. In each case the activity was unchanged or somewhat higher after extraction. In the case of the horse and sheep erythrocyte glycoproteins we have also recovered the major PAS staining band from SDS-polyacrylamide gel electrophoresis, removed the SDS by repeated precipitations from ethanol and shown undiminished activity in the hemagglutination inhibition test.

TABLE 3
Relative Molecular Weights of Bovine Erythrocyte Glycoproteins on SDS-Polyacrylamide Gel Electrophoresis

	BAND	MW_R
LIS-phenol extract	PAS 1	140,000
	2	120,000
	3	81,000
	4	25,500
Hot Aqueous Ethanol Extract	PAS 1	25,500

The hemagglutination of horse erythrocytes by our preparation of Limulin could be blocked by N-acetylneuraminic acid. Its hemagglutination titer against horse erythrocytes was 1:1000. The purified lectin gave a single band in SDS-polyacrylamide gel electrophoresis (Fig. 1a). The apparent molecular weight of this band at a gel concentration of 7.5% was 26,500. This is in agreement with Finstad, Good and

Litman (1974) and Kaplan, Stevens, Li, and Rehoe (1977), who report a subunit size of 25-27,000 and somewhat larger than the value of 19,000 reported by Roche and Monsigny (1974).

TABLE 4
Hemagglutination Inhibition of Limulin by Erythrocyte Glycoprotein

Erythrocytes Used For Preparation of Glycoproteins	Solubilization Method	µg/Ml Inhibiting Agglutination of Horse Erythrocytes by Four Hemagglutining doses of Limulin
Horse	Hot aqueous ethanol extraction	9.0
Sheep	"	0.3
Goat	"	0.2
Bovine	"	4.7
Human	"	227.0
Bovine	LIS-phenol extraction	0.02
Human	"	119.0

By both chlor mine-T and lactoperoxidase methods we got appreciable iodine incorporation (50-80%) into Limulin. The major radio-labeled peak co-electrophoresed with the protein band. There were also some aggregated forms (Fig. 1b). However the binding activity of this labeled Limulin is consistently low (<10% as measured by binding to horse erythrocytes). This led us to consider that there might be a critical tyrocine in the binding site. Thus we have iodinated in the presence of 4-fold excess of sialic acid then dialysed extensively. This preparation still had low binding activity. One explanation for our results is that >90% of the Limulin we have isolated is inactive. The preparation has some activity as was demonstrated by our finding that a small portion of the iodinated material binds to membrane glycoprotein. This binding was shown by the following experiment. We electrophoresed the bovine glycoprotein (LIS-phenol extraction) on SDS-gels. The unstained gel was soaked in several changes of 40% ethanol, 5% acetic acid, to fix the glycoprotein and to remove the SDS. The gel was then soaked in several changes of 0.1M NaCl, 0.01M $CaCl_2$, 0.05 M Tris-HCl, pH 8.5. The gel was incubated 24 hours in the same buffer containing ^{131}I-labeled Limulin, followed by five washes over 24 hours with buffer alone. The gel was sliced into 2 mm segments and radioactivity determined. The total cpm remaining associated with each slice are shown in Fig. 2a. The peak of counts seen at the dye front (India ink stab mark) is also seen in Fig. 2b, which is a control gel without glycoprotein. The other peaks of activity seen in Fig. 2a correspond in relative mobility to the large molecular weight, PAS-staining bands (Table 3).

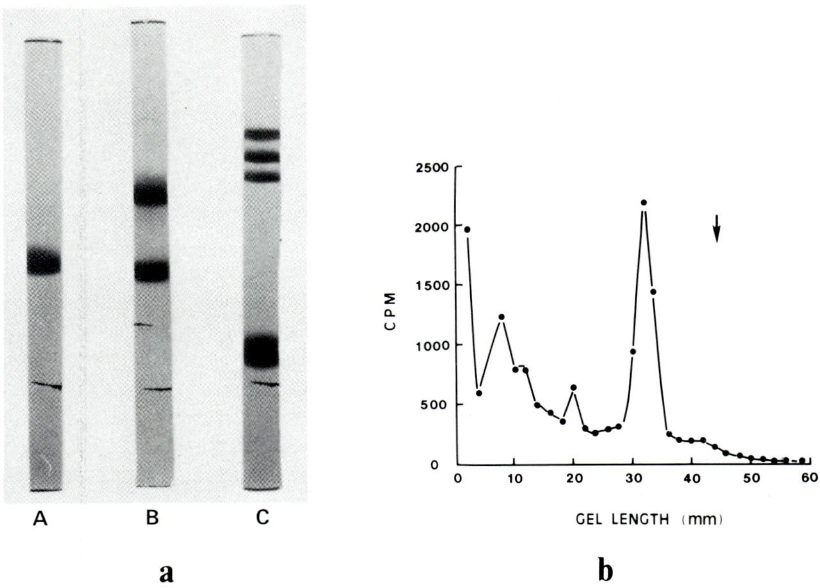

Fig. 1 Electrophoresis of purified Limulin on SDS polyacrylamide gels

(a) gel A-50 mg Limulin; gel B-ovalalbumin and Bence-Jones protein; gel C-fibrinogen and lysozyme gel stained with CB. stab mark indicates bromphenol blue dye migration.

(b) ^{131}I Limulin gel was sliced into 2 mm segments and radioactivity of each slice was determined. Arrow indicates point of bromphenol blue dye migration.

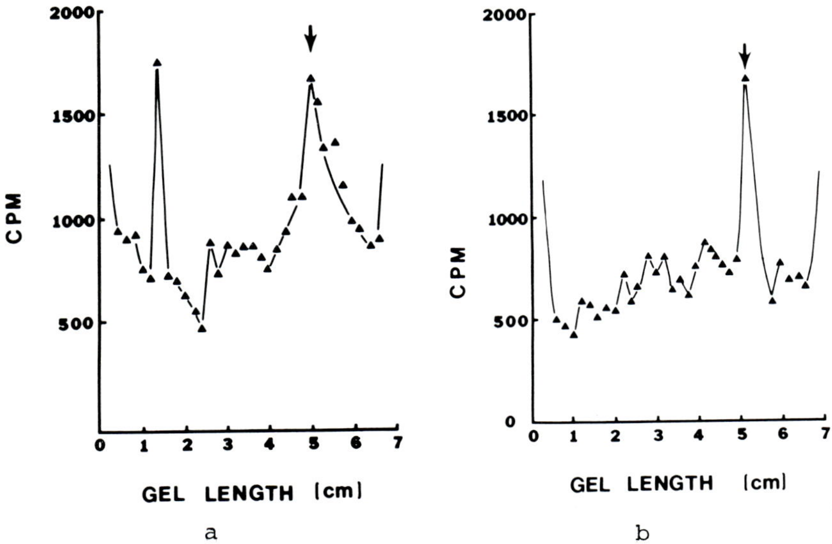

Fig. 2a Electropherogram of 50 µg of bovine erythrocyte glycoprotein (LIS-phenol extract) on SDS polyacrylamide. Gel was fixed and washed, incubated with ^{131}I-Limulin, washed and sliced (see text). The radioactivity was determined in each slice.

Fig. 2b Control electropherogram run exactly as 2a except that bovine erythrocyte glycoprotein was omitted.

Our experiments indicate that Limulin may prove to be a useful probe for the study of the number and distribution of sialic acid containing macromolecules on membrane surfaces. However, isolation of a more active preparation by affinity methods will be necessary. Our studies illustrate also that one cannot predict lectin reactivity with isolated membrane components on the basis of behavior with the intact cell. The bovine erythrocyte does not agglutinate when exposed to Limulin, yet an extremely potent inhibitor of the lectin was extracted from bovine red cell membranes. This is another example of the refractoryness of bovine red cells to agglutination (Fletcher, Lo and Graves, 1977; Uhlenbruck, Seaman, and Coombs, 1967). Of interest in regard to the high activity of bovine erythrocyte glycoproteins and the low activity of the human erythrocyte glycoproteins is that the extract from bovine cells contains only the N-glycolyl derivative of neuraminic acid and the latter only the N-acetyl derivative. Another possible explanation for the high inhibitory activity of the bovine as compared to all of the other preparations is its very large aggregate size which by electronmicroscopy is 3-4 times that of the other glycoproteins. (Fletcher, Lo, and Smith, unpublished observations). A direct relationship frequently exist between molecular size and activity of substances in hemagglutination inhibition tests (Springer, Schwick and Fletcher, 1969).

REFERENCES

Aminoff, D (1961). Method for the quantitative estimation of N-acetylneuraminic acid and their application to hydrolysate of sialomucoids. Biochem J 81:384.

Cohen, E (1968). Immunologic observations of the agglutinins of the hemolymph of Limulus polyphemus and Birgus patro. Trans NY Acad Sci 30:427.

Finstad, CL, Good, RA, and Litman, GW (1974). Erythrocyte agglutinin from Limulus polyphemus hemolymph. Molecular structure and biological function. Ann NY Acad Sci 234:170.

Fletcher, MA, Lo, TM, and Graves,WR (1976). Immunochemical studies of infectious mononucleosis V. Isolation and characterization of a glycoprotein from goat erythrocyte membranes. J Immunol 117:717.

Fletcher, MA, Lo, TM, and Graves,WR (1977). Immunochemical studies of infectious mononucleosis VII. Isolation and partial characterization of a glycopeptide from bovine erythrocytes. Vox Sang 33:150.

Fletcher, MA, and Woolfolk,BJ (1971). Immunochemical studies of infectious mononucleosis. I. Isolation and characterization of heterophile antigens from hemoglobin-free stroma. J Immunol 107:842.

Fletcher, MA and Woolfolk, BJ (1972). Immunochemical studies of infectious mononucleosis II. Sodium dodecyl sulfate gel electrophoresis of membrane glycoprotein antigens. Biochim Biophys Acta 278:163.

Greenwood, FC, Hunter, WM, and Glover, JC (1963). The preparation of ^{131}I-labeled human growth hormone of high specific activity. Biochem J 89:114.

Kaplan, R, Li, S S-L, and Kehoe, JM (1977). Molecular characterization of limulin, a sialic acid binding lectin from the hemolymph of the horse shoe crab, Limulus polyphemus. Biochemistry 16:4297.

Marchalonis, JJ (1969). Enzymatic method for the trace iodination of immunoglobulins and other proteins. Biochem J 113:299.

Marchesi, VT and Andrews, EP (1971). Glycoproteins: Isolation from cell membranes with lithium diido-salicylate. Science 174:1247.

Roche, AC and Monsigny, M (1974). Purification and properties of Limulin: A lectin (agglutinin) from hemolymph of Limulus polyphemus. Biochim Biophys Acta 371:242.

Schwartzweiss, H, and Tomcsik, J (1948). Isolation of the heterogenetic mononucleosis antigen from the stroma of beef erythrocytes. Proc Soc Exp Biol Med. 69:558.

Springer, GF, Schwick, HG, and Fletcher, MA (1969). The relationship of the influenza virus inhibitory activity of glycoproteins to their molecular size and sialic acid contact. Proc Nat Acad Sci USA. 64: 634.

Tuppy, H, and Gottschalk, A (1972). The structure of sialic acids and their quantitation. In Gottschalk, A (Ed): "The Glycoproteins" Amsterdam: Elsevier, p. 404.

Ulhenbruck, G, Seaman, GVF, and Coombs, RRA (1967). Factors influencing the agglutinability of red cells III. Physico chemical studies on ox red cells of different classes of agglutinability. Vox Sang. 12:420.

LIMULUS AGGLUTININS: PAST, PRESENT AND FUTURE.

John J. Marchalonis and Myron J. Waxdal

Cancer Biology Program (JJM), NCI Frederick Cancer Research Center, Frederick, Maryland 21701 and National Institute of Allergy and Infectious Diseases (MJW), National Institutes of Health, Bethesda, Maryland 21701.

The turn of the century was a time of great excitement and progress in immunological research and theory. In 1903, H. Noguchi, a young assistant in Pathology at the University of Pennsylvania, wrote that Bordet and Erhlich were carrying out "epoch making studies." He lamented, however, that all of their work was performed using mammalian systems. For this reason, Noguchi came to the Marine Biological Laboratory to determine whether or not the theories of Paul Erhlich might be extended beyond the mammals to the so-called "cold-blooded" animals. Thus began the association of the horseshoe crab agglutinin with the Marine Biological Laboratory, and with the history of immunology. Noguchi (1903) found powerful agglutinins to vertebrate erythrocytes in the hemolymph of this primitive arachnoid, *Limulus polyphemus*. Sixty years later, Marchalonis (1964), a first year graduate student, also turned to the horseshoe crab in the hopes of finding in this primitive fossil a protein recognition molecule which might be the precursor of immunoglobulins. At the same time, Elias Cohen and his coworkers (Cohen, 1968; Cohen *et al.* 1965) at Rosewell Park were studying agglutinins of the horseshoe crab for use as a means of separating distinct cell populations. The agglutinin reacting with horse erythrocytes was first isolated and partially characterized in 1968 (Marchalonis and Edelman, 1968). This symposium on biochemistry and biophysics of limulus agglutinins bears witness to the fact that substantial progress has been made in elucidating the structure and function of the molecule during the past ten years.

Pages 665–675, Biomedical Applications of the Horseshoe Crab (Limulidae)

REASONS FOR STUDY

Horseshoe crab agglutinins were studied originally with the hope that such molecules would be relevant to the evolution of specific immune recognition. As of this time, it would be naive to retain this hope. The structure of the agglutinin molecule is quite distinct from that of immunoglobulins, both at the level of overall subunit arrangement and at the level of primary amino acid sequence (Marchalonis, 1977; Kehoe, this symposium). Even though the hemagglutinin may not be directly relevant to the search for the beginnings of vertebrate immunity, it is important to ask what the function of the agglutinin is within the animal. It has, for example, been proposed that sugar-binding molecules of this nature might well be involved in the formation of the chitin shell of the creature (Marchalonis, 1977). The agglutinin might, alternatively, function in cell/cell interactions within the horseshoe crab. Experiments must be done to test this hypothesis. Furthermore, it is still a viable hypothesis that such sugar-binding molecules might function in the defense of the animal against infection by bacteria and/or other microorganisms. Experiments to test this hypothesis should involve studies to determine whether or not the agglutinin interacts with the phagocytic amoebocytes of the horseshoe crab and might function as an opsonin.

The structural characterization of the limulus hemaglutinin has provided information which has served as a prototype for the study of invertebrate lectins. It has been found that the agglutinins of the oyster (Acton *et al.*, 1969), the murray molusk (Marchalonis, J. J. and Jenkin, C. R., unpublished observation), and other invertebrates are assembled according to the same general principles that govern the assembly of the limulus lectin. For example, these consist of small subunits (approximate mass 20,000 daltons) which form regular, large aggregates via noncovalent interactions. In many cases, other agglutinins resemble that of the horseshoe crab in requiring the presence of the divalent cation calcium for optimal binding of sugars (Gold and Balding, 1975). We emphasize that the agglutinin is a hemolymph protein. In a series of studies designed to determine whether or not the relative amount of the agglutinin varied with the developmental stage of the animal, hemolymph was taken from animals ranging from 2 cm in size up to the largest adult form. In all cases, the titers of agglutinating activity towards horse erythrocytes were identical (Marchalonis,

J. J., unpublished). Preliminary indications, thus, suggest that the amount of agglutinin in the hemolymph is not a function of the developmental stage of the organism following hatching from the egg. In addition, the possibility was considered that the hemagglutinin might have arisen from lysed amoebocytes in a manner comparable to that of the clotting factor. In fact, Professor Bang has suggested that the agglutinin and the clotting factor might be identical. It was shown, however, by Levin (unpublished observation) and by Marchalonis (unpublished observation) that cell-free plasma, which was incapable of clotting in the presence of endotoxin, was equivalent in its content of hemagglutinin (as measured by agglutination of erythrocytes) to hemolymph prepared in the normal way, that is, by allowing the cells to clump and then removing them by centrifugation. Furthermore, structural studies presented at this meeting indicate that the clotting factors, including the initiating enzyme and the coagulogen, and the hemagglutinin possess quite distinct protein structure.

ISOLATION AND BIOCHEMICAL CHARACTERIZATION

Two approaches have been used to purify limulus agglutinins. The first approach is a strictly biochemical one, which is based upon the electric charge and molecular size of the molecule. This approach was first used by Marchalonis and Edelman (1968) who combined a series of steps in which a) hemocyanin, the bulk hemolymph protein, was removed by ultracentrifugation, b) the remaining clear fluid was separated on the basis of electric charge using zone electrophoresis, and c) the fraction containing agglutinating activity was fractionated further (on the basis of molecular weight) by gel filtration. It was found that the agglutinin had a mobility roughly equivalent to that of a beta globulin and a molecular weight which was between that of IgM (approximate mass 900,000) and IgG (approximate mass 150,000) by gel filtration on Sephadex G200. Figure 1 is an an electropherogram illustrating the electrophoretic mobility of hemagglutinating activity. It can be seen in this Figure that the agglutinin comprises a small percentage of the protein in the serum. It would comprise approximately 2% of hemolymph proteins. Figure 2 is an electronmicrograph of the purified agglutinin (Fernadez-Moran *et al.* 1968). The molecule consists of a cylindrical torus with a diameter of approximately 100 Å and a height of approximately 70 Å. There is a hole in the center of the molecule and the molecule

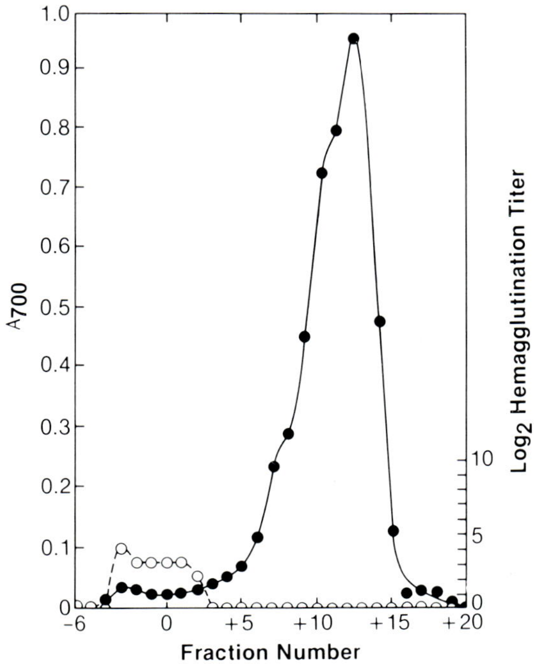

Figure 1. Zone electrophoresis on starch of *Limulus polyphemus* hemolymph illustrating distribution of proteins (●——●) and localization of hemagglutining activity (o---o). (●——●), A_{700} of the Folin protein reagent; o——o, $\log_2$ hemagglutination titer towards horse erythrocytes. The major protein component, fractions +10 to +15 is hemocyanin. (From Marchalonis and Edelman, 1968).

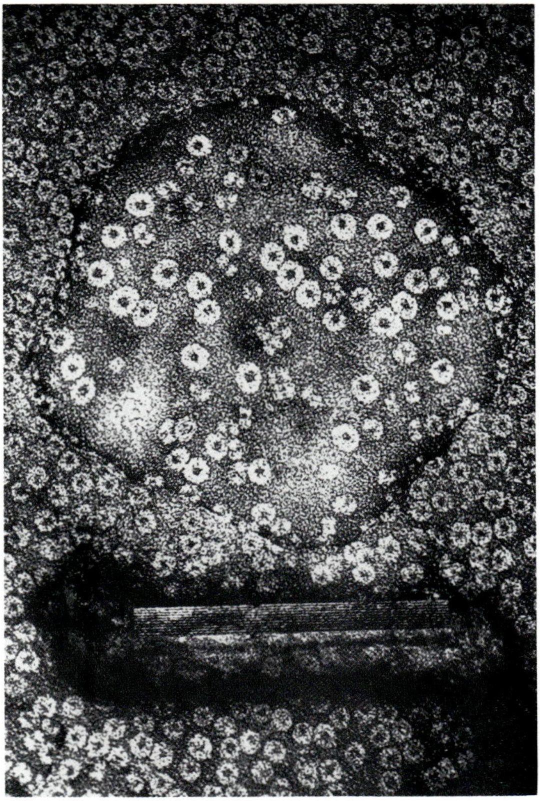

Figure 2. High resolution electronmicrograph of the hemagglutinin of the horseshoe crab. Some of the ring-shaped hemagglutinin molecules in the central area are distended, displaying their fine structure in various orientations (X 444,000). Asbestos filament insert (X 800,000) exhibits a lattice period of 7.3 Å which serves as a calibration standard. (From Fernandez-Moran *et al.*, 1968).

has detectable subunit structure. Figure 3 is a diagramatic model of the hemagglutinin molecule based upon the electronmicrographic and hydrodynamic studies. The intact molecule has a mass of approximately 400,000 daltons and contains subunits of approximate mass 20-25,000 daltons. These are arranged as trimers in subunit structure, and the intact molecule is comprised of 6 of these subunits. The association of the subunits is noncovalent and changes in pH are sufficient to reduce the molecule from the intact unit down to the sixth molecule. Dissociation under more brutal conditions

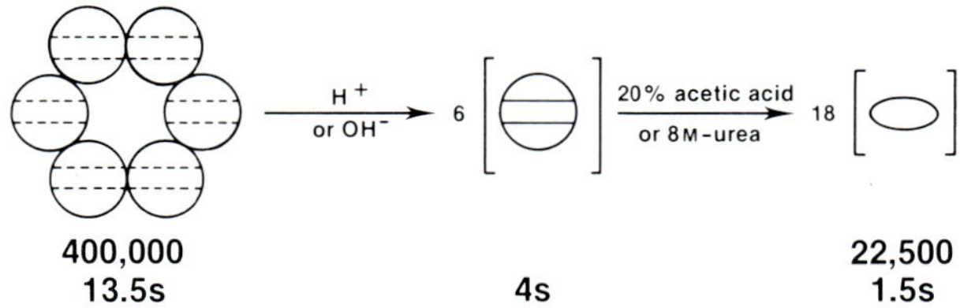

Figure 3. Structural model for the limulus agglutinin based upon physicochemical measurements and electronmicroscopic studies. (From Marchalonis and Edelman, 1968).

such as 6 M urea 1 M acetic acid or the anionic detergent sodium dodecylsulfate will cause dissociation of the molecule into its minimal subunit.

The second major approach to the isolation of the agglutinin utilizes affinity chromatography. It has been possible to use fixed horse erythrocytes (Nowak and Barondes, 1975) or matrices such as mucin or fetuin coupled to Sepharose as solid phase adsorbents (Oppenheim *et al.*, 1974; Roche *et al.*, 1975; Shimizu *et al.*, 1977). At this juncture, it is worth pointing out that the specificity of limulus agglutinins is certainly more complex than a simple affinity for N-acetyl-neuraminic acid (NANA). It might be more correct to conclude that the specificity is directed against oligosachharides expressing terminal N-acetyl-neuraminic acid or o-glycolyl-neuraminic acid residues. The capacity to react with N-acetyl-glucosamine is apparently present also, most probably when this sugar occurs in association with NANA.

Nowak and Barondes (1975), using red cells as affinity reagents, produced a limulus agglutinin with a structure identical to that isolated using strictly biochemical methods (Marchalonis and Edelman, 1968; Roche and Monsigny, 1974). It is interesting also that Niwa *et al.* (this symposium) have used affinity reagents consisting of N-acetyl-glucosamine coupled to Sepharose and have isolated agglutinins from the

Japanese horseshoe crab (*Tachypleus spp.*) which are quite distinct from those isolated from the American species (*Limulus*). Shimiza *et al.* (this symposium) reported data suggesting that two broad catagories of agglutinins occur in the *Tachypleus tridentatus*. One agglutinin has its greatest reactivity with horse erythrocytes and appears to have structural properties similar to those reported for the horseshoe crab limulus. The second has its major reactivity against human erythrocytes; its major specificity is for N-acetyl-glucosamine; it will react with N-acetylated amino acids; and may occur as a disulfide-bonded aggregrate comprised of units, which are significantly larger than those of the limulus agglutinin.

It is of great interest that Monsigny and his associates (this symposium) have fractionated limulus agglutinin into three components by affinity chromatography on mucin-Sepharose. In this case, the agglutinins were eluted by changing the salt concentration and by adding EDTA to the eluent. Although some biological activity was present in all fractions, the most tightly bound component (LIII) was the most potent hemagglutinin. The size of the minimal subunit of LIII was approximately 29,000 daltons, as estimated by SDS-PAGE. Electronmicrographs of LIII are identical to those of agglutinins prepared by strictly biochemical fractionation procedures.

Prior studies have shown that the agglutinin was not mitogenic for murine lymphocytes. However, Monsigny has now demonstrated that LIII can stimulate DNA synthesis and cell division by murine T lymphocytes and that LI also has some activity at high concentrations. This cellular response to LIII is less than that found by stimulation with Con A or PHA and may indicate that only certain subsets of murine T lymphocytes are being activated.

Kehoe has been pursuing the complete amino acid sequence of the limulus agglutinin, and presented substantial sequence data. The material is prepared by biochemical procedures and contains 3 bands by SDS-PAGE. The approximate mol. wts. of these bands are 25,000, 26,000 and 27,000 daltons. The first two are present in about equal amounts and the 27,000-dalton component is rather sparse. However, amino acid sequence analysis on the unfractionated agglutinin yielded a single unique sequence. The recovery of the first residue, leucine, was greater than 80%, suggesting that at least the

25,000- and 26,000-dalton components have the same N-terminal sequence. Cyanogen bromide cleavage yielded 3 fragments. The complete sequences of two of these have been determined. Severe technical problems have hampered studies on the final fragment.

It is clear from the sequence data observed to date (Finstad *et al.*, 1974; Kaplan *et al.*, 1977) that the molecule bears no relationship to vertebrate immunoglobulins, and it does not bear any obvious homology to histocompatibility antigens or to plant lectins. It would be worthwhile to obtain detailed amino acid sequence information on other invertebrate lectins to determine the relationships among these molecules.

Another area of interest is the apparent discrepancy between the biochemical and affinity techniques for the isolation of the agglutinin. Both approaches produce material with approximately the same physicochemical properties: mol. wts., amino acid composition and sequence, distinctive arrangement in electronmicrographs. However, the measurable activities differ. These observations raise several questions, such as: Are some of the binding sites blocked by previously bound carbohydrate? Do some molecules have other specificities similar to the case with Tachypleus? Has the isolation procedure inactivated some of the molecules?

We hope that such studies will become available in the not too distant future. Moreover, it is necessary to determine which portion of the molecule is involved in forming the combining site for sugar. This is a very interesting area, particularly insomuch as it is not even known at this time whether or not there is binding of one mole of oligosaccharide per subunit or whether there is subunit interaction in forming combining sites.

AFFINITY REAGENTS FOR SURFACE, SERUM AND URINE GLYCOPROTEINS

As data accumulate on the structure and sugar specificity of the horseshoe crab agglutinin, it is apparent that this molecule should prove useful as an affinity reagent for the isolation of certain sialoglycoproteins of membranes and biological fluids. Monsigny presented the exciting finding that the purified limulus agglutinin acts as a T cell mitogen. The receptor molecules on T cells which bind the agglutinin should be characterized, and furthermore it is important to determine the manner in which the agglutinin (a homogeneous molecular species) interacts with surface components and causes triggering of the cell. Spivack and Hollemberg (this symposium) show how lectins such as the limulus lectin can be used in the isolation of glycopeptide hormones. In their particular case, the limulus lectin did not bind erythropoietin. This result should not discourage use of limulus lectin affinity columns in purification of sialoglycoproteins, but serves to emphasize the possible selectivity of the reagent. We believe that the use of limulus lectins as probes for the study of surface sialoglycoproteins of mammalian cells may prove a very fruitful avenue of approach and is worthy of continued effort in the future.

CONCLUSIONS

The search for agglutinins in the horseshoe crab was begun by Noguchi (1903) with the overall goal of obtaining information on the mechanisms of immune recognition. Sixty years after his pioneering studies, the limulus agglutinin was approached in a biochemical fashion with essentially the same goal of obtaining information on the evolution of the molecule responsible for specific immune recognition (Marchalonis, 1964; Marchalonis and Edelman, 1968). The data which have accrued to date indicate that the agglutinin is most probably not related to the molecules, which must be the evolutionary precursors of vertebrate immunoglobulins. The function of the agglutinin within the animal is still a subject of conjecture: the hypotheses which must be considered are first, that the molecule serves a defense function, possibly against invasion by infectious bacteria; and second, that the molecule does not serve a defense role at all. Perhaps it serves in a more general physiological sense; for example, it might be required for formation of the chitin shell. Experiments were discussed and outlined to test these possibilities. The horseshoe crab serves as a very important focal point for these studies, because although virtually all living organisms possess lectins of some sort, in no species has the function of the lectin been determined precisely. Continued studies are required to definitively establish the covalent structure of the limulus lectin and to compare it with plant lectins, with the lectins of other species of invertebrates, and with those of the other surviving horseshoe crabs, in particular, the Japanese horseshoe crab, Tachypleus. Further data must be obtained on the definitive oligosaccharide specificity of the agglutinin, the number of combining sites for sugar in the intact molecule, the affinity of binding, and on the structural requirement for calcium for the maintenance of agglutinating potential. Such studies are relevant to the limulus agglutinin, and also as models for studying protein subunit/subunit interaction and in determining general physicochemical parameters underlying the construction of specific sugar-binding molecules. Lastly, the agglutinins present an enormous, virtually untapped potential as tools for selective isolation of different types of cells, in particular lymphocyte subpopulations, and also as probes for investigating the participation of specific cell surface glycoproteins in the initiation of blastogenesis and differentiation. Moreover, the limulus agglutinin remains a potentially useful affinity reagent for the isolation of biologically relevant sialoglycoproteins of cell surfaces and biological fluids.

ACKNOWLEDGEMENTS

This work was partially supported by the National Cancer Institute under contract no. CO-NO1-75380 with Litton Bionetics, Inc. Original work (JJM) cited was supported by grants to Professor G. M. Edelman, The Rockefeller University, from the National Science Foundation and the National Institutes of Health. We thank Ms. Linda Stockman for typing the manuscript.

REFERENCES

Acton RT, Bennet JC, Evans EE and Schrohenloher RE (1969). Physical and chemical characterization of an oyster hemagglutinin. J Biol Chem 4:4128.

Cohen E (1968). Immunologic observations on the agglutinins of the hemolymph of *Limulus polyphemus* and *Birgus latro*. Trans NY Acad Sci 30:427.

Cohen E, Rowe AW and Wissler FC (1965). Heteroagglutinins of the horseshoe crab *Limulus polyphemus*. Life Sciences 4:2009.

Fernandez-Moran H, Marchalonis JJ and Edelman GM (1968). Electron microscopy of a hemagglutinin from *Limulus polyphemus*. J Mol Biol 32:467.

Finstad CL, Good RA and Litman GW (1974). The erythrocyte agglutinin from *Limulus polyphemus* hemolymph: Molecular structure and biological function. Ann NY Acad Sci 234:173.

Gold ER and Balding P (1975). Receptor-specific proteins, plant and animal lectins. Excerpta Medica, Amsterdam.

Kaplan R, Li SS-L, Kehoe M (1977). Molecular characterization of limulin, a sialic acid binding lectin from the hemolymph of the horseshoe crab, *Limulus polyphemus*. Biochemistry 16:4297.

Marchalonis JJ (1964). A natural hemagglutinin from *Limulus polyphemus*. Fed Proc 23:1468.

Marchalonis JJ (1977). *Immunity in Evolution*. Harvard University Press, Cambridge.

Marchalonis JJ and Edelman GM (1968). Isolation and characterization of a natural hemagglutinin from *Limulus polyphemus*. J Mol Biol 32:453.

Noguchi H (1903). On the multiplicity of serum haemagglutinins of cold blooded animals. Zentral Bakt Abt 1 Orig 34:286.

Nowak TP and Barondes SH (1975). Agglutinin from *Limulus polyphemus*. Purification with formalinized horse erythrocytes as the affinity adsorbent. Biochim Biophys Acta 393:115.

Oppenheim JD, Nachbar MS, Salton MRJ, Aull F (1974). Purification of a hemagglutinin from *Limulus polyphemus* by affinity chromatography. Biochem Biophys Res Commun 58:1127.

Roche A-C, Monsigny M (1974). Purification and properties of limulin: A lectin (agglutinin) from hemolymph of *Limulus polyphemus*. Biochim Biophys Acta 371:242.

Roche A-C, Schauer R, Monsigny M (1975). Protein-sugar interactions. Purification by affinity chromatography of limulin on N-acylneuraminyl-binding proteins. FEBS Lett 57:245.

Shimizu S, Ito M, Niwa M (1977). Lectins in the hemolymph of Japanese horseshoe crab, *Tachypleus tridentatus*. Biochim Biophys Acta 500:71.

Dr. Frederik B. Bang

John Hopkins University
Baltimore, Maryland

FINALE

The use of the *Limulus* lysate test for endotoxin, or pyrogen, has now become an accepted procedure in the pharmaceutical industry, and is used routinely in the search for the source of contaminating endotoxin in water, glassware, etc. It seems to be fully acceptable in the diagnosis of gram-negative menigitis and perhaps in the estimation of the number of gram-negative bacteria present in freshly obtained urine. However, disagreement as to the meaning of a positive test in the presence or absence of different kinds of infection and injury (burns) remained. For this reason, instead of attempting a summary of this diversified and stimulating conference, the final session reviewed those points on which there was general agreement as to need for future work. These were as follows:

Section I - *Biology and Ecology of Limulus polyphemus* - *Dr. Carl Shuster Jr.*, (Chairman)

(1) The population dynamics of *Limulus* particularly discreteness of populations, should be studied in order to determine how many animals may be harvested in offshore deep water to prevent depletion and/or extinction of this valuable and ancient species.

Pages 677–680, Biomedical Application of the Horseshoe Crab (Limulidae)

(2) There is a need to study the health of Limulus especially after the impact of repeated bleeding and return to the ocean.

(3) The effects of ecological variables, such as pollution, destruction of critical inshore habitat, present killing by clam fishermen and others, etc., need to be studied.

Section II - Detection of bacterial endotoxin and biomedical applications of Limulus - Dr. Robert Prendergast, (Chairman, Part A, chemical detection of endotoxin); Dr. Jack Levin, (Chairman, Part B, biomedical detection).

(1) A detailed analysis should be done concerning the the coagulation process itself. Some disagreement still exists on the Limulus system. How may antigenically different coagulogens (Tachypleus and Limulus) still interact satisfactorily?

(2) A study of the enzyme kinetics following induction by endotoxin should be undertaken.

(3) The role of inhibitors in the coagulation process is obvious, when comparing in vitro tests with the relatively high resistance of the animal. It needs study.

(4) The exact specificity of the LAL test in its application in human disease is difficult to demonstrate, since it is the most sensitive test available for endotoxin. Does it always detect only endotoxin in human tissues? Correlation with gram-negative infection as demonstrated by positive blood cultures is not high, sometimes not apparent, and perhaps is not to be expected.

(5) Effort should be expended on standardization of the Limulus lysate and of the endotoxin. Evaluation of the test in humans is greatly influenced by variation in the sensitivity of the lysate. Thus, there is a great need for standardization. In turn, the endotoxin

preparations also vary.

Section III - Round Table conclusions and LAL essay of pharmaceutical products - Dr. Stanley Watson, (Chairman)

(1) The role of inhibitors and of modifiers of the LAL tests in pharmaceutical preparations should be investigated.

(2) What is the tolerable level of endotoxin in a preparation as determined by the LAL test?

Section IV - Bacterial agglutinins and protection against bacteria - Dr. Thomas Pistole, (Chairman)

(1) What is the biological significance of antibacterial activity to *Limulus*?

(2) What is the role of endotoxin in disease of *Limulus*, particularly of sublethal amounts?

Section V - Identification of mammalian cell receptors - Dr. Elias Cohen, (Chairman)

(1) How does hemolymph agglutinin (lectin) function in *Limulus*, both in natural situations and with specificity of function? Does endotoxin react with *Limulus* lectin which is probably on the surface of the amebocytes? Does the agglutinin play any role in the change of cell shape induced by endotoxin? What is the specificity of the agglutinin or agglutinins, in relation to human and other mammalian cell populations? What are the cell membrane receptors of *Limulus* or related species agglutinins (lectins), in relation to normal *versus* neoplastic cells?

Section VI - Biochemistry and biophysics of Limulus agglutinins and other hemolymph lectins: Potential applications - Dr. John Marchalonis, (Chairman)

The chairman pointed out that it was Noguchi, the inspired Japanese worker at Rockefeller Institute, who first demonstrated that *Limulus* blood caused agglutination of mammalian red cells, leading much later to the subsequent studies on the nature of this lectin. Thus it was particularly gratifying to have at this meeting a presentation of the excellent Japanese studies on the similar coagulation system present in Tachypleus.

The following questions seem to need early resolution:

(1) What enzymatic changes follow or result from Limulidae agglutinin action?

(2) What is/are the sugar determinant(s) of agglutinin(s)?

(3) With what cells is/are Limulidae agglutinin(s) associated?

(4) What is the optimal purification procedure of Limulus or related species agglutinin(s)?

In closing, the audience was reminded that although the development of the endotoxin/agglutinin story originated at the Marine Biological Laboratory in Woods Hole, it was also here that Leo Loeb worked on the reaction of Limulus cells to a variety of environmental changes and chemical influences. He described in separate work the coagulase activity of staphylococci/on goose plasma, (gram-positive organism), but by bad luck seems to have just missed the effect of gram-negative organisms on Limulus serum.

AFTERWORD

Lewis Thomas, M.D.

Memorial Sloan-Kettering Cancer Center

New York, New York 10021

Although there is a solid conviction within the biomedical science community that basic research is the most central, crucial and indispensable element of the whole enterprise, and whatever benefits society may hope to receive in the future will depend on the quality and scope of the fundamental information being generated today, this point of view is by no means accepted as a given truth, or even well understood, by the political world from which science derives most of its support. There is, indeed, a growing skepticism about the value of basic science and a mounting impatience for practical, useful and usable results.

In this climate, one looks about for evidence to support the scientists' contention. It is there, all over the place, but scattered around in the form of anecdotes, many of them sounding like stories of accidents, information emerging by chance, bits of unaccountable good luck.

This is, in real life, the way it goes. Basic reasearch is an inquiry into unknown mechanisms, and must by its nature be carried out in an atmosphere of high uncertainty and unpredictability. When it works very well, in the sense of producing powerful

Pages 681–683, Biomedical Applications of the Horseshoe Crab (Limulidae)

and meaningful information, the outcome is almost always bound to be a surprise, astounding not only outside observers but (most of all) the investigator himself. It is rather like gambling.

If you were looking about for an anecdote to illustrate this element of unexpectedness and surprise, there is no better story than that of Limulus polyphemus. Who could have guessed, long ago, that this strange, archaic creature would become the source of so much information in so many different areas of modern medicine, illuminating visual science, neurophysiology, cell biology, protein chemistry, immunology and infectious disease? None of these things could have been predicted in advance. How does one explain this amazing productivity?

One answer, of course, is simply that Limulus is such an intrinsically interesting animal. It arouses curiosity, just to look at it.

The great piece of luck has been that this fundamentally engrossing animal has been encountered by investigators with rather more than the usual habit of curiosity. Frederick Bang, a spectacular example of this trait, was as uniquely necessary for the discovery with which much of this conference deals as was Limulus itself. At the outset of the work, Bang was purely a curious observer, with nothing more on his mind, I suspect, than a wish to know more about the way blood circulated through the microvasculature. The sludging then observed in several sick, infected animals might have been overlooked by most people, or at any rate not followed up, but Bang's curiosity was insatiable, and the net result was "limulus extract," one of the handiest tools ever to become available for the study of infection of man by gram-negative bacteria.

For this kind of basic research to succeed, then,

you need an object as interesting as <u>Limulus</u> and an observer as interested as Bang.

The world is filled with creatures to study, all fascinating when approached closely enough. I trust that the papers in this book will demonstrate that skilful observers exist as well, and that more of the latter are needed to deal with more of the former.

Author Index

A

Anderson, Roger L., 465
Armstrong, Peter B., 73, 203
Atkins, E., 485

B

Bang, Frederik B., 109, 677
Bertrand, O., 159
Bird, G. W. G., 559, 601
Blumenson, L. E., 589
Bondar, R. J. L., 435
Bowanko, A., 435

C

Caldwell, Karen E., 655
Carson, Loretta A., 453
Casterlin, Martha Elizabeth, 47
Cohen, Elias, ix, 579, 589
Cooper, James F., 345
Cotter, Richard, 403

D

Dastoli, Frank R., 403
Dixon, Richard E., 465
Donahoe, Robert M., 473
Dubczak, John A., 403

E

Elin, Ronald J., 279
Enzinger, R. M., 353

F

Favero, Martin S., 453
Feinstein, Robert, 265
Ferry, E. W., 339
Fletcher, Mary A., 655
French, Kathleen A., 61
Freudenberg, M. A., 321
Fumarola, D., 379

G

Gaffin, Stephen L., 221
Galanos, C., 321
Galler, Sidney R., 107
Gilbride, Kevin J., 525
Guzman, Ruben J., 333

H

Harada, T., 209
Harris, Nick S., 265
Harrison, Susan J., 353, 367
Highsmith, Anita K., 465
Hochstein, H. D., 501
Hokama, Y., 169
Hollenberg, M. D., 641

I

Ito, Mie, 625
Iwanaga, S., 169, 209

J

Jirillo, E., 379
Jorgensen, James H., 255

K

Kaplan, Ruth, 617
Kehoe, J. Michael, 617
Kelly, Kristine M., 423, 435
Kropach, Chaim, 103
Kuo, Harng-Shen, 333

L

Levin, Jack, 131, 159, 203, 235, 485
Li, Steven S.-L., 617
Liang, Su-Mei, 147
Liehr, Heinrich, 309
Lindberg, Robert B., 275
Liu, Teh-Yung, 147
Lo, Timothy M., 655
Lüderitz, O., 321

M

McManus, Albert T., 275
Maget-Dana, Régine, 567
Marchalonis, John J., 665
Mascoli, Carmine C., 387
Mason, Arthur D., Jr., 275
Minowada, J., 589
Monsigny, Michel, 567, 603
Morita, T., 169, 209
Mosesson, M. W., 159
Müller, Werner E. G., 579

N

Nachum, Ronald, 245, 513
Nakamura, Koichiro, 37
Nakamura, S., 169, 209
Neter, Edwin, 555
Niwa, Makoto, 169, 209, 625

O

Ornberg, R. L., 125

P

Pearson, Frederick C., 93
Peters, Robert L., 473
Petersen, Norman J., 453
Pistole, Thomas G., 525, 537, 547, 557
Pliss, M., 589
Prendergast, Robert A., 229
Pruitt, Basil A., Jr., 275

Q

Quesenberry, P., 485

R

Reese, T. S., 125
Reynolds, William Wallace, 47
Rick, P.D., 203
Rickles, F. R., 203, 485
Rietschel, E. T., 321
Robbins, Joan B., 147
Roche, Annie-Claude, 567, 603
Rostam-Abadi, Homayoun, 537
Rudloe, Anne, 27

S

Sakmar, Thomas P., 147
Seid, Robert C., Jr., 147
Sekiguchi, Koichi, 37, 185
Seligmann, Edward B., Jr., 501
Shimizu, Satoru, 625
Shishikura, Fumio, 185
Shuster, Carl N., Jr., 1, 3
Siegel, Stuart E., 245
Spivak, J. L., 641

T

Tai, Joseph Y., 147
Takagi, T., 169
Takahashi, Noriko, 625
Teller, Joseph D., 423, 435
Thomas, Lewis, 681
Tomasulo, Peter A., 293
Tsuji, K., 353, 367

U

Uhlenbruck, Gerd, 579

V

Vaith, Peter, 579
Valois, Frederica, 415

W

Waxdal, Myron J., 665
Weary, Marlys E., 387
Weiss, Peter J., 507
West, C. Michael, 465
Westphal, O., 321
Wilkinson, S. P., 303
Wolfenstein-Todel, C., 159
Woodland, Ellen, 93

Subject Index

A

Activator, 423
Affinity chromatography, 603, 625, 641
Agglutinin, 525, 537, 547, 567, 589, 603
Agglutination studies, 547
Amebocyte lysate, 131, 169
Amebocytes, 73, 93, 109, 125, 203
Amidase activity
  α-N-acetylglycyl-lysine methyl ester, 209
  α-N-tosylarginine methyl ester (TAME), 209
Amino acid composition, 625
Antibiotic preparations, 353
Anti-galactin, 525
Antitumor agents, 379

B

Bacteremia, 235
Bacterial agglutinin, 525, 537
Bacterial endotoxin, 131, 147, 209
Bacterial meningitis, 245
Bacteriocidal activity, 513
B-cell, 513
Behavior, 27
Bibliography of Limulus literature, 27
Biologic injectables, 387
Biological products, 379
Blood cell, 73
Blood coagulation
  evolution of, 131
  in horseshoe crab, 131
  in invertebrate, 131
Buffers, 423

C

Carboxymethylcellulose, 403
Carcinoscorpius rotundicauda, 37
Cell motility, 73
Cells in culture, 73, 93
Chromogenic substrates, 209
Circadian rhythms, 47
Circular dichroism, 603
Coagulation factors, 379
Coagulogen, 131, 147, 169, 185
Coagulogen clot, 403
^{60}Co irradiation, 367
Concanavalin A, 483
Contact inhibition, 73
Contact paralysis, 73
Creatinine, 465

D

Diel activity, 47
Dry heat, 367

E

Ecology, 27
Embryonic and juvenile Limulus, 61
Endogenous pyrogen, 483
Endotoxemia, 235
Endotoxin, 93, 109, 235, 245, 387, 483
  contamination, 379
  detection of, 235
  determinations, 221
  inactivation, 367
  pyrogenicity, 353
Endotoxin-like substance, 465
Equivalency test, 505
Erythropoietin, 379, 483
  purification, 641
Ethylene oxide (ETO) treatment, 367
Extant Limulidae, 37

F

Fat emulsion, 333
Fluorescence polarization, 567

G

Galactans, 579
Gangliosides, 567
Gel formation, 169
Glucosamine, 537
Glucuronic acid, 579
Glycoproteins, 641, 655
GMP (Good Manufacturing Practices), 339
Gram-negative bacteria, 147, 547
Gram-negative infection, 235, 245
Gram-positive bacteria, 547

H

Health Industry, 339
Hemocytes, 185
Heterogeneity, 625
HIMA (Health Industry Manufacturers Association), 339
Homology of coagulogen and primate fibrinogen, 169
Host defense, 513

I

Immunological reagents, 379
Interference, 505, 221
Intralipid®, 333
Intrathecal drugs, 345
Iron dextran, 221

K

KDO (2-keto-3-deoxy-octonate), 537

L

LAL assay, 454
Lectin, 525, 567, 589, 625
Leukemia, human lymphocytic, 589
Limulin, 567, 579, 603, 617
Limulus agglutinins, 665
Limulus amebocyte lysate (LAL), 131, 245, 345, 367, 387, 403, 415, 423, 463, 473, 483, 499, 505, 513
Limulus distribution, 3
Limulus haemagglutinin, 559, 625
Limulus lysate, 93
Limulus polyphemus, 1, 27, 37, 73, 93, 109, 131, 185, 209, 229, 513, 589
Limulus test, 235
Lipid A, 537
Lipopolysaccharide, 415, 537
Liposomes, 567
LPS contamination, 379
Lymphoblastoid cell line, 589
LVP (Large volume parenterals), 339

M

Membrane fluidity, 567
Minimum pyrogenic dose (MPD), 353
Mitogenicity, 483, 603
Molecular weight, 625

N

N-acetylamino compounds, 625
Nephelometry, 403
Neuraminic acid, 603

O

Ontogeny, 109
Opaque solutions, 221

P

Parenterals, 387
Phagocytosis, 73
Photic responses, 47
Phylogeny, 109
Polymorphism, 625
Pyrogenic reaction, 465
Pyrogen test
  Limulus, 333
  rabbit, 333, 463, 499
  USP, 345, 353, 367, 505
Pyrogens, 387
Pyrostat™ Limulus amebocyte lysate, 435

Q

Quantitation, 403, 415, 423, 435

R

Radioactive lipopolysaccharide, 203
Radiographic contrast media, 379
Radiopharmaceuticals, 339, 345

S

Sialic acid, 559, 579, 655
Sodium dodecylsulfate, 403
Spectrophotometric assay, 415
Standardization, 505
Subunit heterogeneity, 579
Subunits, 603
Sugar specificity, 567, 603

T

Tachypleus gigas, 37
Tachypleus tridentatus, 1, 37, 131, 169, 185, 209, 229, 625
T-cell, 589
T-cell mitogen, 603
Temperature preference, 37
Thermogenic response, 333
Thermoregulation, 47
Tissue factor, 483
Turbidimetric assay, 423, 435
Tubidity, 435

U

Urea, 465
U.S.P. (United States Pharmacopeia), 339

V

Vesicles, 567
Vizualization, 603

[redacted]